Michael Cullinane

Real Analysis - An Introduction

Also of Interest

Applied Nonlinear Functional Analysis
An Introduction
2nd Edition
Nikolaos S. Papageorgiou, Patrick Winkert, 2024
ISBN 978-3-11-128421-7, e-ISBN (PDF) 978-3-11-128695-2,
e-ISBN (EPUB) 978-3-11-128832-1

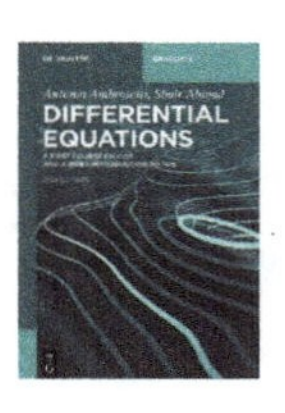

Differential Equations
A First Course on ODE and a Brief Introduction to PDE
2nd Edition
Antonio Ambrosetti, Shair Ahmad, 2023
ISBN 978-3-11-118524-8, e-ISBN (PDF) 978-3-11-118567-5,
e-ISBN (EPUB) 978-3-11-118578-1

Real Analysis
Measure and Integration
Marat V. Markin, 2019
ISBN 978-3-11-060097-1, e-ISBN (PDF) 978-3-11-060099-5,
e-ISBN (EPUB) 978-3-11-059882-7

Elementary Functional Analysis
Marat V. Markin, 2018
ISBN 978-3-11-061391-9, e-ISBN (PDF) 978-3-11-061403-9,
e-ISBN (EPUB) 978-3-11-061409-1

Michael Cullinane

Real Analysis - An Introduction

Mathematical Arguments and Elementary Proof Techniques

DE GRUYTER

Mathematics Subject Classification 2020
26-01

Author
Prof. Michael Cullinane
Keene State College
Department of Mathematics
229 Main Street
Keene NH 03435
U.S.A.
mcullina@keene.edu

ISBN 978-3-11-142928-1
e-ISBN (PDF) 978-3-11-142956-4
e-ISBN (EPUB) 978-3-11-142994-6

Library of Congress Control Number: 2025937540

Bibliographic information published by the Deutsche Nationalbibliothek
The Deutsche Nationalbibliothek lists this publication in the Deutsche Nationalbibliografie; detailed bibliographic data are available on the Internet at http://dnb.dnb.de.

Cover image: benjaminec / iStock / Getty Images Plus
Typesetting: VTeX UAB, Lithuania

www.degruyter.com
Questions about General Product Safety Regulation:
productsafety@degruyterbrill.com

Preface

This book endeavors to present a reasonably comprehensive treatment of undergraduate real analysis that is accessible to students who have successfully completed two semesters of single variable calculus. Because elementary proof writing strategies are developed within the book as they are needed, there is no expectation that a reader must have already completed an introductory course in mathematical reasoning and proof.

The book addresses both the structure of the real number system and the basic theory of calculus. Topics include completeness, sequential convergence, decimal representation of real numbers, a bit of topology, function limits, continuity, the derivative, the Riemann integral, and infinite series, as well as sequences and series of functions. We even demonstrate the irrationality of both e and π.

The pacing of the narrative is somewhat more leisurely at the beginning so as to assist readers for whom the book represents a first venture into the formal study of mathematics. For example, more basic matters such as inequalities, absolute value, and algebraic properties of the natural numbers, integers, and rational numbers are treated in some depth to show how familiar mathematical content can be axiomatized and deductive logic applied to formally derive theorems. Doing so developmentally prepares the reader to apply this more formal methodology when the mathematical content is less familiar or even completely new.

While the book goes into more detail on certain elementary ideas than many other books covering essentially the same content, it still includes enough material for a full year of study. The uncountability of the real numbers, Lebesgue's integrability criterion, rearrangements of series, the Weierstrass approximation theorem, and numerical approximation of integrals are just some of the topics the book addresses that go beyond the essential fundamentals. Thus, a variety of single-semester courses or two-semester course sequences can be fashioned from what is included.

Especially in the first half of the book, proofs incorporate considerable detail to help readers more easily follow the arguments being presented. Quite often, such as with proofs involving convergence of a sequence or function limits, the preliminary thinking needed to discover how the proof can be developed and written is discussed.

Also, in further recognition of the developmental nature of mathematical learning, when calculus topics such as the derivative and the integral are considered, more attention is paid than is usually the case to drawing upon students' prior experiences with them. For instance, readers are reminded of how the integral may be interpreted as measuring signed area, and this understanding is then leveraged to lead them to a fully rigorous definition of the integral.

To promote more active learning, the exercises, of which there are more than 700, are embedded within the narrative rather than being collected at the end of each chapter. They are designed to engage the reader in developing a deeper level of understanding of the ideas under discussion. Some are routine and others more challenging. There

https://doi.org/10.1515/9783111429564-202

are many that relate directly to the narrative, asking the reader to fill in details or justification left out of a proof or to reflect on a particular example. Hints and answers to selected exercises can be found on the publisher's platform at weblink https://www.degruyterbrill.com/document/isbn/9783111429564/html QR code .

With respect to both the content covered within the book and the exercises, I make no claim to originality. Over many decades, the material has become standard, and versions of the proofs and exercises can be found in just about every undergraduate real analysis book. It is only the organization and relative emphasis of ideas that are my own (as well as any errors that remain, for which I apologize in advance).

I want to acknowledge my colleagues at Keene State College who have supported me through the writing of this book, especially Joe Witkowski and Lisha Hunter. Many years ago, my dissertation advisor, Sam Shore, who is Professor Emeritus from the University of New Hampshire, taught me the importance of creating careful, precise, and detailed mathematical arguments, and I know his influence continues to inform my writing. It has also been a pleasure to collaborate with everyone at De Gruyter, and I am particularly grateful to Melanie Götz and Steve Elliot for their guidance and professionalism. Finally, this book is dedicated in loving memory of my parents, Don and Connie, and my uncle, Hubert.

I truly hope this book may help all who read it to develop an understanding of and appreciation for a part of the foundation on which a significant amount of higher mathematics rests.

Contents

called *mathematical analysis* began to develop in the early 19th century in response to such demands.

Within the realm of mathematical analysis sits *real analysis*, the analytic study of the real number system as well as functions for which both the domain and range consist of real numbers. The key descriptor here is *analytic*, which signifies the application of *analytical reasoning*. The philosopher Immanuel Kant described analytical reasoning as the process of making judgments about statements based solely on the content of those statements, rather than by means of observations or experiences external to the statements [1]. In mathematics, analytical reasoning takes the form of building arguments called *proofs* that are based on principles of *deductive logic*.

Logic and proof arose in the study of both number and geometry undertaken by the ancient Greek mathematicians. However, it would be a mistake to think that the development of mathematics occurred within a primarily proof-oriented framework from that time onward. When Newton and Leibniz discovered calculus in the 17th century, and for a period of about 150 years thereafter, mathematics developed largely as a collection of methods and procedures in response to computational questions that arose in scientific fields such as physics. It was not until the early 19th century, when the mathematical community came to realize the necessity of a foundation on which to build the ideas of the calculus, that deductive logic and proof began to serve the characteristic roles they continue to play in mathematics today. In particular, Augustin–Louis Cauchy's successful development in the 1820s of the theory of limits as a basis for the calculus not only gave birth to the subject of mathematical analysis, but contributed immensely to the modern-day deductive approach to mathematical research in general [2].

Thus, our study of real analysis commences with the articulation of a set of assumptions, called *axioms*, about the real numbers, from which we derive other facts, called *theorems*, via the application of deductive logic. By carrying out such an analytic study, we are able to conclusively answer many more questions about the real numbers than is feasible with less formal approaches.

Even though we are taking an analytic approach, we do not mean to suggest that geometric intuition or any other aspects of our prior, mostly informal, work with the real numbers and the calculus should be ignored. Our analytic study should be informed by intuition and experience, but intuition and experience are not permitted to substitute for rigorous deductive arguments. For example, we eventually analytically define the number π so that

$$\pi = 2\int_{-1}^{1} \sqrt{1-x^2}\,dx,$$

but we also show how this definition is consistent with the more common geometric approach to π that relates to circles.

We also demonstrate how an analytic perspective permits us to derive certain intuitive features of the real number system that are manifested in the number line repre-

1 Algebraic and order properties of the real numbers

Through many years of school mathematics, we come to think of the *real numbers* as those numbers associated with points on a number line, which is therefore often referred to as the **real line** (Figure 1.1).

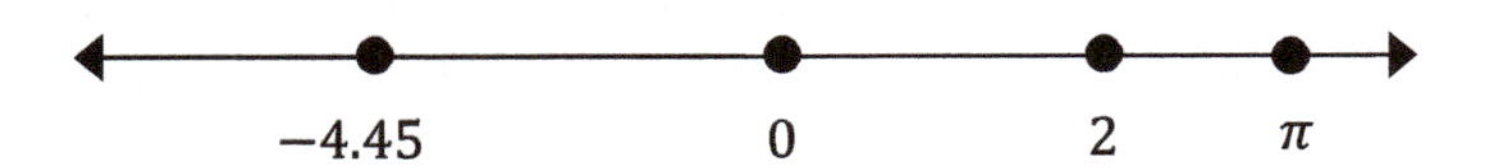

Figure 1.1: The real line.

Real numbers are also informally viewed as those numbers that can be represented symbolically via decimal notation. Thus, for example, each of 2, 0, −4.45, and π (which, as we have all learned, is approximately 3.14) is a real number.

We have learned how to work with the real numbers in various ways, performing computations such as $-4.45 + 2 - \pi \cdot 0$, making comparisons by means of statements like "2 is less than π," and applying labels such as *positive* and *rational* to categorize them. In our study of calculus, we focused on functions whose inputs and outputs are real numbers, learning how to, for example, differentiate and integrate them.

There are, however, questions about the real numbers and the mathematical things we do with them that are not easily resolved using either the number line or decimal notation representations. For instance, exactly what real number is $\sqrt{2} + \pi$? Certainly this question would seem to have an answer, for we can conceive of two segments, one of length $\sqrt{2}$ and the other of length π, and might argue that $\sqrt{2} + \pi$ is the length of the new segment formed by "combining" the original segments. Of course, this approach to resolving the question requires that we agree to what is meant by the "length" of a segment as well as exactly how the segments are to be "combined." The very notion of "segment" is geometric and not intrinsic to the real numbers themselves. For instance, what if the numbers being added instead represented temperatures? How can we be sure that an approach to real number addition based on interpreting the numbers as lengths of segments would legitimately transfer to a setting in which we need to add temperatures?

If we instead draw upon decimal representations, we might note $\sqrt{2}$ is approximately 1.414 and π is approximately 3.142, so it would seem reasonable to conclude that $\sqrt{2} + \pi$ is approximately $1.414 + 3.142 = 4.556$. Here though we obtain only an approximation to the sum, not the exact sum, as a result of inherent limitations with the ability to represent some irrational numbers in decimal form.

The sort of question we have posed here would apparently demand a more thorough understanding of the real numbers, a depth of comprehension that is unlikely to be achieved by appealing to geometric intuition or notational conventions. The subject

https://doi.org/10.1515/9783111429564-001

sentation of the real numbers, and we explain how the analytic approach gives rise to the decimal representations of the real numbers. Thus, both the real line and decimal representation models for the real number system guide our thinking as we carry out a more formal analytic study.

We should also mention that while it is possible to analytically define the real numbers, we shall not do so here. The standard analytic constructions of the real numbers are less than straightforward and better left to a more advanced course. Anyway, what the real numbers are is actually far less important than how they behave.

Any formal mathematical study typically begins with some **undefined notions** together with one or more **axioms** that provide an initial set of assumed facts regarding the undefined notions. In our analytic treatment of the real number system, we take the notions of *real number*, *addition*, *multiplication*, and *positive* as undefined and then put together a system of axioms that describe the most fundamental ways in which these undefined notions are related to one another.

Algebraic assumptions about the real numbers

We begin with an axiom that catalogs our assumptions concerning the *real numbers* relative to the operations of *addition* and *multiplication*.

Axiom 1.1 (The field axiom). *There exists a set* $\mathbb{R}$ *whose elements are called* real numbers. *There are also two functions, called* addition *and* multiplication, *that can be applied to any ordered pair* (a, b) *of real numbers to produce, respectively, a* **sum** $a+b$ *and a* **product** $a \cdot b$ *or* ab, *both of which are themselves real numbers. The real numbers under addition and multiplication satisfy all of the following properties.*

(1) Commutative properties: *For all real numbers a and b, we have*

$$a + b = b + a$$

and

$$ab = ba.$$

(2) Associative properties: *For all real numbers a, b, and c, we have*

$$(a + b) + c = a + (b + c)$$

and

$$(ab)c = a(bc).$$

(3) Distributive property: *For all real numbers a, b, and c, we have*

$$a(b + c) = ab + ac.$$

(4) Existence of an additive identity: *There is a real number* 0, *called* ***zero*** *or the* ***additive identity****, with the property that* $0 + a = a$ *for every real number a.*
(5) Existence of a multiplicative identity: *There is a real number* 1, *called* ***one*** *or the* ***multiplicative identity****, which is distinct from* 0 *(i. e.,* $1 \neq 0$*) and with the property that* $1a = a$ *for every real number a.*
(6) Existence and fundamental property of additive inverses: *Every real number a has an* ***opposite*** *or* ***additive inverse*** $-a$ *that is also a real number and for which* $a + (-a) = 0$.
(7) Existence and fundamental property of multiplicative inverses: *Every nonzero real number a has a* ***reciprocal*** *or* ***multiplicative inverse*** $\frac{1}{a}$ *that is also a real number and for which* $a \cdot \frac{1}{a} = 1$.

The fundamental mathematical notions of *set*, *relation*, and *function* are introduced in Appendix A, which we encourage you to read now and consult as necessary. When a set of objects is accompanied by operations that behave like addition and multiplication, in the specific sense that properties paralleling (1–7) of Axiom 1.1 are satisfied, the resulting algebraic structure is called a *field*, and explains why the axiom is referred to as the field axiom.

We want to make a few comments on the ideas referred to in this axiom. First, we agree that our intention when declaring that real numbers are *equal* is that they are in fact the same real number and may therefore be freely substituted for each other. That is, when a and b are real numbers and $a = b$, we always have the option to replace a with b (or b with a). One consequence of this agreement is that new true equations can be obtained by adding or multiplying existing true equations. Specifically, if $a = b$ and $c = d$, then both $a + c = b + d$ and $ac = bd$.

It is customary to refer to addition and multiplication as *operations*, and we shall follow this custom. Formally, though, each of addition and multiplication is a function whose domain is $\mathbb{R} \times \mathbb{R}$ and whose codomain (actually range) is $\mathbb{R}$. We use the familiar operational notations $a + b$ and $a \cdot b$ in place of, respectively, standard function notation $+(a, b)$ and $\cdot(a, b)$, which is of course in keeping with how we learned to write addition and multiplication calculations in grade school.

Generally, **a set is closed under an operation** applied to the members of the set if the result of applying the operation to members of the set *always* yields a member of the set. For instance, as the sum of even counting numbers is always an even counting number, the set of all even counting numbers is closed under addition. On the other hand, the set of odd counting numbers is not closed under addition as, for example, $1 + 3 = 4$. We are assuming that the set $\mathbb{R}$ of real numbers is closed under both addition and multiplication, meaning that for any real numbers a and b, both $a + b$ and ab are real numbers.

An operation applied to the members of a set is **commutative** if the order in which members of the set are combined under the operation is irrelevant. We have assumed that addition and multiplication of real numbers are commutative operations (and this should, of course, agree with your prior experiences working with these operations). However, not all operations are commutative. Subtraction, for instance, is not commutative as the order in which numbers are subtracted can yield different results, for example, $7 - 4 \neq 4 - 7$.

An operation applied to the members of a set is **associative** if, when three or more elements in the set are combined under the operation, it does not matter where in the calculation we begin computing. For instance, we have assumed that addition of real numbers is associative, meaning the expression

$$a + b + c$$

is unambiguous; whether we view it as $(a + b) + c$ or as $a + (b + c)$ makes no difference to the final computed result. Similarly, our assumption that multiplication of real numbers is associative means that

$$abc$$

is unambiguous. Of course, not all operations are associative. For example, note that $(20 \div 2) \div 2 = 5$, whereas $20 \div (2 \div 2) = 20$, so division is not associative.

The terminology *additive identity* for 0 and *multiplicative identity* for 1 reflects the idea that neither adding 0 nor multiplying by 1 changes the value of a number, meaning the number "retains its identity." Similarly, *additive inverse* and *multiplicative inverse* refer to their roles in "inverting" or "undoing" addition and multiplication.

Example 1.2. To undo addition by 2.5, we add its *additive inverse* -2.5. We apply this idea in "solving" an equation such as

$$x + 2.5 = 15.$$

We add -2.5 to "both sides" to obtain

$$x + 2.5 + (-2.5) = 15 + (-2.5),$$

then use the fact that (-2.5) is the additive inverse of 2.5 to get

$$x + 0 = 12.5,$$

and finally use the fact that 0 is the additive identity to reach the conclusion that

$$x = 12.5.$$

Similarly, to undo multiplication by 2.5, we multiply by its *multiplicative inverse* $\frac{1}{2.5}$. For instance, to "solve" the equation

$$2.5x = 15,$$

we multiply "both sides" by $\frac{1}{2.5}$ to obtain

$$\frac{1}{2.5} \cdot 2.5 \cdot x = \frac{1}{2.5} \cdot 15,$$

then use the fact that $\frac{1}{2.5}$ is the multiplicative inverse of 2.5 to get

$$1 \cdot x = 6,$$

and finally use the fact that 1 is the multiplicative identity to conclude that

$$x = 6.$$

Theorems and proofs

In analytic mathematical studies, we typically want to assume as little as possible and deduce as much as possible from our assumptions. Such an approach helps to more fully reveal how mathematical ideas are interrelated so that we can better understand them.

A **proof** is an argument that uses logical reasoning to deduce a new result from given assumptions. A **theorem** is a mathematical statement for which there is a proof. We are allowed to use axioms, the definitions of specialized terms and notations, and previously established theorems to justify the conclusions we make within a proof.

The assumptions made in Axiom 1.1 are all basic mathematical truths with which we are extremely familiar. We can deduce many other such truths from these assumptions. For instance, we will soon prove the following theorem.

Theorem 1.3 (Cancellation laws). *Let a, b, and c be real numbers.*
(1) *If $a + c = b + c$, then $a = b$.*
(2) *If $ac = bc$ and $c \neq 0$, then $a = b$.*

These cancellation laws are also very familiar to us from high school algebra, but our point here is that they need not be assumed, as they can be shown to follow from some of the assumptions put forth in Axiom 1.1.

Many readers of this book are already well-versed in the fundamentals of logic and basic proof techniques, but we still remind you of some of the standard mathematical proof strategies as we encounter them. Appendix B provides a primer on elementary logic and should be consulted as needed. A list of standard proof strategies used throughout mathematics can be found in Appendix C.

Proving an *if… then…* statement

A generic *if… then…* statement

$$\textit{If } P, \textit{ then } Q,$$

is often symbolized as

$$P \Rightarrow Q,$$

and referred to as the **conditional** or **implication** with **hypothesis** P and **conclusion** Q. The following proof strategy is typically employed when proving an *if… then…* statement and is based on the fact that an *if… then…* statement is false only when its hypothesis is true and its conclusion is false.

Proof Strategy 1.4 (*If… then…* statement). To prove a statement of the form

$$P \Rightarrow Q,$$

assume P and then show how to deduce Q from this assumption.

Logical symbols such as the implication arrow $\Rightarrow$ are not usually employed in formal mathematical writing, unless the subject matter is logic itself. Therefore, we generally do not use such symbols in stating mathematical results or in our written proofs. Because proof strategies are part of the subject matter of logic, we do allow for the use of logical symbols in stating these strategies. Doing so can be helpful as there are often a variety of alternative ways to verbally render a logical operator. For instance, *if* is sometimes conveyed by writing *as long as* or *when*. The use of $\Rightarrow$ in the statement of Proof Strategy 1.4 helps make clear that it is a conditional statement's logical structure, not the specific choice of wording, that is relevant to the proof strategy.

We now apply Proof Strategy 1.4 to establish one of the cancellation laws.

Proof of Theorem 1.3. We prove (1) and leave (2) as an exercise.

Assume $a + c = b + c$. Then as c is a real number, c has an additive inverse $-c$, and it follows from our assumption that

$$a + c + (-c) = b + c + (-c).$$

Then, because $c + (-c) = 0$, we may conclude that

$$a + 0 = b + 0,$$

from which it then follows, as 0 is the additive identity, that

$$a = b. \qquad \square$$

In writing the proof, we have tried to provide suitable justification for the deductions we have made. Here the critical observation required to carry out and justify the proof is that the cancellation of c results from our assumption that every real number has an additive inverse. On the other hand, we did not indicate our use of the associative property of addition in this proof. We employed it, implicitly, when we wrote, for instance, the expression

$$a + c + (-c)$$

rather than first writing

$$(a + c) + (-c)$$

and then indicating that this expression could be rewritten as

$$a + \big(c + (-c)\big).$$

When writing proof, there are always questions of how much detail to provide, both in terms of what explicit deductions are shown and how much information is provided with respect to justifying these deductions. Different people at different times, and writing for different audiences, may include more or fewer deductions, and more or less justification, in writing what is essentially the same proof. As we become more accomplished in our ability to reason mathematically and more experienced with the particular mathematical content we are studying, we often write less than we would have done when we had less accomplishment and experience, especially if we believe our audience should also have acquired sufficient experience to follow a more compressed argument. That is the practice we shall follow in this book, our proofs tending to be more expansive in the first few chapters and more succinct later on. In any case, the reader of a proof should always have paper and pencil at hand to work through details the proof omits, the goal being to confirm for oneself that a valid and complete argument has been presented.

In almost any mathematics textbook, items such as axioms, theorems, and some examples, are numbered so that they may be easily referenced. In our proof above, for example, we could have referred to Axiom 1.1(6) in providing justification for the existence of $-c$ and to support why $c + (-c) = 0$. However, referencing items by number only really makes sense within the written work in which those item numbers are established, so when you write up a proof as part of an exercise, you should not refer to items in this book by their item numbers, but instead should paraphrase the content of a result you are using to justify a particular deduction.

We use the symbol □ to indicate the end of a proof.

Exercise 1.1. Prove the cancellation law stated in Theorem 1.3(2).

Proving a *for all* statement

We are very familiar with the fact that multiplying a number by zero just produces zero.

Theorem 1.5. *For every real number a, we have* $0a = 0$.

This theorem incorporates so-called *universal quantification*, which is signaled by phrases such as *for all*, *for every*, and *for each*, and which may be denoted symbolically by $\forall$. In logic, the statement

$$\forall x \in A, P(x),$$

is a contraction for

$$\forall x, x \in A \Rightarrow P(x).$$

Thus, in purely symbolic form Theorem 1.5 could be stated as

$$\forall a \in \mathbb{R}, 0a = 0,$$

though as we have previously mentioned, logical symbols, among which are quantifiers such as $\forall$, should be avoided in formal mathematical writing where the subject is not logic itself.

Proof Strategy 1.6 (Proving a *for all* statement). To prove a statement of the form

$$\forall x \in A, P(x),$$

consider an anonymous element x of the set A and then show the statement $P(x)$ is true for this anonymous x.

We employ this proof strategy in the following proof of Theorem 1.5.

Proof of Theorem 1.5. Let a be a specific, but anonymous, real number. Note that

$$0a + 0a = (0 + 0)a = 0a = 0a + 0,$$

where the first equality holds because of the distributive property, and the other two equalities because 0 is the additive identity. Thus, we have determined that

$$0a + 0a = 0a + 0.$$

It now follows by additive cancellation of the real number $0a$ that $0a = 0$. □

In applying Proof Strategy 1.6 to prove Theorem 1.5, observe that the role of x is being played by a (the particular choice of letter used to name an anonymous object is not important), the type of object that a represents is that of a real number, and the specific conclusion $P(a)$ we wish to reach about the anonymous real number a is $0a = 0$. Within our proof, we did not assign a an actual numerical value (such as $\frac{5}{3}$) because we want our argument to apply to all real numbers. The first sentence of our proof could be written more succinctly as "Consider any real number a," or "Let a be a real number," or even "Suppose $a \in \mathbb{R}$."

As we come to better understand a particular proof, there is often the potential to gain further insight into the mathematical ideas the proof addresses. For instance, among all of the assumptions compiled in Axiom 1.1, it is only the distributive property that relates the operations of addition and multiplication. Thus, we might expect the proof of a result such as Theorem 1.5 concerning *multiplication* of numbers by the *additive* identity should rely on distributivity. Moreover, it is also now apparent why the real number 0 cannot have a multiplicative inverse: There is no real number r for which $0 \cdot r = 1$.

Exercise 1.2. Prove that for any real numbers a and b, we have $a(b + 1) = a + ba$. Indicate each application of any part of Axiom 1.1 in your proof.

Exercise 1.3. Solve the equation $3x + 2 = 10$, providing justification for each step in the solution process using an appropriate part of Axiom 1.1.

Proving uniqueness

Sometimes a mathematical object is unique with respect to a certain property. To establish the uniqueness of an object with respect to a property, we usually employ the following strategy.

Proof Strategy 1.7 (Proving uniqueness). To prove there is only one object having a certain property, assume there are two objects possessing the property and show they really are the same.

The proof of the following theorem illustrates this strategy, establishing the uniqueness of the additive identity 0 by entertaining the possibility of a "second" additive identity z and then demonstrating that $z = 0$.

Theorem 1.8. *The number* 0 *is the only additive identity for the real numbers.*

Proof. Suppose z is a real number and z is an additive identity. Because z is an additive identity, it follows that $0 + z = 0$. But because 0 is an additive identity, it also follows that $0 + z = z$. As we have shown $0 + z$ is equal to both z and 0, it follows that $z = 0$. □

Exercise 1.4. Prove that 1 is the unique multiplicative identity for the real numbers.

Exercise 1.5. Prove that each real number has a unique additive inverse. Then prove that each nonzero real number has a unique multiplicative inverse.

The use of definitions in proofs

When a statement we want to prove makes use of a certain mathematical concept that has been formally defined, it is nearly always a good idea to consider trying to use the concept's definition within the proof.

For example, we are all familiar with the algebraic property $(-a)b = -(ab)$ that permits us to extract a negative sign from a factor in a product and relocate it in front of the product. To prove this result, we focus on the *definition of additive inverse* rather than the "movement" of the negative sign. By definition, the number q is the additive inverse of the number p, provided that $p + q = 0$. So, as the expression $-(ab)$ represents the additive inverse of ab and we want to prove that $(-a)b = -(ab)$, we need to show $(-a)b$ is the additive inverse of ab, which we can do by showing that when we add $(-a)b$ to ab we get 0. That is precisely what we do in the following proof.

Theorem 1.9. *For all real numbers a and b, we have $(-a)b = -(ab)$.*

Proof. Let a and b be real numbers. We want to show that $(-a)b$ is the additive inverse of ab. It suffices to show that $ab + (-a)b = 0$. Observe that

$$ab + (-a)b = (a + (-a))b = 0b = 0. \qquad \square$$

In this proof, we did not explicitly justify the equalities in the final sentence because we expect that by showing each step in the simplification the reader is able to supply the justification on their own. As we mentioned earlier, the amount of justification given in a proof depends on what the writer of the proof can expect from the reader of the proof.

Exercise 1.6. Use Theorem 1.9 to show that $(-a)(-b) = ab$ for all real numbers a and b. Then use this result to explain why $(-1)(-1) = 1$.

A **corollary** is a theorem that follows from another theorem via a very short proof.

Corollary 1.10. *For every real number a, we have $(-1)a = -a$.*

Exercise 1.7. Show how Corollary 1.10 follows from Theorem 1.9. (In mathematics, *show* means *prove*; other synonyms for *prove* are *demonstrate*, *deduce*, *establish*, etc.).

Exercise 1.8. Explain why the additive inverse of $-a$ is a.

Exercise 1.9. Prove that the additive inverse of a sum is the sum of the additive inverses; that is, prove that $-(a+b) = (-a) + (-b)$ for all real numbers a and b.

Subtraction and division

The operations of **subtraction** $-$ and **division** $\div$ are defined so that for any real numbers a and b we have

$$a - b = a + (-b),$$

and for any real number a and any nonzero real number b we have

$$a \div b = a \cdot \frac{1}{b}.$$

That is, subtraction is the same as "adding the opposite" and division is the same as "multiplying by the reciprocal." The result of a subtraction is referred to as a **difference** and the result of a division is referred to as a **quotient**. We usually write $\frac{a}{b}$ in place of $a \div b$.

Theorem 1.11. *For every real number a, we have $a - 0 = a$.*

Proof. Consider any real number a. Then

$$a - 0 = a + (-0) = a + 0 = a. \qquad \square$$

Exercise 1.10. Justify each equality in the proof of Theorem 1.11.

Exercise 1.11. Use the definition of subtraction and the result from Exercise 1.8 to prove that $a - (-b) = a + b$ for all real numbers a and b.

Exercise 1.12. Use the definition of subtraction and the result from Exercise 1.9 to prove that for all real numbers a, b, and c, we must have $a - (b + c) = a - b - c$.

Positive and negative numbers

Our initial understanding of the undefined notions of *real number*, *addition*, and *multiplication* comes from the assumptions we stated in Axiom 1.1 and the results we have proved based on these assumptions. Another undefined notion in our study of the real numbers is the idea that some of them are *positive*. The following axiom delineates further assumptions concerning the addition and multiplication of positive real numbers, and also tells us that the real numbers may be separated into three nonoverlapping subsets.

Axiom 1.12 (The positivity axiom). *There exists a subset* $\mathbb{R}^+$ *of the set* $\mathbb{R}$ *of real numbers whose elements are called* positive *real numbers. We also make the following assumptions.*

(1) *Closure properties of* $\mathbb{R}^+$: *The set* $\mathbb{R}^+$ *of all positive real numbers is closed under both addition and multiplication, meaning the sum of positive real numbers is always positive, as is the product.*
(2) *Law of trichotomy: For any real number* a, *exactly one of the following is true:* a *is positive,* $a = 0$, *or* $-a$ *is positive.*

A real number is called **negative** if its additive inverse is positive (i. e., a is **negative** if and only if $-a$ is positive). The set of all negative real numbers is denoted $\mathbb{R}^-$; thus,

$$\mathbb{R}^- = \{a \in \mathbb{R} \mid -a \in \mathbb{R}^+\}.$$

The law of trichotomy tells us that a given real number is positive, zero, or negative. Moreover, this law tells us that no real number can be both positive and negative, and that the number 0 is neither positive nor negative.

Theorem 1.13. *If* a *is negative and* b *is negative, then* ab *is positive.*

Proof. Assume a is negative and b is negative; we must show ab is positive. As a and b are negative, we deduce that $-a$ and $-b$ are positive. As $\mathbb{R}^+$ is closed under multiplication, it follows that $(-a)(-b)$ is positive. Then, as $(-a)(-b) = ab$, we may therefore conclude that ab is positive. □

Exercise 1.13. Prove that if a is negative and b is negative, then $a + b$ is negative.

Exercise 1.14. Prove that if a is positive and b is negative, then ab is negative.

Counterexamples

Mathematical exploration naturally leads to the formulation of *conjectures*, statements for which there is some evidence supporting their truth, but for which valid proofs have not yet been found. Inevitably, some conjectures turn out to be false. To establish that a *for all* statement is false, it is only necessary to provide one instance of an object of the type to which the *for all* quantifier applies and for which the underlying statement is false. Such an object, if it exists, is called a **counterexample** to the statement and **disproves** the statement.

Example 1.14. Recall that the **square** x^2 of a real number x is the real number $x \cdot x$. Consider the statement,

The square of any real number is positive.

As

$$0^2 = 0 \cdot 0 = 0,$$

and 0 is not positive, the real number 0 serves as a counterexample to the statement. So it is *not true* that the square of any real number is positive.

Exercise 1.15. Provide a counterexample to the given statement, thus disproving it.
(a) The sum of a positive real number and a negative real number is always positive.
(b) For any real number a, we have $a + 3a = 3a^2$.

The ordering of the real numbers

A real number a is defined to be **less than** a real number b, denoted $a < b$, if $b - a$ is positive (Figure 1.2).

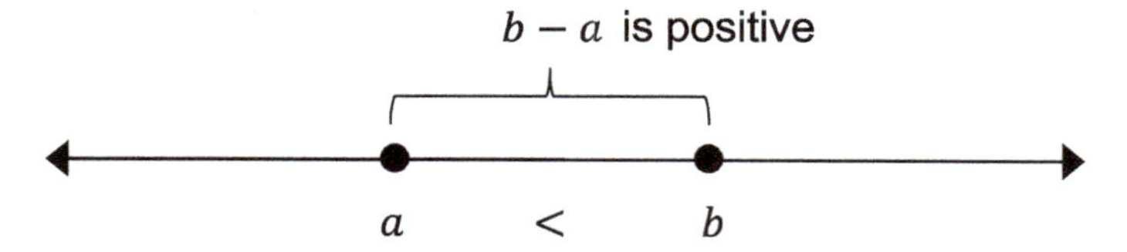

Figure 1.2: Visualizing the meaning of "less than."

When $a < b$ we may also say that b is **greater than** a and write $b > a$. We write $a \leq b$ to mean a is less than or equal to b. When $a \leq b$ we may also say that b is greater than or equal to a and write $b \geq a$.

Exercise 1.16. Use the law of trichotomy to prove that for any real numbers a and b, exactly one of the following is true: $a < b$, $a = b$, or $a > b$. (This fact may also be referred to as the *Law of Trichotomy*.)

Exercise 1.16 establishes the familiar ordering of the real numbers on the number line. In particular, it tells us that, for any two distinct real numbers, one of them "comes before" (i. e., is less than) the other. This fact helps to justify the use of a number line as an intuitive geometric model for the real numbers.

Theorem 1.15. *If $a < b$ and $c < d$, then $a + c < b + d$.*

Proof. Suppose $a < b$ and $c < d$. Then, by the definition of $<$, each of $b - a$ and $d - c$ is positive. So, as the sum of positive real numbers must be positive, it follows that $(b - a) + (d - c)$ is positive. But $(b - a) + (d - c) = (b + d) - (a + c)$, so we may conclude, by the definition of $<$, that $a + c < b + d$. □

Exercise 1.17. Prove each of the following.

(a) If $a < b$ and $b < c$, then $a < c$. (Because of this result, when $a < b$ and $b < c$, we often write $a < b < c$ and say that b is **between** a and c.)

(b) If a is negative and b is positive, then $a < b$.

(c) If $a < b$, then $a + c < b + c$.

(d) If $a < b$ and $c > 0$, then $ac < bc$.

(e) If $a < b$ and $c < 0$, then $ac > bc$.

Exercise 1.18. Suppose $0 < a < b$.

(a) Prove that $a^2 < b^2$.

(b) Prove that $\sqrt{a} < \sqrt{b}$. (Recall that, for any positive number p, the notation $\sqrt{p}$ represents the *positive* number whose square is p. At this stage, we do not necessarily know that $\sqrt{p}$ exists as a real number for every positive number p, but for this exercise, there is no harm in assuming that they do.)

Proof by contradiction

The **(logical) negation** of a statement P is the statement

It is not the case that P,

which can be symbolized as

$$\sim P.$$

A statement's negation is true when the statement itself is false, and false when the statement itself is true. As a result, exactly one of a given statement P and its negation $\sim P$ is true, the other false. Therefore, we can establish the truth of a statement by showing that assuming its negation to be true leads to a contradiction.

Proof Strategy 1.16 (Proof by contradiction). To prove a statement P using the method of proof by contradiction, assume its negation $\sim P$ is true and deduce a contradiction.

We employ proof by contradiction to establish that 1 is positive.

Theorem 1.17. *The multiplicative identity 1 is positive.*

Proof. Suppose 1 is not positive. Then by the law of trichotomy we may conclude that $1 = 0$ or 1 is negative. The former cannot be true, as the multiplicative identity is assumed to be distinct from the additive identity; so 1 must be negative. However, as $1 \cdot 1 = 1$, we have a contradiction, as the product of two negative numbers cannot be negative. □

Note that it is typical to end a proof by contradiction as soon as a specific contradiction has been established.

Exercise 1.19. Use contradiction to prove each of the following.

(a) For every real number a, we have $a^2 \geq 0$. *Hint*: After setting up your proof based on the contradiction strategy, consider separately the possibilities that a is positive, a is negative, and $a = 0$.

(b) If $a > 0$, then $\frac{1}{a} > 0$. *Hint*: When using contradiction to prove an *if... then...* statement, note that we should assume the hypothesis and the negation of the conclusion.

(c) If $0 < a < b$, then $\frac{1}{a} > \frac{1}{b}$.

Proving existence

Sometimes we want to prove that a mathematical object with a certain property actually exists.

Theorem 1.18. *There exists a real number a with the property that* $a + a = a \cdot a$.

This theorem involves so-called *existential quantification*, which is signaled by phrases such as *there exists*, *there is*, and *for some*, and which may be denoted symbolically by $\exists$. In logic, the statement

$$\exists x \in A, P(x),$$

is a contraction for

$$\exists x, x \in A \wedge P(x),$$

where $\wedge$ is the logical operator *and*. Thus, in purely symbolic form, Theorem 1.18 could be stated as

$$\exists a \in \mathbf{R}, a + a = a \cdot a.$$

Proof Strategy 1.19 (Proving existence). To prove a statement of the form

$$\exists x \in A, P(x),$$

identify, define, or construct an element t of the set A and show that the statement $P(t)$ is true for this particular object t.

In the following proof of Theorem 1.18, we observe that the real number 0 possesses the property the theorem conveys.

Proof of Theorem 1.18. The additive identity 0 is a real number. Note that $0 + 0 = 0$ and $0 \cdot 0 = 0$; hence, $0 + 0 = 0 \cdot 0$. □

In many cases, establishing the existence of an object possessing a certain property requires us to define or construct the desired object using objects whose existence is already known or assumed. The proof of the next theorem, which states a very important property of the real numbers, provides an example of such a constructive existence proof.

Theorem 1.20 (Density property of the real numbers). *Given any real numbers a and b for which $a < b$, there exists a real number c such that $a < c < b$. In other words, between any two distinct real numbers there is another real number.*

Proof. Assume a and b are real numbers for which $a < b$. Then

$$a + a < a + b < b + b,$$

which may be rewritten as,

$$2a < a + b < 2b,$$

by writing 2 in place of $1 + 1$. We may then conclude that

$$a < \frac{1}{2}(a + b) < b.$$

Since the set of real numbers is closed under both addition and multiplication, we may conclude that $\frac{1}{2}(a + b)$ is a real number, constructed from the real numbers a, b, and $\frac{1}{2}$, that lies between a and b. □

Exercise 1.20. Use the density property to prove that if $a < b$ there must exist three different real numbers between a and b. Do you think there must be more than three?

Exercise 1.21. Use contradiction and the density property to prove that there is no smallest positive real number.

Exercise 1.22. Use contradiction and the density property to prove that if a is a real number for which $0 \leq a \leq r$ for every positive number r, then $a = 0$. Can you write the statement being proved here without the use of any symbols (words only)?

Exercise 1.23. Use contradiction and the density property to prove that if a and b are real numbers for which $a \le b + r$ for every positive number r, then $a \le b$.

If and only if statements

The **biconditional**

$$P \textit{ if and only if } Q,$$

which can be symbolized as

$$P \Leftrightarrow Q,$$

is shorthand for

$$\textit{If } P \textit{ then } Q, \textit{ and also if } Q \textit{ then } P.$$

It is true precisely when P and Q are both true or both false. Thus, a biconditional may be regarded as expressing a notion of equivalence and we sometimes read $P \Leftrightarrow Q$ as "P is equivalent to Q."

Example 1.21. The biconditional

$$x^2 = 25 \textit{ if and only if } x = 5$$

is false because, by taking the value of x to be -5, we see that it is possible for the statement $x^2 = 25$ to be true at the same time that the statement $x = 5$ is false.

Because mathematical definitions are always intended to represent equivalences, they really are *if and only if* statements even though they are often written with the *and only if* suppressed.

Example 1.22. In the definition of **less than**, we wrote

$$a < b \textit{ if } b - a \textit{ is positive},$$

but what we actually mean, as the statement is intended to define $a < b$, is

$$a < b \textit{ if and only if } b - a \textit{ is positive}.$$

This convention of writing *if* when we really mean *if and only if* is only allowed when formulating a definition. Thus, for instance, since neither of the statements

$$x^2 = 25 \textit{ if } x = 5,$$

and

$$x^2 = 25 \textit{ if and only if } x = 5,$$

represents a definition, neither one can replace the other (note the absurdity of attempting to do so anyway, as the first statement is true while the second is false).

We usually prove a biconditional by independently proving the two conditional statements of which it is comprised.

Proof Strategy 1.23 (Proving an *if and only if* statement). To prove a statement of the form

$$P \Leftrightarrow Q,$$

prove each of $P \Rightarrow Q$ and $Q \Rightarrow P$.

In presenting the proof of an *if and only if* statement $P \Leftrightarrow Q$, it is common to employ the label ($\Rightarrow$) to indicate where $P \Rightarrow Q$ is being proved and the label ($\Leftarrow$) to indicate where $Q \Rightarrow P$ is being proved.

Theorem 1.24. *Let a be a real number. Then $a^2 < a$ if and only if $0 < a < 1$.*

Proof. ($\Rightarrow$) Suppose $a^2 < a$. We must show $a > 0$ and $a < 1$.

Suppose to the contrary that it is not the case that $a > 0$, so that $a = a - 0$ is not positive. By the law of trichotomy, we may conclude that $a = 0$ or a is negative. But if $a = 0$, we would have $a^2 = a$, which contradicts our assumption that $a^2 < a$. Also, if a is negative, then $a^2 = a \cdot a$ would be positive, again contradicting our assumption that $a^2 < a$, as no positive number is less than any negative number. Thus, we have shown that a must be positive.

Now, since a is positive, it follows that $\frac{1}{a}$ is positive. Thus, from our assumption that $a^2 < a$, we may conclude that $\frac{1}{a} \cdot a^2 < \frac{1}{a} \cdot a$, which simplifies to $a < 1$.

($\Leftarrow$) Suppose $0 < a < 1$. As $a > 0$, we know $a = a - 0$ is positive. Thus, from our additional hypothesis that $a < 1$, it follows that $a \cdot a < 1 \cdot a$, that is, $a^2 < a$. □

Exercise 1.24. Sometimes we want to prove more than two statements are equivalent to one another. For instance, suppose we want to prove the following four statements are equivalent:

(i) $a > 0$;
(ii) a is positive;
(iii) $-a$ is negative;
(iv) $-a < 0$.

(a) Prove each of the four implications (i) $\Rightarrow$ (ii), (ii) $\Rightarrow$ (iii), (iii) $\Rightarrow$ (iv), and (iv) $\Rightarrow$ (i).
(b) List the eight other implications that would have to hold in order for all four of the statements (i), (ii), (iii), and (iv) to be equivalent to one another.
(c) Explain how it is that the eight implications you listed in (b) must automatically follow logically having already proved the four implications listed in (a).

The law of trichotomy together with the equivalence established in Exercise 1.24 permit us to reach the following conclusions:
(1) $a > 0$ if and only if a is positive;
(2) $a \geq 0$ if and only if a is nonnegative;
(3) $a < 0$ if and only if a is negative;
(4) $a \leq 0$ if and only if a is nonpositive.

Proving an *or* statement

Another important property of the real numbers is that a product of real numbers can be equal to zero if and only if one of the factors comprising the product is equal to zero.

Theorem 1.25. *For any real numbers a and b, if $ab = 0$, then $a = 0$ or $b = 0$.*

To prove this theorem, we would consider arbitrary real numbers a and b, and assume that $ab = 0$. Our goal would then be to show, based on this assumption, that $a = 0$ or $b = 0$. The logical symbol for *or* is $\vee$, and since an *or* statement is true precisely when at least one of its components is true, we immediately obtain the following standard approach to proving $P \vee Q$.

Proof Strategy 1.26 (Proving an *or* statement). To prove a statement of the form

$$P \vee Q,$$

assume P is not true and then deduce Q.

We apply this strategy in carrying out a proof of Theorem 1.25.

Proof of Theorem 1.25. Let a and b be real numbers, assume $ab = 0$, and also assume $a \neq 0$; we need only show $b = 0$. Since $a \neq 0$, we know a has a multiplicative inverse $\frac{1}{a}$. Then, from our hypothesis that $ab = 0$, it follows that $\frac{1}{a} \cdot (ab) = \frac{1}{a} \cdot 0$. But, because $\frac{1}{a} \cdot (ab) = b$ and $\frac{1}{a} \cdot 0 = 0$, we may then conclude that $b = 0$. □

Exercise 1.25. Prove that if $a^2 = 1$, then $a = 1$ or $a = -1$.

Exercise 1.26. Find all solutions to the equation $x \cdot x = x$, providing justification for your conclusion.

Intervals and interval notation

An **interval** of real numbers is a subset I of $\mathbb{R}$ with the property that any real number lying between two members of I is also a member of I (so an interval has no "gaps");

that is, I is an **interval** if whenever $p < x < q$ and both p and q are in I, it follows that x is in I.

Example 1.27. The set $\{x \in \mathbb{R} \mid \pi \leq x < 5\}$, pictured in Figure 1.3, is an interval because it includes all real numbers lying between any two of its members.

Figure 1.3: The interval $[\pi, 5)$.

Given real numbers a and b, with $a < b$, we employ the following notations for various types of intervals:

$$
\begin{aligned}
(a, b) &= \{x \in \mathbb{R} \mid a < x < b\}, \\
[a, b] &= \{x \in \mathbb{R} \mid a \leq x \leq b\}, \\
(a, b] &= \{x \in \mathbb{R} \mid a < x \leq b\}, \\
[a, b) &= \{x \in \mathbb{R} \mid a \leq x < b\}, \\
(a, \infty) &= \{x \in \mathbb{R} \mid x > a\}, \\
[a, \infty) &= \{x \in \mathbb{R} \mid x \geq a\}, \\
(-\infty, a) &= \{x \in \mathbb{R} \mid x < a\}, \\
(-\infty, a] &= \{x \in \mathbb{R} \mid x \leq a\}.
\end{aligned}
$$

We may also write $(-\infty, \infty)$ for the set $\mathbb{R}$ of all real numbers. Note that $\emptyset$ is an interval, as is any singleton $\{a\}$ where a is a real number.

For each of the intervals (a, b), $[a, b]$, $(a, b]$, and $[a, b)$, the number a is the **left endpoint** of the interval and the number b is the **right endpoint** of the interval. For each of the intervals (a, ∞) and $[a, \infty)$, the number a is the **left endpoint** and there is no right endpoint. For each of the intervals $(-\infty, a)$ and $(-\infty, a]$, the number a is the **right endpoint** and there is no left endpoint.

Example 1.28. Thus,

(1) the interval $[\pi, 5)$ includes all real numbers between π and 5, along with its left endpoint π, but not its right endpoint 5;
(2) the interval $(-\infty, 0)$ is the set $\mathbb{R}^-$ of all negative real numbers;
(3) the intervals $(4, \infty)$ and $[5, \infty)$ do not represent the same set of numbers since $(4, \infty)$ includes among its members all real numbers greater than 4 and less than 5, for instance 4.25, while $[5, \infty)$ does not include any of these numbers;
(4) similarly, $[1, 12] \neq \{1, 2, 3, \ldots, 12\}$, as, for example, $\pi \in [1, 12]$, but $\pi \notin \{1, 2, 3, \ldots, 12\}$.

Exercise 1.27. Compute each of the following.

(a) $(-\infty, 100) \cup [\pi, \infty)$;

(b) $(-\infty, 100) \cap [\pi, \infty)$;

(c) $(-\infty, 100) - [\pi, \infty)$;

(d) $\{4, 5, 6, 7, 8\} - [\pi, 100)$;

(e) $\mathbb{R} - (\mathbb{R}^+ \cup \mathbb{R}^-)$;

(f) $(-4, 10] \cap \mathbb{R}^-$;

(g) $((-\infty, 1] \cap (-\frac{1}{4}, \infty)) - [1, \infty)$.

2 Absolute value and distance

A careful treatment of the real number system depends heavily on the distance between real numbers when they are viewed as points on a line. We first look at the notion of *absolute value*, which analytically measures distance from the origin 0 on the number line, then use absolute value as a means to express the *distance* between any two real numbers.

Absolute value

For each real number a we define the **absolute value** of a, denoted $|a|$, so that

$$|a| = \begin{cases} a, & \text{if } a \geq 0; \\ -a, & \text{if } a < 0. \end{cases}$$

The definition is formulated in such a way that $|a|$ can be interpreted as the distance from 0 to a on the number line (we are thinking of distance from a geometric perspective, so a distance is never negative; see Figure 2.1).

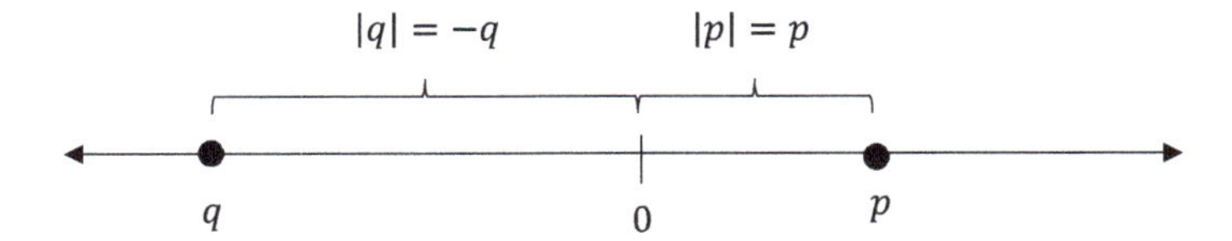

Figure 2.1: Visualizing absolute value as distance from the origin.

For example, $|4| = 4$ and $|-4| = -(-4) = 4$, which tells us that both 4 and -4 are 4 units away from 0 on the real line.

Exercise 2.1. Use the definition of absolute to evaluate each of $|\pi|$, $|-\pi|$, and $|0|$. Then evaluate $-|-t|$ if $t > 2$.

Theorem 2.1 (Fundamental properties of absolute value).

(1) *For any real number a, we have $|a| \geq 0$.*
(2) *For any real number a, we have $|a| = 0$ if and only if $a = 0$.*
(3) *For any real number a, we have $|-a| = |a|$.*
(4) *For any real number a, we have $-|a| \leq a \leq |a|$.*
(5) *For any real number a and any positive real number r, we have $|a| \leq r$ if and only if $-r \leq a \leq r$.*
(6) *For any real numbers a and b, we have $|ab| = |a||b|$.*

https://doi.org/10.1515/9783111429564-002

Observe that (1) of Theorem 2.1 says that the absolute value of a real number is always nonnegative; (2) says that there is precisely one number whose absolute value is 0, namely 0 itself; (3) says that a number and its opposite have the same absolute value; (4) says that every number lies "between" the opposite of its absolute value and its absolute value, inclusive; and (6) says that the absolute value of a product is the product of the absolute values of the factors comprising the product.

The equivalence stated in (5) permits us to rewrite certain inequalities involving absolute value in a way that does not use absolute value. More specifically, it tells us that a real number's distance from 0 is no greater than a specified positive number r, precisely when the number falls between $-r$ and r inclusive.

Exercise 2.2. Draw a picture that incorporates the real line and illustrates what is being conveyed by
(a) Theorem 2.1(3);
(b) Theorem 2.1(4);
(c) Theorem 2.1(5).

Exercise 2.3. Prove Theorem 2.1. *Hint*: The definition of absolute value forces us to consider cases in our arguments here.

The following result, known as the triangle inequality, is of great importance and will be used often in our future work. A bit later in this chapter, we will provide some insight as to where the name for this property comes from.

Theorem 2.2 (The triangle inequality). *For all real numbers a and b, we have*

$$|a + b| \leq |a| + |b|.$$

In other words, the absolute value of a sum is always less than or equal to the sum of the absolute values of the terms comprising the sum.

Proof. Consider any real numbers a and b. Since

$$|0 + 0| = |0| = 0 = 0 + 0 = |0| + |0|,$$

the desired result holds if $a = b = 0$. Now assume at least one of a or b is nonzero. Then from Theorem 2.1(4) we have $-|a| \leq a \leq |a|$ and $-|b| \leq b \leq |b|$. Adding yields

$$-(|a| + |b|) \leq a + b \leq |a| + |b|.$$

Now as a and b are not both zero, we see that $|a| + |b|$ is a positive number. Thus, by applying Theorem 2.1(5) to the previous inequality, we may conclude that

$$|a + b| \leq |a| + |b|.$$

□

Exercise 2.4. Show that for all real numbers a and b, we have $|a - b| \leq |a| + |b|$.

Theorem 2.3 (The reverse triangle inequality). *For all real numbers a and b, we have*

$$\big||a| - |b|\big| \leq |a - b|.$$

Proof. Let a and b be real numbers. Using the triangle inequality, we have

$$|a| = \big|(a - b) + b\big| \leq |a - b| + |b|$$

and

$$|b| = \big|(b - a) + a\big| \leq |b - a| + |a|.$$

From the first of these statements, we may conclude that

$$|a| - |b| \leq |a - b|$$

and from the second that

$$|b| - |a| \leq |b - a|.$$

This last inequality can be rewritten as

$$-\big(|a| - |b|\big) \leq |a - b|,$$

since $b - a$ and $a - b$ are opposites of each other, hence, have the same absolute value.

Now a number's absolute value is either equal to itself or to its opposite, so we know that $||a| - |b||$ is equal to $|a| - |b|$ or to $-(|a| - |b|)$. But as we have shown that these two quantities are less than or equal to $|a - b|$, it follows that $||a| - |b|| \leq |a - b|$. □

The distance between real numbers

We can represent the larger of two real numbers a and b by $\max\{a, b\}$, and the smaller by $\min\{a, b\}$, also observing that $\max\{a, b\} = \min\{a, b\}$ when $a = b$. To measure how close one real number is to another, we use the usual distance between them on the number line. By definition, the **distance** between real numbers a and b is

$$\max\{a, b\} - \min\{a, b\}.$$

Example 2.4. The distance between the real numbers 3 and 8 is

$$\max\{3, 8\} - \min\{3, 8\} = 8 - 3 = 5,$$

which reflects how far we must travel on the number line to get from 3 to 8 (Figure 2.2).

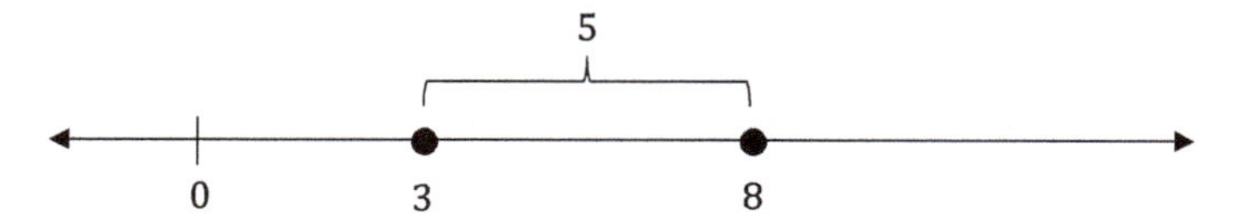

Figure 2.2: The distance between 3 and 8 is 5.

The ability to measure the distance between real numbers plays a central role in real analysis. Absolute value can be used to express the distance between real numbers.

Theorem 2.5 (Representing distance using absolute value). *Given real numbers a and b, the expression $|a - b|$ represents the distance between a and b.*

Proof. Consider any real numbers a and b.

Case: $a < b$.

Then $a - b < 0$ so that $|a - b| = -(a - b) = b - a$. Hence, the distance between a and b is

$$\max\{a, b\} - \min\{a, b\} = b - a = |a - b|.$$

Case: $a \geq b$.

Then $a - b \geq 0$ so that $|a - b| = a - b$. Hence, the distance between a and b is

$$\max\{a, b\} - \min\{a, b\} = a - b = |a - b|. \qquad \square$$

Example 2.4 (Continued). Thus, the distance between the real numbers 3 and 8, which is defined to be

$$\max\{3, 8\} - \min\{3, 8\} = 8 - 3 = 5,$$

can also be calculated as

$$|3 - 8| = |-5| = 5.$$

Whenever we are faced with an expression whose form is that of the absolute value of a difference, the expression may, and usually should, be interpreted as the distance between the real number quantities being subtracted. For instance, assuming t is a real number, the expression $|t^2 - 1|$ represents the distance between t^2 and 1.

Exercise 2.5. Interpret each expression as a distance and compute the distance.

(a) $|-3 - 11|$;

(b) $|11 + 3|$;

(c) $|\pi - \pi|$;

(d) $|x - (-x)|$ if x is negative.

Exercise 2.6. Locate the two numbers on the real line whose distance from the number 7.5 is 3. Relative to these two numbers, where are the real numbers whose distance from 7.5 is less than 3? Where are the real numbers whose distance from 7.5 is greater than 3?

Exercise 2.7. Suppose $r > 0$. Show that $|x - p| < r$ if and only if $p - r < x < p + r$. Also draw a picture to convince yourself of the truth of this equivalence.

Exercise 2.8. Solve the given inequality, justifying each step.
(a) $|2x - 3| < 8$;
(b) $\left|x^2 - 4\right| < 3$.

Exercise 2.9. Show that if $x < -5$ and $y > -2$, then $|x - y| > 3$.

More about the triangle inequality and distance

There is a result called the triangle inequality that you probably learned in high school geometry. It states that the sum of the lengths of any two sides of a triangle is always greater than the length of the third side. Drawing a picture makes this intuitively clear.

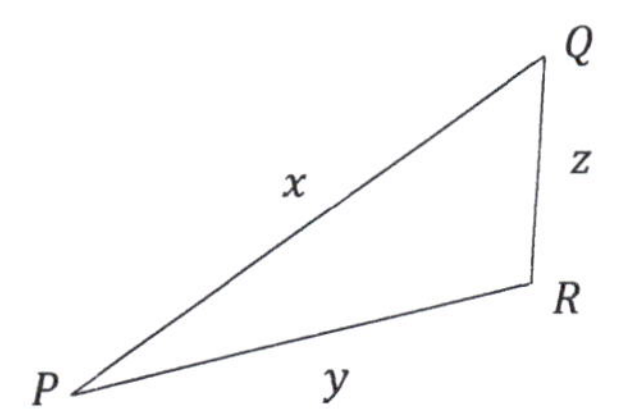

Figure 2.3: The triangle inequality for triangles: $x < y + z$.

For instance, if x, y, and z are the lengths, respectively, of sides PQ, PR, and RQ in the triangle shown in Figure 2.3, the triangle inequality from geometry tells us that

$$x < y + z,$$

which is based on the intuitive notion that the "shortest distance between two points is the straight-line distance." In other words, there is no advantage, from the perspective of minimizing distance, to go out of our way to visit point R if we are trying to get from point P to point Q.

Of course, the real line is not "spacious" enough to contain actual triangles, but interpreting the absolute value expressions in the inequality stated in the following theorem as distances helps to explain why Theorem 2.2 is referred to as the triangle inequality (for real numbers).

Theorem 2.6. *For all real numbers a, b, and c, we have*

$$|a-b| \leq |a-c| + |c-b|.$$

This theorem tells us that the distance between real numbers a and b on the number line is no larger than the sum of the distance between a and a "third" number c and the distance between this number c and b, meaning there is no advantage, from the perspective of minimizing distance, in going out of our way to visit any particular real number c if we are trying to get from the real number a to the real number b.

Exercise 2.10. Use the triangle inequality to prove Theorem 2.6.

Exercise 2.11. Consider again the reverse triangle inequality from Theorem 2.3.
(a) What is this inequality saying in terms of distance between numbers?
(b) Use the interpretation you gave in (a) to build an argument for the truth of the inequality.

Neighborhoods of a point

It is important for us to be able to analytically describe "relative closeness" of real numbers to a specified real number. The mathematical notion of *neighborhood* provides one means for doing so.

We call any interval of the form $(p-\varepsilon, p+\varepsilon)$, where ε is a positive number, a **neighborhood** of p (Figure 2.4).

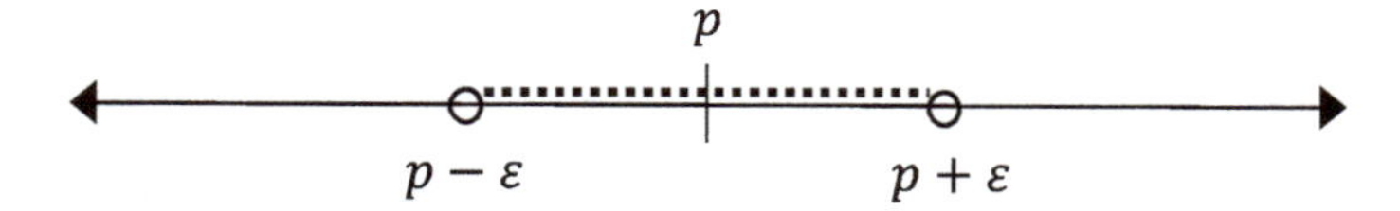

Figure 2.4: The neighborhood $(p-\varepsilon, p+\varepsilon)$ of p.

For a specific choice of ε the neighborhood

$$(p-\varepsilon, p+\varepsilon)$$

can be referred to as the **neighborhood of** p **with radius** ε or as, simply, the ε**-neighborhood** of p. The ε-neighborhood of p consists of all real numbers whose distance from p is less than ε.

The symbol ε is the lowercase Greek letter *epsilon* and is employed a great deal in mathematical analysis. Below we discuss why this symbol was chosen back when analysis was first developed.

Example 2.7. The neighborhoods of the real number 4 are all the intervals

$$(4 - \varepsilon, 4 + \varepsilon),$$

where $\varepsilon > 0$. In particular, the 0.5-neighborhood of 4, that is, the neighborhood of 4 having radius 0.5, is

$$(4 - 0.5, 4 + 0.5) = (3.5, 4.5),$$

and consists of all real numbers whose distance from 4 is less than 0.5.

Exercise 2.12. Which of the numbers

$$-10, \quad 1.8, \quad 10, \quad -9, \quad -11, \quad -8, \quad -8.5, \quad -8.2, \quad -11.8$$

are in the 1.8-neighborhood of -10?

Exercise 2.13. Find the radius of the neighborhood $(5, 13)$ of 9.

As $|x - p|$ represents the distance between x and p, we see that x is in $(p - \varepsilon, p + \varepsilon)$, the ε-neighborhood of p, if and only if $|x - p| < \varepsilon$ (see Exercise 2.7 above).

Example 2.7 (Continued). For instance, x is in the 0.5-neighborhood

$$(4 - 0.5, 4 + 0.5) = (3.5, 4.5)$$

of 4 if and only if the distance between x and 4 is less than 0.5, that is, if and only if $|x - 4| < 0.5$.

In Figure 2.5, both p and q are in the 0.5-neighborhood of 4, as each has distance from 4 that is less than 0.5, while r is not in the 0.5-neighborhood of 4 as its distance from 4 is greater than 0.5.

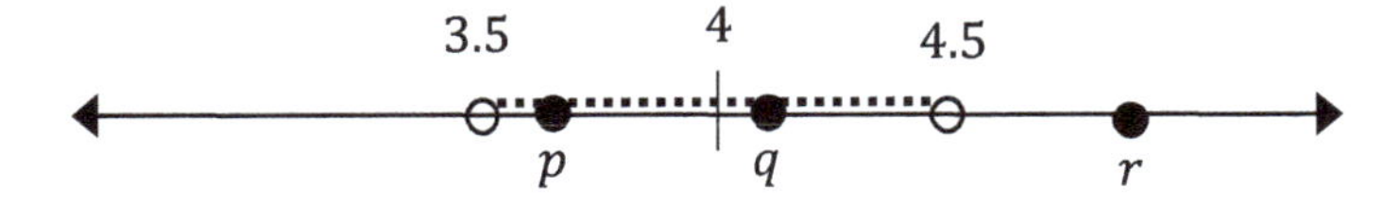

Figure 2.5: The 0.5-neighborhood of 4 includes p and q, but not r.

In symbols, then, we have

$$p \in (4 - 0.5, 4 + 0.5)$$

and

$$q \in (4 - 0.5, 4 + 0.5),$$

because $|p - 4| < 0.5$ and $|q - 4| < 0.5$, but

$$r \notin (4 - 0.5, 4 + 0.5)$$

because $|r - 4| > 0.5$.

Exercise 2.14. Suppose a and b are real numbers for which $|a - b| < 0.1$.
(a) Must a be in the 0.2-neighborhood of b?
(b) Must b be in the 0.02-neighborhood of a?
(c) Must b be in the 0.1-neighborhood of a?
(d) Must a be in the 0.00001-neighborhood of a?

A particular choice of a positive value for ε specifies what is meant in a given situation by numbers being "relatively close" to each other.

Example 2.7 (Continued). If we are told that numbers are to be considered "close" if they are less than 0.5 units from each other, we have been told to take $\varepsilon = 0.5$. In this case, the 0.5-neighborhood of 4 is the set $(3.5, 4.5)$ of all real numbers that can be considered "close" to 4 according to the prescribed measure of closeness determined by $\varepsilon = 0.5$.

It is conceivable for any positive number ε to be specified as a measure of relative closeness. The choice of a particular value really depends on the circumstances. For instance, sometimes we might need to measure length only so that we are sure the measured value is less than 0.5 inches from the true value, but in other situations it could be important that the measurement is less than 0.01 inches from the true value.

Historically, the choice to use the symbol ε came about as it was intended to signify "error." If we are trying to measure a quantity whose actual value is the real number p and our measurement error must be less than a specified positive number ε, then the neighborhood

$$(p - \varepsilon, p + \varepsilon)$$

can be viewed as the set of all acceptable measurements for the quantity. That is, while we "aim for" the particular number p, inherent limitations in whatever measuring device (ruler, thermometer, clock, and so on) we are employing typically force us to consider an "acceptable target range" $(p - \varepsilon, p + \varepsilon)$.

This perspective provides us with an initial sense of why neighborhoods are important in our study of the real number system. Later, we shall see that many fundamental calculus concepts, for example the limit of a function or sequence, can be precisely formulated by essentially measuring the "error" resulting from how two relevant quantities differ from each other.

Because the number line forms a geometric model for the set $\mathbb{R}$ of real numbers, with real numbers corresponding to points on the line, a real number is itself sometimes referred to as a *point*. We may then speak of, for example, a *neighborhood of a point p* rather than a *neighborhood of a real number p*.

The neighborhoods of a point fit inside one another based on their radii; when one neighborhood of a point has a smaller radius than another neighborhood of that same point, the former is a subset of the latter.

Example 2.7 (Continued). As $0.2 < 0.5$, the 0.2-neighborhood

$$(4 - 0.2, 4 + 0.2) = (3.8, 4.2)$$

of 4 is contained within the 0.5-neighborhood

$$(4 - 0.5, 4 + 0.5) = (3.5, 4.5)$$

of 4.

Exercise 2.15. Let A be the neighborhood of 2 having radius 2, let B be the 0.5-neighborhood of 2, and let C be the 0.5-neighborhood of 1. Determine all the subset relationships among A, B, and C.

---

3 The natural numbers, the integers, and the rational numbers

Among the subsets of the set of real numbers that play a significant role in our work are the sets of *natural numbers*, *integers*, and *rational numbers*. We use this chapter to formally introduce these types of real numbers and some of the special properties they possess, along with the important notions of *proof by induction* and *recursive definition*.

The natural numbers

We are already familiar with the real number 1 and its role as the multiplicative identity. Through iterated addition by 1, the other numbers used for counting can be obtained:

$$\begin{aligned}2 &= 1+1,\\3 &= 2+1 = 1+1+1,\\4 &= 3+1 = 1+1+1+1,\end{aligned}$$

and so on. These numbers are the *natural numbers*, which we formally introduce via the following axiom.

Axiom 3.1 (The natural numbers). *There exists a subset $\mathbb{N}$ of the set $\mathbb{R}$ of real numbers whose elements are called* ***natural numbers****. The natural numbers satisfy all of the following properties.*

(1) *The number 1 is a natural number.*
(2) *Given any natural number n, the number $n+1$ is also a natural number and is sometimes referred to as the* ***successor*** *of n.*
(3) *Axiom of induction: When a subset of the set $\mathbb{N}$ of all natural numbers includes 1 and includes the successor of each of its members, that subset must actually be the entire set $\mathbb{N}$ of natural numbers.*

Taken together, (1) and (2) of Axiom 3.1 tell us that all of the numbers $1, 2, 3, 4, \dots$ are natural numbers. Thus, $\{1, 2, 3, 4, \dots\}$ is certainly a subset of the set of all natural numbers. However, as $\{1, 2, 3, 4, \dots\}$ includes 1 and includes the successor of each of its members (i. e., includes one more than each of its members), we can use (3) to conclude that $\{1, 2, 3, 4, \dots\}$ is the entire set of natural numbers.

Proof by induction

The following proof strategy follows directly from the axiom of induction.

https://doi.org/10.1515/9783111429564-003

Proof Strategy 3.2 (Proof by induction). Suppose that for each natural number n we have a statement $S(n)$. To prove that all of the statements

$$S(1), \quad S(2), \quad S(3), \quad S(4), \quad \ldots,$$

are true, it is enough to do both of the following:

Base step: Prove that $S(1)$ is true.

Inductive step: Prove that whenever $S(k)$ is true, it follows that $S(k+1)$ is true.

The intuition behind proof by induction is often portrayed by the following "thought experiment." Imagine an infinite sequence of dominos numbered $1, 2, 3, 4, \ldots$. If we are convinced that both (1) the first domino in the sequence is knocked over (*base step*) and (2) for each domino that is knocked over, the one immediately following it is also knocked over (*inductive step*), it would seem to follow that all of the dominos are knocked over. Proof by induction simply generalizes the logic underlying this situation, replacing "domino is knocked over" with "statement is true."

In establishing the inductive step within a proof by induction, the assumption that $S(k)$ is true is referred to as the **inductive hypothesis**. It is important to keep in mind that the inductive hypothesis assumes only the truth of a single statement, because k is a fixed, though anonymous, natural number.

The proof of the following theorem provides a straightforward application of proof by induction.

Theorem 3.3. *Every natural number is positive. That is, for each n in $\mathbb{N}$, the number n is positive.*

Proof. We apply induction.

Base step: By the law of trichotomy, the number 1 must be positive, negative, or 0. But one of our basic assumptions about the real number system is that $1 \neq 0$. Moreover, it is impossible for 1 to be negative as, if 1 were negative, the true equation $1 \cdot 1 = 1$ would contradict the result we proved earlier that a negative number multiplied by a negative number is positive (Theorem 1.13). Hence, 1 must be positive. (This is essentially the same argument we gave in proving Theorem 1.17.)

Inductive step: Let k be a fixed but anonymous natural number and suppose that k is positive. As we have also demonstrated, in carrying out the base step, that 1 is positive, it follows that $k + 1$ is positive because the sum of positive numbers is always positive according to Axiom 1.12(1). □

In applying proof by induction in the above proof, note that $S(n)$, in this particular case, is the statement

n is positive,

which means the sequence

$$S(1), \quad S(2), \quad S(3), \quad S(4), \quad \dots,$$

in this case is

$$1 \textit{ is positive}, \quad 2 \textit{ is positive}, \quad 3 \textit{ is positive}, \quad 4 \textit{ is positive}, \quad \dots.$$

Our proof by induction establishes that all of the infinitely many statements in this sequence are true.

Exercise 3.1. Use Axiom 3.1 to prove Proof Strategy 3.2, proof by induction.

Exercise 3.2. Use induction to prove that for each natural number n, we have $n \geq 1$ (so, as expected, 1 is the least natural number).

We can use induction to establish that the set $\mathbb{N}$ of natural numbers is closed under addition and multiplication.

Theorem 3.4 (Closure of $\mathbb{N}$ under addition and multiplication). *The sum of natural numbers is always a natural number, as is the product.*

Proof. We use induction to show that $\mathbb{N}$ is closed under addition, leaving the proof that $\mathbb{N}$ is closed under multiplication as an exercise.

Consider any natural number m, which we treat as fixed throughout our proof. We use induction to prove that for any natural number n, the sum $m+n$ is a natural number.

Base step: We must show that $m + 1$ is a natural number. This is indeed the case, because $m+1$ is the successor of the natural number m, and by Axiom 3.1(2), the successor of any natural number is always a natural number.

Inductive step: Consider a fixed but anonymous natural number k and assume $m + k$ is a natural number; we must show that $m + (k + 1)$ is a natural number. Using the associativity of addition, we see that $m + (k + 1) = (m + k) + 1$, which is the successor of the natural number $m + k$, hence, is itself a natural number. □

Exercise 3.3. Prove that the set $\mathbb{N}$ of natural numbers is closed under multiplication. *Hint*: Take an approach similar to that used above to establish that $\mathbb{N}$ is closed under addition by fixing m as an arbitrary natural number and then using induction to show that for every natural number n, the product mn is a natural number.

The natural numbers are generated by starting with the number 1 and then iteratively adding one, thus producing 2, then 3, then 4, and so on. Our intuition is that this process never comes to an end, meaning there must be infinitely many natural numbers. The following theorem shows that this is the case by demonstrating that each natural number generated by adding one is different from the natural numbers previously generated.

Our proof is by induction, but within the base step we employ proof by contradiction, and within the inductive step we actually employ contradiction twice.

Theorem 3.5. *For each natural number n, the natural number $n+1$ is not in $\{1,2,\ldots,n\}$. Hence, there are infinitely many natural numbers.*

Proof. We apply induction.

Base step: We must show $1+1 \notin \{1\}$. Suppose to the contrary that $1+1 \in \{1\}$. Then $1+1=1$ and it follows that $1=0$, a contradiction.

Inductive step: Assume k is a natural number for which $k+1 \notin \{1,2,\ldots,k\}$. We must show $(k+1)+1 \notin \{1,2,\ldots,k,k+1\}$.

If $(k+1)+1=1$, then $k+1=0$, which is impossible as $k+1$ is a natural number but 0 is not.

If $(k+1)+1 \in \{2,\ldots,k,k+1\}$, then $k+1 \in \{1,2,\ldots,k\}$, contradicting our inductive hypothesis.

Thus, $(k+1)+1 \notin \{1,2,\ldots,k,k+1\}$. □

The integers

The **integers** are the real numbers

$$\ldots, \quad -3, \quad -2, \quad -1, \quad 0, \quad 1, \quad 2, \quad 3, \quad \ldots.$$

That is, the integers comprise the natural numbers, their additive inverses, and the additive identity 0. The set of all integers is denoted by $\mathbb{Z}$, the set of all positive integers by $\mathbb{Z}^+$, and the set of all negative integers by $\mathbb{Z}^-$. Note that the set of all positive integers is really just the set $\mathbb{N}$ of natural numbers, making the terms *positive integer* and *natural number* synonymous.

We now prove that the sum of integers is always an integer. One of the cases within the proof involves showing that whenever a positive integer is added to a negative integer, the resulting sum is an integer. The argument for this case is a bit delicate, involving the use of “induction within induction.”

Specifically, the statement we prove in this case is the following:

For every positive integer m and every negative integer n, the sum $m+n$ is an integer.

This statement has the form

$$\forall m \in \mathbb{Z}^+, P(m), \tag{*}$$

where $P(m)$ is the statement

For every negative integer n, the sum $m+n$ is an integer.

To prove (*) via induction, in the base step we must show (*) is true for $m = 1$, in other words, that

For every negative integer n, the sum $1 + n$ is an integer.

Note that this statement is equivalent to the statement

For every positive integer n, the sum $1 + (-n)$ is an integer,

which we shall prove by induction. Thus, we are going to use induction as the means through which we establish the base step of another induction argument.

Then, in order to carry out the inductive step for proving (*), we assume k is a positive integer and also that it is true that for every negative integer n, the sum $k + n$ is an integer; we then need to show that

For every negative integer n, the sum $(k + 1) + n$ is an integer,

or, equivalently,

For every positive integer n, the sum $(k + 1) + (-n)$ is an integer.

We then go on to use induction to establish the truth of this statement, another instance of applying induction to verify a particular result needed within another induction proof.

Theorem 3.6 (Closure of $\mathbb{Z}$ under addition and multiplication). *The sum of integers is always an integer, as is the product.*

Proof. We show that $\mathbb{Z}$ is closed under addition. You will show that $\mathbb{Z}$ is closed under multiplication in Exercise 3.4. Let m and n be integers. We must show $m+n$ is an integer. We consider cases based on whether each of m and n are zero, positive, or negative.

Case: At least one of m or n is 0.

As 0 is the additive identity, $m+n$ must be equal to one of m or n, which are integers, so in this case, $m + n$ is an integer.

Case: Both m and n are positive.

Then $m + n$ is a positive integer as we have already shown that the sum of natural numbers (positive integers) is also a natural number. Therefore, in this case, $m + n$ is an integer.

Case: Both m and n are negative.

Then each of $-m$ and $-n$ is a positive integer, so as the sum of positive integers is a positive integer, we may conclude that $(-m) + (-n)$ is a positive integer. As the additive inverse of a positive integer (natural number) is also an integer, it follows that $-((-m) + (-n)) = m + n$ is an integer.

Case: One of m or n is positive and the other is negative.

Without loss of generality, assume that m is positive and n is negative. We use induction to prove that for each positive integer m the statement

For every negative integer n, the sum $m + n$ is an integer,

is true.

Base step: Here we must show that for every negative integer n, the sum $1 + n$ is an integer or, equivalently, that for every positive integer n, the sum $1 + (-n)$ is an integer, which we now do by induction.

For the base step we must show that $1 + (-1)$ is an integer. But $1 + (-1) = 0$ and 0 is an integer, so $1 + (-1)$ is an integer.

For the inductive step, we assume k is a positive integer and $1 + (-k)$ is an integer; we must show $1+(-(k+1))$ is an integer. Note that $1+(-(k+1)) = -k$, so as k is a positive integer, it follows that $-k$ is a negative integer. Thus, $1 + (-(k + 1))$ is an integer.

Inductive step: Here we assume that k is a positive integer and for every negative integer n, the sum $k + n$ is an integer; we must show that for every negative integer n, the sum $(k + 1) + n$ is an integer, or equivalently, that for every positive integer n, the sum $(k + 1) + (-n)$ is an integer, which we now demonstrate via induction.

For the base step, we must show that $(k+1)+(-1)$ is an integer. Since $(k+1)+(-1) = k$ and k is a positive integer, it follows that $(k + 1) + (-1)$ is an integer.

For the inductive step, we assume j is a positive integer and $(k+1)+(-j)$ is an integer; we must show $(k + 1) + (-(j + 1))$ is an integer. Observe that $(k + 1) + (-(j + 1)) = k + (-j)$. Now, recall that in the inductive hypothesis for the "outer" induction, we assumed that whenever a negative integer is added to k, the result is an integer. Since $-j$ is a negative integer, we may therefore conclude that $k+(-j)$ is an integer, which means $(k+1)+(-(j+1))$ is an integer. □

It is customary to use the phrase *without loss of generality* in a proof when there really are multiple possibilities to consider, but the other instances can be handled by interchanging variable names in the explicitly presented argument for one of the possibilities. We did this in the above proof when we wrote, "Without loss of generality, assume that m is positive and n is negative." The argument for the other possibility, that m is negative and n is positive, can be obtained by using the commutativity of addition to rewrite $m + n$ as $n + m$, then interchange the names n and m, so that the argument we did write for the situation where m is positive and n is negative would now apply.

Exercise 3.4. Prove that the set $\mathbb{Z}$ of integers is closed under multiplication. *Hint*: Consider cases; you do not need to use induction.

We learn in elementary school that when an integer n is divided by a positive integer m, we can obtain an integer quotient and a nonnegative integer remainder that is less than m. For example, when 47 is divided by 9, the quotient is 5 and the remainder is 2, when −47 is divided by 9, the quotient is −6 and the remainder is 7, and when 72 is divided by 9, the quotient is 8 and the remainder is 0.

The following theorem provides the foundation for performing division in this way. In the proof, note that we apply induction with a base of $n = 0$ rather than $n = 1$. In general, we may apply induction to establish that all of the statements

$$S(n^*), \quad S(n^*+1), \quad S(n^*+2), \quad S(n^*+3), \quad \ldots,$$

are true, where the initial index n^* is any integer, not necessarily 1, but this requires that we adjust the base step accordingly.

Theorem 3.7 (The division algorithm). *For any integer n and any positive integer m, there exist unique integers q, called the* **quotient**, *and r, called the* **remainder**, *for which $n = mq + r$ and $0 \leq r < m$.*

Proof. Let m be any positive integer, which remains fixed through the entire proof. We begin by using induction to prove the existence of a quotient and remainder when a nonnegative integer n is "divided by" m.

The base step is established by observing that $0 = m \cdot 0 + 0$, where 0 is a nonnegative integer that is necessarily less than the positive integer m.

Then, for the inductive step, assume that k is a nonnegative integer for which there exist integers q and r such that $k = mq + r$ and $0 \leq r < m$. If $r < m - 1$, observe that $k + 1 = mq + (r + 1)$, where m and $r + 1$ are integers with $0 \leq r + 1 < m$. If $r = m - 1$, observe that $k+1 = m(q+1)+0$, where $q+1$ and 0 are integers, and of course $0 \leq 0 < m$. So in either case, we have obtained an integer quotient and suitable integer remainder when $k + 1$ is "divided by" m.

Having established the base and inductive steps, we may now conclude that a quotient and remainder exist whenever a nonnegative integer n is "divided by" m. In Exercise 3.5, we ask you to show that a quotient and remainder exist for the situation in which n is a negative integer.

To demonstrate the uniqueness of the quotient and remainder for a given integer n, assume $n = mq + r$ and $n = mq' + r'$, where all of q, r, q', and r' are integers, with $0 \leq r < m$ and $0 \leq r' < m$. We show $r' = r$, from which it follows that $q' = q$. Assume to the contrary that $r' \neq r$ and, without loss of generality, further assume that $r' < r$. It follows from the inequalities $0 \leq r < m$ and $0 \leq r' < m$ that $r - r' < m$. But as $r' < r$, we really have $0 < r - r' < m$. Note also that as $mq + r = mq' + r'$, we are able to conclude that $r - r' = m(q' - q)$. Thus, $0 < m(q' - q) < m$, from which it follows that $0 < q' - q < 1$, an impossibility because $q' - q$ is an integer and, as you demonstrated in Exercise 3.2 that 1 is the least positive integer, there are no integers between 0 and 1. ☐

Exercise 3.5. Show that the existence portion of the Division Algorithm holds for any negative integer n. *Hint*: Let n be a negative integer and consider the positive integer $-n$. There is no need to use induction in your argument.

An integer m is said to be a **divisor** of an integer n provided there is an integer k such that $n = mk$ (in other words, the remainder is 0 when the division algorithm is applied to divide n by m). When m is a divisor of n, we also say that n is a **multiple** of m and that n is **divisible by** m.

Example 3.8. Since $40 = 5 \cdot 8$, we can say that 5 is a divisor of 40 or, equivalently, that 40 is a multiple of 5.

Similarly, as $-40 = 5 \cdot (-8)$, we can say that 5 is a divisor of -40, so that -40 is a multiple of 5.

On the other hand, 5 is not a divisor of 32 because $32 = 5 \cdot 6 + 2$, so by the uniqueness of the quotient and remainder determined via the division algorithm, there is no integer k for which $32 = 5k$. Hence, 32 is not a multiple of 5.

Theorem 3.9. *If m and n are positive integers, and m is a divisor of n, then $m \leq n$.*

Proof. Assume m and n are positive integers, and suppose m is a divisor of n; then there is a positive integer k such that $n = mk$. As $k \geq 1$ and m is positive, it follows that $n = mk \geq m \cdot 1 = m$. □

An integer n is **even** if $n = 2k$ for some integer k and is **odd** if $n = 2k + 1$ for some integer k.

Exercise 3.6. Use the division algorithm to prove that every integer is either even or odd, and no integer is both even and odd.

Proof by contraposition

The logical equivalence of a conditional statement $P \Rightarrow Q$ and its contrapositive $\sim Q \Rightarrow \sim P$ (see Example B.9 in Appendix B) provides us with another method for proving an *if... then...* statement.

Proof Strategy 3.10 (Proof by contraposition). To prove the *if... then...* statement

$$P \Rightarrow Q$$

by contraposition, assume $\sim Q$ and then deduce $\sim P$.

We use this method of proof to show that if the square of an integer is even, then the integer itself must be even.

Theorem 3.11. *Let n be an integer. If n^2 is even, then n is even.*

Proof. We proceed by contraposition. Assume the integer n is not even. Thus, n is odd, so $n = 2k + 1$ for some integer k. It follows that

$$n^2 = (2k + 1)^2 = 4k^2 + 4k + 1 = 2(2k^2 + 2k) + 1,$$

where $2k^2 + 2k$ is an integer as both the product and the sum of integers is an integer. Thus, n^2 is odd, hence, not even. □

Exercise 3.7. Let n be an integer. Use contraposition to prove that if n^2 is odd, then n is odd.

Decimal representation of positive integers

The ten **decimal digits** are 0, 1, 2, 3, 4, 5, 6, 7, 8, and 9, where 0 and 1 are, respectively, the additive and multiplicative identities in the real number system, and where we define

$$\begin{aligned}\mathbf{2} &= 1 + 1,\\ \mathbf{3} &= 2 + 1,\\ \mathbf{4} &= 3 + 1,\\ \mathbf{5} &= 4 + 1,\\ \mathbf{6} &= 5 + 1,\\ \mathbf{7} &= 6 + 1,\\ \mathbf{8} &= 7 + 1,\\ \mathbf{9} &= 8 + 1.\end{aligned}$$

Integers larger than 9 can be represented by forming an appropriate string of decimal digits, leading to the *base ten* or *decimal* representation of the integer.

Example 3.12. We understand the mathematical meaning of 3482 to be

$$3482 = 3 \cdot 10^3 + 4 \cdot 10^2 + 8 \cdot 10 + 2.$$

The digits in this representation are the remainders obtained when we apply “dividing by 10” via the division algorithm to 3482 and the resulting quotients:

$$\begin{aligned}3482 &= 10 \cdot 348 + 2,\\ 348 &= 10 \cdot 34 + 8,\\ 34 &= 10 \cdot 3 + 4,\\ 3 &= 10 \cdot 0 + 3.\end{aligned}$$

The remainders are the digits in 3482 because

$$\begin{aligned}3482 &= 10 \cdot 348 + 2\\ &= 10 \cdot (10 \cdot 34 + 8) + 2\\ &= 10 \cdot \big(10 \cdot (10 \cdot 3 + 4) + 8\big) + 2\\ &= 3 \cdot 10^3 + 4 \cdot 10^2 + 8 \cdot 10^1 + 2.\end{aligned}$$

The procedure described in this example is generalized in the following theorem.

Theorem 3.13. *Given an integer b greater than 1, each positive integer n can be uniquely expressed in the form*

$$n = a_k \cdot b^k + a_{k-1} \cdot b^{k-1} + \cdots + a_2 \cdot b^2 + a_1 \cdot b + a_0,$$

where k is a nonnegative integer and where, for each nonnegative integer i for which $i \leq k$, the integer a_i is between 0 and $b - 1$, inclusive, with $a_k \neq 0$.

The expression

$$a_k \cdot b^k + a_{k-1} \cdot b^{k-1} + \cdots + a_2 \cdot b^2 + a_1 \cdot b + a_0$$

for a positive integer n is called the **base b representation of n** and is usually written as

$$(a_k a_{k-1} \quad \cdots \quad a_2 a_1 a_0)_b.$$

When $b = 10$, we have the **base ten** or **decimal representation** of n, and in this case, we typically leave out the parentheses and subscript, writing for instance 3482 rather than $(3482)_{\text{ten}}$. Base two representations are also called **binary representations** and base three representations **ternary representations**.

Example 3.14. When the division algorithm is applied to divide an integer by 3, the only possible remainders are 0, 1, and 2, which are the three **ternary digits**. Every positive integer can be expressed as an appropriate string of these digits. For example, the base three representation of 33 is $(1020)_{\text{three}}$ as $33 = 1 \cdot 3^3 + 0 \cdot 3^2 + 2 \cdot 3 + 0$.

Exercise 3.8. Use the division algorithm to prove Theorem 3.13. You should find it helpful to study Example 3.12.

Exercise 3.9. Find the binary representation of 40.

Recursive definition

The axiom of induction enables us to define a function f having domain $\mathbb{N}$ by a process known as **recursive definition**. The idea is to directly specify $f(1)$ and then, for each natural number n for which $f(n)$ has been defined, uniquely define $f(n+1)$ by means of either $f(n)$ or a combination of n and $f(n)$.

Example 3.15. Given any real number a, the expression a^n, called the ***n*th power of** a, comprised of the **base** a and the **exponent** n, is defined recursively on the set of natural numbers as follows:

$$a^1 = a,$$

and

$$a^{n+1} = a^n \cdot a,$$

for each natural number n. For example,

$$\pi^3 = \pi^2 \cdot \pi = \pi^1 \cdot \pi \cdot \pi = \pi \cdot \pi \cdot \pi.$$

The following theorem formally expresses the process through which a function is defined recursively. After proving the theorem, we illustrate how it is applied in the context of Example 3.15, where we defined natural number powers of a real number.

Theorem 3.16 (Principle of recursive definition). *Given a set A, an element a in A, and a function $g : \mathbb{N} \times A \to A$, there is a unique function $f : \mathbb{N} \to A$ for which both $f(1) = a$ and $f(n+1) = g(n, f(n))$ for each natural number n. This function f is said to have been **defined recursively**.*

Proof. We show that under the stated hypotheses a function $f : \mathbb{N} \to A$ with the properties listed has been defined and is unique. Let

$$K = \{n \in \mathbb{N} \mid f(n) \text{ is a uniquely defined element in } A\}.$$

We need only show $K = \mathbb{N}$. First, as $f(1)$ has been uniquely defined to be the element a in A, we know that $1 \in K$. Now consider any n in K; then we know $f(n)$ is a uniquely defined element in A. So, as g is a function with domain $\mathbb{N} \times A$ and codomain A, the value $g(n, f(n))$ is a uniquely determined element in A. But as $f(n+1)$ is defined to be $g(n, f(n))$, we may now conclude that $n+1 \in K$. Thus, as K is a subset of $\mathbb{N}$ that includes 1 and includes the successor of each of its members, it follows by the axiom of induction 3.1(3), that $K = \mathbb{N}$, that is, the function f is uniquely defined for all natural numbers. □

Example 3.15 (Continued). In applying Theorem 3.16 to define natural number powers of a real number, we are taking A to be the set $\mathbb{R}$ and the distinguished element a in this set to be whatever real number a is to serve as the base, then defining the function

$g\colon \mathbb{N}\times\mathbb{R} \to \mathbb{R}$ so that $g(n,x) = x \cdot a$. The theorem then tells us there is a unique function $f : \mathbb{N} \to \mathbb{R}$ for which

$$f(1) = a$$

and

$$f(n+1) = g(n, f(n)) = f(n) \cdot a$$

for each natural number n.

In this example, the value of the function g at (n, x) depends only on x, so that the value of $f(n+1)$ depends only on the "previous" value $f(n)$, along with the given real number a, and does not involve the value of n itself. It is quite often the case that a recursively defined function's value $f(n+1)$ depends only on $f(n)$.

We now use induction to show that for every natural number n, the number $f(n)$ is actually a^n. The base step is established by observing that $f(1) = a = a^1$. The inductive step is established by noting that, under the assumption that $f(k) = a^k$, it follows that $f(k+1) = f(k) \cdot a = a^k \cdot a = a^{k+1}$.

With the recursive definition from Example 3.15 in hand, we can use induction to prove various properties of natural number powers.

Theorem 3.17. *Let a and b be any real numbers. Then for any natural number n, we have* $(ab)^n = a^n b^n$.

Proof. We apply induction. First, note that $(ab)^1 = ab = a^1 b^1$, which establishes the base step. Now consider any fixed natural number k and assume $(ab)^k = a^k b^k$; it follows that

$$(ab)^{k+1} = (ab)(ab)^k = (ab)(a^k b^k) = (a \cdot a^k)(b \cdot b^k) = a^{k+1} b^{k+1},$$

which establishes the inductive step. □

Exercise 3.10. Justify each equality stated in establishing the inductive step in the proof of Theorem 3.17 above. (As we have pointed out before, justification of each step in a proof is not always provided explicitly. But we should always be ready to supply the justification if asked to do so.)

Exercise 3.11. Let a be a real number. Use induction to prove each of the following.

(a) For any natural numbers m and n, we have $a^m a^n = a^{m+n}$.

(b) For any natural numbers m and n, we have $(a^m)^n = a^{mn}$.

Exercise 3.12. Assuming $a > 1$, use induction to prove that $a^n > a$ for every natural number n for which $n \geq 2$.

Exercise 3.13. Assuming $a \geq 2$, use induction to prove that $a^n > n$ for every natural number n.

Exercise 3.14. Suppose $a > -1$. Use induction to prove that for any natural number n, we have $(1 + a)^n \geq 1 + na$. (This inequality is known as *Bernoulli's inequality*.)

In Example 3.15, we defined a^n for any real number a and any positive integer n. If a is nonzero, we can also define a^n, the **nth power of a**, for any nonpositive integer n as follows:

$$a^0 = 1$$

and, when n is a negative integer,

$$a^n = \left(\frac{1}{a}\right)^{-n}.$$

Thus, for example,

$$5^0 = 1$$

and

$$5^{-3} = \left(\frac{1}{5}\right)^3 = \frac{1}{125}.$$

Exercise 3.15. Given below are several familiar properties of exponents. Prove each of them under the assumption that a and b are real numbers, and m and n are integers (as necessary, you may also assume a and/or b is nonzero). If you prove the properties in the order listed, you can use earlier properties in the list within your proofs of later ones. Also keep in mind that (b), (c), and (d) were already established for positive integer exponents (see Theorem 3.17 and Exercise 3.11).

(a) $a^{-n} = \frac{1}{a^n}$.
(b) $(ab)^n = a^n b^n$.
(c) $a^m a^n = a^{m+n}$.
(d) $(a^m)^n = a^{mn}$.
(e) $\frac{a^m}{a^n} = a^{m-n}$.
(f) $\left(\frac{a}{b}\right)^n = \frac{a^n}{b^n}$.

In the next example, we use recursive definition to define factorials. This time the value of $f(n + 1)$ depends on both $f(n)$ and n.

Example 3.18. In Theorem 3.16, take $A = \mathbb{N}$, set $a = 1$, and define the function $g : \mathbb{N} \times \mathbb{N} \to \mathbb{N}$ so that $g(x, y) = x \cdot y$. Then there is a unique function $f : \mathbb{N} \to \mathbb{N}$ for which

$$f(1) = 1$$

and

$$f(n+1) = g(n, f(n)) = n \cdot f(n)$$

for each natural number n. Observe that

$$\begin{aligned} f(2) &= 1 \cdot f(1) = 1 \cdot 1 = 1, \\ f(3) &= 2 \cdot f(2) = 2 \cdot 1 = 2, \\ f(4) &= 3 \cdot f(3) = 3 \cdot 2 = 6, \\ f(5) &= 4 \cdot f(4) = 4 \cdot 6 = 24, \end{aligned}$$

and so forth.

For each nonnegative integer n, we define n **factorial**, denoted $n!$, to be the number $f(n+1)$. Thus,

$$\begin{aligned} 0! &= f(1) = 1, \\ 1! &= f(2) = 1, \\ 2! &= f(3) = 2 = 2 \cdot 1, \\ 3! &= f(4) = 6 = 3 \cdot 2 \cdot 1, \\ 4! &= f(5) = 24 = 4 \cdot 3 \cdot 2 \cdot 1, \end{aligned}$$

and so forth. In Exercise 3.16, you use induction to show that, in general, when n is a natural number, $n!$ is the product of the least n natural numbers.

Exercise 3.16. Use induction to prove that for every natural number n, the number $n!$ defined recursively in Example 3.18, is actually equal to $1 \cdot 2 \cdot 3 \cdot \dots \cdot n$, the product of the least n natural numbers.

Exercise 3.17. Use induction to prove that for every natural number n, we have $2^{n-1} \leq n!$.

Sigma notation for sums

If m and n are integers for which $m \leq n$, and $a_m, a_{m+1}, a_{m+2}, \dots, a_n$ are real numbers, the sum

$$a_m + a_{m+1} + a_{m+2} + \dots + a_n$$

can be represented in what is referred to as **Σ-notation** or **summation notation** as

$$\sum_{i=m}^{n} a_i.$$

The number m is called the **lower index**, the number n the **upper index**, and the letter i the **index of summation**. A different letter can be used for the index of summation; for instance, $\sum_{k=m}^{n} a_k$ represents the same sum. The symbol Σ is the uppercase Greek letter *sigma* and is used in mathematics to signify "summation."

Example 3.19. In the summation

$$\sum_{i=2}^{5} i^3,$$

the lower index is 2 and the upper index is 5, so the index i of summation in turn takes on the values 2, 3, 4, and 5. For each of these values of i, the expression i^3 appearing after the symbol Σ is evaluated, yielding 2^3, 3^3, 4^3, and 5^3. These quantities are then added, so that

$$\sum_{i=2}^{5} i^3 = 2^3 + 3^3 + 4^3 + 5^3.$$

Exercise 3.18. Write out $\sum_{i=10}^{15}(6 - i)$ in expanded form and then find the numerical value of this sum.

Exercise 3.19. Express

$$[(-2)^2 + (-3)^3] + [(-2)^3 + (-3)^4] + \cdots + [(-2)^{12} + (-3)^{13}]$$

using Σ-notation.

When the value of a mathematical expression does not change, we say the expression is **constant**. For our purposes, a *constant* usually refers to a **real number constant**, a mathematical expression representing a fixed real number.

The following properties of summations follow immediately from the commutative and associative properties of addition, and the distributive property.

Theorem 3.20 (Fundamental properties of summations).

(1) *The summation of a sum is the sum of the summations. That is,*

$$\sum_{i=m}^{n} (a_i + b_i) = \sum_{i=m}^{n} a_i + \sum_{i=m}^{n} b_i.$$

(2) *A constant factor can be factored out of a summation. That is, assuming c is a real number constant,*

$$\sum_{i=m}^{n} ca_i = c \sum_{i=m}^{n} a_i.$$

There are a number of other summation formulas that we occasionally require.

Theorem 3.21 (Some useful summation formulas).
(1) Summing a constant: *Assuming c is a real number constant,* $\sum_{i=1}^{n} c = nc$.
(2) Summing consecutive positive integers: $\sum_{i=1}^{n} i = \frac{1}{2}n(n+1)$.
(3) Summing consecutive odd positive integers: $\sum_{i=1}^{n}(2i-1) = n^2$.
(4) Summing consecutive squares: $\sum_{i=1}^{n} i^2 = \frac{1}{6}n(n+1)(2n+1)$.
(5) Summing consecutive cubes: $\sum_{i=1}^{n} i^3 = \frac{1}{4}n^2(n+1)^2$.
(6) Summing consecutive powers: *Assuming r is a real number constant for which* $r \neq 0$ *and* $r \neq 1$, *we have* $\sum_{i=0}^{n} r^i = \frac{1-r^{n+1}}{1-r}$.

Formula (1) of this theorem holds because repeated addition can be performed as multiplication:

$$\sum_{i=1}^{n} c = \underbrace{c + c + \cdots + c}_{n \text{ summands}} = nc.$$

The other formulas can be proved using induction.

Example 3.22. Using several of the results stated in Theorems 3.20 and 3.21, we may observe that

$$\begin{aligned}
&\sum_{n=10}^{22}\left(\frac{1}{2}n + n^3 - 2\right) \\
&= \frac{1}{2}\sum_{n=10}^{22} n + \sum_{n=10}^{22} n^3 - \sum_{n=10}^{22} 2 \\
&= \frac{1}{2}\left(\sum_{n=1}^{22} n - \sum_{n=1}^{9} n\right) + \sum_{n=1}^{22} n^3 - \sum_{n=1}^{9} n^3 - \left(\sum_{n=1}^{22} 2 - \sum_{n=1}^{9} 2\right) \\
&= \frac{1}{2}\left(\frac{1}{2}\cdot 22\cdot 23 - \frac{1}{2}\cdot 9\cdot 10\right) + \frac{1}{4}\cdot 22^2\cdot 23^2 - \frac{1}{4}\cdot 9^2\cdot 10^2 - (22\cdot 2 - 9\cdot 2) \\
&= 62{,}062.
\end{aligned}$$

Exercise 3.20. Use the appropriate summation properties and formulas to evaluate each of the following.
(a) $\sum_{i=1}^{36}(i+2)$;
(b) $\sum_{k=1}^{100}(k^2-k)$;
(c) $\sum_{k=20}^{86} 2k^3$;
(d) $\sum_{j=1}^{10} 2^{j+1}$.

Exercise 3.21. Use induction arguments to prove each of (2), (3), (4), (5), and (6) of Theorem 3.21.

The binomial theorem

The following famous result provides a formula for the expanded form of a natural number power of a binomial.

Theorem 3.23 (The binomial theorem). *Given nonzero real numbers a and b, for any natural number n, we have*

$$(a+b)^n = \sum_{k=0}^{n} \binom{n}{k} a^k b^{n-k},$$

where, by definition, $\binom{n}{k} = \frac{n!}{k!(n-k)!}$ *and is referred to as a* ***binomial coefficient****.*

Proof. We are taking a and b to be nonzero to avoid the undefined expression 0^0; besides, it is straightforward to simplify $(a+b)^n$ if either a or b is zero. Our proof is by induction.

For the base step, note that since $\binom{1}{0} = \frac{1!}{0!(1-0)!} = 1$ and $\binom{1}{1} = \frac{1!}{1!(1-1)!} = 1$, it follows that

$$(a+b)^1 = b + a = \binom{1}{0} a^0 b^{1-0} + \binom{1}{1} a^1 b^{1-1} = \sum_{k=0}^{1} \binom{1}{k} a^k b^{1-k},$$

so the desired result holds when $n = 1$.

For the inductive step, assume that $(a+b)^n = \sum_{k=0}^{n} \binom{n}{k} a^k b^{n-k}$ for some natural number n. In our calculations here, we use the fact, easily verified, that each of the binomial coefficients $\binom{n}{0}$, $\binom{n+1}{0}$, $\binom{n}{n}$, and $\binom{n+1}{n+1}$ is equal to 1, as well as the fact (see Exercise 3.22 below) that $\binom{n}{k-1} + \binom{n}{k} = \binom{n+1}{k}$. Keeping all this in mind, we have

$$\begin{aligned}
(a+b)^{n+1} &= (a+b)(a+b)^n \\
&= (a+b) \sum_{k=0}^{n} \binom{n}{k} a^k b^{n-k} \\
&= a \sum_{k=0}^{n} \binom{n}{k} a^k b^{n-k} + b \sum_{k=0}^{n} \binom{n}{k} a^k b^{n-k} \\
&= \sum_{k=0}^{n} \binom{n}{k} a^{k+1} b^{n-k} + \sum_{k=0}^{n} \binom{n}{k} a^k b^{n-k+1} \\
&= \sum_{k=1}^{n+1} \binom{n}{k-1} a^k b^{(n+1)-k} + \sum_{k=0}^{n} \binom{n}{k} a^k b^{(n+1)-k} \qquad \text{(see Exercise 3.23 below)} \\
&= \binom{n}{0} a^0 b^{n+1} + \sum_{k=1}^{n} \left[\binom{n}{k-1} + \binom{n}{k}\right] a^k b^{(n+1)-k} + \binom{n}{n} a^{n+1} b^0 \\
&= \binom{n+1}{0} a^0 b^{n+1} + \sum_{k=1}^{n} \binom{n+1}{k} a^k b^{(n+1)-k} + \binom{n+1}{n+1} a^{n+1} b^0 \\
&= \sum_{k=0}^{n+1} \binom{n+1}{k} a^k b^{(n+1)-k}.
\end{aligned}$$

□

Example 3.24. Observe that

$$
\begin{aligned}
(x^2+2)^3 &= \sum_{k=0}^{3} \binom{3}{k} \cdot (x^2)^k \cdot 2^{3-k} \\
&= \binom{3}{0} \cdot (x^2)^0 \cdot 2^{3-0} + \binom{3}{1} \cdot (x^2)^1 \cdot 2^{3-1} \\
&\quad + \binom{3}{2} \cdot (x^2)^2 \cdot 2^{3-2} + \binom{3}{3} \cdot (x^2)^3 \cdot 2^{3-3} \\
&= \frac{3!}{0!(3-0)!} \cdot 8 + \frac{3!}{1!(3-1)!} \cdot x^2 \cdot 4 + \frac{3!}{2!(3-2)!} \cdot x^4 \cdot 2 + \frac{3!}{3!(3-3)!} \cdot x^6 \\
&= 1 \cdot 8 + 3 \cdot x^2 \cdot 4 + 3 \cdot x^4 \cdot 2 + 1 \cdot x^6 \\
&= 8 + 12x^2 + 6x^4 + x^6.
\end{aligned}
$$

Exercise 3.22. Prove that if k and n are natural numbers for which $k \le n$, then

$$\binom{n}{k-1} + \binom{n}{k} = \binom{n+1}{k}.$$

Exercise 3.23. In the proof of the binomial theorem, the summation

$$\sum_{k=0}^{n} \binom{n}{k} a^{k+1} b^{n-k}$$

is rewritten as

$$\sum_{k=1}^{n+1} \binom{n}{k-1} a^k b^{(n+1)-k}.$$

By writing out the first few terms of each of these summations, we can easily see that they agree. This is an example of "re-indexing" a summation. Show that if we let $j = k+1$, the original summation $\sum_{k=0}^{n} \binom{n}{k} a^{k+1} b^{n-k}$ can be expressed as $\sum_{j=1}^{n+1} \binom{n}{j-1} a^j b^{(n+1)-j}$. Why can the latter summation then be rewritten as $\sum_{k=1}^{n+1} \binom{n}{k-1} a^k b^{(n+1)-k}$?

The rational numbers

A real number that can be expressed in the form $\frac{m}{n}$ for some integer m and some nonzero integer n is called a **rational number**. Thus, rational numbers are quotients (i. e., ratios) of integers. The set of all rational numbers is denoted by $\mathbb{Q}$, the set of all positive rational numbers by $\mathbb{Q}^+$, and the set of all negative rational numbers by $\mathbb{Q}^-$.

A given rational number can always be represented by a fraction in "reduced" form. For instance, the reduction of $\frac{15}{20}$ to $\frac{3}{4}$ can be justified because

$$\frac{15}{20} = 15 \cdot \frac{1}{20} = (3 \cdot 5) \cdot \left(\frac{1}{4} \cdot \frac{1}{5}\right) = \left(3 \cdot \frac{1}{4}\right) \cdot \left(5 \cdot \frac{1}{5}\right) = \frac{3}{4} \cdot 1 = \frac{3}{4}.$$

The calculation in this example is generalized as follows.

Theorem 3.25. *If m and n are integers with $n \neq 0$, then there exist integers j and k having no common divisor greater than 1, and for which $k \neq 0$ and $\frac{m}{n} = \frac{j}{k}$.*

Proof. Let m and n be integers with $n \neq 0$. Take d to be the greatest divisor that is common to m and n. Then there exist integers j and k such that $m = dj$ and $n = dk$, where $d \neq 0$ and $k \neq 0$ because $n \neq 0$. As $d \neq 0$, the multiplicative inverse $\frac{1}{d}$ of d exists. It follows that

$$\frac{m}{n} = m \cdot \frac{1}{n} = dj \cdot \frac{1}{dk} = d \cdot j \cdot \frac{1}{d} \cdot \frac{1}{k} = j \cdot \frac{1}{k} = \frac{j}{k}$$

(*can you justify each step in this calculation?*). Note that j and k have no common divisor greater than 1, for otherwise we could argue that d is not the greatest divisor common to m and n, which would be a contradiction. □

The next theorem provides us with the basis for determining when two quotients of integers represent the same rational number, and for calculating products and sums of rational numbers.

Theorem 3.26. *Let a, b, c, and d be real numbers, with b and d both nonzero.*
(1) $\frac{a}{b} = \frac{c}{d}$ *if and only if* $ad = bc$.
(2) $\frac{a}{b} \cdot \frac{c}{d} = \frac{ac}{bd}$.
(3) $\frac{a}{b} + \frac{c}{d} = \frac{ad+bc}{bd}$.

Exercise 3.24. Prove Theorem 3.26.

Corollary 3.27 (Closure of $\mathbb{Q}$ under addition and multiplication). *The sum of rational numbers is always a rational number, as is the product.*

Exercise 3.25. Explain how Corollary 3.27 follows from relevant parts of Theorem 3.26 and the closure of $\mathbb{Z}$ under addition and multiplication.

Not all real numbers are rational. A real number that is not rational is called **irrational.** Famous examples of irrational numbers include $\sqrt{2}$, π, and e. The ancient Greeks "constructed" $\sqrt{2}$ as the length of the hypotenuse of a right triangle whose legs both have unit length, but as discussed near the beginning of Chapter 1, our analytic approach to the real number system does not permit such geometric constructions. In the next chapter, we take a different conceptual approach to demonstrating the existence of $\sqrt{2}$. For now, assuming there is indeed a positive real number $\sqrt{2}$ whose square is 2, we show that this number cannot be rational.

Theorem 3.28. *The real number $\sqrt{2}$ is irrational.*

Proof. Suppose by way of contradiction that $\sqrt{2}$ is rational. Then $\sqrt{2} = \frac{m}{n}$, where m is an integer and n is a nonzero integer. Assume the fraction $\frac{m}{n}$ has been reduced to lowest terms so that m and n have no common divisor greater than 1. Squaring both sides of the equation $\sqrt{2} = \frac{m}{n}$ yields

$$2 = \frac{m^2}{n^2},$$

from which it follows that

$$m^2 = 2n^2.$$

Since any product of integers is itself an integer, we see that m^2 and n^2 are both integers. It then follows from our equation $m^2 = 2n^2$ that m^2 is even. But as m^2 is even, we may conclude, using Theorem 3.11, that m is even, so that $m = 2k$ for some integer k. Thus, we have

$$2n^2 = m^2 = (2k)^2 = 2 \cdot 2k^2,$$

from which it follows that

$$n^2 = 2k^2.$$

Thus, as k^2 is an integer, we may conclude that n^2 is even, from which it follows that n is even. But now we have deduced that both m and n are even, which means they are both divisible by 2, a contradiction to our assumption that m and n have no common divisor greater than 1. □

In Chapter 27, we prove the number e is irrational and in Chapter 28, we prove the number π is irrational.

Exercise 3.26. Prove that if a is rational and b is irrational, then $a + b$ must be irrational. *Hint*: Try proof by contradiction.

Exercise 3.27. If a is rational and b is irrational, must ab be irrational? Justify your conclusion.

Exercise 3.28. Assuming $\sqrt{3}$ and $\sqrt[3]{2}$ both exist within the real number system, prove that they are irrational.

More about set inclusion and set equality

We have defined the set of integers so that it includes among its members all natural numbers; thus, $\mathbb{N} \subseteq \mathbb{Z}$. Also, the rational numbers have been defined as those real numbers that are ratios of integers; thus, $\mathbb{Q} \subseteq \mathbb{R}$. It may not be immediately clear that the set inclusion $\mathbb{Z} \subseteq \mathbb{Q}$ also holds, so we use the following strategy for proving that one set is a subset of another set to show that this is the case.

Proof Strategy 3.29 (Proving set inclusion). To prove the subset relationship

$$A \subseteq B,$$

consider an anonymous member x of the set A and show x is a member of the set B.

Thus, to prove $\mathbb{Z} \subseteq \mathbb{Q}$, we should begin with an anonymous member of the set $\mathbb{Z}$ and then show this object is actually a member of the set $\mathbb{Q}$. Our proof also makes use of the familiar fact, which you are asked to prove below, that dividing by 1 leaves a number unchanged.

Theorem 3.30. *The set $\mathbb{Z}$ of integers is a subset of the set $\mathbb{Q}$ of rational numbers; that is, $\mathbb{Z} \subseteq \mathbb{Q}$.*

Proof. Consider any n in the set $\mathbb{Z}$. Then $n = \frac{n}{1}$, because dividing by 1 does not change a number's value. Now, as both n and 1 are integers, we see that we have been able to express n as a ratio of integers. Thus, using the definition of rational number, we may conclude that n is in the set $\mathbb{Q}$. Thus, we have shown that $\mathbb{Z} \subseteq \mathbb{Q}$. □

Exercise 3.29. Prove that for every real number a, we have $\frac{a}{1} = a$.

We can also apply Proof Strategy 3.29 to establish more general subset relationships.

Theorem 3.31. *For any sets A and B,*
(1) $A \cap B \subseteq A$;
(2) $A - B \subseteq A$;
(3) $A \subseteq A \cup B$.

Proof. Let A and B be sets. We prove $A \cap B \subseteq A$, leaving the proofs of the other results as exercises.

Suppose $x \in A \cap B$. Then the definition of intersection of sets allows us to conclude that $x \in A$, so our proof is complete. □

Exercise 3.30. Prove that for any sets A and B, we have $A - B \subseteq A$ and $A \subseteq A \cup B$.

By definition, sets are equal provided they have exactly the same members, meaning each set includes all the members of the "other" set. We thus have the following strategy, sometimes referred to as **double containment**, for demonstrating a set equality.

Proof Strategy 3.32 (Proving sets are equal). To prove the set equality

$$A = B,$$

prove both $A \subseteq B$ and $B \subseteq A$.

When A and B are sets and we prove $A = B$, it is customary within the proof to use the label ($\subseteq$) to indicate where we prove the "forward" containment $A \subseteq B$ and the label ($\supseteq$) to indicate where we prove the "backward" containment $B \subseteq A$.

Example 3.33. A number line sketch (you might want to make one) suggests that

$$(-4, 10] \cap \mathbb{R}^- = (-4, -1) \cup [-1, 0).$$

We use Proof Strategy 3.32 to prove this set equality.

Proof. ($\subseteq$) Assume $x \in (-4, 10] \cap \mathbb{R}^-$. Then, by definition of intersection, $x \in (-4, 10]$ and $x \in \mathbb{R}^-$, that is, $-4 < x \leq 10$ and $x < 0$. Thus, $-4 < x < 0$, so that either $-4 < x < -1$ or $-1 \leq x < 0$. Hence, $x \in (-4, -1)$ or $x \in [-1, 0)$, so we may now conclude, using the definition of union, that $x \in (-4, -1) \cup [-1, 0)$.

($\supseteq$) Assume $x \in (-4, -1) \cup [-1, 0)$. Then, by definition of union, $x \in (-4, -1)$ or $x \in [-1, 0)$, that is, $-4 < x < -1$ or $-1 \leq x < 0$. It follows that $-4 < x < 0$. Since $-4 < x < 0$ and $0 \leq 10$, we may conclude that $-4 < x \leq 10$, in other words, $x \in (-4, 10]$. Also, since $x < 0$, we may conclude that $x \in \mathbb{R}^-$. Hence, using the definition of intersection, we have $x \in (-4, 10] \cap \mathbb{R}^-$. □

Exercise 3.31. Prove that for any sets A and B, we have $A - (A \cap B) = A - B$.

Exercise 3.32. Prove that the intersection of two neighborhoods of a real number p is also a neighborhood of p.

---

4 The completeness of the real numbers

In this chapter, we develop the final assumption, the completeness axiom, we shall make about the real number system. This assumption, which is built upon the notion of *upper bounds* for a subset of the set of real numbers, has far-reaching consequences and plays a critical role throughout the rest of the book.

Taking stock and looking forward

The field axiom, Axiom 1.1, and the positivity axiom, Axiom 1.12, set forth the fundamental assumptions we have made thus far concerning the *real numbers*, their *addition* and *multiplication*, and their classification into the three types we have designated *positive*, *negative*, and *zero*. They also give rise to the ordering of the real numbers via the *less than* relation. Taken together, these axioms make the resulting real number system into what is called an **ordered field**. Precisely the same properties, however, are satisfied by the rational numbers, thus making the rational number system an ordered field, too.

Exercise 4.1. Verify that all the properties listed in Axioms 1.1 and 1.12, are satisfied if we replace the set $\mathbb{R}$ of real numbers with the set $\mathbb{Q}$ of rational numbers, and the set $\mathbb{R}^+$ of positive real numbers with the set $\mathbb{Q}^+$ of positive rational numbers, while also restricting the operations of addition and multiplication so that they are applied only to rational numbers. Thus, $\mathbb{Q}$ is an ordered field, just as $\mathbb{R}$ is.

So we may legitimately ask whether it is possible to study *rational* analysis in place of *real* analysis. Is there anything that structurally distinguishes the real number system from the rational number system? Would what we learned in our calculus courses still hold if we were to work with functions whose inputs and outputs were restricted to being rational numbers?

To explore this question, we recall the notion of *continuity of a function* and the *intermediate value theorem*, which states that a continuous function cannot "skip over" numbers lying between two of its values.

Theorem 4.1 (The intermediate value theorem). *Let I be an interval, let the function $f : I \to \mathbb{R}$ be continuous, and let a and b be numbers in I for which $a < b$. If r is any number between $f(a)$ and $f(b)$, then there exists a number c such that $a < c < b$ and $f(c) = r$.*

This theorem provides the basis for the commonly encountered geometric characterization of continuous functions as those whose graphs are "unbroken," meaning the graph can be drawn without lifting pencil from paper. The diagram in Figure 4.1 visualizes what the intermediate value theorem is telling us.

Note that the value of r shown in the diagram can be shifted to any position between $f(a)$ and $f(b)$ and there is still a corresponding number c between a and b for

https://doi.org/10.1515/9783111429564-004

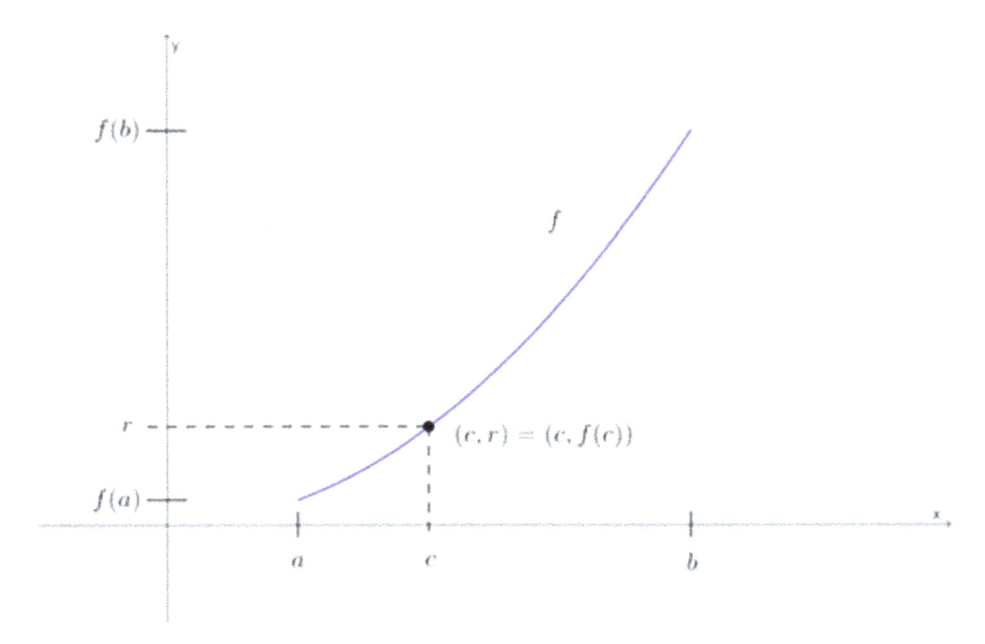

Figure 4.1: The intermediate value theorem.

which $f(c) = r$; without continuity of the function f, such a conclusion would not be guaranteed.

We now show how the intermediate value theorem can be used to argue for the existence of $\sqrt{2}$ as a real number. To that end, consider the function $f : \mathbb{R} \to \mathbb{R}$ defined so that $f(x) = x^2 - 2$. The graph of f is displayed in Figure 4.2.

In Chapter 13, we provide an analytic definition of continuity of a function and use it to prove that all polynomial functions are continuous, from which it follows that this particular function f is continuous, though this conclusion should also seem credible as the graph of f appears to be unbroken. Observe that

$$f(1) = -1 < 0$$

and

$$f(2) = 2 > 0,$$

so by the intermediate value theorem it should follow that there is a real number c with

$$1 < c < 2$$

and

$$f(c) = c^2 - 2 = 0,$$

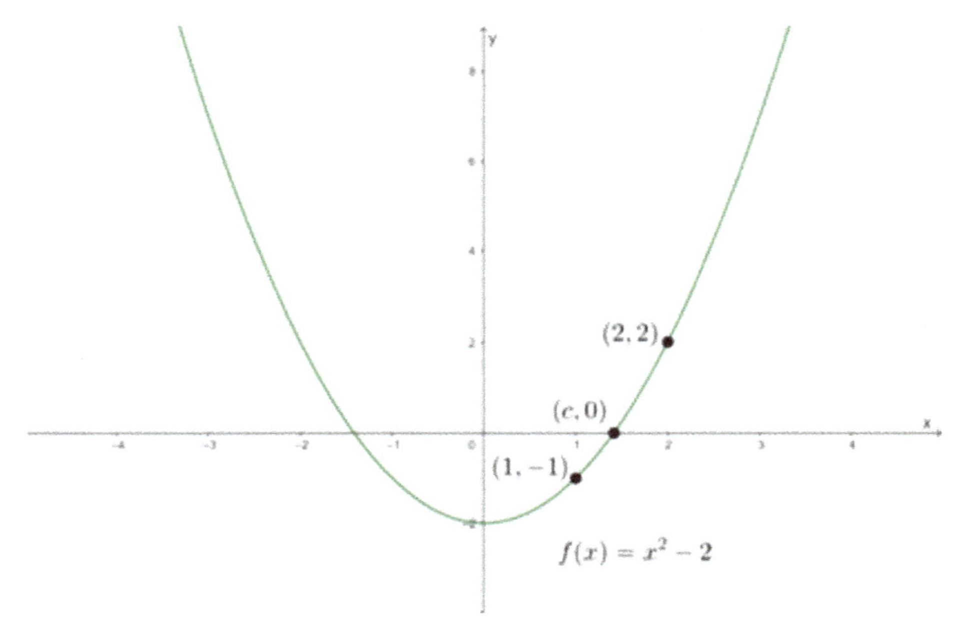

Figure 4.2: The graph of the function f where $f(x) = x^2 - 2$.

that is, $c^2 = 2$. But as we proved in Theorem 3.28 there is no rational number c such that $c^2 = 2$, we have therefore demonstrated the existence of an irrational number, namely, $c = \sqrt{2}$, within the set of real numbers. Essentially, we are showing here that if we did not have the irrational numbers present as elements of $\mathbb{R}$, then the intermediate value theorem would fail to hold and continuous functions would not behave as we intuitively expect them to.

We therefore really have no choice but to bring forth an additional axiom that permits us to eventually prove the intermediate value theorem and thereby provide for the existence within the real number system of irrational numbers such as $\sqrt{2}$. The irrationals, whose existence cannot be established on the basis of the field and positivity axioms alone, "complete" the real number system.

To motivate the formulation of this final axiom, observe that the irrational number $\sqrt{2}$ is the smallest real number that is at least as big as all the members of the set $\{x \in \mathbb{R}^+ \mid x < \sqrt{2}\}$. As our final axiom is intended to guarantee the existence of such a "least upper bound" for certain sets of real numbers, it is advantageous for us to first gain some experience with this notion of *upper bound*.

Exercise 4.2. What is it specifically that permits us to conclude that $\sqrt{2}$ is the least real number that is greater than every member of the set $\{x \in \mathbb{R}^+ \mid x < \sqrt{2}\}$?

Upper bounds and lower bounds

A real number m that is in a subset A of $\mathbb{R}$ is the **maximum** (or **greatest element**) of A if $a \leq m$ for all a in A, and is the **minimum** (or **least element**) of A if $m \leq a$ for all a in A.

A subset A of $\mathbb{R}$ need not have a maximum, but if it does, the maximum is unique (see Exercise 4.3) and is denoted by $\max(A)$. Likewise, A need not have a minimum, but if it does, the minimum is unique and is denoted by $\min(A)$.

Example 4.2. The set $\{0, 2, -\pi\}$ has maximum 2 and minimum $-\pi$. The minimum of the singleton set $\{0.87\}$ is the same number as the maximum, namely, 0.87.

Exercise 4.3. Show that if the maximum of a subset of $\mathbb{R}$ exists, it is unique.

Example 4.3. The interval $[0, 1]$ has maximum 1 and minimum 0, but the interval $(0, 1)$ has neither a maximum nor a minimum.

To see why $(0, 1)$ has no maximum, suppose to the contrary that m is the maximum of $(0, 1)$. As $m \in (0, 1)$, it follows that $m < 1$. But the density property, Theorem 1.20, tells us that between any two distinct real numbers we can find another, so there exists a real number c for which $m < c < 1$. Observe that $c > 0$ as well, so that $c \in (0, 1)$ with $m < c$, contradicting our assumption that m is the maximum of $(0, 1)$.

Exercise 4.4. Show that the interval $(0, 1)$ has no minimum.

Example 4.4. The interval $(0, 1]$ has maximum 1, but no minimum. The interval $[0, 1)$ has minimum 0, but no maximum.

Example 4.5. For each natural number n, note that $n + 1$ is a larger natural number; hence, the set $\mathbb{N}$ of all natural numbers has no maximum (though it has minimum 1).

The set $\mathbb{R}$ of all real numbers has neither a minimum nor a maximum. Note also that the set $\mathbb{R}^+ = (0, \infty)$ of positive real numbers also has neither a minimum nor a maximum.

Exercise 4.5. Explain why $\mathbb{R}$ has no maximum. Then explain how the fact that $\mathbb{R}$ has no maximum implies that $\mathbb{R}$ has no minimum.

Given a nonempty subset A of $\mathbb{R}$, a real number b is an **upper bound** of A if $a \leq b$ for all a in A, and is a **lower bound** of A if $b \leq a$ for all a in A. An upper bound (respectively, lower bound) *of* a set is also sometimes said to be an upper bound (respectively, lower bound) *for* the set.

Example 4.6. The real number 10.3 is an upper bound of $\{0, 2, -\pi\}$ because $0 \leq 10.3$, $2 \leq 10.3$, and $-\pi \leq 10.3$.

The real number 2 is also an upper bound of $\{0, 2, -\pi\}$ because $0 \leq 2$, $2 \leq 2$, and $-\pi \leq 2$.

There are many other upper bounds of $\{0, 2, -\pi\}$. Specifically, any real number that is greater than or equal to 2 is an upper bound of $\{0, 2, -\pi\}$.

Exercise 4.6. Find three different lower bounds for the set $\{0, 2, -\pi\}$. What is the largest real number that is a lower bound for $\{0, 2, -\pi\}$?

A subset of $\mathbb{R}$ need not have an upper bound, and it may not have a lower bound. If an upper bound exists, any larger real number is also an upper bound, so upper bounds, if they exist, are not unique. Similarly, if a lower bound exists, any smaller real number is also a lower bound, so lower bounds, if they exist, are not unique.

A subset of $\mathbb{R}$ having an upper bound is said to be **bounded above**, while a subset of $\mathbb{R}$ having a lower bound is said to be **bounded below**. A subset of $\mathbb{R}$ having both a lower bound and an upper bound is said to be **bounded** and a subset of $\mathbb{R}$ that is not bounded is said to be **unbounded**.

Example 4.7. We saw in Example 4.6 that the set $\{0, 2, -\pi\}$ has an upper bound; hence, $\{0, 2, -\pi\}$ is bounded above. In Exercise 4.6, you showed that $\{0, 2, -\pi\}$ has a lower bound, so $\{0, 2, -\pi\}$ is bounded below. Because $\{0, 2, -\pi\}$ has both a lower bound and an upper bound, the set $\{0, 2, -\pi\}$ is bounded.

The set $\mathbb{R}$ itself has no upper bound, so it is not bounded above. Simply knowing that $\mathbb{R}$ is not bounded above is enough for us to conclude that $\mathbb{R}$ is unbounded, but it is also the case that as $\mathbb{R}$ has no lower bound, $\mathbb{R}$ is not bounded below.

The set $\mathbb{N}$ of natural numbers is bounded below, but we eventually show $\mathbb{N}$ is not bounded above (see Theorem 4.18); thus, $\mathbb{N}$ is unbounded.

Exercise 4.7. Determine whether the given set is bounded below, whether it is bounded above, and whether it is bounded or unbounded.

(a) $\mathbb{Z}$;
(b) $\mathbb{R}^+$;
(c) $\{x \in \mathbb{R} \mid x^2 > 1\}$;
(d) $\{x \in \mathbb{R} \mid x^2 < 1\}$.

When a subset A of $\mathbb{R}$ has a maximum, the maximum is an upper bound of A. Similarly, when a subset A of $\mathbb{R}$ has a minimum, the minimum is a lower bound of A. However, a subset of $\mathbb{R}$ need not have a maximum in order to have upper bounds, and need not have a minimum in order to have lower bounds.

Example 4.8. The interval $(0, 1)$ has neither a maximum nor a minimum. But any real number greater than or equal to 1 is an upper bound of $(0, 1)$, and any real number less than or equal to 0 is a lower bound of $(0, 1)$.

Suprema and infima

Given a nonempty subset A of $\mathbb{R}$, a real number b is the

(1) **supremum** or **least upper bound** of A if b is an upper bound of A and $b \leq u$ for every upper bound u of A;

(2) **infimum** or **greatest lower bound** of A if b is a lower bound of A and $l \leq b$ for every lower bound l of A.

A subset A of $\mathbb{R}$ need not have a supremum, but if it does, the supremum is unique and is denoted by $\sup(A)$; similarly, A need not have an infimum, but if it does, the infimum is unique and is denoted by $\inf(A)$.

Example 4.9. Since 1 is greater than or equal to every member of the interval $(0,1]$, we see that 1 is an upper bound of $(0,1]$. Moreover, since $1 \in (0,1]$, no number less than 1 can be an upper bound of $(0,1]$. Therefore, as indicated in Figure 4.3, the number 1 is the least upper bound of $(0,1]$.

Also, since 0 is less than or equal to every member of $(0,1]$, we see that 0 is a lower bound of $(0,1]$. Moreover, since you show in Exercise 4.8 that there is no lower bound of $(0,1]$ that is greater than 0, it follows that 0 is the greatest lower bound of $(0,1]$, as is also indicated in Figure 4.3.

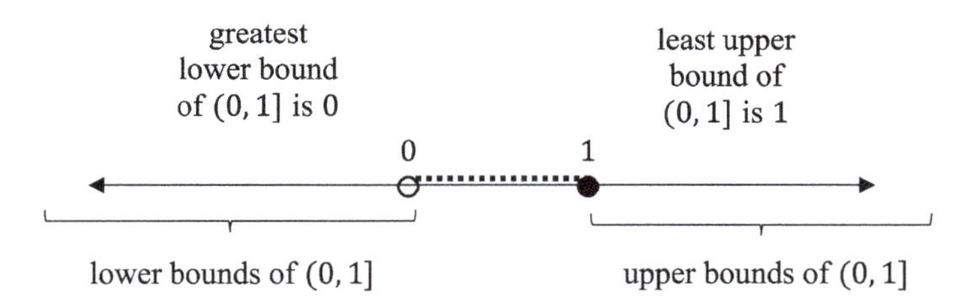

Figure 4.3: Visualizing the upper bounds and lower bounds of $(0,1]$.

When it exists, the least upper bound of a set may or may not be a member of the set. Similarly, when it exists, the greatest lower bound of a set may or may not be a member of the set.

Example 4.9 (Continued). Note that the least upper bound 1 of $(0,1]$ is a member of $(0,1]$, whereas the greatest lower bound 0 is not.

To prove a real number is the least upper bound of a subset of $\mathbb{R}$, we usually just apply the definition of least upper bound, first showing the number is an upper bound, then showing it is no larger than any upper bound.

Example 4.10. Here we actually prove that 1 is the supremum of the interval $(0,1)$ by establishing that 1 is an upper bound of $(0,1)$ and that 1 is no larger than any upper bound of $(0,1)$.

Proof. First, we show that 1 is an upper bound of $(0,1)$. To do so, consider any x in $(0,1)$. Then $x < 1$. Thus, as every element of $(0,1)$ is less than 1, we may conclude that 1 is an upper bound of $(0,1)$.

Now we show that for every upper bound u of $(0,1)$, we have $1 \leq u$; it then immediately follows that 1 is the least upper bound of $(0,1)$. Consider any upper bound u of $(0,1)$ and suppose by way of contradiction that $u < 1$. As $0.5 \in (0,1)$ and u is an upper bound of $(0,1)$, it follows that $0.5 \leq u$. Then the density property tells us there exists a real number a such that $0 < 0.5 \leq u < a < 1$. So as $a \in (0,1)$ and u is an upper bound of $(0,1)$, we must have $a \leq u$, which contradicts our earlier deduction that $u < a$. As the assumption that $u < 1$ has led to a contradiction, we may conclude that $1 \leq u$, the desired result. □

Our argument here uses the fact that 0.5 is an element of $(0,1)$ in order to be sure that the number a obtained via the density property is also an element of $(0,1)$. Any element of $(0,1)$ could have been chosen in place of 0.5.

Similarly, we usually prove a real number is the greatest lower bound of a subset of $\mathbb{R}$ by first showing it is a lower bound, then showing it is no smaller than any lower bound.

Exercise 4.8. Prove that 0 is the infimum of the interval $(0,1]$.

Exercise 4.9. Show that if a nonempty subset of $\mathbb{R}$ has a supremum, then this supremum is unique.

The notion of supremum (respectively, infimum) is a generalization of maximum (respectively, minimum).

Theorem 4.11. *If a subset A of $\mathbb{R}$ has a maximum m, then* $\sup(A) = m$*. Similarly, if a subset A of $\mathbb{R}$ has a minimum m, then* $\inf(A) = m$.

Proof. We prove that the maximum of a set, when it exists, is the supremum of the set, leaving the other result as an exercise. Suppose $A \subseteq \mathbb{R}$ and $m = \max(A)$. Then for every a in A, we have $a \leq m$. Thus, m is an upper bound of A. Now consider any upper bound u of A. As u is an upper bound and $m \in A$, it follows that $m \leq u$. Therefore, m is the least upper bound of A. □

Exercise 4.10. Prove that if a subset A of $\mathbb{R}$ has a minimum m, then $\inf(A) = m$.

Exercise 4.11. Suppose A is a subset of $\mathbb{R}$ for which $\sup(A)$ exists. Let $s = \sup(A)$ and $B = \{-a \mid a \in A\}$. Show that $\inf(B)$ exists and $\inf(B) = -s$.

Exercise 4.12. Let $S = \{\frac{(-1)^n}{n} \mid n \in \mathbb{N}\}$. Find $\sup(S)$ and $\inf(S)$, justifying your conclusions.

When something is not always true, it fails at least once, and when there does not exist an instance for which something is true, it fails always. We invariably apply these principles, articulated in Axiom B.14 in Appendix B, when we need to form the negation of a statement that includes one or more of the quantifiers *for all* and *there exists*.

Example 4.12. The negation of the false statement

Every real number is positive,

is the true statement

Some real numbers are not positive.

Example 4.13. The negation of the statement

For every positive number ε, there exists a in the set A for which $u - \varepsilon < a$.

is the statement

There exists a positive number ε such that for each a in the set A, we have $u - \varepsilon \geq a$.

Here we do not know what set A is and we do not know what number u is, so we do not know which of these two statements is false and which is true, though once values have been assigned to A and u, exactly one of the statements is true and exactly one is false.

Note the several uses of Axiom B.14, including the specific illustration given in Example 4.13, within the proof of the following alternate method for establishing that an upper bound of a subset of $\mathbb{R}$ is the least upper bound of the subset.

Theorem 4.14 (Alternate characterization of the least upper bound). *Let u be an upper bound of a nonempty subset A of $\mathbb{R}$. Then $u = \sup(A)$ if and only if, for every positive number ε, there exists a number a in A for which $u - \varepsilon < a$.*

Proof. ($\Rightarrow$) Suppose $u = \sup(A)$ and consider any positive number ε. Then $u - \varepsilon < u$ and as u is the *least* upper bound of A, it follows that $u - \varepsilon$ is not an upper bound of A. Hence, there exists a in A such that $u - \varepsilon < a$.

($\Leftarrow$) We apply contraposition. Suppose $u \neq \sup(A)$. Then, as we are assuming u is an upper bound of A, there exists an upper bound b of A for which $b < u$. Take $\varepsilon = u - b$, which is positive. Then as b is an upper bound of A, we have

$$a \leq b = u - (u - b) = u - \varepsilon$$

for all a in A. Thus, for the particular positive number $\varepsilon = u - b$, there is no member of A for which $u - \varepsilon < a$. □

In the statement of this theorem, it is imperative to be aware of the dependency of a on the choice of ε. In other words, the element a of A whose existence is asserted can be different for different values of ε.

Example 4.10 (Continued). Earlier we demonstrated that 1 is the supremum of the interval $(0, 1)$. According to Theorem 4.14 this means that for any positive number ε we should be able to find an element a of $(0, 1)$ that is greater than $1 - \varepsilon$; the particular element a can vary, though, with the particular value of ε.

For example, if we take $\varepsilon = 0.4$, we may take $a = 0.7$ since $0.7 \in (0, 1)$ and

$$1 - \varepsilon = 1 - 0.4 = 0.6 < 0.7 = a.$$

But if we take $\varepsilon = 0.01$, we cannot take $a = 0.7$ since now

$$1 - \varepsilon = 1 - 0.01 = 0.99 > 0.7 = a.$$

Instead, for $\varepsilon = 0.01$ we must choose a number between 0.99 and 1, which is of course feasible via the density property. For instance, letting $a = 0.999$ we have

$$1 - \varepsilon = 1 - 0.01 = 0.99 < 0.999 = a < 1.$$

When the value of one thing, say a, can vary with the value of another thing, say ε, we sometimes write a_ε in place of just a to emphasize the dependency of a on ε. For instance, we can use this convention to restate Theorem 4.14 as follows.

Theorem 4.14 (Restated). *Let u be an upper bound of a nonempty subset A of $\mathbb{R}$. Then $u = \sup(A)$ if and only if for every positive number ε, there exists a number a_ε in A for which $u - \varepsilon < a_\varepsilon$ (Figure 4.4).*

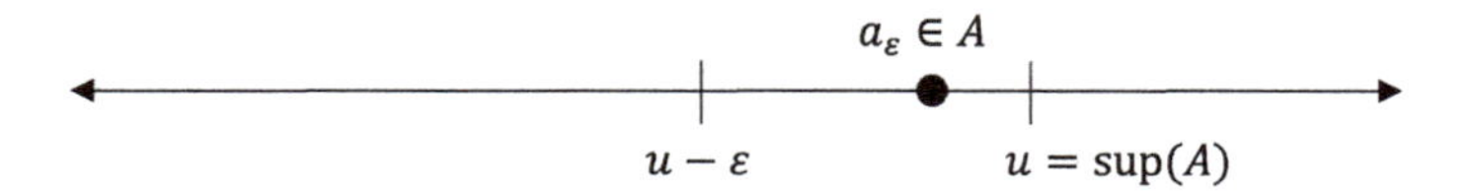

Figure 4.4: Visualizing Theorem 4.14.

Again, the point here is to realize that to each different positive number ε there correspond potentially different numbers a_ε in the set A.

Example 4.10 (Continued). We can use Theorem 4.14 to prove that the upper bound 1 of the interval $(0,1)$ is the supremum of $(0,1)$. All we need to do is show that for every positive number ε, there exists a number a_ε in $(0,1)$ for which $1-\varepsilon < a_\varepsilon$.

Consider any positive number ε. By the density property, there exists a real number a_ε such that $\max\{0, 1-\varepsilon\} < a_\varepsilon < 1$. Thus, $0 < a_\varepsilon < 1$ so that $a_\varepsilon \in (0,1)$, and $1-\varepsilon < a_\varepsilon$.

Exercise 4.13. In the argument presented in the last appearance of Example 4.10 immediately above,
(a) How do we know that $\max\{0, 1-\varepsilon\} < 1$?
(b) Why did we use $\max\{0, 1-\varepsilon\}$ instead of $1-\varepsilon$?

The notion of greatest lower bound can also be characterized in a manner similar to that provided in Theorem 4.14 for the least upper bound.

Theorem 4.15 (Alternate characterization of the greatest lower bound). *Let l be a lower bound of a nonempty subset A of $\mathbb{R}$. Then $l = \inf(A)$ if and only if for every positive number ε, there exists a number a_ε in A for which $a_\varepsilon < l + \varepsilon$.*

Exercise 4.14. Prove Theorem 4.15.

Exercise 4.15. Use Theorem 4.15 to establish that the lower bound 0 of the interval $(0,1)$ is the infimum of $(0,1)$.

The completeness axiom

We are now ready to state our final assumption about the real number system, the *completeness axiom*, which gets its name from its role in guaranteeing the existence of irrational numbers, thereby "completing" the real number system. The axiom accomplishes this objective by asserting the existence of a real number least upper bound (supremum) for any nonempty subset of $\mathbb{R}$ that is bounded above.

Axiom 4.16 (The completeness axiom). *Every nonempty subset of $\mathbb{R}$ that is bounded above has a least upper bound that is itself a real number.*

The completeness axiom is essential to much of our work in the remainder of this book. That every nonempty subset of $\mathbb{R}$ having a lower bound must have a *greatest* lower bound (infimum) is a consequence of the axiom.

Theorem 4.17 (Existence of infima). *Every nonempty subset of $\mathbb{R}$ that is bounded below has a greatest lower bound that is itself a real number.*

Proof. Consider any nonempty subset A of $\mathbb{R}$ that is bounded below and let $B = \{-a \mid a \in A\}$. Since $x \le y$ if and only if $-y \le -x$, it follows that b is a lower bound for A if and only if $-b$ is an upper bound for B. Thus, as A is bounded below, we may conclude that B is bounded above and so, by the completeness axiom, has a real number supremum s. It then follows that $-s$ is the real number infimum of A (see Exercise 4.11). □

The suprema and infima whose existence is provided for by Axiom 4.16 and Theorem 4.17 are real numbers, and therefore "live within" the real number system whose structure we are analyzing. In most instances, then, making an explicit statement that a supremum or infimum is a real number is not necessary, as it is understood that this must be the case.

By way of contrast, we shall soon see that, when they exist, suprema and infima for subsets of the set $\mathbb{Q}$ of rational numbers may not themselves be rational numbers, and this fact finally reveals exactly how the real number system and the rational number system are different from each other (our work at the beginning of this chapter, in which we used the intermediate value theorem to provide for the existence of $\sqrt{2}$ as a real number, foreshadows this outcome).

Exercise 4.16. Show that if A is a nonempty bounded subset of $\mathbb{R}$, then $\inf(A) \le \sup(A)$. Under what circumstances does equality hold?

Exercise 4.17. Suppose A and B are subsets of $\mathbb{R}$ for which A is nonempty, B is bounded, and $A \subseteq B$.
(a) Explain how we know $\inf(B)$, $\sup(B)$, $\inf(A)$, and $\sup(A)$ all exist.
(b) Prove that $\inf(B) \le \inf(A)$ and $\sup(A) \le \sup(B)$.
(c) Find specific nonempty subsets A and B of $\mathbb{R}$ for which $A \subset B$ and also both $\inf(B) = \inf(A)$ and $\sup(A) = \sup(B)$.

Exercise 4.18. Let A be a nonempty subset of $[0, \infty)$ that is bounded above, let r be a nonnegative real number, and let $rA = \{ra \mid a \in A\}$. Prove that rA is bounded above and that $\sup(rA) = r\sup(A)$.

Exercise 4.19. Suppose A and B are nonempty subsets of $\mathbb{R}$ that are bounded above. Let $A + B = \{a + b \mid a \in A \text{ and } b \in B\}$. Prove that $\sup(A + B) = \sup(A) + \sup(B)$.

Exercise 4.20. Suppose A and B are nonempty subsets of $[0, \infty)$ that are bounded above. Let $AB = \{ab \mid a \in A \text{ and } b \in B\}$. Prove that $\sup(AB) = \sup(A)\sup(B)$.

The Archimedean property

One very important consequence of the completeness of $\mathbb{R}$ is the fact that, given any real number, *no matter how large*, we can always find a natural number that is even larger. In other words, the set $\mathbb{N} = \{1, 2, 3, \dots\}$ of natural numbers is not bounded above.

Theorem 4.18 (The Archimedean property). *For every real number a, there exists a natural number N_a such that $N_a > a$.*

Proof. Otherwise, there exists a real number a for which $n \leq a$ for all natural numbers n, thus making a an upper bound for $\mathbb{N}$. By the completeness of $\mathbb{R}$, it would then follow that there is a supremum s for $\mathbb{N}$. Hence, as $s - 1$ is not an upper bound for $\mathbb{N}$ (*why?*), there would exist a natural number K such that $s - 1 < K$, which would then imply that $s < K + 1$. Since K is a natural number, so is $K + 1$, and we have reached a contradiction, as the status of s as an upper bound for $\mathbb{N}$ means no member of $\mathbb{N}$ can be larger than s. □

We wrote N_a in the above statement of the Archimedean Property to emphasize that the natural number whose existence is being asserted depends on the value of the real number a. For instance, if $a = -5.7$ we can take N_a to be any natural number, but as the smallest natural number larger than $a = \sqrt{101}$ is 11, in this case we could take N_a to be 11, but not 10. There is, however, no requirement that we incorporate the subscript a here; we could have written N instead of N_a.

Exercise 4.21. The Archimedean Property can be traced back to the ancient Greek mathematician Archimedes, who formulated a version of the property stating that, given positive numbers a and b for which $a < b$, there is a natural number N for which $Na > b$. Prove that this version of the property is equivalent to the one given in Theorem 4.18.

The following corollary to the Archimedean property is actually equivalent to it and is itself, therefore, often referred to as the Archimedean property. We use the Archimedean Property in the form of this corollary quite often.

Corollary 4.19 (The Archimedean property). *For every positive number ε, there exists a natural number N such that $\frac{1}{N} < \varepsilon$.*

This corollary tells us that given any positive real number, *no matter how small*, we can find a suitably large natural number for which the reciprocal of this natural number is even smaller than the original positive real number. Since the natural number N, whose existence is being asserted in the corollary, depends on the value of the positive number ε, we could have written N_ε in place of N.

Exercise 4.22. Find the smallest natural number N for which

a. $\frac{1}{N} < 10$;
b. $\frac{1}{N} < 1$;
c. $\frac{1}{N} < 0.003$.

Exercise 4.23. Prove Corollary 4.19.

Example 4.20. The Archimedean property permits us to establish 0 as the infimum of

$$A = \left\{ \frac{1}{n} \;\middle|\; n \in \mathbb{N} \right\}.$$

As every member of A is positive, 0 is a lower bound of A. Suppose to the contrary that 0 is not the greatest lower bound of A; then the infimum of A is some positive number l. As $l > 0$, the Archimedean property allows us to conclude that there exists a natural number N such that $\frac{1}{N} < l$, from which it follows that l cannot be a lower bound for A, a contradiction. Thus, $\inf\{\frac{1}{n} \mid n \in \mathbb{N}\} = 0$.

Exercise 4.24. Let $V = \{5 - \frac{2}{n} \mid n \in \mathbb{N}\}$.
(a) Does $\min(V)$ exist? If so, find $\min(V)$ and justify your conclusion. If not, explain why not.
(b) Does $\inf(V)$ exist? If so, find $\inf(V)$ and justify your conclusion. If not, explain why not.
(c) Does $\sup(V)$ exist? If so, find $\sup(V)$ and justify your conclusion. If not, explain why not.
(d) Does $\max(V)$ exist? If so, find $\max(V)$ and justify your conclusion. If not, explain why not.

Exercise 4.25. Let A be a nonempty subset of $\mathbb{R}$ that is bounded above. Show that $s = \sup(A)$ if and only if for every natural number n, the number $s - \frac{1}{n}$ is not an upper bound of A and the number $s + \frac{1}{n}$ is an upper bound of A.

Exercise 4.26. Find the least upper bound of $\{\frac{1}{j} - \frac{1}{k} \mid j, k \in \mathbb{N}\}$. Justify your conclusion.

Exercise 4.27. Find the greatest lower bound of $\{\frac{1}{j} + \frac{1}{k} \mid j, k \in \mathbb{N}\}$. Justify your conclusion.

Exercise 4.28. Show that for every positive number ε, there exists a natural number n such that $\frac{1}{2^n} < \varepsilon$. *Hint*: Make use of the result from Exercise 3.13.

Roots and non-integer powers

We saw near the beginning of this chapter that the intermediate value theorem, may be applied to assert the existence within the set $\mathbb{R}$ of the irrational number $\sqrt{2}$. A modification of that argument shows how the theorem actually provides for roots of positive numbers more generally.

Theorem 4.21 (Existence of roots). *For each positive number a and each natural number n, there exists a unique positive number c for which $c^n = a$. When $n \geq 2$, this number c is usually denoted by either $\sqrt[n]{a}$ or $a^{1/n}$, and is called the **nth root** of a when n is odd and the **principal** nth **root** of a when n is even. We interpret $\sqrt[1]{a}$ as simply representing a.*

Proof. We establish the existence portion of the result and ask you to establish uniqueness in Exercise 4.29. Let a be any positive number and let n be any natural number. First note that as $1^n = 1$, the result holds if $a = 1$, and since $a^1 = a$, the result holds if $n = 1$. So we now assume that $a \neq 1$ and $n \neq 1$. We show in Chapter 13 that all polynomial functions are continuous, so here we assume that the polynomial function $f : \mathbb{R} \to \mathbb{R}$ defined by $f(x) = x^n - a$ is continuous. Observe that $f(0) = -a < 0$.

Case: $a > 1$.

From Exercise 3.12 it follows that $a^n > a$, so that $f(a) = a^n - a > 0$. Hence, by the intermediate value theorem there exists a real number c for which $0 < c < a$ and $f(c) = c^n - a = 0$, which implies $c^n = a$.

Case: $0 < a < 1$.

Then $\frac{1}{a} > 1$ and from Exercise 3.12 it follows that $(\frac{1}{a})^n > \frac{1}{a} > a$, so that $f(\frac{1}{a}) = (\frac{1}{a})^n - a > 0$. Hence, by the intermediate value theorem, there exists a real number c for which $0 < c < \frac{1}{a}$ and $f(c) = c^n - a = 0$, which implies $c^n = a$. □

Example 4.22. Because 7.28 is a positive number, there is a positive number c for which $c^3 = 7.28$. This number c can be written as either $\sqrt[3]{7.28}$ or $7.28^{1/3}$, and is called the cube root of 7.28.

There is also a positive number d for which $d^8 = 7.28$. This number d can be written as either $\sqrt[8]{7.28}$ or $7.28^{1/8}$, and is called the principal eighth root of 7.28.

Exercise 4.29. Show that if c and d are positive real numbers for which $c^n = d^n$ for some natural number n, we must have $c = d$.

For each negative number a and each odd natural number n we also define the ***n*th root** of a so that $\sqrt[n]{a} = a^{1/n} = -\sqrt[n]{-a}$.

Example 4.23. Thus, $\sqrt[3]{-8} = -\sqrt[3]{-(-8)} = -\sqrt[3]{8} = -2$.

When a is a natural number and $\sqrt[n]{a}$ is not, we can use arguments similar to the one employed in the proof of Theorem 3.28 that $\sqrt{2}$ irrational to show that $\sqrt[n]{a}$ is irrational.

Example 4.24. As $\sqrt{16} = 4$ and $\sqrt[4]{16} = 2$, each of $\sqrt{16}$ and $\sqrt[4]{16}$ is rational. However, $\sqrt[n]{16}$ is not a natural number when $n \neq 2$ and $n \neq 4$, which implies $\sqrt[n]{16}$ is irrational for every natural number n other than 2 and 4.

Keep in mind that we have not yet proved the intermediate value theorem. When the proof is presented in Chapter 14, we make sure to indicate how it depends on the completeness axiom. But since we have used the intermediate value theorem to argue why those irrational numbers that are roots of positive real numbers must be present in the real number system, the existence of these irrational numbers as real numbers can therefore be traced back to the completeness axiom. We want to stress that the primary motivation for including the completeness axiom among our assumptions about the real numbers is its role in guaranteeing the presence of the irrational numbers among the real numbers.

It is also possible to directly prove via the completeness axiom, and without first establishing the intermediate value theorem, the existence within the real number system of roots of positive real numbers, but the arguments tend to be quite technical. For instance, to establish the existence of $\sqrt{2}$ among the real numbers, it is possible to argue using the completeness axiom that the set

$$\{x \in \mathbb{R}^+ \mid x^2 < 2\}$$

must have a supremum s and that s is a positive number satisfying the equation $s^2 = 2$, from which it follows that $s = \sqrt{2}$. We do not carry out this construction here, but we do want to demonstrate that under the assumption that $\sqrt{2}$ exists as a real number, it follows that

$$\sqrt{2} = \sup\{x \in \mathbb{R}^+ \mid x^2 < 2\},$$

as doing so helps to make clear the role of the completeness axiom in bringing forth irrational numbers.

So let $A = \{x \in \mathbb{R}^+ \mid x^2 < 2\}$. If $x > \sqrt{2}$, it follows using that $x^2 > 2$, in which case x is not in A. Hence, all elements of A are less than or equal to $\sqrt{2}$, thus making $\sqrt{2}$ an upper bound for A. Now suppose u is an upper bound for A and $u < \sqrt{2}$. As $1 \in A$, we have $1 \le u$, so from the density property there is a positive number t such that $u < t < \sqrt{2}$. We may then conclude that $t^2 < 2$, which implies that $t \in A$, thereby contradicting the assumption that u is an upper bound for A. Thus, $\sqrt{2}$ is the least upper bound of A, that is, $\sqrt{2} = \sup\{x \in \mathbb{R}^+ \mid x^2 < 2\}$.

We may similarly represent any root of a positive number as the supremum or infimum of a well-chosen subset of $\mathbb{R}$. For example, $\sqrt[3]{7.28} = \sup\{x \in \mathbb{R}^+ \mid x^3 < 7.28\}$.

Exercise 4.30. Represent $\sqrt[5]{24}$ as the least upper bound of an appropriate subset of $\mathbb{R}$.

We are now in a position to give meaning to non-integer powers of positive real numbers, and in some cases, negative real numbers. Specifically, for each positive real number a and each real number r, we define the **rth power of a**, denoted a^r, as follows:

(1) If $r = 0$, then $a^r = a^0 = 1$.
(2) If r is a positive rational number, then $a^r = (\sqrt[n]{a})^m$, where m and n are positive integers with $r = \frac{m}{n}$.
(3) If r is a positive irrational number, then

$$a^r = \sup\{a^q \mid q \in \mathbb{Q}^+ \text{ and } q < r\} \text{ when } a > 1$$

and

$$a^r = \inf\{a^q \mid q \in \mathbb{Q}^+ \text{ and } q < r\} \text{ when } 0 < a < 1.$$

(4) If r is negative, then $a^r = \frac{1}{a^{-r}}$.

Also, for each positive real number r, we define 0^r, the r**th power of** 0, so that $0^r = 0$.

Finally, when a is a negative real number, and m and n are natural numbers with n odd, we define the $\frac{m}{n}$**th** and $-\frac{m}{n}$**th powers** of a so that $a^{m/n} = \sqrt[n]{a^m}$ and $a^{-m/n} = \frac{1}{\sqrt[n]{a^m}}$.

Exercise 4.31. Evaluate each of the following according to our definitions.
(a) $64^{-2/3}$;
(b) $0^{\sqrt{3}}$;
(c) $\sqrt[5]{-32}$;
(d) $(\sqrt{3})^0$.

Exercise 4.32. The property $a^r a^s = a^{r+s}$ holds for all real numbers for which the exponential expressions in this equation are defined. You have already verified the property in the situation where r and s are integers. Prove the property holds in the following special cases.
(a) a is a positive real number and both r and s are positive rational numbers.
(b) a is a positive real number, r is a positive rational number, and s is a positive irrational number greater than 1.
(c) a is a positive real number and both r and s are positive irrational numbers greater than 1.

Exercise 4.33. Indicate how the property $\frac{a^r}{a^s} = a^{r-s}$ can be deduced from the property $a^r a^s = a^{r+s}$ and the definitions of division, subtraction, and negative exponents.

Exercise 4.34. Prove that if $b > 1$ and $r > 0$, then $b^r \geq 1$. *Hint*: Consider in turn the cases where the exponent r is a natural number, a positive rational number, and a positive irrational number.

5 Finite sets and infinite sets

We now explore functions in more depth, in particular the notions of one-to-one function, onto function, and bijection. Bijections are used to make precise the distinction between a finite set and an infinite set. Along the way we establish that every real number that is not an integer falls between consecutive integers and demonstrate that between any two distinct real numbers there must exist both a rational number and an irrational number.

Recall that a **function** is a set of ordered pairs for which each first coordinate has exactly one second coordinate paired with it; the set of all first coordinates of the ordered pairs is the **domain** of the function and the set of all second coordinates is the **range**. A member of the domain of a function can be referred to as an **input** of the function, and a member of the range as an **output**. For a function f, it is customary to write $f(x)$ for the output of f corresponding to the input x.

To indicate that f is a function with domain X and range a subset of Y, we may say f is a **function from** X **into** Y, which can be expressed in symbols by writing $f : X \to Y$, in which case the set Y is referred to as the **codomain**.

Example 5.1. Consider the function $F : \{1, 2\} \to \mathbb{R}$ defined so that $F(x) = 4x^2$. The domain of F is $\{1, 2\}$ and the codomain is $\mathbb{R}$. Because $F(1) = 4$ and $F(2) = 16$, the range of F is $\{4, 16\}$. As a set of ordered pairs, $f = \{(1, 4), (2, 16)\}$.

Exercise 5.1. Define the function $f : \mathbb{N} \to \mathbb{Q}$ so that $f(n) = \frac{n}{n+1}$.

(a) Identify the domain and codomain of f.

(b) Evaluate $f(1)$ and $f(100)$.

(c) Identify the range of f.

As sets are *equal* precisely when they have the same elements, it follows that functions are *equal* when they consist of precisely the same ordered pairs. This observation leads immediately to the following proof strategy.

Proof Strategy 5.2 (Proving functions are equal). To prove that functions f and g are equal, show f and g have the same domain, and also show that $f(x) = g(x)$ for each x in this shared domain.

Evidently, functions are not equal if either their domains are unequal or there is an input for which the functions produce different outputs.

Example 5.3. The function $f : \mathbb{R} \to \mathbb{R}$, defined so that $f(x) = x^3$, and the function $g : \mathbb{R}^+ \to \mathbb{R}$, defined so that $g(x) = x^3$, are not equal because the domain of f is $\mathbb{R}$, but the domain of g is $\mathbb{R}^+$.

The function $\alpha : \{1, 2, 3\} \to \mathbb{R}$, defined so that $\alpha(t) = t - 2$, and the function $\beta : \{1, 2, 3\} \to \mathbb{R}$, defined so that $\beta(t) = t^2 - 2t$, are not equal because $\alpha(3) = 1$, but $\beta(3) = 3$ (note carefully that there is no other way to justify that $\alpha \neq \beta$).

https://doi.org/10.1515/9783111429564-005

On the other hand, the function $F : \{1, 2\} \to \mathbb{R}$, where $F(x) = 4x^2$ and the function $G : \mathbb{Z} \cap (0, 3) \to \mathbb{R}$, where $G(x) = 4^x$ are equal. Since the only integers in the interval $(0, 3)$ are 1 and 2, we see that F and G have the same domain. Also,

$$F(1) = 4 \cdot 1^2 = 4 = 4^1 = G(1)$$

and

$$F(2) = 4 \cdot 2^2 = 16 = 4^2 = G(2).$$

Composition of functions

Given functions $f : X \to Y$ and $g : Y \to Z$, the **composite function** $g \circ f : X \to Z$ is defined so that

$$(g \circ f)(x) = g(f(x)).$$

The operation $\circ$ is referred to as **function composition**. Function composition permits us to build more elaborate functions from simpler functions.

Example 5.4. If the function $f : [0, 100] \to [0, 10]$ is defined so that $f(x) = \sqrt{x}$ and the function $g : [0, 10] \to [5, 15]$ is defined so that $g(x) = x + 5$, it follows that the function $g \circ f : [0, 100] \to [5, 15]$ is defined so that

$$(g \circ f)(x) = g(f(x)) = g(\sqrt{x}) = \sqrt{x} + 5.$$

Exercise 5.2. Reordering a composition usually produces different functions. Let the functions $f : \mathbb{R} \to \mathbb{R}$ and $g : \mathbb{R} \to \mathbb{R}$ be defined so that $f(x) = 3x$ and $g(x) = x^2$. Show that $f \circ g \neq g \circ f$.

One-to-one functions, onto functions, and bijections

A function is **one-to-one** if distinct inputs always yield distinct outputs. A function with a specified codomain is **onto** provided that this codomain is the range. A one-to-one function is also called an **injection**, and is said to be **injective**, while an onto function is also called a **surjection**, and is said to be **surjective**. A function that is both an injection and a surjection is called a **bijection**, and is said to be **bijective**.

Example 5.5. Consider the function $f : \{1, 2, 3\} \to \mathbb{R}$ defined so that $f(1) = 2$, $f(2) = 3$, and $f(3) = 2$. This function is *not* one-to-one, since as $f(1) = 2 = f(3)$ we see that distinct inputs do not always yield distinct outputs. This function is also *not* onto, since the codomain is not the same set as the range; for instance, 0 is in the codomain $\mathbb{R}$, but not in the range $\{2, 3\}$.

Exercise 5.3. Determine whether the given function is one-to-one.
(a) $A : \{-1, 0, 1\} \to \{0, 1\}$ defined by $A(x) = |x|$;
(b) $i : \{1, 2\} \to \{2, 3\}$ defined by $i(x) = x + 1$.

Exercise 5.4. Determine whether each function from Exercise 5.3 is onto.

Exercise 5.5. Based on your conclusions in Exercises 5.3 and 5.4, which of the functions, if any, are bijections?

Proof Strategy 5.6 (Proving a function is one-to-one, onto, or a bijection). To prove a function $f : X \to Y$ is
(1) *one-to-one*, assume x_1 and x_2 are in X with $f(x_1) = f(x_2)$ and show $x_1 = x_2$;
(2) *onto*, assume y is in Y and show there exists x in X with $f(x) = y$;
(3) a *bijection*, prove f is one-to-one and prove f is onto.

The approach to proving a function is one-to-one given in this strategy is obtained by forming the logically equivalent contrapositive of *if the inputs are different, then the outputs are different.*

Example 5.7. Define the function $h : \mathbb{Z}^- \to \mathbb{Z}^+$ so that $h(x) = -2x$. Here is a proof that h is one-to-one.

Proof. Consider any negative integers n_1 and n_2, and suppose $h(n_1) = h(n_2)$ (*we begin our argument with "two" arbitrary members of the domain* $\mathbb{Z}^-$ *of* h, *and assume* h *generates the same output from "both" of these inputs*). Then $-2n_1 = -2n_2$ and upon dividing through by -2 we obtain $n_1 = n_2$ (*so we have shown that the "two" arbitrary inputs that generated the same output are actually the same*). □

In the write-up of the proof, we included two parenthetical remarks designed to help you see how the standard strategy for proving a function is one-to-one is being carried out; these types of remarks would not usually be included in the proof.

Exercise 5.6. Provide evidence that the function h of Example 5.7 is not a surjection.

A function's range is always a subset of its codomain, so the approach given in Proof Strategy 5.6 for proving a function is onto is based on the observation that the codomain and range are equal precisely when the codomain is a subset of the range.

Example 5.8. Consider the function $g : \mathbb{R} \to \mathbb{Z}$, defined so that $g(x)$ is the greatest integer less than or equal to x. Observe that $g(3.8) = 3$, $g(-3.8) = -4$, and $g(3) = 3$. We now prove g is onto.

Proof. Consider any integer n (*we are beginning our argument with an arbitrary member n of the codomain $\mathbb{Z}$ of g*). As any integer is also a real number, n is a real number (*thus we have shown that n is in the domain $\mathbb{R}$ of g*). Also, the greatest integer less than or equal to the integer n is n itself, so $g(n) = n$ (*thus we have shown there is a member of the domain that is mapped to the original codomain member n, thus making n a member of the range*). □

The three parenthetical remarks are not part of the actual proof; we have included them only for the purpose of helping you to see how the standard strategy for proving a function is onto is being implemented in this particular situation.

When a function $f : X \to Y$ is onto, we may say that f is a **function from X onto Y**.

Exercise 5.7. Provide evidence showing the function g of Example 5.8 is *not* an injection.

Exercise 5.8. Show that the composition of one-to-one functions is one-to-one and the composition of onto functions is onto. That is, given functions $f : X \to Y$ and $g : Y \to Z$, show each of the following.

(a) If f and g are both one-to-one, then so is the function $g \circ f : X \to Z$.

(b) If f and g are both onto, then so is the function $g \circ f : X \to Z$.

It follows from these results that the composition of bijections is a bijection.

Exercise 5.9. Let m and n be natural numbers for which $m < n$. Prove that there is no one-to-one function from $\{1, 2, \ldots, n\}$ into $\{1, 2, \ldots, m\}$. *Hint*: Apply induction. In the base step, prove the result for $m = 1$. In the inductive step, assume the result holds for $m = k$ and then show the result holds for $m = k + 1$.

Finite and infinite sets

When the elements of a set have been listed, we can count them to determine the number of elements in the set. For instance, we can quickly decide that the set

$$\{\text{north}, \text{south}, \text{east}, \text{west}\}$$

has four elements by pointing to "north" and saying "one," pointing to "south" and saying "two," pointing to "east" and saying "three," and pointing to "west" and saying "four." In carrying out this counting procedure we have formed the bijection

$$\begin{array}{lll} 1 & \mapsto & \text{north} \\ 2 & \mapsto & \text{south} \\ 3 & \mapsto & \text{east} \\ 4 & \mapsto & \text{west} \end{array}$$

from the set $\{1, 2, 3, 4\}$ onto the set {north, south, east, west}. That we needed to count the natural numbers up through 4 is what tells us that the set has four elements.

Informally, a set is *finite* if the process of counting the members of the set necessarily must come to an end. This idea is captured more formally by defining a set A to be **finite** if A is empty or if there exists a bijection from $\{1, 2, \dots, n\}$ onto A for some natural number n. When A is empty, we say A **has zero elements** and when there is a bijection from $\{1, 2, \dots, n\}$ onto A for the natural number n we say that A **has n elements**. When a set has n elements, we may also say **there are n elements in** A and that n is the **number of elements in** A.

Example 5.9. Let D = {north, south, east, west} and define $f : \{1, 2, 3, 4\} \to D$ so that $f(1)$ = north, $f(2)$ = south, $f(3)$ = east, and $f(4)$ = west. As f is a bijection, we may conclude that the set D is a finite set and D has 4 elements.

The number of elements in a nonempty finite set is always some specific natural number.

Theorem 5.10. *If A is a finite set having n elements, then for every natural number m different from n, the set A does not have m elements.*

Exercise 5.10. Prove Theorem 5.10.

Exercise 5.11. Prove that if A is a nonempty finite set that has n elements and the function $f : A \to B$ is a bijection, then B has n elements.

We now want to establish that every nonempty finite subset of the set of real numbers has a maximum element (one can similarly show every nonempty finite subset of $\mathbb{R}$ has a minimum element). Our strategy is to use induction to prove that for any natural number n, each subset of $\mathbb{R}$ having n elements has a maximum. This requires, in the base step, that we show that every subset of $\mathbb{R}$ having just one element has a maximum and, in the inductive step, that if each subset of $\mathbb{R}$ having k elements, for some fixed natural number k, has a maximum, it must follow that each subset of $\mathbb{R}$ having $k+1$ elements also has a maximum. In carrying out these arguments, we apply what it means, by definition, for a set to have a certain number n of elements, namely, that there is a bijection from $\{1, 2, \dots, n\}$ onto the set.

Theorem 5.11. *Every nonempty finite subset of $\mathbb{R}$ has both a maximum and a minimum.*

Proof. We apply induction to show that every nonempty finite subset of $\mathbb{R}$ has a maximum; the proof that every such subset of $\mathbb{R}$ must also have a minimum is similar.

Base step: Consider any nonempty finite subset A of $\mathbb{R}$ having just one element, that is, for which there exists a bijection $f : \{1\} \to A$. Then $A = \{f(1)\}$. Since $f(1) \geq f(1)$ we see that $f(1)$ is an element of A that is greater than or equal to every member of A. Thus, $f(1)$ is the maximum element of A.

Inductive step: Let k be a fixed natural number and assume that every finite subset of $\mathbb{R}$ having k elements, that is, for which there exists a bijection from $\{1, 2, \ldots, k\}$ onto the finite subset, has a maximum. Consider any nonempty finite subset A of $\mathbb{R}$ with $k+1$ elements, that is, for which there exists a bijection $f : \{1, 2, \ldots, k, k+1\} \to A$; we show A has a maximum.

Define $g : \{1, 2, \ldots, k\} \to A - \{f(k+1)\}$ so that $g(i) = f(i)$ for each $i \in \{1, 2, \ldots, k\}$. As f is a bijection, so is g (*do you see why?*). Thus, by our inductive hypothesis, $\{g(1), g(2), \ldots, g(k)\}$, which is really the set $\{f(1), f(2), \ldots, f(k)\}$, has a maximum element, say m. As $f(k+1) \neq m$ (*do you see why?*), either $f(k+1) < m$ or $f(k+1) > m$. If the former is true, then $\max(A) = m$; if the latter is true, then $\max(A) = f(k+1)$. □

Exercise 5.12. Referring back to the inductive step in the proof of Theorem 5.11,
(a) argue in more detail as to why the function g must be a bijection;
(b) tell why $f(k+1) \neq m$.

A set that is not finite is called **infinite**.

Example 5.12. The set $\mathbb{N} = \{1, 2, 3, 4, \ldots\}$ must be infinite, for if $\mathbb{N}$ were finite, being a subset of $\mathbb{R}$, it would have a maximum element, which it does not.

A **lemma** is a theorem that is used to help us prove another theorem. Typically, the information content conveyed by a lemma is either not of great importance in itself or is subsumed by the content of the theorem the lemma is being used to prove.

In order to prove that all subsets of a finite set are themselves finite, we first prove a lemma demonstrating that all subsets of the set $\{1, 2, \ldots, n\}$, where n is a natural number, are finite. Doing so facilitates the more general proof, since the definition of finite set involves sets of the form $\{1, 2, \ldots, n\}$.

Lemma 5.13. *For each natural number n, every subset of* $\{1, 2, \ldots, n\}$ *is finite.*

Proof. We proceed by induction.

Base step: Here we must show that every subset of $\{1\}$ is finite. The only subsets of $\{1\}$ are $\emptyset$ and $\{1\}$. By definition, $\emptyset$ is finite. Since the function $f : \{1\} \to \{1\}$ defined by $f(1) = 1$ is a bijection, $\{1\}$ is finite.

Inductive step: Here we must show that if every subset of $\{1, 2, \ldots, k\}$ is finite, then it follows that every subset of $\{1, 2, \ldots, k, k+1\}$ is finite. So assume that every subset of $\{1, 2, \ldots, k\}$ is finite and let A be a subset of $\{1, 2, \ldots, k, k+1\}$; we show A is finite. If A is empty, A is finite, so we now assume $A \neq \emptyset$.

Case 1: $k+1 \notin A$.

Then $A \subseteq \{1, 2, \ldots, k\}$ and our inductive hypothesis allows us to conclude that A is finite.

Case 2: $k+1 \in A$.

Then $A = A^* \cup \{k+1\}$ for some subset A^* of $\{1,2,\dots,k\}$. By our inductive hypothesis, A^* is finite. If A^* is empty, then $A = \{k+1\}$ and the function $f : \{1\} \to \{k+1\}$ where $f(1) = k+1$ is a bijection; hence, A is finite. Otherwise, there exists a natural number m for which the function $f : \{1,2,\dots,m\} \to A^*$ is a bijection. Then the function $g : \{1,2,\dots,m,m+1\} \to A$ defined so that $g(i) = f(i)$ for all i in $\{1,2,\dots,m\}$ and $g(m+1) = k+1$ is a bijection, thus making A finite. □

We now argue more generally that if F is a finite set and A is a subset of F, then A must be finite. The central idea, under the assumption that there is a bijection from $\{1,2,\dots,n\}$ onto F, is to pick out the members of $\{1,2,\dots,n\}$ that correspond to the members of the subset A, thereby creating a subset K of $\{1,2,\dots,n\}$ to which Lemma 5.13 may be applied. Function composition is then used to produce the bijection needed to show, by definition, that A is finite.

Theorem 5.14. *Every subset of a finite set is finite.*

Proof. Suppose $A \subseteq F$ and F is finite. If F is empty, so is A, thus making A finite. Otherwise, there exists a natural number n such that $f : \{1,2,\dots,n\} \to F$ is a bijection. If A is empty, then A is finite, so we assume now that $A \neq \emptyset$.

Let $K = \{i \in \mathbb{N} \mid i \leq n, f(i) \in A\}$ and note that K is a subset of $\{1,2,\dots,n\}$, hence, is finite by Lemma 5.13. As $A \neq \emptyset$ and f is onto, it follows that $K \neq \emptyset$. So there exists a bijection $g : \{1,2,\dots,m\} \to K$ for some natural number m. Now define $h : K \to A$ so that $h(i) = f(i)$ for each i in K, and observe that h is a bijection. Then the function $h \circ g : \{1,2,\dots,m\} \to A$, being a composition of bijections, is also a bijection. Hence, A is finite. □

The following exercise illustrates the idea underlying the proof of Theorem 5.14 via a concrete example.

Exercise 5.13. Let $F = \{a,b,c,d,e\}$ and $A = \{b,d,e\}$, and note that $A \subseteq F$.

(a) Define a bijection $f : \{1,2,\dots,n\} \to F$ using the appropriate value of n.

(b) For the bijection f you created in (a), list the members of the set $K = \{i \in \mathbb{N} \mid i \leq n, f(i) \in A\}$, where n is the same natural number you used in defining f.

(c) Define a bijection $g : \{1,2,\dots,m\} \to K$ using the appropriate value of m.

(d) Define $h : K \to A$ so that $h(i) = f(i)$ for each i in K. Write out the individual values $h(i)$ for each i in K and be sure you agree that h is a bijection.

(e) The function $h \circ g : \{1,2,\dots,m\} \to A$, where m is the same natural number you used in defining g, is a composition of bijections, hence, a bijection itself. Write out the individual values $(h \circ g)(i)$ for each i in $\{1,2,\dots,m\}$. How do we now know that A is finite?

(f) If you had difficulty following the proof of Theorem 5.14, re-read it now in light of what you have already done in this exercise.

Corollary 5.15. *Every superset of an infinite set is infinite.*

Exercise 5.14. Prove Corollary 5.15.

It is often possible to tell that a set is finite by direct inspection. For instance, even before we created a formal definition, we were certainly aware that the set $\{6, 8, 10, \ldots, 20\}$ is finite. The formal definition has helped to put the notion of a set being finite on a firm foundation. Having done so, we usually just state that a given set is finite when it is clear that an appropriate bijection could be constructed.

Exercise 5.15. We recognize each of the following sets as being finite just by looking at them. For each set, construct an appropriate bijection to show it really is finite.

(a) $\{4\}$;

(b) $\{-1, 1\}$;

(c) $\{6, 8, 10, \ldots, 20\}$.

Writing proofs: audience and purpose matter

In the early chapters of this book, we have discussed a variety of strategies for developing and writing proofs, for example, how to prove an *If... then...* statement and how to prove sets are equal. We have used these strategies in subsequent proofs we have presented and you have used them in proofs requested in the exercises. Now we briefly digress in order to share with you some general advice about proofs that goes beyond specific proof-writing strategies.

Whenever we write a proof, we should take into account the sophistication of the audience who will be reading it. Proofs written for a more experienced audience tend to omit details that would be included for beginners. The goal is to maintain a clear line of reasoning, pointing out the major deductions and providing some evidence as to why they are valid, while at the same time allowing for the ability of the reader to fill in more routine or elementary steps. Almost every mathematical proof we could imagine writing necessarily leaves out things that could have been made more explicit; the writer of the proof has a responsibility to take some care in deciding what to actually include in the write-up so that the argument can be agreed to by a knowledgeable reader.

One specific area of concern in writing a proof is how much explicit justification of conclusions to include. Less justification is included when it is expected readers of a proof should be able to supply it for themselves without particular difficulty. This is why, in our proofs of Theorem 5.11, Lemma 5.13, and Theorem 5.14, we only stated, rather than gave detailed evidence, that certain functions are bijections; we believe you should recognize that in each case it would be straightforward to develop explicit justification of these assertions.

Exercise 5.16. Provide detailed evidence as to why each function is a bijection:
(a) the function g that appeared in the proof of Lemma 5.13;
(b) the function h that appeared in the proof of Theorem 5.14.

Closely related to the *audience* for whom a proof is written is the *purpose* for which it is written. For example, the proofs appearing in this textbook are written for an audience consisting of advanced undergraduate mathematics students and for the purpose of *helping them to learn elementary real analysis*. This purpose influences how the proofs are written, the approaches taken in their development, and the degree of detail provided in them. Along with the proofs themselves, we often elaborate on how they are structured in light of the standard proof strategies we have introduced. In some cases, there are homework exercises where the reader is asked to justify claims made in a proof or to modify a given argument to prove another result. All of these choices reflect the context in which the proofs appear, namely, as part of a textbook from which students are learning specific subject matter.

In your role as a student, the ways in which you write proofs also tend to vary with the purposes underlying them. Sometimes you are just writing for yourself in order to make sure you understand a certain concept, so your proof might be more of a sketch that leaves out things you are already comfortable with. But if you are writing up a proof as part of an exam or a graded homework assignment, the purpose is different as you are now trying to demonstrate the depth of your understanding, and so it is to your advantage to provide as much detail as possible, carefully justifying your conclusions and ensuring they are sequenced appropriately.

As is the case with writing in general, both audience and purpose influence the written presentation of a mathematical proof.

The Archimedean property revisited

The number line model of the real number system suggests that every real number is either an integer or else lies between consecutive integers.

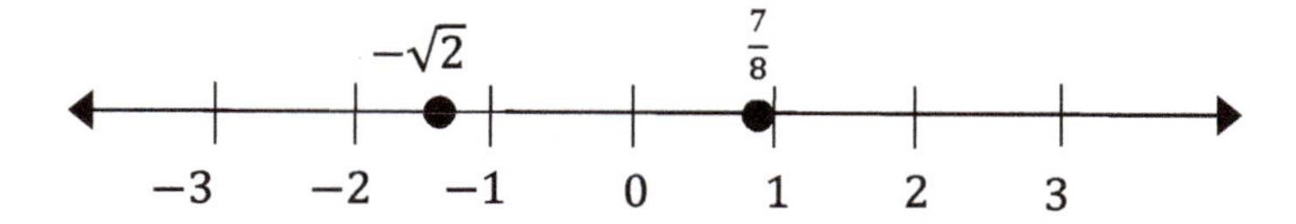

Figure 5.1: A real number that is not an integer lies between consecutive integers.

For example, Figure 5.1 shows several integers along with the non-integer real numbers $\frac{7}{8}$, which lies between the consecutive integers 0 and 1, and $-\sqrt{2}$, which lies between the consecutive integers −2 and −1.

That every non-integer real number does indeed lie between consecutive integers is a consequence of the Archimedean property. Note how our proof makes use of the already proven results that a subset of a finite set is also finite (Theorem 5.14) and a nonempty finite subset of $\mathbb{R}$ has a maximum (Theorem 5.11).

Theorem 5.16. *For every real number a, there exists a unique integer N such that $N \leq a < N + 1$.*

Proof. We establish the existence portion of this result and ask you to establish the uniqueness portion in Exercise 5.17.

Consider any real number a. If a is an integer, then $a \leq a < a + 1$ and our proof is complete. Thus, for the remainder of the proof we assume a is not an integer.

Case 1: $a > 0$.

If $a < 1$, then $0 \leq a < 1$ and the desired result holds. So suppose $a > 1$. By the Archimedean property there exists a natural number M such that $a < M$. Let $S = \{k \in \mathbb{N} \mid k < a\}$, and note that, as $1 \in S$, the set S is nonempty, and S is a subset of the finite set $\{1, 2, 3, \ldots, M-1\}$. Thus, S is a nonempty finite subset of $\mathbb{R}$ and, therefore, has a maximum N, which is necessarily an integer. It follows that $N < a < N + 1$.

Case 2: $a < 0$.

Then $-a > 0$, so our work in Case 1 allows us to assert the existence of an integer K for which $K < -a < K + 1$. It follows that $-(K + 1) < a < -K$, so letting $N = -(K + 1)$, we have $-K = N + 1$ so that $N < a < N + 1$. As N is an integer (*why?*), the proof for this case is now complete. □

In the statement of Theorem 5.16, the value of the integer N depends on the value of the real number a we start with; thus, we could have written N_a instead of n to emphasize this dependency. Whether to use such notation or not is often just a matter of preference on the part of the writer. To reduce the apparent complexity of symbolic expressions, from now on we intend not to employ notation that makes such dependencies explicit, unless doing so would be likely to cause confusion or we are trying to make a specific pedagogical point.

For instance, we are quite confident that no one would read Theorem 5.16 and think we are stating that there is a single (magical!) integer N with the property that *all* real numbers lie inside the interval $[N, N + 1)$. Clearly that would be nonsensical, and the actual intent is to convey the notion that each real number fits into an interval $[N, N+1)$ for some integer N, but that the value of N can be different for different real numbers. For example, our earlier number line diagram shows that for $\frac{7}{8}$ we would take $N = 0$, whereas for $-\sqrt{2}$ we would take $N = -2$.

Exercise 5.17. Establish the uniqueness portion of Theorem 5.16.

The density property revisited

We can improve upon the result stated in Theorem 1.20, that between any two distinct real numbers there is another real number, by showing that both a rational number and an irrational number can be found between any two distinct real numbers.

The key idea employed in the following proof is to use the Archimedean Property to find a fraction $\frac{1}{N}$ with natural number denominator N that is less than the distance $b-a$ between a and b. We then know, essentially because of Theorem 5.16, that if we start at 0 on the number line and move to the right in increments of $\frac{1}{N}$, we eventually reach a position M increments of $\frac{1}{N}$ that, since $\frac{1}{N} < b - a$, lies between a and b (Figure 5.2). This position is the rational number $M \cdot \frac{1}{N} = \frac{M}{N}$.

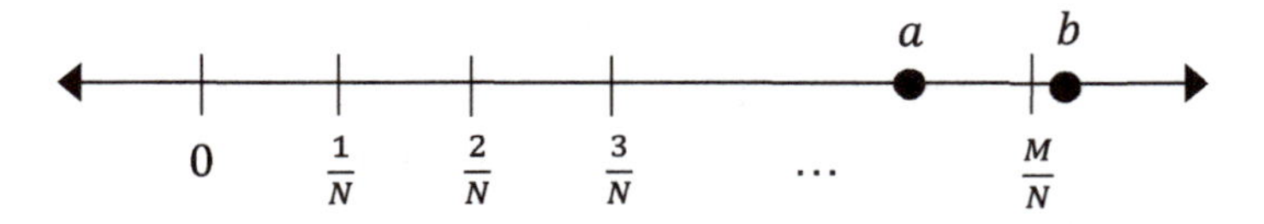

Figure 5.2: Finding a rational number between a and b.

Theorem 5.17 (The density property). *Between any two distinct real numbers there is both a rational number and an irrational number.*

Proof. Suppose $a < b$. We find a rational number c such that $a < c < b$, leaving the existence of an irrational number between a and b as an exercise. Note that if $a = 0$, the Archimedean property guarantees the existence of a natural number N for which $0 < \frac{1}{N} < b$, and our proof is complete.

We now assume that $a > 0$; the argument for the case in which $a < 0$ is similar. By the Archimedean property, because $b - a$ is positive, there exists a natural number N such that $\frac{1}{N} < b - a$. Then, since $Na > 0$, by Theorem 5.16 there exists a natural number M for which $M - 1 \leq Na < M$, which implies that both $M \leq Na + 1$ and $a < \frac{M}{N}$. From $\frac{1}{N} < b - a$, it follows that $Na + 1 < Nb$. Now, as $M \leq Na + 1$ and $Na + 1 < Nb$, we may conclude that $M < Nb$, so that $\frac{M}{N} < b$. Thus, we have shown that $a < \frac{M}{N} < b$, where $\frac{M}{N}$ is a rational number, as M and N are natural numbers. □

Exercise 5.18. Complete the proof of Theorem 5.17 by showing that if $a < b$, there exists an irrational number c with $a < c < b$. *Hint*: Observe that if $a < b$, then $a\sqrt{2} < b\sqrt{2}$; what do we already know, from the part of the theorem already proved, must lie between $a\sqrt{2}$ and $b\sqrt{2}$?

Exercise 5.19. Explain how it follows from Theorem 5.17 that there are actually infinitely many rational numbers and also infinitely many irrational numbers between any two distinct real numbers.

Exercise 5.20. Any real number of the form $\frac{m}{2^n}$, where m is an integer and n is a nonnegative integer, is called a **dyadic rational number**. Prove that between any two distinct real numbers there is a dyadic rational number.

Theorems 5.16 and 5.17 help to reinforce the number line as a model for the real number system, because they analytically bring into play properties that appear intuitively evident from the number line perspective. Of course, the "rational number line," consisting of points corresponding only to rational numbers, also possesses these properties, which explains why the number line remains just an informal model and cannot serve as the foundation for a critical study of the real number system.

Completeness revisited

We say a subset S of $\mathbb{R}$ is **complete** if every nonempty subset A of S that has an upper bound in S has a least upper bound in S.

Example 5.18. The set $\mathbb{R}$ of all real numbers is complete, because the completeness axiom, Axiom 4.16, tells us that every nonempty subset of $\mathbb{R}$ that is bounded above has a real number least upper bound.

Example 5.19. There are (many) subsets of the set $\mathbb{Q}$ of all rational numbers that are bounded above, but which do not have least upper bounds that are rational numbers. Hence, $\mathbb{Q}$ is not complete.

For instance, we can show that the subset $B = \{x \in \mathbb{Q}^+ \mid x^2 < 2\}$ of $\mathbb{Q}$ has upper bounds in $\mathbb{Q}$ but does not have a least upper bound in $\mathbb{Q}$. Note that 2 is a rational number upper bound of B (verification left to Exercise 5.21). Now if B were to have a least upper bound in $\mathbb{Q}$, this rational number would also be the least upper bound of B in $\mathbb{R}$, since a subset of $\mathbb{R}$ cannot have more than one supremum. We can show, however, that when B is viewed as a subset of $\mathbb{R}$, its supremum is $\sqrt{2}$, thereby making it impossible for B to have a rational number supremum.

First, we show that $\sqrt{2}$ is an upper bound of B. Let x be any positive rational number for which $x > \sqrt{2}$. It follows that $x^2 > 2$, so $x \notin B$. Thus, all elements of B are less than or equal to $\sqrt{2}$, which makes $\sqrt{2}$ an upper bound of B.

Suppose now that $0 < y < \sqrt{2}$. Because there is a rational number between any two distinct real numbers, we can find a rational number q for which $y < q < \sqrt{2}$. Thus, q is a positive rational number and $q^2 < 2$, so it follows that $q \in B$. Hence, as $y < q$, the number y cannot be an upper bound of B. So all upper bounds of B are greater than or equal to $\sqrt{2}$, and since we already showed that $\sqrt{2}$ is an upper bound of B, it follows that $\sqrt{2}$ is the least upper bound of B.

Example 5.19 shows that the rational number system does not possess the property of completeness, which the real number system does possess. It is, therefore, the completeness property of $\mathbb{R}$ that sets apart the real number system from the system of rational numbers.

Exercise 5.21. Verify that 2 is an upper bound for the set $B = \{x \in \mathbb{Q}^+ \mid x^2 < 2\}$.

Exercise 5.22. Determine whether the set $\mathbb{Z}$ of all integers is a complete subset of $\mathbb{R}$, justifying your conclusion.

Countably infinite sets

A **sequence** is a function whose domain is the set $\mathbb{N}$ of natural numbers. For a sequence s, it is customary to write s_n in place of $s(n)$ and to then think of the sequence as an unending list

$$s = (s_1, s_2, s_3, \dots)$$

in which, for every natural number n, the output s_n appears as the ***n*th term** in the list.

Example 5.20. The sequence

$$a = (1, -1, 1, -1, 1, -1, \dots)$$

is defined so that $a_n = 1$ when n is odd and $a_n = -1$ when n is even. Thus, for instance, $a_{35} = 1$, whereas $a_{1000} = -1$.

The sequence

$$b = \left(1, \frac{1}{2}, \frac{1}{3}, \frac{1}{4}, \dots\right)$$

is defined so that $b_n = \frac{1}{n}$.

When the elements of an infinite set A can be arranged into a sequence, with each element of A appearing exactly once as a term of the sequence, this arrangement represents a bijection from $\mathbb{N}$ onto the set A, and the set A is said to be **countably infinite**.

Example 5.21. The set $\mathbb{Z}$ of all integers is countably infinite as the integers can be arranged into the sequence

$$(0, 1, -1, 2, -2, 3, -3, \dots),$$

which represents the bijection

$$\begin{array}{rcr} 1 & \mapsto & 0 \\ 2 & \mapsto & 1 \\ 3 & \mapsto & -1 \\ 4 & \mapsto & 2 \\ 5 & \mapsto & -2 \\ 6 & \mapsto & 3 \\ 7 & \mapsto & -3 \\ & \vdots & \end{array}$$

from $\mathbb{N}$ onto $\mathbb{Z}$.

Exercise 5.23. If we give the bijection in Example 5.21 the name c, find the values of $c_9 = c(9)$ and $c_{100} = c(100)$.

It may appear to you that every infinite set is countably infinite, but in Chapter 6, we demonstrate that there are infinite sets whose elements cannot be arranged into a sequence.

Exercise 5.24. Show that the given set is countably infinite by describing a way to put all of its members into a sequence.

(a) The set of all rational numbers that can be expressed as fractions in which the numerator is an integer and the denominator is one more than the square of the numerator.

(b) The set of all finite-length strings of letters of the alphabet.

Exercise 5.25. Prove that the union of two countably infinite sets is countably infinite. *Hint*: Use the same idea employed in Example 5.21.

Theorem 5.22. *Every infinite subset of a countably infinite set is countably infinite.*

Proof. Suppose A is an infinite subset of the countably infinite set B. Since B is countably infinite, its members can be arranged in a sequence

$$(b_1, b_2, b_3, \ldots).$$

By deleting anything in this list that is not a member of A, we are left with a list of the members of A. As A is infinite, the resulting list is still unending, thereby representing a bijection from $\mathbb{N}$ onto A, thus making A countably infinite. □

Exercise 5.26. Show that the union of two finite sets is always finite. *Hint*: Begin with sets A and B, each of which is finite. Note that $A \cup B = A \cup (B - A)$, where A and $B - A$ are disjoint. The set $B - A$, being a subset of the finite set B, must be finite.

Exercise 5.27. Use Exercise 5.26 to explain why, if A is an infinite set and B is a finite subset of A, it necessarily follows that $A - B$ is infinite.

Exercise 5.28. Use Exercise 5.27 to prove that every infinite set has a countably infinite subset.

We can form the union of any number of sets, even infinitely many. By definition, an object is in the **union** of sets exactly when it is a member of at least one of the sets over which the union is formed. We still use the symbol $\cup$ to denote the operation of union, no matter how many sets are involved.

Example 5.23. Observe that

$$\{1\} \cup \{2\} \cup \{3\} \cup \cdots = \bigcup_{n=1}^{\infty} \{n\} = \mathbb{N}.$$

The notation

$$\bigcup_{n=1}^{\infty} \{n\}$$

being employed here is similar to Σ-notation for summations. The idea is that for each of the infinitely many natural numbers $n = 1, n = 2, n = 3, \ldots$, we create the singleton sets $\{1\}, \{2\}, \{3\}, \ldots$, and then form their union

$$\{1\} \cup \{2\} \cup \{3\} \cup \cdots.$$

The expression $\bigcup_{n=1}^{\infty} \{n\}$ is just a more compact way of expressing this infinite union.

When we form the union of a countably infinite number of sets, each of these sets also being countably infinite, the resulting set is still countably infinite. In the following proof of this result, we essentially show that it is possible to "weave together" an unending list of unending lists into a single unending list.

Theorem 5.24. *A countably infinite union of countably infinite sets is countably infinite.*

Proof. Consider sets $A_1, A_2, A_3, \ldots$, each countably infinite, with

$$\begin{aligned} A_1 &= \{a_{11}, a_{12}, a_{13}, \ldots\}, \\ A_2 &= \{a_{21}, a_{22}, a_{23}, \ldots\}, \\ A_3 &= \{a_{31}, a_{32}, a_{33}, \ldots\}, \end{aligned}$$

and so forth, so that a_{ij} is the jth element in the listing of all the elements of A_i. Note that all of the elements in these sets appear in the following array.

$$\begin{array}{ccccccc}
a_{11} & a_{12} & a_{13} & a_{14} & a_{15} & a_{16} & \cdots \\
a_{21} & a_{22} & a_{23} & a_{24} & a_{25} & a_{26} & \cdots \\
a_{31} & a_{32} & a_{33} & a_{34} & a_{35} & a_{36} & \cdots \\
a_{41} & a_{42} & a_{43} & a_{44} & a_{45} & a_{46} & \cdots \\
a_{51} & a_{52} & a_{53} & a_{54} & a_{55} & a_{56} & \cdots \\
a_{61} & a_{62} & a_{63} & a_{64} & a_{65} & a_{66} & \cdots \\
\vdots & \vdots & \vdots & \vdots & \vdots & \vdots & \vdots
\end{array}$$

We now put all the entries in this array into the single sequence

$$(a_{11}, a_{21}, a_{12}, a_{31}, a_{22}, a_{13}, a_{41}, a_{32}, a_{23}, a_{14}, \dots)$$

by beginning at the upper left of the array and moving along the diagonals that extend up and to the right. As some of the sets $A_1, A_2, A_3, \dots$ might have elements in common, once an element has been listed in the sequence, we do not list it again. The sequence therefore includes each element of $B = A_1 \cup A_2 \cup A_3 \cup \cdots$ exactly once, and so represents a bijection from $\mathbb{N}$ onto B, thereby establishing that B is countably infinite. □

Corollary 5.25. *The set $\mathbb{Q}$ of all rational numbers is countably infinite.*

Exercise 5.29. Use Theorem 5.24 to prove Corollary 5.25. *Hint*: First put all the integers into a sequence, then put all the rational numbers having denominator 2 into a sequence, then put all the rational numbers having denominator 3 into a sequence, and so forth.

A set is **countable** if it is either finite or countably infinite. Thus, not only is every finite set countable, but so are the sets $\mathbb{N}$, $\mathbb{Z}$, and $\mathbb{Q}$. From Theorems 5.14 and 5.22 we immediately deduce the following result.

Theorem 5.26. *Every subset of a countable set is countable.*

6 The nested intervals property

The categorization of sets by size is taken one step further in this chapter when infinite sets are separated into two classes, *countably infinite* and *uncountable*. In particular, we show the set $\mathbb{Q}$ of rational numbers is countably infinite, then use the *nested intervals property* to show the set $\mathbb{R}$ of real numbers is uncountable. We also indicate how to apply the nested intervals property to obtain a decimal representation for each nonzero real number.

Collections of sets

By a **collection of sets** we simply mean a set whose elements are sets.

Example 6.1. As each element of the set

$$\mathcal{S} = \{(a, \infty) \mid a \in \mathbb{R}\}$$

is a set (more specifically, an interval), $\mathcal{S}$ is a collection of sets. Note that the collection $\mathcal{S}$ is "indexed" by the set $\mathbb{R}$ of real numbers; that is, for each real number a, there is a set, namely, (a, ∞), in $\mathcal{S}$. For instance, corresponding to the real number $-\sqrt{\pi}$ we have the element $(-\sqrt{\pi}, \infty)$ of $\mathcal{S}$; thus, we may write $(-\sqrt{\pi}, \infty) \in \mathcal{S}$.

Exercise 6.1. Consider the collection of sets $\mathcal{V} = \{\{-n, 0, n\} \mid n \in \mathbb{N}\}$.
(a) What set "indexes" the collection $\mathcal{V}$?
(b) Which member of $\mathcal{V}$ includes the number 3?

In Example 5.23, and also in the proof of Theorem 5.24, we formed the union of infinitely many sets. Generally, we can form both the *union* and the *intersection* of any nonempty collection of sets. The **union** is the set of all elements that are in at least one of the sets in the collection and the **intersection** is the set of all elements that are in every one of the sets in the collection.

Example 6.1 (Continued). The union of $\mathcal{S} = \{(a, \infty) \mid a \in \mathbb{R}\}$ is

$$\bigcup_{a\in\mathbb{R}} (a, \infty) = \mathbb{R},$$

as, given any real number a, the interval $(a - 1, \infty)$ includes a, which is one of the sets in $\mathcal{S}$.

Also, the intersection of the collection $\mathcal{S}$ is

$$\bigcap_{a\in\mathbb{R}} (a, \infty) = \emptyset,$$

https://doi.org/10.1515/9783111429564-006

since, given any real number a, the interval (a, ∞) does not include a, so a is not in every one of the sets in $\mathcal{S}$.

Exercise 6.2. Determine each of $\bigcap_{n=1}^{\infty}\{-n, 0, n\}$ and $\bigcup_{n=1}^{\infty}\{-n, 0, n\}$.

Example 6.2. Theorem 5.16 tells us that

$$\mathbb{R} = \cdots \cup [-2, -1) \cup [-1, 0) \cup [0, 1) \cup [1, 2) \cup \cdots = \bigcup_{n \in \mathbb{Z}} [n, n + 1).$$

Example 6.3. We can show that

$$\bigcap_{n=1}^{\infty}\left[0, \frac{1}{n}\right] = \{0\},$$

by demonstrating that $0 \in \bigcap_{n=1}^{\infty}[0, \frac{1}{n}]$ and that $\bigcap_{n=1}^{\infty}[0, \frac{1}{n}]$ includes no negative and no positive numbers.

First, observe that since $0 \in [0, \frac{1}{n}]$ for every natural number n, it follows, by the definition of intersection, that $0 \in \bigcap_{n=1}^{\infty}[0, \frac{1}{n}]$.

Next, suppose x is negative. Then x is not in $[0, \frac{1}{n}]$ for any natural number n, hence, cannot be in $\bigcap_{n=1}^{\infty}[0, \frac{1}{n}]$.

Finally, suppose x is positive. Then, by the Archimedean property, there exists a natural number m such that $\frac{1}{m} < x$. Thus, $x \notin [0, \frac{1}{m}]$ and so x cannot be in $\bigcap_{n=1}^{\infty}[0, \frac{1}{n}]$.

Exercise 6.3. Compute each of

$$\bigcup_{n=1}^{\infty}\left[1 + \frac{1}{n}, 2 + \frac{1}{n}\right],$$

and

$$\bigcap_{n=1}^{\infty}\left[1 + \frac{1}{n}, 2 + \frac{1}{n}\right],$$

justifying your conclusions.

The British mathematician Augustus De Morgan (1806–1871), who is also credited for the early development of mathematical induction, formalized many of the fundamental principles of deductive reasoning. The results stated in the following theorem are known as DeMorgan's laws for sets because they parallel DeMorgan's laws of logic (see Theorem B.10(6, 7) in Appendix B). The first equality says that the complement of the intersection of sets is the union of the complements of these sets, while the second equality says that the complement of the union of sets is the intersection of the complements of these sets.

Theorem 6.4 (DeMorgan's laws for sets). *Suppose that A is a set and also that B_i is a set for each i in some set I. Then*
(1) $A - \bigcap_{i \in I} B_i = \bigcup_{i \in I}(A - B_i)$;
(2) $A - \bigcup_{i \in I} B_i = \bigcap_{i \in I}(A - B_i)$.

In this theorem, the statement "B_i is a set for each i in some set I," means we have a collection of sets $\{B_i \mid i \in I\}$. Think of I as an arbitrary *indexing* set: Corresponding to each member i of I (that is, to each index value i), there is a set named B_i. Thus, if $I = \{1, 2, 3\}$, we have sets B_1, B_2, and B_3, which together comprise the collection $\{B_1, B_2, B_3\}$, and in this case the two DeMorgan's laws are

$$A - (B_1 \cap B_2 \cap B_3) = (A - B_1) \cup (A - B_2) \cup (A - B_3)$$

and

$$A - (B_1 \cup B_2 \cup B_3) = (A - B_1) \cap (A - B_2) \cap (A - B_3).$$

But if $I = \mathbb{Q}$, we have a set B_q for each rational number q, hence a collection $\{B_q \mid q \in \mathbb{Q}\}$, and DeMorgan's laws become

$$A - \bigcap_{q \in \mathbb{Q}} B_q = \bigcup_{q \in \mathbb{Q}} (A - B_q)$$

and

$$A - \bigcup_{q \in \mathbb{Q}} B_q = \bigcap_{q \in \mathbb{Q}} (A - B_q).$$

Proof of Theorem 6.4. We prove (1) and leave the proof of (2) as an exercise.

First, suppose $x \in A - \bigcap_{i \in I} B_i$. Then $x \in A$ and $x \notin \bigcap_{i \in I} B_i$. As x is not in every one of the sets in the collection $\{B_i \mid i \in I\}$, there exists an index k in I for which $x \notin B_k$. So as $x \in A$ and $x \notin B_k$, it follows that $x \in A - B_k$. Now, as x is in at least one of the sets in the collection $\{A - B_i \mid i \in I\}$, it follows that $x \in \bigcup_{i \in I}(A - B_i)$. We have thus shown that $A - \bigcap_{i \in I} B_i \subseteq \bigcup_{i \in I}(A - B_i)$.

Now suppose $x \in \bigcup_{i \in I}(A - B_i)$. Then there exists an index k in I, for which $x \in A - B_k$. We then deduce that $x \in A$ and $x \notin B_k$. As $x \notin B_k$, we see that x is not in every one of the sets in the collection $\{B_i \mid i \in I\}$, so $x \notin \bigcap_{i \in I} B_i$. Since we have $x \in A$ and $x \notin \bigcap_{i \in I} B_i$, it follows that $x \in A - \bigcap_{i \in I} B_i$. We have thus shown that $\bigcup_{i \in I}(A - B_i) \subseteq A - \bigcap_{i \in I} B_i$.

Having shown that each of the sets $A - \bigcap_{i \in I} B_i$ and $\bigcup_{i \in I}(A - B_i)$ is a subset of the other, we may conclude they are equal. □

Exercise 6.4. Prove (2) of Theorem 6.4.

A collection of sets is called **pairwise disjoint** if each pair of distinct sets in the collection is disjoint (have no elements in common).

Example 6.5. The collection $\{(n, n+1] \mid n \in \mathbb{N}\}$ is pairwise disjoint, since for distinct natural numbers j and k, the intervals $(j, j+1]$ and $(k, k+1]$ have no elements in common.

Note, however, that even though there is no real number that is in every one of the intervals in the collection $\{(a, \infty) \mid a \in \mathbb{R}\}$, this collection is not pairwise disjoint, because any two of the intervals have points in common.

Exercise 6.5. Determine which of the following collections of sets are pairwise disjoint.
(a) $\{\{-n, 0, n\} \mid n \in \mathbb{N}\}$;
(b) $\{\{-n, n\} \mid n \in \mathbb{N}\}$;
(c) $\{\{n, n^2\} \mid n \in \mathbb{N}\}$;
(d) $\{(a - 0.5, a + 0.5) \mid a \in \mathbb{R}\}$.

The nested intervals property

Recall that an interval is a subset of $\mathbb{R}$ that includes all real numbers between any two of its members. An interval that includes all of its endpoints is said to be **closed**, while an interval that includes none of its endpoints is said to be **open**. An interval is **bounded** if it is a bounded subset of $\mathbb{R}$; otherwise, an interval is **unbounded**.

Example 6.6. The interval $[0, 1]$ is closed and bounded. The interval $(0, 1)$ is open and bounded. The intervals $(0, 1]$ and $[0, 1)$ are both bounded, but neither is closed and neither is open; thus, an interval may be neither open nor closed. The interval $[0, \infty)$ is closed and unbounded, while the interval $(0, \infty)$ is open and unbounded.

The nested intervals property tells us that whenever we have a sequence of closed bounded intervals that are "nested" in the sense that the next one in the sequence is always a subset of the previous one, the intersection of all the intervals must include at least one real number.

Theorem 6.7 (The nested intervals property). *If, for each natural number n, the interval $I_n = [a_n, b_n]$ is closed and bounded, with $I_1 \supseteq I_2 \supseteq I_3 \supseteq \cdots$, then*

$$\bigcap_{n=1}^{\infty} I_n \neq \emptyset.$$

Moreover, if $\inf\{b_n - a_n \mid n \in \mathbb{N}\} = 0$, *then*

$$\bigcap_{n=1}^{\infty} I_n = \{p\}$$

for some real number p.

Since $I_n = [a_n, b_n]$, the statement

$$I_1 \supseteq I_2 \supseteq I_3 \supseteq \cdots$$

means

$$[a_2, b_2] \subseteq [a_1, b_1],$$
$$[a_3, b_3] \subseteq [a_2, b_2],$$
$$[a_4, b_4] \subseteq [a_3, b_3],$$

and so forth. In other words, the intervals $I_1, I_2, I_3, \ldots$ are "collapsing inward" on one another, as shown in Figure 6.1.

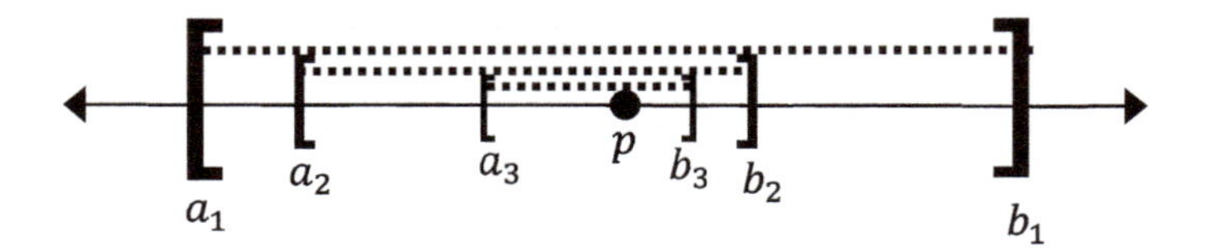

Figure 6.1: Visualizing the nested intervals property.

The theorem says that because the intervals are also bounded, and include both their left and right endpoints, this "nesting effect" results in a nontrivial "collapse" in which the intersection of the intervals contains at least one number p.

Because the length of the interval $I_n = [a_n, b_n]$ is $b_n - a_n$, the supplementary condition

$$\inf\{b_n - a_n \mid n \in \mathbb{N}\} = 0$$

would force the lengths of the intervals to become arbitrarily close to 0, in which case the intersection of the intervals consists of a single point p.

Proof of the nested intervals property, Theorem 6.7. For each natural number n, let the interval $I_n = [a_n, b_n]$ be closed and bounded, and assume that $I_1 \supseteq I_2 \supseteq I_3 \supseteq \cdots$. We first show that the right endpoint b_n of each interval is an upper bound for the set $A = \{a_n \mid n \in \mathbb{N}\}$ of all left endpoints of the intervals.

Consider any natural number n. To show b_n is an upper bound for A, we must show $a_m \leq b_n$ for every natural number m. So let m be a natural number. If $m \geq n$, we have $I_m \subseteq I_n$ so that $a_n \leq a_m \leq b_m \leq b_n$. If $m < n$, we have $I_n \subseteq I_m$ so that $a_m \leq a_n \leq b_n \leq b_m$. Hence, regardless of whether $m \geq n$ or $m < n$, we now know that $a_m \leq b_n$. Thus, b_n is an upper bound for A.

Since A is bounded above, the completeness axiom tells us that A has a supremum, which we shall denote by p. Consequently, for each natural number n, knowing p is an upper bound for A allows us to conclude that $a_n \leq p$, and knowing that p is the *least* upper bound for A and that b_n is an upper bound for A allows us to conclude that $p \leq b_n$. Thus, as $a_n \leq p \leq b_n$ for every natural number n, it follows that $p \in \bigcap_{n=1}^{\infty} I_n$, which means $\bigcap_{n=1}^{\infty} I_n \neq \emptyset$.

Now let us make the further assumption that $\inf\{b_n - a_n \mid n \in \mathbb{N}\} = 0$, and suppose to the contrary that p and q are both in $\bigcap_{n=1}^{\infty} I_n$ with $p \neq q$. Without loss of generality, assume $p < q$. It follows, as each I_n is an interval, that $[p, q] \subseteq \bigcap_{n=1}^{\infty} I_n$, which implies $[p, q] \subseteq [a_n, b_n]$ for every natural number n. Hence, $q - p \leq b_n - a_n$ for every natural number n, in which case it follows that $\inf\{b_n - a_n \mid n \in \mathbb{N}\} \geq q - p > 0$, a contradiction. □

Example 6.3 (Continued). For each natural number n, the interval $[0, \frac{1}{n}]$ is closed and bounded. Also,

$$[0, 1] \supseteq \left[0, \frac{1}{2}\right] \supseteq \left[0, \frac{1}{3}\right] \supseteq \cdots.$$

Thus, the nested intervals property assures us that $\bigcap_{n=1}^{\infty}[0, \frac{1}{n}] \neq \emptyset$. Because the lengths $\frac{1}{n} - 0 = \frac{1}{n}$ of the intervals $[0, \frac{1}{n}]$ are becoming arbitrarily small as n becomes ever larger, as $\inf\{\frac{1}{n} | n \in \mathbb{N}\} = 0$, we would anticipate the intersection consists of a single real number. In fact, we established earlier that $\bigcap_{n=1}^{\infty}[0, \frac{1}{n}] = \{0\}$.

The conclusion of a theorem may not hold if all of the hypotheses of the theorem are not met. For instance, a nested sequence of intervals may have *empty* intersection if the intervals are not all closed or if they are not all bounded.

Exercise 6.6. Give an example of a sequence of *open* bounded intervals $I_1, I_2, I_3, \ldots$ for which $I_1 \supseteq I_2 \supseteq I_3 \supseteq \cdots$ and $\bigcap_{n=1}^{\infty} I_n = \emptyset$.

Exercise 6.7. Give an example of a sequence of closed *unbounded* intervals $I_1, I_2, I_3, \ldots$ for which $I_1 \supseteq I_2 \supseteq I_3 \supseteq \cdots$ and $\bigcap_{n=1}^{\infty} I_n = \emptyset$.

The uncountability of the real numbers

We have not as yet exhibited an infinite set that is not countable, but such sets do exist. We need look no further than $\mathbb{R}$ itself, which we shall show is not countably infinite.

First, though, we use the nested intervals property to demonstrate that the closed bounded interval $[0, 1]$ is not countably infinite. In the proof, we make reference to a closed bounded interval being **nontrivial**, by which we simply mean it is not a singleton (that is, its left and right endpoints are different). The proof makes use of the observation that, given any particular number in a nontrivial closed bounded interval, we may choose a nontrivial closed bounded subinterval of the interval that does not include the particular number.

Exercise 6.8. Show that if $a < b$, then for any point p in the interval $[a, b]$, there exist real numbers c and d for which $c < d$, the interval $[c, d]$ is a subset of $[a, b]$, and p is not in $[c, d]$.

Theorem 6.8. *The interval* $[0, 1]$ *is not countably infinite.*

Proof. Suppose to the contrary that $[0, 1]$ is countably infinite. Then we may put the members of $[0, 1]$ into an infinite sequence $(a_1, a_2, a_3, \ldots)$ so that

$$[0, 1] = \{a_1, a_2, a_3, \ldots\}.$$

Use the result from Exercise 6.8 to choose a nontrivial closed subinterval I_1 of $[0, 1]$ that does not include a_1. Also, whenever n is a natural number and we have a nontrivial closed subinterval I_n of $[0, 1]$ that does not include a_n, again apply Exercise 6.8 to choose a nontrivial closed subinterval I_{n+1} of I_n that does not include a_{n+1}. The result, via the process of recursive definition (Theorem 3.16), is a sequence $(I_1, I_2, I_3, \ldots)$ of closed bounded intervals for which $I_1 \supseteq I_2 \supseteq I_3 \supseteq \cdots$.

Therefore, by the nested intervals property, there must exist a point p in $\bigcap_{n=1}^{\infty} I_n$. Then, as $\bigcap_{n=1}^{\infty} I_n \subseteq [0, 1]$, it follows that $p \in [0, 1]$. But, by hypothesis, $[0, 1] = \{a_1, a_2, a_3, \ldots\}$, so we may conclude that $p = a_k$ for some natural number k. But, by the way in which I_k was chosen, $a_k \notin I_k$, which implies that $p = a_k \notin \bigcap_{n=1}^{\infty} I_n$, a contradiction. □

Thus, not all infinite sets are "listable," even from the perspective of an unending list. We view countable sets as those sets whose elements can be put into a terminating (if the set is finite) or unending (if the set is countably infinite) list. A set that is not countable is called **uncountable**, so uncountable sets can be viewed as those infinite sets whose elements cannot be arranged into a list (sequence).

Having shown that the interval $[0, 1]$ is uncountable, it actually follows that all intervals on the real line that include more than one point are uncountable.

Theorem 6.9. *The set* $\mathbb{R}$ *of real numbers is uncountable, as is the set* $\mathbb{R} - \mathbb{Q}$ *of irrational numbers.*

Exercise 6.9. Prove Theorem 6.9. *Hint*: Make use of Theorem 6.8, Corollary 5.25, Theorem 5.26, and Exercise 5.25.

Our work here tells us there are at least two "sizes" an infinite set can exhibit, countably infinite and uncountable. In truth, the uncountable sets can be further subcategorized, and there are actually infinitely many different possible sizes an infinite set may have. For our work analyzing the real number system, though, it is sufficient to be able to distinguish between countably infinite and uncountable subsets of $\mathbb{R}$.

It was Georg Cantor (1845–1918) who developed the theory of *set cardinality* as a means for distinguishing among the various sizes of infinite sets. His work on this and related problems helped to create the foundation on which mathematics from the early 20th century onward has been built.

Decimal representation of real numbers

We use the shorthand

$$N.d_1d_2\ldots d_n,$$

called a **terminating decimal representation**, for the positive real number

$$N + d_1 \cdot \frac{1}{10} + d_2 \cdot \frac{1}{10^2} + \cdots + d_n \cdot \frac{1}{10^n},$$

where N is a nonnegative integer and each of $d_1, d_2, \ldots, d_n$ is a decimal digit, that is, a member of the set $\{0, 1, 2, 3, 4, 5, 6, 7, 8, 9\}$. For example,

$$0.75 = 0 + 7 \cdot \frac{1}{10} + 5 \cdot \frac{1}{10^2}$$

and

$$120.4026 = 120 + 4 \cdot \frac{1}{10} + 0 \cdot \frac{1}{10^2} + 2 \cdot \frac{1}{10^3} + 6 \cdot \frac{1}{10^4}.$$

If a positive number p can be represented as $N.d_1d_2\ldots d_n$, we may write the negative number $-p$ as $-N.d_1d_2\ldots d_n$. Thus, as $\frac{3}{4} = 0.75$, we also have $-\frac{3}{4} = -0.75$.

Exercise 6.10. Assuming N is a nonnegative integer and each of $d_1, d_2, \ldots, d_n$ is a decimal digit, explain why

$$N + d_1 \cdot \frac{1}{10} + d_2 \cdot \frac{1}{10^2} + \cdots + d_n \cdot \frac{1}{10^n}$$

must be a rational number. Then explain why this rational number, when expressed as a "reduced" fraction, must have a denominator of the form $2^i \cdot 5^j$, where i and j are nonnegative integers.

The previous exercise reveals that no irrational numbers and only certain rational numbers have terminating decimal representations. However, we can show that every nonzero real number has a *decimal representation* of the form

$$\pm N.d_1d_2d_3\ldots$$

that includes infinitely many decimal digits to the right of the decimal point, infinitely many of which are also nonzero, and where N is a nonnegative integer.

We begin by using the nested intervals property to obtain a decimal representation for each number in the interval $(0,1]$. Consider any number p in $(0,1]$. Partition $(0,1]$ into the ten subintervals

$$\left(0,\frac{1}{10}\right],\ \left(\frac{1}{10},\frac{2}{10}\right],\ \left(\frac{2}{10},\frac{3}{10}\right],\ \left(\frac{3}{10},\frac{4}{10}\right],\ \left(\frac{4}{10},\frac{5}{10}\right],\ \left(\frac{5}{10},\frac{6}{10}\right],$$
$$\left(\frac{6}{10},\frac{7}{10}\right],\ \left(\frac{7}{10},\frac{8}{10}\right],\ \left(\frac{8}{10},\frac{9}{10}\right],\ \text{and}\ \left(\frac{9}{10},1\right].$$

The number p is a member of precisely one of these intervals, so let $(\frac{d_1}{10},\frac{d_1+1}{10}]$ be the one that includes p, and define $I_1=[\frac{d_1}{10},\frac{d_1+1}{10}]$.

Example 6.10. Consider $p=\frac{3}{4}$, which is in $(0,1]$. As shown in Figure 6.2, the number $\frac{3}{4}$ is in $(\frac{7}{10},\frac{8}{10}]$, so $d_1=7$ and $I_1=[\frac{7}{10},\frac{8}{10}]$.

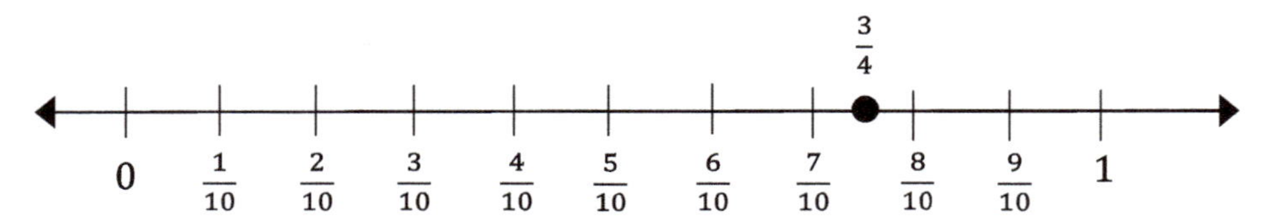

Figure 6.2: The number $\frac{3}{4}$ is in the interval $(\frac{7}{10},\frac{8}{10}]$.

For convenience, we let $D=\{0,1,2,3,4,5,6,7,8,9\}$, the set of decimal digits. Now for any natural number n for which $I_n=[a_n,b_n]$ has been defined so as to include p, partition $(a_n,b_n]$ into the ten subintervals of the form

$$\left(a_n+\frac{i}{10^{n+1}},a_n+\frac{i+1}{10^{n+1}}\right],$$

where $i\in D$. The number p is a member of precisely one of these intervals, so let $(a_n+\frac{d_{n+1}}{10^{n+1}},a_n+\frac{d_{n+1}+1}{10^{n+1}}]$ be the one that includes p, and define

$$I_{n+1}=\left[a_n+\frac{d_{n+1}}{10^{n+1}},a_n+\frac{d_{n+1}+1}{10^{n+1}}\right].$$

According to the principle of recursive definition, Theorem 3.16, we have now defined two sequences, a sequence $(I_1,I_2,I_3,\dots)$ of closed bounded intervals and a sequence $(d_1,d_2,d_3,\dots)$ of decimal digits. The sequence

$$(d_1,d_2,d_3,\dots)$$

is usually written as

$$0.d_1d_2d_3\dots$$

and is called the **nonterminating decimal representation** of the number p. By the manner in which it is created, this representation for a particular number in $(0,1]$ is unique.

Example 6.10 (Continued). From $I_1 = [\frac{7}{10}, \frac{8}{10}]$, we now determine d_2 and I_2 as follows. Partition $(\frac{7}{10}, \frac{8}{10}]$ into the ten subintervals $(\frac{7}{10} + \frac{i}{10^2}, \frac{7}{10} + \frac{i+1}{10^2}]$, where $i \in D$. Figure 6.3 displays these intervals.

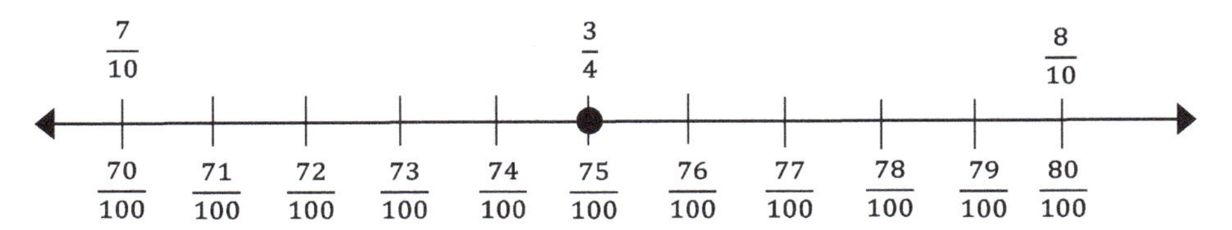

Figure 6.3: The number $\frac{3}{4}$ is in the interval $(\frac{74}{100}, \frac{75}{100}]$.

Note that it is the interval

$$\left(\frac{7}{10} + \frac{4}{10^2}, \frac{7}{10} + \frac{4+1}{10^2}\right] = \left(\frac{74}{100}, \frac{75}{100}\right]$$

that includes the number $\frac{3}{4}$, so $d_2 = 4$ and $I_2 = [\frac{74}{100}, \frac{75}{100}]$.

We can now obtain d_3 and I_3 from I_2. Partition $(\frac{74}{100}, \frac{75}{100}]$ into the ten subintervals $(\frac{74}{100} + \frac{i}{10^3}, \frac{74}{100} + \frac{i+1}{10^3}]$, where $i \in D$, as illustrated in Figure 6.4.

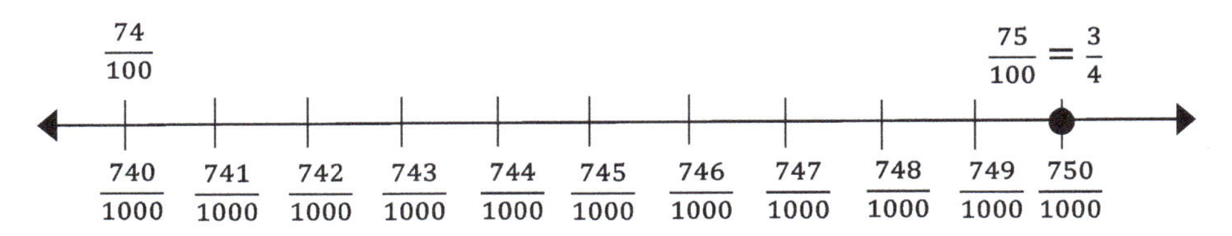

Figure 6.4: The number $\frac{3}{4}$ is in the interval $(\frac{749}{1000}, \frac{750}{1000}]$.

Of these intervals, it is the final one

$$\left(\frac{74}{100} + \frac{9}{10^3}, \frac{74}{100} + \frac{9+1}{10^3}\right] = \left(\frac{749}{1000}, \frac{750}{1000}\right]$$

that includes the number $\frac{3}{4}$, so $d_3 = 9$ and $I_3 = [\frac{749}{1000}, \frac{750}{1000}]$.

Exercise 6.11. Continue the process begun in Example 6.10 to obtain d_4 and I_4.

Example 6.10 (Continued). The number $\frac{3}{4}$, for which we are creating a decimal representation, is the right endpoint of the interval $I_2 = [\frac{74}{100}, \frac{75}{100}]$, and this location in the interval actually forces $\frac{3}{4}$ to be the right endpoint of each subsequent interval I_n that is selected (we saw this to be the case for I_3 and you should have noticed the same thing for I_4 when you completed Exercise 6.11). The result is that $d_n = 9$ for all n larger than 2. Thus, the nonterminating decimal representation of $\frac{3}{4}$ is

$$0.74\overline{9},$$

where the "bar" over the 9 indicates this digit endlessly repeats.

Exercise 6.12. Find $d_1, I_1, d_2, I_2, d_3, I_3, d_4$, and I_4 in the process of obtaining the nonterminating decimal representation of $\frac{1}{27}$.

Exercise 6.13. Find the nonterminating decimal representation of 1.

By the way in which the intervals in the sequence $(I_1, I_2, I_3, \dots)$ are determined, we see they are nested inside one another, they are closed and bounded, and their lengths are becoming arbitrarily small, so the nested intervals property, Theorem 6.7, tells us that their intersection includes precisely one number. Thus, as the original number p chosen from $(0, 1]$ is in every one of the intervals, we may conclude that

$$\bigcap_{n=1}^{\infty} I_n = \{p\}.$$

The fact that the number p is the only number in all of the intervals indicates that different numbers in $(0, 1]$ must have different nonterminating decimal representations.

Furthermore, the construction employed clearly identifies the endpoints a_n and b_n of $I_n = [a_n, b_n]$ as

$$a_n = \sum_{k=1}^{n} d_k \cdot \frac{1}{10^k}$$

and

$$b_n = \sum_{k=1}^{n} d_k \cdot \frac{1}{10^k} + \frac{1}{10^n}.$$

Exercise 6.14. It is more common to express $\frac{3}{4}$ in decimal form as 0.75 rather than $0.74\overline{9}$. Explain why the process we have described for obtaining nonterminating decimal representations could never lead to the sequence $(7, 5, 0, 0, 0, 0, \dots)$.

Exercise 6.15. Show that if a sequence $(d_1, d_2, d_3, \ldots)$ of decimal digits has the property that there exists a natural number m for which $d_n = 0$ for every n greater than m, then the sequence is not the nonterminating decimal representation of any real number in $(0, 1]$.

Every nonzero real number has a nonterminating decimal representation, though in the case of an integer we usually just employ the representation provided for by Theorem 3.13 (for example, 3482 rather than $3481.\overline{9}$). Therefore, we concentrate here on real numbers that are not integers.

Given a positive real number p that is not an integer, Theorem 5.16 tells us there is a unique nonnegative integer N for which $N < p < N+1$. As it follows that $0 < p-N < 1$, we see that $p-N$ has a unique nonterminating decimal representation $0.d_1d_2d_3\ldots$. We then define the **nonterminating decimal representation** of the positive number p to be

$$N.d_1d_2d_3\ldots.$$

We also define the **nonterminating decimal representation** of a negative real number q that is not an integer to be

$$-N.d_1d_2d_3\ldots,$$

where $N.d_1d_2d_3\ldots$ is the nonterminating decimal representation of the positive number $-q$.

Example 6.11. The nonterminating decimal representation of $\frac{15}{4}$ is $3.74\overline{9}$ and the nonterminating decimal representation of $-\frac{15}{4}$ is $-3.74\overline{9}$.

What mathematicians call *real numbers* are often referred to as "decimals" by the general public, probably in large part to help distinguish them from "whole numbers." The decimal representations we have developed here provide some validation for this point of view. We discuss decimal representations further in Chapter 22, where we study infinite summations and make the anticipated connection that the expression $N.d_1d_2d_3\ldots$ is actually mathematical shorthand for

$$N + d_1 \cdot \frac{1}{10} + d_2 \cdot \frac{1}{10^2} + d_3 \cdot \frac{1}{10^3} + \cdots.$$

Exercise 6.16. Modify the process we used to create nonterminating decimal representations to develop a method for obtaining a unique nonterminating binary representation $(b_1, b_2, b_3, \ldots)$, where each b_n is in $\{0, 1\}$, for an arbitrary number in $(0, 1]$.

Exercise 6.17. Repeat Exercise 6.16 to show how to obtain a unique nonterminating ternary representation $(t_1, t_2, t_3, \ldots)$, where each t_n is in $\{0, 1, 2\}$, for an arbitrary number in $(0, 1]$.

7 Limit points, closed sets, and open sets

During the latter half of the 19th century, mathematical analysis began to grow beyond the study of the real number system, and mathematicians began to more formally employ structures such as the plane $\mathbb{R}^2 = \mathbb{R}\times\mathbb{R}$, three-dimensional space $\mathbb{R}^3 = \mathbb{R}\times\mathbb{R}\times\mathbb{R}$, and function spaces representing the solutions to certain differential equations. The mathematical subject of *topology* emerged from this activity and offers an extremely general setting for the study of concepts such as *limits*, *convergence*, and *continuity*. As these concepts are of fundamental importance in the analysis of the real number system, we use this chapter and the next to introduce several relevant topological notions in the context of the real numbers. They include *limit point*, *open set*, *closed set*, *connected set*, and *compact set*, each of which is governed by the distance between real numbers we defined back in Chapter 2. The number line model proves particularly helpful in developing an intuitive feel for the topological notions we discuss.

Limit points

The notion of *limit point*, as the choice of the terminology itself suggests, is closely related to fundamental limiting processes, such as *convergence of a sequence* and *limit of a function* that are first studied in calculus and that we pursue in more depth later in this book.

Recall that a **neighborhood** of a real number p is any interval of the form

$$(p - \varepsilon, p + \varepsilon),$$

where ε is a positive number.

A real number p is a **limit point** of a subset A of $\mathbb{R}$ if every neighborhood of p includes infinitely many points of A, that is, if for every positive number ε, the set $A \cap (p - \varepsilon, p + \varepsilon)$ is infinite.

Intuitively, when p is a limit point of A, there is a "pile up" or "accumulation" of infinitely many points of A that are "arbitrarily close" to p. In other words, within *any* positive distance, *no matter how small*, of a limit point of a set, we are able to find infinitely many points of the set.

Example 7.1. The number 0.25 is a limit point of the interval $(0, 1)$ because we can show that, for every positive number ε, the neighborhood $(0.25 - \varepsilon, 0.25 + \varepsilon)$ of 0.25 includes infinitely many points of $(0, 1)$. We consider separately the two possibilities $\varepsilon \leq 0.25$ and $\varepsilon > 0.25$.

If $\varepsilon \leq 0.25$, then $(0.25 - \varepsilon, 0.25 + \varepsilon)$ is a subset of $(0, 1)$, so $(0.25 - \varepsilon, 0.25 + \varepsilon)$ includes infinitely many points of $(0, 1)$, namely, *all* the points of $(0.25 - \varepsilon, 0.25 + \varepsilon)$ (Figure 7.1).

https://doi.org/10.1515/9783111429564-007

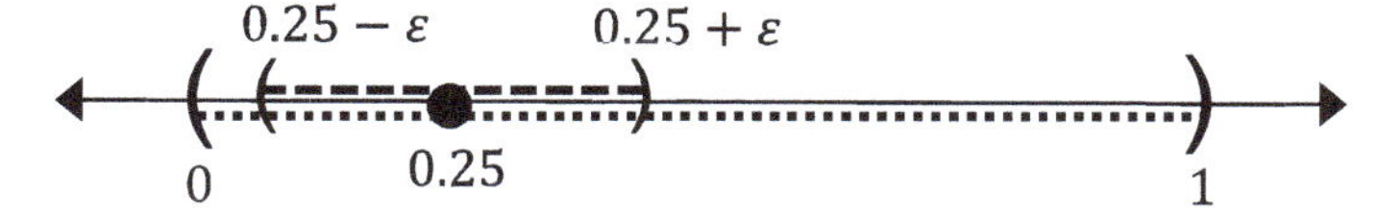

Figure 7.1: Any ε-neighborhood of 0.25 for which $\varepsilon \leq 0.25$ is a subset of $(0, 1)$.

If $\varepsilon > 0.25$, then $(0, 0.5)$ is a subset of both $(0.25 - \varepsilon, 0.25 + \varepsilon)$ and $(0, 1)$, so $(0.25 - \varepsilon, 0.25 + \varepsilon)$ includes infinitely many points of $(0, 1)$, namely, all of the infinitely many points of $(0, 0.5)$ (Figure 7.2).

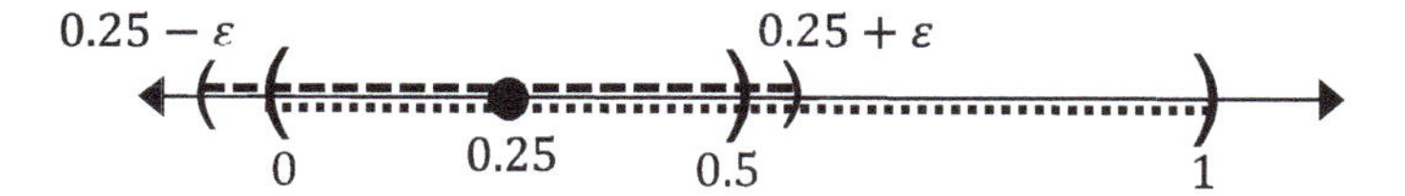

Figure 7.2: Any ε-neighborhood of 0.25 for which $\varepsilon > 0.25$ contains $(0, 0.5)$.

It is important to be aware that a limit point of a set need not be a member of the set.

Example 7.1 (Continued). The number 0 is a limit point of the interval $(0, 1)$, even though 0 is not in $(0, 1)$, as given any positive number ε, the neighborhood $(0 - \varepsilon, 0 + \varepsilon)$, that is, $(-\varepsilon, \varepsilon)$, of 0 includes infinitely many members of $(0, 1)$, either all infinitely many points of $(0, \varepsilon)$, if $\varepsilon \leq 1$, or all infinitely many points of $(0, 1)$, if $\varepsilon > 1$. Figure 7.3 illustrates the situation for a choice of ε where $\varepsilon < 1$.

Figure 7.3: The number 0 is a limit point of $(0, 1)$ even though 0 is not in $(0, 1)$.

To demonstrate that a number p is not a limit point of a set, we need only find a *single* neighborhood of p that does not include infinitely many members of the set in question.

Example 7.1 (Continued). The number 1.1 is not a limit point of the interval $(0, 1)$, as the particular neighborhood $(1.1 - 0.1, 1.1 + 0.1)$, that is, $(1, 1.2)$, of 1.1 includes no members of $(0, 1)$ (Figure 7.4).

Note also that any neighborhood $(1.1 - \varepsilon, 1.1 + \varepsilon)$, where $\varepsilon < 0.1$, also includes no members of the set $(0, 1)$, and so could be cited as evidence as to why 1.1 is not a limit point of $(0, 1)$. But, generally speaking, it is sufficient to identify just one neighborhood

Figure 7.4: The number 1.1 is not a limit point of $(0, 1)$.

of p that does not include infinitely many elements of A in order to conclude that p is not a limit point of A.

Exercise 7.1. Convince yourself that every point of the interval $[0, 1]$ is a limit point of the interval $(0, 1)$, and that $(0, 1)$ has no other limit points (drawing pictures should be helpful). Then give explicit arguments justifying each of the following:

(a) The number 1 is a limit point of $(0, 1)$.

(b) The number 0.9 is a limit point of $(0, 1)$.

(c) The number -0.03 is not a limit point of $(0, 1)$.

It is possible for a set to have no limit points.

Example 7.2. It is impossible for a finite subset of $\mathbb{R}$ to have any limit points, as no neighborhood of any real number could include infinitely many points of a set that has only finitely many members.

In the next example, we determine the limit points of the set

$$\left\{\frac{1}{n} \;\middle|\; n \in \mathbb{N}\right\} = \left\{1, \frac{1}{2}, \frac{1}{3}, \frac{1}{4}, \ldots\right\}$$

of reciprocals of the natural numbers. We encourage you to draw pictures to visualize what is being described.

Example 7.3. For convenience, let $B = \{\frac{1}{n} \mid n \in \mathbb{N}\}$. First, we show the number 0 is a limit point of B. Note that for any positive number ε, the Archimedean property, in the form given in Corollary 4.19, can be used to find a natural number N such that $\frac{1}{N} < \varepsilon$. It then follows that

$$B \cap (0 - \varepsilon, 0 + \varepsilon) = \left\{\frac{1}{N}, \frac{1}{N+1}, \frac{1}{N+2}, \ldots\right\},$$

which is an infinite set. Hence, we may conclude that 0 is a limit point of B.

On the other hand, the number -0.5 is not a limit point of B, as the neighborhood $(-0.5 - 0.1, -0.5 + 0.1)$, that is, $(-0.6, -0.4)$, of -0.5 contains no points of B (all members of $(-0.6, -0.4)$ are negative, while all members of B are positive). In a similar fashion, one can show that no negative number is a limit point of B and no positive number larger than 1 is a limit point of B.

Now observe that each of the three numbers 1, $\frac{1}{2} = 0.5$, and $\frac{1}{3} = 0.\overline{3}$ is greater than 0.3, and when n is a natural number for which $n \geq 5$, the number $\frac{1}{n}$ is less than or equal to $\frac{1}{5} = 0.2$. Hence,

$$\left(\frac{1}{4} - 0.05, \frac{1}{4} + 0.05\right) = (0.25 - 0.05, 0.25 + 0.05) = (0.2, 0.3)$$

includes no point of B other than $\frac{1}{4}$. Thus, as the neighborhood $(\frac{1}{4} - 0.05, \frac{1}{4} + 0.05)$ of $\frac{1}{4}$ does not include infinitely members of B, the number $\frac{1}{4}$ is not a limit point of B. In a similar way one can show that no member of B is a limit point of B.

Finally, observe that each of the four numbers 1, $\frac{1}{2} = 0.5$, $\frac{1}{3} = 0.\overline{3}$, and $\frac{1}{4} = 0.25$ is greater than 0.23, and when n is a natural number for which $n \geq 5$, the number $\frac{1}{n}$ is less than or equal to $\frac{1}{5} = 0.2$, which is in turn less than 0.21. Thus, the neighborhood

$$(0.23 - 0.02, 0.23 + 0.02) = (0.21, 0.25)$$

of 0.23 contains no points of B, so 0.23 is not a limit point of B. In a similar way, one can show that no number between 0 and 1 that is not in B is a limit point of B.

Given the entirety of our analysis in this example, we conclude that 0 is the only limit point of the set $\{\frac{1}{n} \mid n \in \mathbb{N}\}$. Thus, it is possible for the only limit points a set possesses to be non-members of the set.

Exercise 7.2. Let $B = \{\frac{1}{n} \mid n \in \mathbb{N}\}$.
(a) Provide details showing that 2 is not a limit point of B.
(b) Provide details showing that $\frac{1}{100}$ is not a limit point of B.
(c) Provide details showing that $\frac{2}{5}$ is not a limit point of B.

Exercise 7.3. Identify all the limit points of the following subsets of $\mathbb{R}$.
(a) $(-5, 5]$;
(b) $\{\frac{1-n}{n} \mid n \in \mathbb{N}\}$.

By definition, to establish that p is a limit point of A requires us to show that *every* neighborhood of p *includes infinitely many members* of A. However, it turns out that as long as we are able to show that *every* neighborhood of p *includes at least one member of A that is distinct from p*, we are guaranteed that p is a limit point of A.

This comes about because we can use the (necessarily positive) distance between p and the single member of A distinct from p that is assumed to be in a given neighborhood of p as the radius of another neighborhood of p sitting inside the original. The new neighborhood does not include the member of A found in the original neighborhood, but must, by hypothesis, include some member of A distinct from p. The nesting of the new neighborhood inside the original neighborhood means the new member of A also lies in the original. This process can be repeated indefinitely to generate infinitely many points of A that are in the original neighborhood of p.

Figure 7.5 illustrates visually what we have just described. The notation incorporated in it is identical to that employed at the beginning of the argument for (ii) ⇒ (i) in the formal proof presented below.

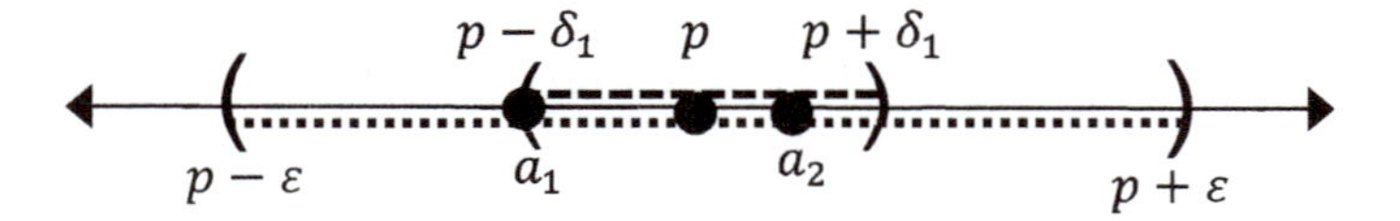

Figure 7.5: Generating infinitely many points of a set that are also in a neighborhood of a limit point of the set.

Theorem 7.4. *Let A be a subset of $\mathbb{R}$ and let p be a real number. The following are equivalent:*
(i) *p is a limit point of A;*
(ii) *every neighborhood of p includes a member of A different from p.*

Proof. (i) ⇒ (ii) Assume that p is a limit point of A and let ε be any positive number. The neighborhood $(p - \varepsilon, p + \varepsilon)$ of p includes infinitely many points of A, so it must include members of A that are different from p.

(ii) ⇒ (i) Assume that every neighborhood of p includes a member of A different from p. Consider any positive number ε; we must show $(p - \varepsilon, p + \varepsilon)$ includes infinitely many members of A.

By assumption, $(p - \varepsilon, p + \varepsilon)$ includes a point a_1 of A different from p. As $a_1 \neq p$, the distance $|a_1 - p|$ between a_1 and p must be some positive number δ_1. So, again applying our assumption that every neighborhood of p has a member of A different from p, we conclude that $(p - \delta_1, p + \delta_1)$ includes a point a_2 of A different from p and, necessarily, different from a_1 as $a_1 \notin (p - \delta_1, p + \delta_1)$. If we let $\delta_2 = |a_2 - p|$, the distance between a_1 and p, which is a positive number, we can once more apply our assumption that every neighborhood of p includes a member of A different from p to obtain an element a_3 of A that is in $(p - \delta_2, p + \delta_2)$, that is different from p, and, necessarily, also different from both a_1 and a_2.

This process may be continued indefinitely, via the principle of recursive definition, to obtain infinitely many distinct points $a_1, a_2, a_3, \ldots$ of A and infinitely many positive numbers $\delta_1, \delta_2, \delta_3, \ldots$ for which $\varepsilon > \delta_1 > \delta_2 > \delta_3 > \cdots$. It then follows that for each natural number n, we have

$$a_n \in (p - \delta_{n-1}, p + \delta_{n-1}) \subseteq (p - \varepsilon, p + \varepsilon).$$

Thus, the original neighborhood $(p - \varepsilon, p + \varepsilon)$ of p includes the infinitely many members $a_1, a_2, a_3, \ldots$ of A. □

The way in which we indicated the application of the principle of recursive definition, Theorem 3.16, in the previous proof is a bit less formal than what we have done before. Formally, after identifying a_1 and δ_1, we would have assumed a_n and δ_n had been identified for some arbitrary natural number and then indicated how to use them to generate a_{n+1} and δ_{n+1}. In the less formal approach, which is actually more typical of the way arguments using recursive definition tend to be structured, we instead indicated how to generate a_2 and δ_2 from a_1 and δ_1, then how to generate a_3 and δ_3 from a_2 and δ_2, and then noted that the same process can be carried out indefinitely to generate a_{n+1} and δ_{n+1} from a_n and δ_n, no matter what the value of n.

Exercise 7.4. Show that every real number is a limit point of the set $\mathbb{Q}$ of all rational numbers.

The Bolzano–Weierstrass theorem

Our next result, the *Bolzano–Weierstrass theorem*, tells us that whenever an infinite subset of $\mathbb{R}$ is bounded, the elements of the subset must accumulate ("pile up") around one or more real numbers. Bernard Bolzano (1781–1848) was the first mathematician to prove this theorem, but in a different form (see Theorem 11.5). In the mid-19th century, Karl Weierstrass (1815–1897) recognized the importance of the theorem to a rigorous foundation for the calculus and offered a different proof of it.

Weierstrass' proof uses the nested intervals property, with the intervals created through iterative bisection of an initial closed bounded interval that contains the given bounded infinite set. Each time we bisect an interval containing infinitely many points of this given set, it follows that at least one of the two subintervals created must contain infinitely many points of the set. Because the lengths of the intervals shrink toward zero, the nested intervals theorem permits us to conclude that the intersection of all the intervals consists of a single point, a point that we can show must be a limit point of the original bounded infinite set.

Theorem 7.5 (The Bolzano–Weierstrass theorem). *Any bounded infinite subset of* $\mathbb{R}$ *has at least one limit point.*

Proof. Consider any bounded infinite subset A of $\mathbb{R}$. Then $A \subseteq [a, b]$ for some numbers a and b for which $a < b$. Let $I_1 = [a, b]$. Bisect I_1 into two closed subintervals $[a, \frac{1}{2}(a+b)]$ and $[\frac{1}{2}(a+b), b]$. At least one of these intervals includes infinitely many points of A; choose such an interval and call it I_2. Then bisect I_2 and obtain I_3 in the same way as I_2 was obtained from I_1, so that I_3 includes infinitely many points of A.

Continuing this process, we obtain a nested collection $\{I_n \mid n \in \mathbb{N}\}$ of closed bounded intervals, each of which contains infinitely many points of A. By the nested intervals property, as the lengths of the intervals I_n are becoming arbitrarily small, $\bigcap_{n=1}^{\infty} I_n = \{p\}$

for some real number p. To complete our proof, we need only show that p is a limit point of A.

So, consider any positive number ε and use the Archimedean property to choose a natural number N large enough so that $\frac{b-a}{2^{N-1}} < \varepsilon$. Then, as $\frac{b-a}{2^{N-1}}$ is the length of the interval I_N, it follows that $I_N \subseteq (p - \varepsilon, p + \varepsilon)$. But, since I_N includes infinitely many points of A, it follows that $(p - \varepsilon, p + \varepsilon)$ does, too, making p a limit point of A. □

Exercise 7.5. Referring to the last paragraph of the proof of the Bolzano–Weierstrass theorem,
(a) provide the details of how the Archimedean property is being used to obtain the natural number N;
(b) and explain why the length of the interval I_N is $\frac{b-a}{2^{N-1}}$.

Example 7.3 (Continued). The set $\{\frac{1}{n} \mid n \in \mathbb{N}\}$ is bounded, as all of its members are greater than 0 and less than or equal to 1, and infinite. Hence, by the Bolzano–Weierstrass theorem, this set must have a limit point. We in fact demonstrated earlier that this set has only one limit point, namely 0.

Example 7.6. In general an infinite subset of $\mathbb{R}$ need not have a limit point. For instance, the set $\mathbb{N}$ of natural numbers has no limit points, as you show in Exercise 7.6, even though it is infinite. This observation is not in conflict with the Bolzano–Weierstrass theorem, however, as $\mathbb{N}$ is not bounded.

Exercise 7.6. Show that the set $\mathbb{N}$ of natural numbers has no limit points.

Exercise 7.7. Consider the set $A = \{\frac{1}{m} - \frac{1}{n} \mid m, n \in \mathbb{N}\}$.
(a) Can the Bolzano–Weierstrass theorem be applied to argue that A must have a limit point? If so, how precisely? If not, why not?
(b) Find all the limit points of A and verify your conclusions.

Closed sets

A subset A of $\mathbb{R}$ is said to be **closed** if A contains all of its limit points.

Example 7.7. The interval $[0, 1]$ is closed, as is any bounded interval $[a, b]$ that includes its endpoints.

To demonstrate that $[0, 1]$ is closed, consider any real number p that is not in $[0, 1]$. Then either $p < 0$ or $p > 1$. You may find it helpful to draw pictures to help guide you through the following arguments.

If $p < 0$, then let $\varepsilon = -p$, which is a positive number, and note that $(p - \varepsilon, p + \varepsilon) = (2p, 0)$, which is a neighborhood of p that includes no points of $[0, 1]$. Hence, p is not a limit point of $[0, 1]$.

If $p > 1$, then let $\varepsilon = p - 1$, which is a positive number, and note that $(p - \varepsilon, p + \varepsilon) = (1, 2p - 1)$, which is a neighborhood of p that includes no points of $[0, 1]$. Hence, p is not a limit point of $[0, 1]$.

Since no real number in the complement $[0, 1]$ is a limit point of $[0, 1]$, we deduce that $[0, 1]$ contains all of its limit points, hence, is closed.

Example 7.8. The interval $(0, 1)$ is not closed, and neither is any bounded interval (a, b) that does not include its endpoints.

To see that $(0, 1)$ is not closed, note that we determined in Example 7.1 that 0 is a limit point of $(0, 1)$. Thus, as $0 \notin (0, 1)$, we see that $(0, 1)$ does not contain all of its limit points, so is not closed.

Exercise 7.8. Determine whether the given set is closed, justifying your conclusion.
(a) $(3, \infty)$;
(b) $(-\infty, 3]$;
(c) $\mathbb{Q}$.

Exercise 7.9. Prove that if A is a closed subset of $\mathbb{R}$ that has a supremum, then $\sup(A)$ is in A. (It is also true that if A is a closed subset of $\mathbb{R}$ that has an infimum, then $\inf(A)$ is in A.)

Closed sets can be understood by means of the notion of the distance between real numbers. Specifically, we may view a closed set as one that contains all real numbers that could be considered "close" to the set, meaning all real numbers whose distance from the set is zero. So how do we measure the distance between a real number and a nonempty subset of $\mathbb{R}$?

The set of distances between a real number p and the numbers in a nonempty subset A of $\mathbb{R}$ is always bounded below by zero and, hence, must have a greatest lower bound (infimum), so we define the **distance between p and A** to be

$$\inf\{|p - a| \mid a \in A\},$$

the greatest lower bound of the distances between p and the points of A.

Example 7.9. Figure 7.6 suggests that the distance between the real number 3 and the set $[0, 1)$ is 2.

Note that there is no "closest" point to 3 in the set $[0, 1)$, since $[0, 1)$ does not include its right endpoint 1. So even though we expect the distance between 3 and $[0, 1)$ to be 2, there is no number in the set $[0, 1)$ whose distance from 3 is 2.

We now verify that it is actually the case that the distance between 3 and $[0, 1)$ is 2. Let $D = \{3 - a \mid a \in [0, 1)\}$, the set of distances between 3 and the points of $[0, 1)$. By definition, the distance between 3 and $[0, 1)$ is $\inf(D)$, so we need to show that $\inf(D) = 2$.

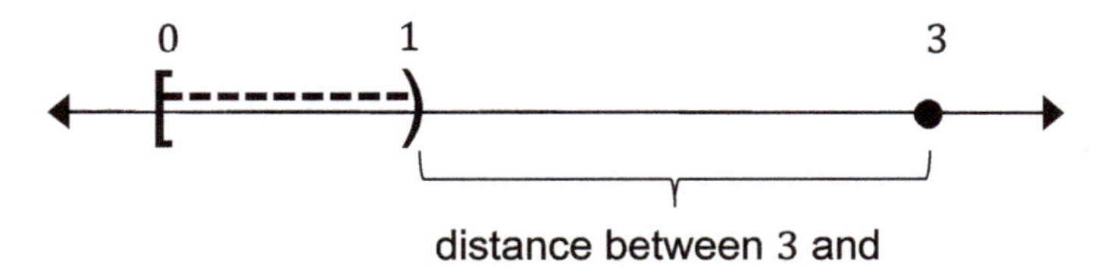

Figure 7.6: The distance between 3 and $[0,1)$ is 2.

Consider an arbitrary point a in the set $[0,1)$. Then $a < 1$, so that $3 - a > 2$. Thus, 2 is a lower bound for D and it follows that $2 \leq \inf(D)$.

Suppose there is a lower bound l for D such that $l > 2$. Then, being a lower bound, we have $l \leq 3 - a$ for all points a in $[0,1)$, so in particular $l \leq 3 - 0 = 3$. Thus, we now know that $2 < l \leq 3$, from which it follows that $0 \leq 3 - l < 1$. By the density property, we may choose a real number p for which $3 - l < p < 1$. Then $p \in [0,1)$, so that $3 - p \in D$. As l is a lower bound for D, it follows that $l \leq 3 - p$, which implies that $p \leq 3 - l$, contradicting our earlier conclusion that $3 - l < p$.

Hence, D has no lower bound greater than the lower bound 2, thus making 2 the greatest lower bound of D.

Exercise 7.10. Find the distance between -7 and the given subset of $\mathbb{R}$.
(a) $\{1.5\}$;
(b) $\mathbb{N}$;
(c) $\mathbb{Z}$;
(d) $\{\frac{1}{n} \mid n \in \mathbb{N}\}$.

We can now establish that closed sets are precisely those sets that include all points that are zero distance from them.

Theorem 7.10. *Let A be a subset of $\mathbb{R}$. The following are equivalent:*
(i) *A is closed;*
(ii) *Whenever the distance between a real number p and the set A is 0, it follows that $p \in A$.*

Proof. (i) $\Rightarrow$ (ii) We proceed by contraposition. Suppose there is a real number p that is not in A, but for which the distance between p and A is 0. Consider any positive number ε. As $\inf\{|p - a| \mid a \in A\} = 0$, it follows that there exists a in A for which $|p - a| < \varepsilon$, which implies $a \in (p - \varepsilon, p + \varepsilon)$. Since $p \notin A$, we may conclude that $a \neq p$. Thus, $(p - \varepsilon, p + \varepsilon)$ includes a point of A different from p, and since ε was arbitrarily chosen, we have shown that every neighborhood of p includes a point of A different from p. Hence, p is a limit point of A that is not a member of A, so by definition A is not closed.

(ii) $\Rightarrow$ (i) We again apply contraposition. Suppose A is not closed. Then A has a limit point p for which $p \notin A$. Thus, for every positive number ε, no matter how small, the interval $(p - \varepsilon, p + \varepsilon)$ includes a point of A necessarily different from p, and it follows that $\inf\{|p - a| \mid a \in A\} = 0$. Thus, the distance between p and A is 0, and we see that condition (ii) fails. □

Open sets

A subset A of $\mathbb{R}$ is said to be **open** if for every point p in A, there exists a positive number ε such that $(p - \varepsilon, p + \varepsilon) \subseteq A$ (the value of ε can, and often will, be different for different values of p). In other words, by definition, an open set contains a neighborhood of each of its points (members).

Example 7.11. The interval $(0, 1)$ is open, as is any bounded interval (a, b) that does not include its endpoints.

To see that $(0, 1)$ is open, consider any point p in $(0, 1)$ and let $\varepsilon = \min\{p, 1 - p\}$, which by drawing a picture you should see is the smaller of the distances from p to the endpoints of $(0, 1)$. Note that $\varepsilon > 0$ and $(p - \varepsilon, p + \varepsilon) \subseteq (0, 1)$. Thus, $(0, 1)$ contains a neighborhood of each of its points.

Example 7.12. The interval $[0, 1]$ is not open, and neither is any bounded interval $[a, b]$ that includes its endpoints.

To see that $[0, 1]$ is not open, note that for any choice of a positive number ε, the density property guarantees the existence of a real number c for which $-\varepsilon < c < 0$, which means $c \in (0 - \varepsilon, 0 + \varepsilon)$, but $c \notin [0, 1]$. Thus, for each positive number ε, it follows that $(0 - \varepsilon, 0 + \varepsilon) \nsubseteq [0, 1]$. Hence, $[0, 1]$ contains no neighborhood of 0 (similarly, there is no neighborhood of 1 that is contained in $[0, 1]$).

Exercise 7.11. Determine whether the given set is open, justifying your conclusion.

(a) $(3, \infty)$;

(b) $(-\infty, 3]$;

(c) $\mathbb{Q}$.

Exercise 7.12. Suppose $n \in \mathbb{N}$ and let $G_1, G_2, \ldots, G_n$ be open sets in $\mathbb{R}$. Show that $\bigcap_{i=1}^{n} G_i$ is also open. Thus, the intersection of finitely many open sets in $\mathbb{R}$ is always open.

Exercise 7.13. For each i in I, let G_i be an open set in $\mathbb{R}$. Show that $\bigcup_{i \in I} G_i$ is also open. Thus, the union of open sets in $\mathbb{R}$ is always open, even if there are infinitely many open sets over which the union is being formed.

The result stated in Exercise 7.12 cannot be extended to conclude that the intersection of infinitely many open sets must be open.

Example 7.13. Even a countably infinite intersection of open sets may not be open. For instance, note that $\bigcap_{n=1}^{\infty}(-\frac{1}{n}, \frac{1}{n}) = \{0\}$, which is not open.

Exercise 7.14. Verify that $\bigcap_{n=1}^{\infty}(-\frac{1}{n}, \frac{1}{n}) = \{0\}$ and explain why $\{0\}$ is not open.

The way in which we measure the distance between real numbers provides some insight into what makes a set open. For any point in an open set, the existence (by definition) of a neighborhood about this point that is entirely contained in the open set means that all points "sufficiently close" to the original point (i. e., within some positive distance ε of the point) are still in the open set. In other words, if we are located at any particular point in an open set, we can move some positive distance from the point and stay within the open set during our entire journey.

Exercise 7.15. In Example 7.11, we showed that the set $(0, 1)$ is open. For each point of $(0, 1)$ given below, find the maximum value of a positive number ε for which the ε-neighborhood of the point is contained in $(0, 1)$.

(a) 0.22;

(b) 0.671;

(c) $p \in (0, 1)$ with $p \leq 0.5$;

(d) $p \in (0, 1)$ with $p > 0.5$.

In fact, every open set can be expressed as a union of neighborhoods of the points in the open set. Furthermore, because of the density of the rational numbers among the real numbers, the union can always be taken over at most countably many points/neighborhoods.

Theorem 7.14. *Let A be a subset of $\mathbb{R}$. The following are equivalent:*

(i) *A is open;*

(ii) *A is the union of countably many open bounded intervals having rational endpoints.*

Proof. (i) $\Rightarrow$ (ii) Assume A is open. Thus, for every point p in A, there exists a positive number ε_p such that $(p - \varepsilon_p, p + \varepsilon_p) \subseteq A$. It follows that

$$A = \bigcup_{p \in A} (p - \varepsilon_p, p + \varepsilon_p).$$

Now, for each point p in A, by the density of $\mathbb{Q}$ in $\mathbb{R}$, there exist rational numbers a_p and b_p for which

$$p - \varepsilon_p < a_p < p < b_p < p + \varepsilon_p.$$

It follows that

$$A = \bigcup_{p \in A} (a_p, b_p).$$

As there are only countably many rational numbers, and each a_p and b_p is rational, the number of intervals (a_p, b_p) we have created is countable. Thus, A is the union of countably many open bounded intervals having rational endpoints.

(ii) $\Rightarrow$ (i) This implication immediately follows from the fact that any union of open sets is open (see Exercise 7.13). □

Exercise 7.16. Referring to the first part, (i) $\Rightarrow$ (ii), of the proof of Theorem 7.14, provide detailed arguments showing that $A = \bigcup_{p \in A}(p - \varepsilon_p, p + \varepsilon_p)$ and $A = \bigcup_{p \in A}(a_p, b_p)$.

The relationship between closed sets and open sets

Just because a set is not closed (respectively, not open) does not mean that the set must be open (respectively, closed). Mathematicians like to say that, unlike a door, a set may be closed, open, both closed and open, or neither closed nor open!

Exercise 7.17. Show that each of the sets $\mathbb{R}$ and $\emptyset$ is both open and closed. 

Example 7.15. We can show the set $B = \{\frac{1}{n} \mid n \in \mathbb{N}\}$ is neither open nor closed.

Because we have already observed in Example 7.3 that 0 is a limit point of B, we see that B does not include all of its limit points, hence, is not closed.

To show that B is not open, note that $1 \in B$. We now show B contains no neighborhood of 1. To do so, consider any positive number ε. Observe that $(1 - \varepsilon, 1 + \varepsilon)$ includes members of the interval $(0.5, 1)$, none of which are in B. Hence, the set $(1 - \varepsilon, 1 + \varepsilon)$ is not contained in B.

Example 7.16. The set $C = \{\frac{1}{n} \mid n \in \mathbb{N}\} \cup \{0\}$ is closed but not open. Observe that the only limit point of C is 0, which is an element of C. Thus, C contains all its limit points and, hence, is closed. The same argument presented in Example 7.15 showing the set B from that example is not open shows that C is not open.

Exercise 7.18. Show that the interval $[0, 1)$ is neither open nor closed. 

Exercise 7.19. Show that any nonempty finite subset of $\mathbb{R}$ is closed, but is not open. 

The following theorem tells us how open sets and closed sets are related to each other. The arguments we employ really just require us to apply the definitions of open set and closed set.

Theorem 7.17. *A subset A of $\mathbb{R}$ is closed if and only if its complement $\mathbb{R} - A$ is open.*

Proof. ($\Rightarrow$) Suppose A is closed. To show $\mathbb{R} - A$ is open we show $\mathbb{R} - A$ contains a neighborhood of each of its points. Consider any point p in $\mathbb{R} - A$. As A is closed, the number p cannot be a limit point of A. Thus, there exists a positive number ε such that the interval $(p - \varepsilon, p + \varepsilon)$ does not include any points of A other than perhaps p itself. However, as $p \in \mathbb{R} - A$, we see that $p \notin A$. Thus, $(p - \varepsilon, p + \varepsilon)$ does not contain any points of A, which means $(p - \varepsilon, p + \varepsilon) \subseteq \mathbb{R} - A$.

($\Leftarrow$) Suppose $\mathbb{R} - A$ is open. To show A is closed, we show A contains all of its limit points. Let p be any limit point of A. Then for every positive number ε, the interval $(p - \varepsilon, p + \varepsilon)$ includes an element of A different from p. Hence, for no positive number ε is it the case that $(p - \varepsilon, p + \varepsilon) \subseteq \mathbb{R} - A$, which means that $p \notin \mathbb{R} - A$ as, otherwise, $\mathbb{R} - A$ could not be an open set, a contradiction. But having deduced that $p \notin \mathbb{R} - A$ permits us to conclude that $p \in A$. □

Theorem 7.17 also tells us that a set is open if and only if its complement is closed. In other words, the open sets in $\mathbb{R}$ are precisely the complements of the closed sets in $\mathbb{R}$, and the closed sets in $\mathbb{R}$ are precisely the complements of the open sets in $\mathbb{R}$.

Example 7.18. In Example 7.7, we showed that the set $[0, 1]$ is closed. Hence, its complement, $\mathbb{R} - [0, 1] = (-\infty, 0) \cup (1, \infty)$ is open.

In Example 7.11 we showed that the set $(0, 1)$ is open. Hence, its complement, $\mathbb{R} - (0, 1) = (-\infty, 0] \cup [1, \infty)$ is closed.

Exercise 7.20. Use your conclusions in Exercises 7.8(c) and 7.11(c) to determine whether the set $\mathbb{R} - \mathbb{Q}$ of all irrational numbers is closed and whether it is open.

Exercise 7.21. Use the results from Exercises 7.12 and 7.13, together with DeMorgan's laws for sets, Theorem 6.4, to explain each of the following.

(a) The intersection of closed sets in $\mathbb{R}$ is always closed.

(b) The union of finitely many closed sets in $\mathbb{R}$ is always closed.

Exercise 7.22. Illustrate why the result stated in Exercise 7.21(b) cannot be extended to conclude that the union of a countably infinite number of closed sets must be closed.

8 Connected sets and compact sets

For pedagogical reasons, we do not emphasize the topological notions of *connectedness* and *compactness* in later chapters of this book, so the content developed here may be viewed as optional, at least upon a first reading of the text. Because there are relatively simple characterizations of connectedness and compactness in the setting of the real line, the true scope and significance of topology does not really become apparent until one works with the ideas in other settings. But some experience with topological ideas in a specific concrete setting, such as is provided in this chapter, can be invaluable in preparing for a dedicated topology course.

Connected and disconnected sets

Intuitively, the set $\mathbb{R}$ appears to be "connected" because it is all in "one piece" (Figure 8.1).

Figure 8.1: The set $\mathbb{R}$ is connected.

On the other hand, the set $\mathbb{R} - \{0\}$ appears to be "disconnected" as it is in "two pieces," the set $(-\infty, 0)$ of all negative real numbers and the set $(0, \infty)$ of all positive real numbers (Figure 8.2).

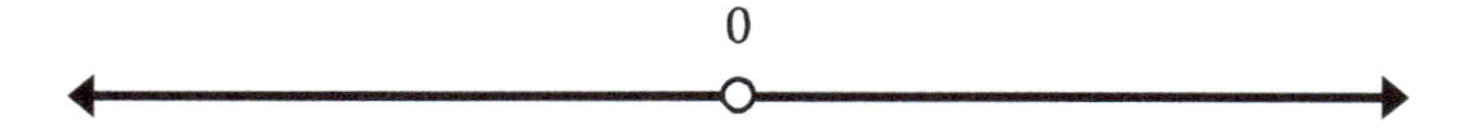

Figure 8.2: The set $\mathbb{R} - \{0\}$ is disconnected.

It also appears that anytime we try to "separate" $\mathbb{R}$ into two disjoint nonempty subsets, at least one of the subsets includes points that are "really close to" the other set (i. e., limit points of the other set). For example, expressing $\mathbb{R}$ as $(-\infty, 0) \cup [0, \infty)$, we see that while the two intervals $(-\infty, 0)$ and $[0, \infty)$ are nonempty and disjoint, $[0, \infty)$ includes the limit point 0 of $(-\infty, 0)$. Soon we shall see that it is actually impossible to "break apart" $\mathbb{R}$ into two disjoint nonempty subsets, neither of which is "really close to" the other.

In contrast, the subset $\mathbb{R} - \{0\}$ of $\mathbb{R}$ is capable of being "separated" into two disjoint nonempty subsets, namely $(-\infty, 0)$ and $(0, \infty)$, neither of which includes a limit point of the other.

These examples help to motivate the following definitions. Two disjoint subsets of $\mathbb{R}$ are said to be **separated** if neither includes a limit point of the other. A **separation** of a

https://doi.org/10.1515/9783111429564-008

subset A of $\mathbb{R}$ is a pair $\{X, Y\}$ of nonempty separated subsets of $\mathbb{R}$ whose union is A. A subset A of $\mathbb{R}$ is **disconnected** if it has a separation and **connected** if it has no separation.

Example 8.1. The limit points of $(-\infty, 0)$ are the elements of $(-\infty, 0]$, while the limit points of $(0, \infty)$ are the elements of $[0, \infty)$. Thus, the sets $(-\infty, 0)$ and $(0, \infty)$ are not only disjoint, but neither includes a limit point of the other one, so they are separated. As they are also nonempty and their union is $\mathbb{R} - \{0\}$, it follows that $\{(-\infty, 0), (0, \infty)\}$ is a separation of $\mathbb{R} - \{0\}$. Having determined that $\mathbb{R} - \{0\}$ has a separation, we may now conclude that $\mathbb{R} - \{0\}$ is disconnected.

Example 8.2. Any singleton subset $\{p\}$ of $\mathbb{R}$ is connected, as a singleton cannot be expressed as the union of two disjoint nonempty sets, hence, cannot have a separation.

Exercise 8.1. Demonstrate that the given set is disconnected by exhibiting a separation of it.
(a) $(-1, 1] \cup \{5\}$;
(b) $\mathbb{Q}$.

Exercise 8.2. Let A be a nonempty subset of $\mathbb{R}$. Prove that A is disconnected if and only if there exist open subsets G and H of $\mathbb{R}$ for which the sets $A \cap G$ and $A \cap H$ are nonempty, the set $A \cap G \cap H$ is empty, and $A = A \cap (G \cup H)$.

Exercise 8.3. Find open sets G and H for which $X = G \cap A$ and $Y = H \cap A$, where $\{X, Y\}$ is the separation of $A = (-1, 1] \cup \{5\}$ you found in Exercise 8.1(a).

Recall that an **interval** of real numbers is a subset of $\mathbb{R}$ that includes all real numbers lying between any two of its members. Once we begin to realize that connected subsets of $\mathbb{R}$ would appear to be what we think they should be "geometrically," it is reasonable to conjecture that it is precisely the intervals in $\mathbb{R}$ that are connected. We now prove this fact.

Theorem 8.3. *A subset of* $\mathbb{R}$ *is connected if and only if it is an interval.*

Proof. ($\Rightarrow$) We employ contraposition. Suppose A is a subset of $\mathbb{R}$ and A is not an interval. Then there exist real numbers a, b, and c such that $a < b < c$ and for which a and c are in A, but b is not in A. We claim that $\{A \cap (-\infty, b), A \cap (b, \infty)\}$ is a separation of A, thereby making A disconnected.

As $a \in A \cap (-\infty, b)$ and $c \in A \cap (b, \infty)$, we see that the sets $A \cap (-\infty, b)$ and $A \cap (-\infty, b)$ are nonempty. They are disjoint because the intervals $(-\infty, b)$ and (b, ∞) are disjoint. We must now show that neither $A \cap (-\infty, b)$ nor $A \cap (b, \infty)$ includes a limit point of the other.

Consider any point p in $A \cap (b, \infty)$. Then $p > b$, and it follows that $\varepsilon = p - b$ is a positive number. Note that the neighborhood $(p - \varepsilon, p + \varepsilon)$ of p is really the interval

$(b, 2p - b)$, which is a subset of (b, ∞). Hence, $(p - \varepsilon, p + \varepsilon)$ is a neighborhood of p that includes no points of $(-\infty, b)$, hence, no points of $A \cap (-\infty, b)$, so it is impossible for p to be a limit point of $A \cap (-\infty, b)$. As p was taken to be an arbitrary point in $A \cap (-\infty, b)$, we may conclude that $A \cap (b, \infty)$ includes no limit points of $A \cap (-\infty, b)$.

A similar argument shows that $A \cap (-\infty, b)$ includes no limit points of $A \cap (b, \infty)$. Thus, we have now demonstrated that $\{A \cap (-\infty, b), A \cap (b, \infty)\}$ is a separation of A, thus making A disconnected.

($\Leftarrow$) Let I be a nonempty interval of real numbers and suppose to the contrary that I is disconnected. Then I has a separation $\{X, Y\}$. As X and Y are nonempty, there exist numbers p in X and q in Y; without loss of generality, assume $p < q$. Note that because I is an interval and we now have $I = X \cup Y$ with both p and q in I, it follows that $[p, q] \subseteq X \cup Y$. Let $X^* = X \cap [p, q]$ and $Y^* = Y \cap [p, q]$. We now show that $\{X^*, Y^*\}$ is a separation of $[p, q]$.

As $p \in X^*$ and $q \in Y^*$, we see that X^* and Y^* are nonempty. Also, X^* and Y^* are disjoint because X and Y are disjoint. We now show that Y^* includes no limit points of X^* (a similar argument demonstrates that X^* includes no limit points of Y^*). Suppose to the contrary that r is a limit point of X^* and $r \in Y^*$. Then $r \in Y$ and every neighborhood of r includes infinitely many points of X^*, hence as X^* is a subset of X, infinitely many points of X. We may then conclude that r is a limit point of X, which is impossible as X and Y are separated. We have now established all the requirements for $\{X^*, Y^*\}$ to be a separation of $[p, q]$.

Next, we show that X^* is a closed set. Consider any limit point r of X^*; we must show r is in X^*. Note that r must be in $[p, q]$, because otherwise r would be in the open set $\mathbb{R} - [p, q]$, which implies there exists a neighborhood of r that includes no points of $[p, q]$, hence, as $X^* \subseteq [p, q]$, no points of X^*, contradicting the assumption that r is a limit point of X^*. Now, as $\{X^*, Y^*\}$ is a separation of $[p, q]$, if r is not in X^*, then r would be in Y^*, hence, in Y. But then, as r is a limit point of X^*, every neighborhood of r includes infinitely many points of X^*, hence, of X, thereby making r a limit point of X, contradicting the fact that, as X and Y are separated, no limit point of X can be an element of Y. We therefore may conclude that r is in X^*. Having shown that X^* contains all of its limit points, X^* is closed.

Observe that X^* is a nonempty subset of $\mathbb{R}$ that is bounded above by q, so by completeness, X^* has a supremum s. In Exercise 7.9 you proved that a closed set having a supremum must include the supremum as one of its members. Thus, as X^* is closed, it follows that $s \in X^*$.

Since no point of X^* is a limit point of Y^*, for each x in X^*, there is a positive number ε_x for which $(x - \varepsilon_x, x + \varepsilon_x) \cap Y^* = \emptyset$. Then $G = \bigcup_{x \in X^*} (x - \varepsilon_x, x + \varepsilon_x)$, being a union of open sets, is open and, by construction, includes every point of X^* and no points of Y^*. Thus, as $[p, q] = X^* \cup Y^*$, it follows that $X^* = G \cap [p, q]$.

Therefore, as $s \in X^*$ we may conclude that $s \in G$, and as G is open, there is a positive number α for which $(s - \alpha, s + \alpha) \subseteq G$. Also, as $q \in Y^*$ and $X^* \cap Y^* = \emptyset$, we see that $s \neq q$. Therefore, $s < q$ and so the set $(s, s + \alpha) \cap [p, q]$ is a nonempty subset of $G \cap [p, q] = X^*$, meaning there exists t in X^* for which $t > s$, contradicting the status of s as an upper bound for X^*. This contradiction is the result of our initial assumption that the interval I is disconnected, so it must actually be the case that I is connected. □

Exercise 8.4. Referring to the last paragraph of the proof of Theorem 8.3, explain how we are able to conclude that $s < q$ and how it then follows that $(s, s + a) \cap [p, q]$ is nonempty.

It may seem that we expended considerable effort in the above proof only to demonstrate the truth of something that is already intuitively plausible. As we mentioned in the chapter opening, the true significance of topological methods do not really become clear until one moves beyond the real line. And, as we have regularly remarked, an intuitive explanation is not the same as an analytic proof.

Example 8.4. Define F_1 to be the closed interval $[0, 1]$ (Figure 8.3).

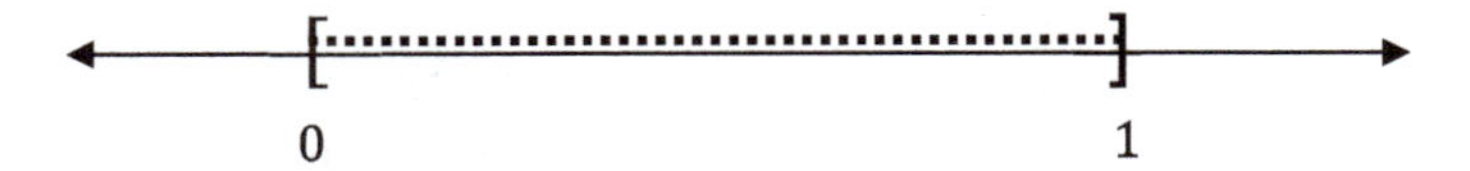

Figure 8.3: In Example 8.4, $F_1 = [0, 1]$.

Now obtain F_2 from F_1 by deleting from F_1 its open "middle third" $(\frac{1}{3}, \frac{2}{3})$. This produces $F_2 = [0, \frac{1}{3}] \cup [\frac{2}{3}, 1]$, a union of closed intervals, which is shown in Figure 8.4.

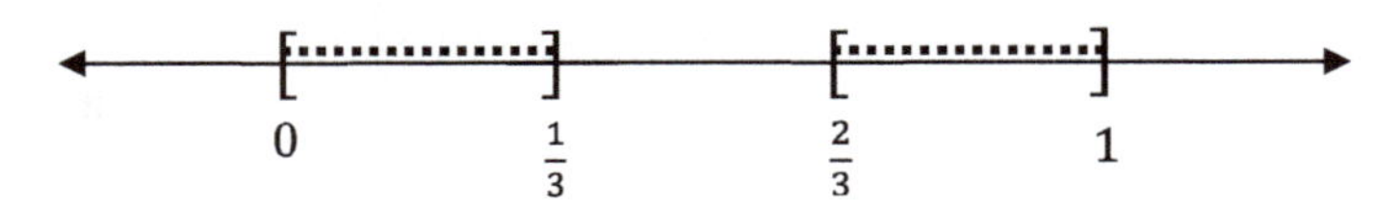

Figure 8.4: In Example 8.4, $F_2 = [0, \frac{1}{3}] \cup [\frac{2}{3}, 1]$.

Next, obtain F_3 from F_2 by deleting from F_2 the open "middle third" of each of the intervals of which F_2 is comprised. This produces

$$F_3 = \left[0, \frac{1}{9}\right] \cup \left[\frac{2}{9}, \frac{1}{3}\right] \cup \left[\frac{2}{3}, \frac{7}{9}\right] \cup \left[\frac{8}{9}, 1\right],$$

another union of closed intervals. The set F_3 is pictured in Figure 8.5.

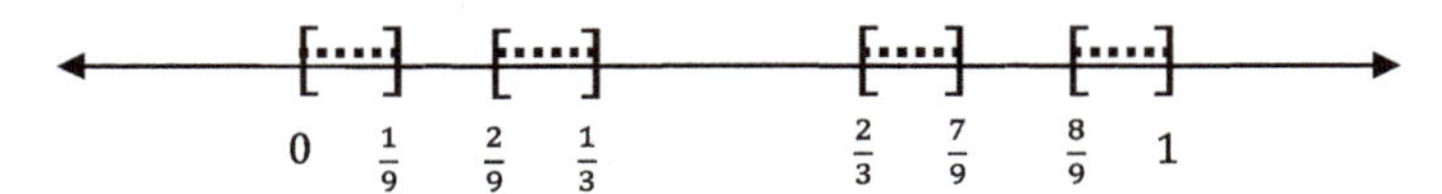

Figure 8.5: In Example 8.4, $F_3 = [0, \frac{1}{9}] \cup [\frac{2}{9}, \frac{1}{3}] \cup [\frac{2}{3}, \frac{7}{9}] \cup [\frac{8}{9}, 1]$.

Continue in this way, obtaining F_{n+1} from F_n by removing from each interval comprising F_n its open "middle third." The result is a recursively defined sequence

$$(F_1, F_2, F_3, F_4, \dots)$$

of sets. The **Cantor set** is defined to be the intersection

$$C = \bigcap_{n=1}^{\infty} F_n$$

of all of the sets in this sequence. Note that as C is not an interval, it follows that C is disconnected.

Exercise 8.5. Let C be the Cantor set from Example 8.4.

(a) Show that C is a closed set.

(b) In Exercise 6.17, you showed that each number in $(0, 1]$ has a unique nonterminating ternary representation. Describe the ternary representations of the numbers in the Cantor set.

(c) Use ternary representations to argue why the Cantor set contains no intervals consisting of more than one point. (This feature makes the Cantor set *totally disconnected*.)

Exercise 8.6. Show that the topology of the plane $\mathbb{R}^2$ is fundamentally different from the topology of the line $\mathbb{R}$ by considering the action of removing a single point from each of them. Then find a way to argue that the topology of three-dimensional space $\mathbb{R}^3$ is fundamentally different from the topologies of the line and the plane.

Compact sets

Let D be a subset of $\mathbb{R}$. We say a function $f : D \to \mathbb{R}$

(1) **achieves a maximum (value)** if there exists M in D for which $f(x) \le f(M)$ for all x in D;

(2) **achieves a minimum (value)** if there exists m in D for which $f(m) \le f(x)$ for all x in D.

A function need not achieve either a maximum or a minimum. For instance, the function $f : \mathbb{R} \to \mathbb{R}$ defined by $f(x) = x^2$ does not achieve a maximum value, and the function $g : (0, 3) \to \mathbb{R}$ defined by $g(x) = x + 2$ achieves neither a maximum value nor a minimum value.

Exercise 8.7. Explain why

(a) the function $f : \mathbb{R} \to \mathbb{R}$ defined by $f(x) = x^2$ does not achieve a maximum value;

(b) the function $g : (0, 3) \to \mathbb{R}$ defined by $g(x) = x + 2$ achieves neither a maximum value nor a minimum value.

However, an important theorem we first encounter in our study of calculus, the *extreme value theorem*, states that if the real-valued function f is continuous and the domain D of f is a closed bounded interval $[a, b]$, then f must achieve both a maximum and a minimum. We prove this result in Chapter 14, but here we use it to provide motivation for another topological concept, that of *compactness*.

During the late 19th and early 20th centuries, mathematicians began to search for properties of the interval $[a, b]$ that, when combined with the continuity of the function f, necessitate the existence of maximum and minimum values for f. For a period of time it was thought that, as $[a, b]$ is both closed and bounded, these were the required attributes of the domain. In fact, this conclusion is correct, if the domain of the function is taken to be a subset of $\mathbb{R}$ or some other Euclidean space.

Eventually, though, as the field of topology matured, the study of continuity grew to include functions whose domains had no notion of distance imposed on them, so the property of "boundedness," which requires the ability to measure distance, would not be able to serve as part of the general characterization of domains on which continuous functions must achieve extreme values. After several false starts, a suitably general characterization, formulated below in our definition of *compact set*, was discovered. As it involves the idea of a *cover* of a set by other sets, we first discuss this notion.

Given a subset A of $\mathbb{R}$, a collection $\mathcal{S} = \{S_i \mid i \in I\}$ of subsets of $\mathbb{R}$ is called a **cover** of A if $A \subseteq \bigcup_{i \in I} S_i$. When $\mathcal{S}$ is a cover of A, we also say that $\mathcal{S}$ **covers** A. The terminology was chosen because, as each element of A must be in at least one of the sets in the collection $\mathcal{S}$, there is a sense that the elements in A are "covered by" $\mathcal{S}$.

Example 8.5. Given any negative real number p, note that $p \in \{a, -a\}$ if we take a to be the positive real number $-p$. Hence, $\mathbb{R}^- \subseteq \bigcup_{a \in \mathbf{R}^+} \{a, -a\}$ and we may conclude that $\mathcal{A} = \{\{a, -a\} \mid a \in \mathbb{R}^+\}$ is a cover of $\mathbb{R}^-$.

Note, however, that as 0 is not in $\{a, -a\}$ for any positive real number a, the collection $\mathcal{A}$ is not a cover of $\mathbb{R}$.

Example 8.6. We can show that the collection $\mathcal{S} = \{(\frac{1}{n}, 2) \mid n \in \mathbb{N}\}$ covers $(0, 2)$. Consider any number x in $(0, 2)$. As $x > 0$, by the Archimedean Property, there exists a natural number N for which $\frac{1}{N} < x$; hence, $x \in (\frac{1}{N}, 2)$. We have thus shown that every member of $(0, 2)$ is a member of at least one of the sets in $\mathcal{S}$, so $(0, 2) \subseteq \bigcup_{n=1}^{\infty} (\frac{1}{n}, \infty)$, and therefore $\mathcal{S}$ covers $(0, 2)$.

Exercise 8.8. Show that $\{(0, n] \mid n \in \mathbb{N}\}$ is a cover of $\mathbb{Q}^+$.

Exercise 8.9. How do we know that $\{(\frac{1}{n}, 1] \mid n \in \mathbb{N}, n > 1\}$ does not cover $[0, 1]$?

If all the sets in a cover are open sets, the cover is called an **open cover**.

Example 8.5 (Continued). For any positive real number a, the doubleton set $\{a, -a\}$ is not open. Thus, the cover $\{\{a, -a\} \mid a \in \mathbb{R}^+\}$ of $\mathbb{R}^-$ is not an open cover.

Example 8.6 (Continued). For each natural number n, the interval $(\frac{1}{n}, 2)$ is an open set, so the cover $\{(\frac{1}{n}, 2) \mid n \in \mathbb{N}\}$ of $(0, 2)$ is an open cover.

Exercise 8.10. A set usually has many covers, even many open covers. Show that each of the collections

$$\mathcal{A} = \left\{\left(-1, \frac{1}{2}\right), \left(\frac{1}{4}, 2\right)\right\},$$
$$\mathcal{B} = \left\{(n, \infty) \mid n \in \mathbb{Z}\right\},$$

and

$$\mathcal{C} = \left\{\left(\frac{1}{n}, \frac{2n-1}{n}\right) \middle| n \in \mathbb{N} \text{ and } n > 1\right\}$$

is an open cover of $(0, 2)$.

A **finite cover** of a set is a cover of the set consisting of finitely many sets.

Example 8.5 (Continued). The cover $\mathcal{A} = \{\{a, -a\} \mid a \in \mathbb{R}^+\}$ of $\mathbb{R}^-$ is not finite as it consists of infinitely many sets (there is one set for each of the infinitely many positive real numbers).

Example 8.7. The collection $\{\mathbb{R}^-, \{0\}, \mathbb{R}^+\}$ is a cover of $\mathbb{R}$ that consists of only three sets, so it is a finite cover of $\mathbb{R}$.

A **subcover** of a cover of A is a subcollection of the cover that also covers A.

Example 8.6 (Continued). The collection $\mathcal{T} = \{(\frac{1}{2n}, 2) \mid n \in \mathbb{N}\}$ is subcollection of the cover $\mathcal{S} = \{(\frac{1}{n}, 2) \mid n \in \mathbb{N}\}$ of $(0, 2)$. Because $\mathcal{S}$ covers $(0, 2)$, given any number x in $(0, 2)$, there is a natural number N for which $x \in (\frac{1}{N}, 2)$. As $2N > N$, it follows that $\frac{1}{2N} < \frac{1}{N}$, so that $(\frac{1}{N}, 2) \subseteq (\frac{1}{2N}, 2)$, and we may conclude that $x \in (\frac{1}{2N}, 2)$. Having shown that each member of $(0, 2)$ is in some member of the subcollection $\mathcal{T}$ of $\mathcal{S}$, it follows that $\mathcal{T}$ is a subcover of $\mathcal{S}$.

Example 8.5 (Continued). As the sets in the cover $\mathcal{A} = \{\{a, -a\} \mid a \in \mathbb{R}^+\}$ of $\mathbb{R}^-$ are pairwise disjoint and each includes a negative real number, there is no proper subcollection of $\mathcal{A}$ that covers $\mathbb{R}^-$. In other words, the cover $\mathcal{A}$ of $\mathbb{R}^-$ has no proper subcover.

Exercise 8.11. For each of the covers $\mathcal{A}$, $\mathcal{B}$, $\mathcal{C}$, $\mathcal{S}$, and $\mathcal{T}$ of $(0, 2)$ given in Exercise 8.10 and Example 8.6, determine whether the cover has a finite subcover.

It can sometimes happen that *every* open cover of a subset A of $\mathbb{R}$ has a finite subcover; when this is the case, the set A is called **compact**. In checking to determine whether a set is compact, note that it is only covers of the set by open sets that are relevant. Also, compactness requires not just that *some* open cover has a finite subcover, but that *every* open cover has a finite subcover.

Example 8.8. Every finite subset of $\mathbb{R}$ is compact. To see why, consider a finite set $A = \{a_1, a_2, \ldots, a_n\}$ consisting of n real numbers. Let $\mathcal{G}$ be an open cover of A, meaning each element of A is in at least one of the open sets in the collection $\mathcal{G}$. Thus, for each natural number i less than or equal to n, there is a set G_i in $\mathcal{G}$ for which $a_i \in G_i$. It follows that $A \subseteq \bigcup_{i=1}^{n} G_i$, thus making $\mathcal{G}_0 = \{G_1, G_2, \ldots, G_n\}$ a cover of A. As all the sets in $\mathcal{G}_0$ are in the original cover $\mathcal{G}$, it follows that $\mathcal{G}_0$ is a subcover of $\mathcal{G}$. Since $\mathcal{G}$ is an arbitrary open cover of A, we have therefore shown that every open cover of A has a finite subcover, so A is compact.

Note also that $\emptyset$ is compact simply because there are no elements to "cover," so any finite subcollection of an open cover of $\emptyset$ necessarily contains $\emptyset$.

To demonstrate that a subset A of $\mathbb{R}$ is not compact, it is enough to find a single open cover of A that has no finite subcover.

Example 8.9. Here we show the set $B = \{\frac{1}{n} \mid n \in \mathbb{N}\}$ is not compact. Let $a_1 = 2$ and for each natural number n that is greater than 1, use the density property to select a real number a_n for which $\frac{1}{n} < a_n < \frac{1}{n-1}$. It follows that, for each n in $\mathbb{N}$, the number $\frac{1}{n}$ is in the open set (a_{n+1}, a_n), but (a_{n+1}, a_n) includes no other members of B. Thus, $\mathcal{H} = \{(a_{n+1}, a_n) \mid n \in \mathbb{N}\}$ is an open cover of B. However, since each set in $\mathcal{H}$ includes just one member of B and B is an infinite set, we see that $\mathcal{H}$ has no finite subcover. Hence, B is not compact.

Exercise 8.12. Prove $C = \{\frac{1}{n} \mid n \in \mathbb{N}\} \cup \{0\}$ is a compact set by showing that every open cover of C has a finite subcover.

Exercise 8.13. Find an open cover of $\mathbb{Z}$ that has no finite subcover, thereby proving that $\mathbb{Z}$ is not compact.

We now show that every closed bounded interval is compact.

Theorem 8.10 (Closed bounded intervals are compact). *Let a and b be real numbers for which $a < b$. Then $[a, b]$ is compact.*

Proof. Consider any open cover $\mathcal{G}$ of $[a, b]$. Define A to be the set of all points x in $[a, b]$ such that $[a, x]$ is covered by a finite subcollection of $\mathcal{G}$. To complete the proof, we must show $b \in A$.

As $a \in A$ we see that A is nonempty. Also, A is bounded above by b. So by completeness, A has a supremum s. We now show that $s \in A$.

As $s \in [a, b]$, there is some open set G in the open cover $\mathcal{G}$ of $[a, b]$ such that $s \in G$. Since G is open, $(s - \varepsilon, s] \subseteq G$ for some positive number ε. Also, because s is the least upper bound of A, the interval $[a, s - \varepsilon]$ can be covered by a finite subcollection $\mathcal{G}_0$ of $\mathcal{G}$. Then $\mathcal{G}_0 \cup \{G\}$ is a finite subcollection of $\mathcal{G}$ that covers $[a, s]$, and it follows that $s \in A$.

We now argue that $s = b$. Otherwise, $s < b$ and any finite subcollection of $\mathcal{G}$ that covers $[a, s]$ actually covers $[a, s + \alpha]$ for some positive number α, which contradicts the fact that s is an upper bound of A. Hence, $s = b$.

Therefore, as $b \in A$, it follows from the way A is defined that $[a, b]$ can be covered by a finite subcollection of $\mathcal{G}$. Having shown that every open cover of $[a, b]$ has a finite subcover, we conclude that $[a, b]$ is compact. □

Exercise 8.14. Explain why, in the next to last paragraph of the proof of Theorem 8.10, if $s = \sup(A)$ is less than b, a finite subcollection of $\mathcal{G}$ that covers $[a, s]$ must actually cover $[a, s + \alpha]$ for some positive number α.

Exercise 8.15. Prove that if A is a compact subset of $\mathbb{R}$ and B is a closed subset of A, then B must be compact.

As we mentioned earlier, the notion of a subset of $\mathbb{R}$ being both closed and bounded was initially thought to be what compactness should mean in general settings that go beyond the real line. Even though this turned out to be incorrect generally, it does provide a straightforward characterization of the compact subsets of $\mathbb{R}$.

Eduard Heine (1821–1881) was one of the first mathematicians to employ the technique of reducing a cover to a finite subcover, but it was Émile Borel (1871–1956) who in 1895 proved a version of the Heine–Borel theorem limited to covers consisting of countably many open sets. Be aware that the Heine–Borel theorem is more general than Theorem 8.10, as the earlier theorem only dealt with closed bounded *intervals*.

Theorem 8.11 (The Heine–Borel theorem). *A subset of $\mathbb{R}$ is compact if and only if it is both closed and bounded.*

Proof. ($\Rightarrow$) Assume A is a compact subset of $\mathbb{R}$. Note that $\mathcal{G} = \{(-r, r) \mid r \in \mathbb{R}^+\}$ is an open cover of A, hence, must have a finite subcover,

$$\mathcal{G}_0 = \{(-r_1, r_1), (-r_2, r_2), \dots, (-r_n, r_n)\}.$$

Let $B = \max\{r_1, r_2, \dots, r_n\}$, which exists as $\{r_1, r_2, \dots, r_n\}$ is a finite set and is positive as all the members of $\{r_1, r_2, \dots, r_n\}$ are positive. Then $(-B, B)$ contains $\bigcup_{i=1}^{n}(-r_i, r_i)$, which in turn contains A, as $\mathcal{G}_0$ covers A. Therefore, A is bounded.

To show A is closed, suppose to the contrary that p is a limit point of A and $p \notin A$. For each a in A, since $a \neq p$, we may choose a positive number ε_a small enough so that the open sets $(a - \varepsilon_a, a + \varepsilon_a)$ and $(p - \varepsilon_a, p + \varepsilon_a)$ are disjoint. The result is an open cover $\mathcal{H} = \{(a - \varepsilon_a, a + \varepsilon_a) \mid a \in A\}$ of A. Since A is compact, $\mathcal{H}$ has a finite subcover

$$\mathcal{H}_0 = \{(a_1 - \varepsilon_{a_1}, a_1 + \varepsilon_{a_1}), (a_2 - \varepsilon_{a_2}, a_2 + \varepsilon_{a_2}), \ldots, (a_m - \varepsilon_{a_m}, a_m + \varepsilon_{a_m})\}.$$

Let $\alpha = \min\{\varepsilon_{a_1}, \varepsilon_{a_2}, \ldots, \varepsilon_{a_m}\}$, which is positive. Then $(p - \alpha, p + \alpha)$ is a neighborhood of p that includes no points of any of the sets in $\mathcal{H}_0$, hence, no points of A, since $\mathcal{H}_0$ covers A. We now have a contradiction, for the assumption that p is a limit point of A means every neighborhood of p should include points of A. Hence, A must contain all of its limit points, and so A is closed.

($\Leftarrow$) Assume A is a subset of $\mathbb{R}$ that is closed and bounded. Being bounded, $A \subseteq [a, b]$ for some closed bounded interval $[a, b]$. By Theorem 8.10 we know $[a, b]$ is compact. Thus, A is a closed subset of a compact set, so by Exercise 8.15, A itself is compact. □

Example 8.4 (Continued). In Exercise 8.5(a), you proved that the Cantor set is closed. Since the Cantor set is a subset of $[0, 1]$ it is bounded. Hence, being closed and bounded, the Cantor set is compact.

Exercise 8.16. According to the Heine–Borel theorem, which of the following sets are compact? In those cases where we were able to reach a conclusion by other means, do the results agree with what the Heine–Borel theorem tells us?

(a) $[0, 1]$;
(b) $(0, 1)$;
(c) $[0, 1)$;
(d) $[0, \infty)$;
(e) $\mathbb{R}$;
(f) $\{\frac{1}{n} \mid n \in \mathbb{N}\}$;
(g) $\{\frac{1}{n} \mid n \in \mathbb{N}\} \cup \{0\}$.

Exercise 8.17. Suppose A is a nonempty compact subset of $\mathbb{R}$.

(a) Must A have a supremum? Must A have an infimum? If so, must either of these numbers be in A?
(b) Let p be a real number. Show that there exists a number r in A for which $|r - p| = \inf\{|a - p| \mid a \in A\}$.

Exercise 8.18. Prove that any infinite compact subset A of $\mathbb{R}$ must have a limit point that is in A.

Exercise 8.19. Prove that any intersection of compact subsets of $\mathbb{R}$ must be compact.

Exercise 8.20. Prove that a finite union of compact subsets of $\mathbb{R}$ must be compact. Then provide an example to show that an infinite union of compact subsets of $\mathbb{R}$ need not be compact.

9 An introduction to sequences and sequential convergence

We now take up the idea of a *sequence* of real numbers, exploring what it means for a sequence to *converge*. The notion of sequential convergence reveals how the completeness of the real numbers leads to *limiting processes* and provides a valuable tool for gaining insight into just about every other area of analysis we shall survey. In particular, our later work with *infinite series of real numbers* as well as *sequences and series of functions* builds upon what we develop in this chapter.

Sequences of real numbers

Somewhat informally, a **sequence of real numbers** is simply an infinite ordered list of real numbers. The nth number in the list is called the nth **term** of the sequence.

Example 9.1. Each of the following is a sequence:
(1) the sequence $(1, 2, 3, 4, \ldots)$ of natural numbers;
(2) the sequence $(1, \frac{1}{2}, \frac{1}{3}, \frac{1}{4}, \ldots)$ of reciprocals of the natural numbers;
(3) the sequence $(0, \frac{3}{4}, \frac{8}{9}, \frac{15}{16}, \ldots)$ whose nth term is $1 - \frac{1}{n^2}$;
(4) the sequence $(5, 4, 5, 8, 5, 12, 5, 16, \ldots)$ whose nth term is 5 if n is odd, but is $2n$ if n is even.

A **constant sequence** is a sequence for which all of the terms are the same number.

Example 9.2. The sequence $(7, 7, 7, 7, \ldots)$ is the constant sequence all of whose terms are 7.

More formally, a **sequence of real numbers** or **real sequence** is a function whose domain is the set $\mathbb{N}$ of natural numbers and whose range consists of real numbers. When a is a sequence it is customary to write a_n in place of $a(n)$. As a sequence is often identified with its terms, in the order in which they appear, we also write (a_n) to represent the sequence a.

We use the terminology **index** to refer to the "input variable" n for a sequence (a_n) as well as to a particular value of the index variable, and we think of the natural numbers as "indexing" the terms of the sequence so that we know "where" in the sequence each term occurs. Thus, for example, the fifth term a_5, that is, the term of the sequence with index 5, immediately "follows" the fourth term a_4, the term with index 4.

Example 9.3. If we let a represent the sequence whose nth term is $1 - \frac{1}{n^2}$, we are really defining the function $a : \mathbb{N} \to \mathbb{R}$ so that $a_n = a(n) = 1 - \frac{1}{n^2}$. Thus,

https://doi.org/10.1515/9783111429564-009

$$a_1 = 1 - \frac{1}{1^2} = 0,$$
$$a_2 = 1 - \frac{1}{2^2} = \frac{3}{4},$$
$$a_3 = 1 - \frac{1}{3^2} = \frac{8}{9},$$

and so forth. The sequence a is identified with its terms as

$$(a_1, a_2, a_3, a_4, \dots)$$

that is, in this case,

$$\left(0, \frac{3}{4}, \frac{8}{9}, \frac{15}{16}, \dots\right).$$

An **alternating sequence** is a sequence whose terms “alternate” between positive values and negative values, meaning either all the odd-indexed terms are positive with all the even-indexed terms negative, or vice-versa.

Example 9.4. The sequence $(1, -\frac{1}{2}, \frac{1}{3}, -\frac{1}{4}, \dots)$ whose nth term is $\frac{(-1)^{n+1}}{n}$ is an alternating sequence, as is the sequence $(-1, 1, -1, 1, \dots)$ whose nth term is $(-1)^n$.

Note that there is a distinction between a sequence's terms and its range: Two different sequences can have the same range.

Example 9.5. The alternating sequences

$$(-1, 1, -1, 1, \dots)$$

and

$$(1, -1, 1, -1, \dots)$$

have the same range $\{-1, 1\}$, but the sequences are not the same as, for instance, their first terms are different.

Any function whose domain is $\mathbb{N}$ can be referred to as a sequence. For instance,

$$\left(\left[0, \frac{1}{n}\right]\right) = \left([0, 1], \left[0, \frac{1}{2}\right], \left[0, \frac{1}{3}\right], \left[0, \frac{1}{4}\right], \dots\right)$$

is a sequence of intervals. However, as we will for now be working almost exclusively with sequences of real numbers, we usually take the term *sequence* to mean *sequence of real numbers* in what follows.

Exercise 9.1. Write out the first four terms of the given sequence.
(a) The sequence x defined so that $x_n = \frac{n}{(n+1)(n+2)}$.
(b) The sequence y defined so that $y_n = (-1)^n - (-1)^{n+1}$.
(c) The sequence a defined so that $a_n = |\pi - n|$.
(d) The sequence b defined so that $b_n = \frac{n}{-|-n|}$.

Exercise 9.2. Is any sequence from Exercise 9.1 a constant sequence? An alternating sequence?

Exercise 9.3. Consider again the sequence y from Exercise 9.1(b). When viewed as a function, what is the domain of y? What is the range of y?

Recursively defined sequences

Theorem 3.16 shows how the axiom of induction, Axiom 3.1(3), provides a mechanism for defining a function with domain $\mathbb{N}$, what we would now call a sequence, by a process called *recursive definition*. The idea is to directly specify the first term of the sequence and then define each subsequent term using the term immediately preceding it. For instance, in Example 3.15, we recursively defined the sequence (a^n) of natural number powers of a given real number a.

The process of recursively defining a sequence can be extended to include situations where any finite number of the initial terms of the sequence are directly specified and a procedure for obtaining the remaining terms from those already defined is provided. The famous *Fibonacci sequence* provides an example of this more general model for recursion.

Example 9.6. The **Fibonacci sequence**

$$f = (1, 1, 2, 3, 5, 8, 13, \ldots),$$

is defined recursively so that

$$f_1 = 1,$$
$$f_2 = 1,$$

and

$$f_{n+1} = f_n + f_{n-1},$$

when $n \geq 2$.

Here it is not just the first term that is directly specified, but the first two terms, f_1 and f_2. Also, subsequent terms depend not just on the immediately preceding term, but the two immediately preceding terms; for instance, f_3 is obtained by adding f_1 and f_2.

Exercise 9.4. Let f be the Fibonacci sequence. Find the numerical values of f_8 and f_9.

Exercise 9.5. Recursively define the sequence s so that $s_1 = 0.5$, $s_2 = -0.5$, $s_3 = 2$, and

$$s_{n+1} = 3s_{n-2} + s_{n-1}s_n$$

when n is a natural number for which $n \geq 3$. Find the numerical values of s_4 and s_5.

Proof by induction provides a means for establishing results about recursively defined sequences, but when the recursion involves looking back further than the immediately preceding term, we usually need to apply so-called *strong induction.*

Theorem 9.7 (Proof by strong induction). *Suppose that for each natural number n we have a statement $S(n)$. To prove that all of the statements*

$$S(1), \quad S(2), \quad S(3), \quad S(4), \quad \ldots,$$

are true, it is enough to do both of the following:

Base step: *Prove that for some natural number m, all of $S(1), S(2), \ldots, S(m)$ are true.*

Inductive step: *Prove that whenever k is a natural number greater than or equal to m and all of $S(1), S(2), \ldots, S(k)$ are true, it follows that $S(k+1)$ is true.*

The assumption in the inductive step of a proof via strong induction that all of $S(1), S(2), \ldots, S(k)$ are true is called the **inductive hypothesis**.

A fundamental question about the Fibonacci sequence is how quickly the terms grow. We can use strong induction to establish lower and upper bounds on the terms of the Fibonacci sequence f. In the next theorem, we prove that $f_n \geq (\sqrt{2})^{n-2}$ for every natural number n, which provides a lower bound for f_n, and in Exercise 9.6, you show that $f_n < 2^n$ for every natural number n, which gives us an upper bound for f_n.

Theorem 9.8. *Let f be the Fibonacci sequence. Then $f_n \geq (\sqrt{2})^{n-2}$ for every natural number n.*

Proof. Observe that

$$(\sqrt{2})^{1-2} = \frac{1}{\sqrt{2}} < 1 = f_1$$

and

$$(\sqrt{2})^{2-2} = 1 = f_2,$$

so the desired result $f_n \geq (\sqrt{2})^{n-2}$ holds for both $n = 1$ and $n = 2$; this establishes the base step.

Now assume $f_i \geq (\sqrt{2})^{i-2}$ for all i in $\{1, 2, \ldots, k\}$, where $k \geq 2$. Observe that since $\frac{1+\sqrt{2}}{2} > 1$, we have

$$\begin{aligned} f_{k+1} &= f_{k-1} + f_k \\ &\geq (\sqrt{2})^{k-3} + (\sqrt{2})^{k-2} \\ &\quad \text{(by our inductive hypothesis)} \\ &= (\sqrt{2})^{k-1}(\sqrt{2})^{-2} + (\sqrt{2})^{k-1}(\sqrt{2})^{-1} \\ &= (\sqrt{2})^{k-1}\left(\frac{1}{2} + \frac{1}{\sqrt{2}}\right) \\ &= (\sqrt{2})^{k-1}\left(\frac{1+\sqrt{2}}{2}\right) \\ &> (\sqrt{2})^{k-1}, \end{aligned}$$

which establishes the inductive step. □

The string of equalities and inequalities presented in the inductive step of the proof of Theorem 9.8 has the specific form

$$p = q \geq r = s = t = u > v,$$

where f_{k+1} is playing the role of p, and $f_{k-1} + f_k$ is playing the role of q, and so forth. The string is a condensed way of stating

$$p = q \quad \text{and} \quad q \geq r \quad \text{and} \quad r = s \quad \text{and} \quad s = t \quad \text{and} \quad t = u \quad \text{and} \quad u > v,$$

from which it follows, because of the strict inequality $u > v$, that

$$p > v,$$

that is,

$$f_{k+1} > (\sqrt{2})^{k-1},$$

which is what we wanted to show. These sorts of strings of equalities and inequalities appear regularly in analysis, so we want to be sure you understand how to read them and what they imply.

Exercise 9.6. Let f be the Fibonacci sequence. Use strong induction to prove that $f_n < 2^n$ for every natural number n.

Exercise 9.7. Prove Theorem 9.7. *Hint*: Apply Proof Strategy 3.2, proof by induction.

Sequential convergence: a case study

We are interested in situations in which the terms of a sequence become “arbitrarily close” to a specific real number, without necessarily ever being equal to this number.

Consider, for instance, the sequence $s = (\frac{3n+2\cdot(-1)^n}{n})$ and observe that

$$s_1 = \frac{3\cdot 1 + 2\cdot(-1)^1}{1} = 1,$$
$$s_2 = \frac{3\cdot 2 + 2\cdot(-1)^2}{2} = 4,$$
$$s_3 = \frac{3\cdot 3 + 2\cdot(-1)^3}{3} = 2.\overline{3},$$
$$s_4 = \frac{3\cdot 4 + 2\cdot(-1)^4}{4} = 3.5,$$

and

$$s_5 = \frac{3\cdot 5 + 2\cdot(-1)^5}{5} = 2.6.$$

Thus, we see that the sequence begins as

$$s = (s_1, s_2, s_3, s_4, s_5, \dots) = (1, 4, 2.\overline{3}, 3.5, 2.6, \dots).$$

If we evaluate s_n for a couple of somewhat larger values of n, say $n = 99$ and $n = 100$, we get

$$s_{99} = \frac{3\cdot 99 + 2\cdot(-1)^{99}}{99} = 2.\overline{97}$$

and

$$s_{100} = \frac{3\cdot 100 + 2\cdot(-1)^{100}}{100} = 3.02.$$

The numerical evidence we have compiled suggests the terms of the sequence are becoming “ever closer” to a number that is either 3 or else very close to 3. Algebraically rewriting the expression for s_n provides more evidence for this conjecture, as

$$s_n = \frac{3n + 2\cdot(-1)^n}{n} = \frac{3n}{n} + \frac{2\cdot(-1)^n}{n} = 3 + (-1)^n\cdot\frac{2}{n}$$

so that

$$s_n = \begin{cases} 3 - \frac{2}{n} & \text{if } n \text{ is odd;} \\ 3 + \frac{2}{n} & \text{if } n \text{ is even.} \end{cases}$$

Since the index n is a natural number, the quantity $\frac{2}{n}$ is positive and becomes smaller as n becomes larger. Thus, the odd-indexed terms of the sequence are all less than 3 but appear to be increasing toward 3, while the even-indexed terms are all greater than 3, but decreasing toward 3.

These observations, while not conclusive, do provide significant evidence to support the belief that the terms of the sequence s are becoming arbitrarily close to 3 as we continue to look further out in the sequence.

Where the analysis we have carried out so far is problematic is the use of phrases such as "ever closer," "arbitrarily close," and "continue to look further out," that have not yet been made precise. But "ever closer" and "arbitrarily close" suggest that certain distances should be "small," so it would seem reasonable to represent these distances using expressions of the form

$$|x - y|,$$

which, as we know from Theorem 2.5, can be interpreted as the distance between the real numbers x and y.

For our sequence s, this process involves measuring the distance

$$|s_n - 3|$$

between terms s_n of the sequence and the conjectured limit 3. Since we have already determined that $s_n = 3 - \frac{2}{n}$ when n is odd and $s_n = 3 + \frac{2}{n}$ when n is even, we see that for any index n, the distance between s_n and 3 is

$$|s_n - 3| = \left|\left(3 \pm \frac{2}{n}\right) - 3\right| = \left|\pm\frac{2}{n}\right| = \frac{2}{n}.$$

Now, if the terms of the sequence s are truly becoming arbitrarily close to 3, it would seem that if someone specifies a positive number representing "how close" to 3 the terms must be, we should then be able to find an index so that from that index onward the terms stay "that close" to 3.

For example, if the specified positive "closeness" is, say, 0.7, since

$$|s_1 - 3| = \frac{2}{1} = 2 \nless 0.7$$

the distance $|s_1 - 3|$ between the first term $s_1 = 1$ of the sequence and 3 is larger than the specified closeness 0.7, and since

$$|s_2 - 3| = \frac{2}{2} = 1 \nless 0.7,$$

the distance $|s_2 - 3|$ between the second term $s_2 = 4$ of the sequence and 3 is also larger than the specified closeness. But, since

$$|s_3 - 3| = \frac{2}{3} = 0.\overline{6} < 0.7,$$

we see that the distance $|s_3 - 3|$ between the third term $s_3 = 2.\overline{3}$ and 3 is less than the specified closeness of 0.7. We may then argue that for any index n where $n \geq 3$, we have

$$|s_n - 3| = \frac{2}{n} \leq \frac{2}{3} < 0.7,$$

so that all the terms of the sequence s from the third term onward are less than 0.7 units from 3.

Of course, as the closeness that is specified is made smaller, we anticipate having to move further out in the sequence to observe terms that close to 3. For instance, if a closeness of 0.1 is specified, since

$$|s_{20} - 3| = \frac{2}{20} = 0.1 \not< 0.1$$

but when $n \geq 21$, it follows that

$$|s_n - 3| = \frac{2}{n} \leq \frac{2}{21} < \frac{2}{20} = 0.1,$$

it is not until the twenty-first term of the sequence that the terms stay less than 0.1 units from 3.

We do anticipate, though, that no matter what positive closeness is specified, we can find an index beyond which all the terms of the sequence are less than the specified closeness from 3.

Exercise 9.8. For the sequence s of the case study we are undertaking, represent the distance between the tenth term s_{10} and the number 3 using absolute value. Then calculate this distance.

The definition of sequential convergence

Generalizing from our case study, when we suspect a number p to be the "limit" of a sequence (a_n), we expect to be able to make the distance

$$|a_n - p|$$

between a_n and p smaller than any specified positive closeness for all terms a_n whose index n is sufficiently large. If the positive closeness is denoted by ε, then we are requiring that

$$|a_n - p| < \varepsilon$$

for index values n such that $n \geq N$, where N is some particular index that depends on the positive closeness ε that has been specified.

For example, considering once more the sequence s from the case study, we found that for the positive closeness $\varepsilon = 0.7$, taking the index $N = 3$ resulted in

$$|s_n - 3| < 0.7$$

for all index values n for which $n \geq 3$. Making the positive closeness ε smaller, though, forced us to move further out in the sequence before we saw the distance between the terms and the anticipated limit 3 become less than ε. Specifically, when $\varepsilon = 0.1$, we needed to take $N = 21$, for when $n \geq 21$, we determined that

$$|s_n - 3| < 0.1.$$

Exercise 9.9. For the sequence s of the case study, take $\varepsilon = 0.01$ and find the least natural number index N for which $|s_n - 3| < \varepsilon$ whenever $n \geq N$.

We are now in a position to formulate a precise definition. In this book, we generally do not set apart mathematical definitions and give them reference numbers, but for a few of the most important ones we shall do so. The definition of sequential convergence is one of these definitions.

Definition 9.9 (The definition of sequential convergence). A sequence (a_n) **converges to a real number** p if for every positive number ε, there is a natural number index N such that when $n \geq N$, it follows that $|a_n - p| < \varepsilon$. If (a_n) converges to p, we call p a **limit** of the sequence (a_n).

The statement that (a_n) converges to p is often rendered symbolically as either

$$a_n \to p$$

or

$$\lim a_n = p.$$

When a sequence has a limit, we say the sequence **converges** and that it is a **convergent sequence**.

It is always a good idea to reconcile a formal mathematical definition with the more intuitive notion it is intended to capture. For the definition of sequential convergence, we interpret $|a_n - p|$ as the distance between the nth term a_n of a sequence and the sequence's (proposed) limit p. We also interpret ε as a "tolerance" or "threshold" or "error allowance" we want the distance between a_n and p to stay below. With this perspective, it is only when, for *any* given positive tolerance, the terms of the sequence from some index value onward can be made to stay less than this tolerance from the proposed limit, that the proposed limit is confirmed as the sequence's actual limit.

Be very aware that the index N beyond which we are able to guarantee the distance between a_n and p is less than ε can, and usually does, vary with the value of ε. We saw this happen with our case study sequence s, where we could take $N = 3$ for $\varepsilon = 0.7$, but needed a minimum value of $N = 21$ for $\varepsilon = 0.1$. To emphasize this dependency, we sometimes write N_ε in place of just N.

Exercise 9.10. Consider the sequence b defined so that $b_n = \frac{2-n}{n+3}$. This sequence converges to -1. For the specific value of ε, find the least natural number index N_ε for which

$$\left|b_n - (-1)\right| < \varepsilon$$

whenever $n \geq N_\varepsilon$.

(a) $\varepsilon = 1$
(b) $\varepsilon = 0.5$
(c) $\varepsilon = 0.1$

To conclusively establish the limit of a convergent sequence, it is not enough to find a suitable index N for one or even a few selected values of the positive error tolerance ε. The definition of sequential convergence requires that we demonstrate the ability to find such an index no matter what positive number is assigned to ε. So a proof that employs this definition to verify convergence of a sequence typically leaves ε unassigned, that is, treats ε as an *arbitrary* positive number, and then proceeds to indicate how an index N can be obtained regardless of what the actual numerical value of ε happens to be. As we have seen, the value of N tends to vary with the value of ε, so we do not usually see an assignment such as "take $N = 37$" in the proof, but rather a discussion of how generally one would obtain a value of N if a value were assigned to ε.

Example 9.10. We now use the definition of sequential convergence to prove that our case study sequence s defined so that

$$s_n = \frac{3n + 2 \cdot (-1)^n}{n}$$

converges to 3.

To do so, we begin with an arbitrary positive number ε. We then need to find a natural number N for which, whenever the index n is at least as large as N, the distance

$$\left|\frac{3n + 2 \cdot (-1)^n}{n} - 3\right|$$

between the nth term $\frac{3n+2\cdot(-1)^n}{n}$ of the sequence and the sequence's proposed limit 3 is less than the given ε. We include in our proof the algebraic simplification of $|\frac{3n+2\cdot(-1)^n}{n}-3|$ to $\frac{2}{n}$, which means we are looking for N so that when $n \geq N$, it follows that $\frac{2}{n} < \varepsilon$. To achieve this goal, we apply the Archimedean property to obtain a natural number N for which $\frac{1}{N} < \frac{\varepsilon}{2}$, as it then follows that when $n \geq N$ we have $\frac{1}{n} \leq \frac{1}{N} < \frac{\varepsilon}{2}$ so that $\frac{2}{n} < \varepsilon$.

Here is how we could write up the argument, based on the plan just discussed, that the sequence $(\frac{3n+2\cdot(-1)^n}{n})$ converges to 3.

Proof. Suppose $\varepsilon > 0$. Then $\frac{\varepsilon}{2}$ is also a positive number, so by the Archimedean property, there exists a natural number N for which $\frac{1}{N} < \frac{\varepsilon}{2}$. Then, assuming $n \geq N$, it follows that

$$\left|\frac{3n + 2\cdot(-1)^n}{n} - 3\right| = \left|\frac{3n}{n} + \frac{2\cdot(-1)^n}{n} - 3\right| = \left|\frac{2\cdot(-1)^n}{n}\right| = \frac{2}{n} \leq \frac{2}{N} < 2\cdot\frac{\varepsilon}{2} = \varepsilon.$$

Thus, $\frac{3n+2\cdot(-1)^n}{n} \to 3$, that is, $\lim \frac{3n+2\cdot(-1)^n}{n} = 3$. □

Sequential convergence is our first example of a limiting process, and it is important to be aware of how this notion relies on the completeness axiom. Most often it is via the Archimedean property, which is a consequence of the completeness axiom, that the completeness of the real number system makes its appearance in proofs of sequential convergence. This was, for instance, the case with the proof in Example 9.10 that the sequence $(\frac{3n+2\cdot(-1)^n}{n})$ converges to 3.

In the next example, we prove the sequence $(\frac{1}{n})$ of reciprocals of the natural numbers converges to 0, which is essentially the Archimedean property recast into the language of sequential convergence.

Example 9.11. We claim the sequence $(\frac{1}{n})$ converges to 0. To see why, consider any positive number ε. By the Archimedean property, there exists a natural number N such that $\frac{1}{N} < \varepsilon$. Now, assuming $n \geq N$, it follows that

$$\left|\frac{1}{n} - 0\right| = \frac{1}{n} \leq \frac{1}{N} < \varepsilon.$$

Hence, according to the definition of sequential convergence, $(\frac{1}{n})$ converges to 0.

There is at least one situation in which we are attempting to apply the definition of convergence of a sequence that the same value of N can be chosen even as the value of the positive number ε is allowed to vary. Because a constant sequence converges to the constant that is the value of every term of the sequence, the distance between such a sequence's terms and its limit is actually zero.

Theorem 9.12. *Given any real number c, the constant sequence (c), all of whose terms are equal to c, converges to c.*

Proof. Consider any real number c and define $a_n = c$ for every natural number n. Let ε be any positive number and let $N = 1$. Assuming $n \geq N$, that is, assuming n is any natural number,

$$|a_n - c| = |c - c| = 0 < \varepsilon.$$

Thus, $a_n \to p$. □

Exercise 9.11. Identify the limit of the sequence from Exercise 9.1(d).

In using the definition of sequential convergence to prove that a sequence converges to a certain number, we usually need to do some preliminary analysis, perhaps to conjecture a value for the limit, but almost always in order to figure out how to obtain the natural number index N from the positive distance threshold ε.

Example 9.13. Consider the sequence $(\frac{6n^2+1}{4+3n^2})$. The reader can check that the sequence begins as

$$(1, 1.56, 1.77, 1.87, 1.91, \dots),$$

where we have approximated the values of the second through fifth terms to two decimal places. Based on this numerical evidence we might conjecture that the sequence converges to 2, which is also suggested by the fact that $\frac{6n^2+1}{4+3n^2} \approx \frac{6n^2}{3n^2} = 2$ for "large" values of n (recall that the symbol $\approx$ means *is approximately equal to*). So we now attempt to prove that the sequence actually does converge to 2.

We start with an anonymous positive number ε and we must find an index N so that when $n \geq N$ we have

$$\left|\frac{6n^2+1}{4+3n^2} - 2\right| < \varepsilon.$$

To achieve this goal, we examine the desired inequality to see if it reveals to us something about how n might relate to ε. Simplifying the expression on the left side should help us gain some insight. Note that

$$\left|\frac{6n^2+1}{4+3n^2} - 2\right| = \left|\frac{6n^2+1}{4+3n^2} - \frac{2(4+3n^2)}{4+3n^2}\right| = \left|\frac{-7}{4+3n^2}\right| = \frac{7}{4+3n^2},$$

so the desired inequality is actually

$$\frac{7}{4+3n^2} < \varepsilon.$$

Furthermore, since

$$\frac{7}{4+3n^2} < \frac{7}{3n^2}$$

and

$$\frac{7}{3n^2} \leq \frac{7}{3n},$$

if we can get $\frac{7}{3n} < \varepsilon$, it would follow that $\frac{7}{4+3n^2} < \varepsilon$. Note that the inequality $\frac{7}{3n} < \varepsilon$ is equivalent to the inequality $\frac{1}{n} < \frac{3}{7}\varepsilon$. Since ε is a positive number, so is $\frac{3}{7}\varepsilon$, and we can apply the Archimedean property to obtain a natural number N for which $\frac{1}{N} < \frac{3}{7}\varepsilon$, which then leads to the inequality $\frac{1}{n} < \frac{3}{7}\varepsilon$ holding whenever $n \geq N$.

The preliminary analysis we have just carried out suggests how we can obtain the natural number index N from the given positive number ε via the Archimedean property. The analysis also indicates many of the steps we employ in our written proof below, but notice that the ordering of these steps is nearly reversed from how they came about when we were "thinking backward" in our work above because, by its very nature, a proof is a record of the deductions made that lead us from hypothesis to our ultimate conclusion.

Here is a write-up of our proof that the sequence $(\frac{6n^2+1}{4+3n^2})$ converges to 2.

Proof. Consider any positive number ε. Then $\frac{3}{7}\varepsilon > 0$ and the Archimedean property tells us there exists a natural number N for which $\frac{1}{N} < \frac{3}{7}\varepsilon$. Now suppose $n \geq N$. It follows that

$$\left|\frac{6n^2+1}{4+3n^2} - 2\right| = \left|\frac{6n^2+1}{4+3n^2} - \frac{2(4+3n^2)}{4+3n^2}\right| = \left|\frac{-7}{4+3n^2}\right| = \frac{7}{4+3n^2} < \frac{7}{3n^2} \leq \frac{7}{3n} \leq \frac{7}{3N} < \frac{7}{3} \cdot \frac{3}{7}\varepsilon = \varepsilon.$$

Therefore, $\frac{6n^2+1}{4+3n^2} \to 2$. □

Note that the proof itself is written in a style that aligns very closely with the formal logical structure underlying what it means, by definition, for a sequence (a_n) to converge to the number p:

(1) In the first sentence of the proof, we introduce an arbitrary positive number ε.
(2) We then use ε to get ahold of the natural number N.
(3) Then, we show how it comes about that the inequality $|a_n - p| < \varepsilon$ must be true when $n \geq N$.

We want to emphasize, though, that the decision in this particular proof to apply the Archimedean property with the positive number $\frac{3}{7}\varepsilon$ is made only after we investigated the quantity $|\frac{6n^2+1}{4+3n^2} - 2|$ we needed to make less than ε for suitably large natural number index values n. Thus, a reader of the proof who had not seen the preliminary analysis would probably only be convinced that the number $\frac{3}{7}\varepsilon$ is relevant once they finish reading the entire proof and see that this choice was helpful.

The situation presented in Example 9.13 is typical of many we shall encounter in the sense that we are looking for a value of a quantity, in that case n, that makes a related quantity, in that case $\frac{7}{4+3n^2}$, suitably small, which here means less than the positive number ε we started with. Except in the most straightforward of situations, we should anticipate having to do some preliminary investigation before trying to write up such a proof.

Exercise 9.12. Use the definition of limit of a sequence to prove each of the following.
(a) The sequence $(\frac{3n}{n^2+5})$ converges to 0.
(b) The sequence $(\frac{2-n}{n+3})$ converges to -1.
(c) The sequence $(\frac{\sqrt{n}}{n^2+10})$ converges to 0.
(d) The sequence $(\frac{1}{n+1} - \frac{1}{n})$ converges to 0.
(e) The sequence $(\frac{n^2+1}{2n^2+n})$ converges to $\frac{1}{2}$.

Exercise 9.13. Show that if $a_n \to p$, then $|a_n| \to |p|$. *Hint*: Use the reverse triangle inequality, Theorem 2.3.

Uniqueness of sequential limits

As you would probably imagine, it is impossible for a sequence to converge to more than one real number. In the following proof, we assume a sequence (a_n) converges to both p and q, then show that $p = q$ by showing that the distance $|p - q|$ between p and q is less than every positive number ε, hence, must be zero.

Theorem 9.14 (Uniqueness of sequential limits). *A sequence of real numbers cannot have more than one limit.*

Proof. Suppose $a_n \to p$ and $a_n \to q$. Let ε be an arbitrary positive number. Because $a_n \to p$, there exists an index N_1 such that $|a_n - p| < \frac{\varepsilon}{2}$ whenever $n \geq N_1$. Because $a_n \to q$, there exists an index N_2 such that $|a_n - q| < \frac{\varepsilon}{2}$ whenever $n \geq N_2$. Let $N = \max\{N_1, N_2\}$. Then

$$|p - q| = |p - a_N + a_N - q| \leq |p - a_N| + |a_N - q| < \frac{\varepsilon}{2} + \frac{\varepsilon}{2} = \varepsilon.$$

Thus, $|p - q|$ is less than every positive real number, from which it follows that $|p - q| = 0$, so that we must have $p = q$. □

In this proof, the use of the maximum of the indices N_1 and N_2 permits us to locate an index N beyond which both of the conditions $|a_n - p| < \frac{\varepsilon}{2}$ and $|a_n - q| < \frac{\varepsilon}{2}$ hold. We can employ this strategy any time we need multiple conditions to be simultaneously satisfied for the terms of a sequence from some index onward.

Also, the hypotheses that $a_n \to p$ and $a_n \to q$ tell us that each of the quantities $|a_n - p|$ and $|a_n - q|$ can be made arbitrarily small for sufficiently large n. To relate $|p - q|$

to these quantities, we rewrote $|p - q|$ as $|p - a_N + a_N - q|$ and then applied the triangle inequality. It is useful to keep in mind the strategy of adding zero in the form of $x - x$ for a suitable choice of x when we are trying to relate various quantities.

Exercise 9.14. Write a proof by contradiction for Theorem 9.14.

Characterizing sequential convergence via neighborhoods

Because

$$|a_n - p| < \varepsilon$$

is equivalent to the statement

$$a_n \in (p - \varepsilon, p + \varepsilon),$$

another way of expressing the notion that the sequence (a_n) converges to p is to say that (a_n) is eventually in (and "stays in") every neighborhood of p. This observation establishes the following theorem.

Theorem 9.15. *Given a sequence (a_n) and a real number p, the following are equivalent:*
(i) *(a_n) converges to p;*
(ii) *(a_n) is eventually in every neighborhood of p, that is, for every positive number ε, there exists an index N for which a_n is in $(p - \varepsilon, p + \varepsilon)$ when $n \geq N$.*

Example 9.10 (Continued). When we proved the sequence

$$s = \left(\frac{3n + 2 \cdot (-1)^n}{n} \right)$$

of our case study converges to 3, we essentially showed that the sequence is eventually in every neighborhood of 3. Beginning with an arbitrary positive number ε is equivalent to beginning with an arbitrary neighborhood $(3 - \varepsilon, 3 + \varepsilon)$ of 3. We then found an index N for which whenever $n \geq N$, we have

$$\left| \frac{3n + 2 \cdot (-1)^n}{n} - 3 \right| < \varepsilon,$$

which is really just another way of saying that, from the index N onward, the terms of the sequence a stay within the neighborhood $(3 - \varepsilon, 3 + \varepsilon)$ of 3 (Figure 9.1).

In other words, because the positive number ε was arbitrary, we showed that the sequence a is eventually in every neighborhood of 3.

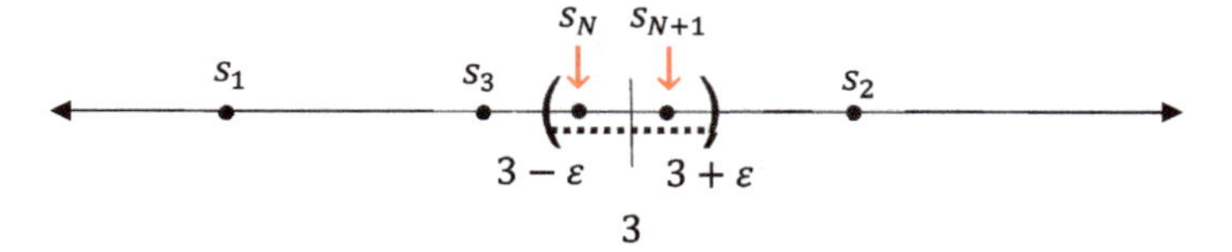

Figure 9.1: The terms of the sequence s from Example 9.10 eventually stay within any specified neighborhood of 3.

Exercise 9.15. Show that if $a_n \to p$ and $p < 0$, then there exists an index N such that when $n \geq N$ we have $a_n < 0$.

Showing a sequence does not converge to a specified real number

There are times when we want to establish that a certain sequence (a_n) does not converge to a certain real number p. Now to say (a_n) converges to p means the terms a_n of the sequence from some index value onward must stay within any positive distance of p that has been prescribed. Therefore, to say (a_n) does not converge to p means that the terms a_n cannot be made to stay within some particular positive distance of p no matter how far out in the sequence we look. The following theorem states this observation more formally. Its proof is the result of applying Axiom B.14, concerning the logical negation of statements involving quantifiers, to Definition 9.9, the formal definition of $a_n \to p$.

Theorem 9.16. *Given a sequence (a_n) and a real number p, the following are equivalent:*
(i) *(a_n) does not converge to p;*
(ii) *there exists a positive number ε such that for every natural number N, there exists an index n for which $n \geq N$ and $|a_n - p| \geq \varepsilon$.*

Example 9.17. We now show the alternating sequence

$$a = (-1, 1, -1, 1, \ldots)$$

does not converge to 1. To do so, we must find a specific positive number ε that we believe provides a tolerance within which we are never able to make the terms of the sequence, from some index onward, stay less than ε away from 1. As every other term of the sequence is -1, and the distance between -1 and 1 is 2, we are led to consider a choice of ε less than or equal to 2. We will try $\varepsilon = 2$.

Now consider any natural number N. Then $2N + 1$ is an odd natural number that is larger than N. Because $2N + 1$ is odd, it follows that $a_{2N+1} = -1$. Then

$$|a_{2N+1} - 1| = |-1 - 1| = 2 \geq 2 = \varepsilon.$$

Thus, the sequence $(-1, 1, -1, 1, \ldots)$ does not converge to 1 (Figure 9.2).

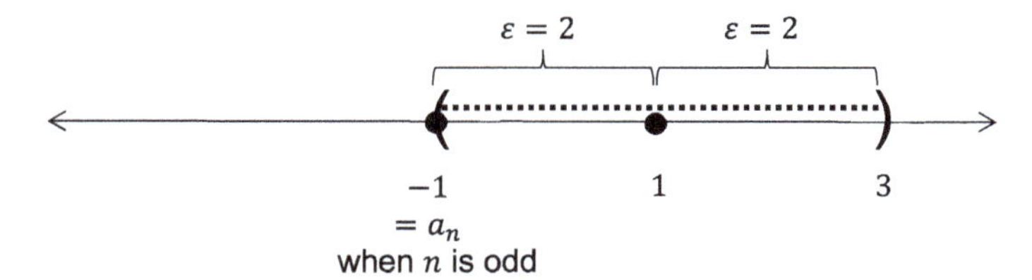

Figure 9.2: Terms of the sequence from Example 9.17 never stay, from some index onward, in the 2-neighborhood of 1.

Exercise 9.16. Use Axiom B.14 to carefully work out, in a step-by-step fashion, the formal negation of Definition 9.9, the definition of sequential convergence, in order to derive condition (ii) of Theorem 9.16.

Exercise 9.17. Use Theorem 9.16 to show that the sequence $(\frac{2n+1}{n})$ does not converge to 3.

Divergence of a sequence

We do not want to give the impression that all sequences converge. That is, not only may a given sequence not converge to some specified real number, it is possible that there is no real number to which the sequence converges, in which case we say the sequence **diverges**.

Example 9.17 (Continued). We have shown that the sequence

$$a = (-1, 1, -1, 1, \dots)$$

does not converge to 1. In fact, this sequence does not converge to any real number. We can show this to be the case by demonstrating that for any real number p, it follows that $a_n \nrightarrow p$.

Let p be any real number and take $\varepsilon = 1$. Note that all points in the interval $(p-\varepsilon, p+\varepsilon) = (p-1, p+1)$ are less than $(p+1)-(p-1) = 2$ units apart, so this particular neighborhood of p, the 1-neighborhood of p, cannot include both -1 and 1, which are exactly 2 units apart. Hence, the 1-neighborhood of p cannot contain all terms of the sequence $a = (-1, 1, -1, 1, \dots)$ from any index value onward, as there is a continual alternation between -1 and 1 in the values of the terms. That is, taking $\varepsilon = 1$, for any natural number N, at least one of a_N or a_{N+1}, one of which is equal to -1 and the other of which is equal to 1, is not in the 1-neighborhood of p. Thus, $a_n \nrightarrow p$.

Having shown that the alternating sequence $(-1, 1, -1, 1, \dots)$ does not converge to any real number, we conclude that it diverges.

Exercise 9.18. Show the sequence $(n \cdot (-1)^n)$ diverges.

Exercise 9.19. Suppose (a_n) and (b_n) are sequences that agree with each other for all but a finite number of indices. Is it possible for one of these sequences to converge and the other to diverge? Justify your conclusion.

Exercise 9.20. Show that $a_n \to 0$ if and only if $|a_n| \to 0$. Then give an example to show that it is possible for the sequence $(|a_n|)$ to converge even if the sequence (a_n) does not.

10 More about sequences and convergence

Having defined what it means for a sequence to converge, we now investigate various means for determining whether a sequence might or must converge, as well as some methods for evaluating limits of convergent sequences. We also introduce some special types of sequences, including *bounded sequences*, *monotone sequences*, and *infinite series*.

Sequences, limit points, and closed sets

Given a nonempty subset A of $\mathbb{R}$, we say that a **sequence is in** A if all its terms are members of the set A.

Example 10.1. All of the terms of the sequence $(1, \frac{1}{2}, \frac{1}{3}, \frac{1}{4}, \ldots)$ of reciprocals of the natural numbers are positive, so we can say this sequence is in $\mathbb{R}^+$.

The notions of limit point and closed set can both be characterized via sequential convergence.

Theorem 10.2. *Given a subset A of $\mathbb{R}$ and a real number p, the following are equivalent:*
(i) *p is a limit point of A;*
(ii) *there exists a sequence (a_n) in $A - \{p\}$ for which $a_n \to p$.*

Informally, the conditions (i) and (ii) of Theorem 10.2 are just different ways of expressing the idea that there are points of A different from p that are arbitrarily close to p.

Exercise 10.1. Prove Theorem 10.2.

Theorem 10.3. *Given a subset A of $\mathbb{R}$, the following are equivalent:*
(i) *A is closed;*
(ii) *whenever (a_n) is a sequence in A and $a_n \to p$, it follows that $p \in A$.*

Informally, the conditions (i) and (ii) of Theorem 10.3 are just different ways of expressing the idea that A contains all real numbers that are arbitrarily close to A.

Exercise 10.2. Prove Theorem 10.3.

Exercise 10.3. A subset A of $\mathbb{R}$ is said to be **perfect** if every point of A is a limit point of A. Show that the Cantor set, defined in Example 8.4, is a perfect set.

---

https://doi.org/10.1515/9783111429564-010

Bounded and unbounded sequences

A sequence is **bounded** if its range is a bounded subset of $\mathbb{R}$. A sequence that is not bounded is said to be **unbounded**.

Example 10.4. The alternating sequence

$$((-1)^n) = (-1, 1, -1, 1, \dots)$$

is bounded because its range, $\{-1, 1\}$, has a lower bound, for instance, -1, and an upper bound, for instance, 1.

The sequence

$$\left(\frac{1}{n}\right) = \left(1, \frac{1}{2}, \frac{1}{3}, \frac{1}{4}, \dots\right)$$

is bounded because its range is bounded below by 0 and bounded above by 1.

The sequence

$$(n) = (1, 2, 3, 4, \dots)$$

is not bounded because its range is the set $\mathbb{N}$, which the Archimedean property tells us has no upper bound.

The sequence

$$((-n)^n) = (-1, 4, -27, 256, \dots)$$

is not bounded because $|(-n)^n| = n^n \geq n$ for every natural number n (see Exercise 10.4(b)).

Exercise 10.4. Use induction to prove each of the following.
(a) If $0 < a < b$, then for every natural number n, it follows that $a^n < b^n$.
(b) For every natural number n, we have $n^n \geq n$.

Theorem 10.5. *A sequence (a_n) is bounded if and only if there exists a positive number B for which $|a_n| \leq B$ for all natural numbers n.*

Exercise 10.5. Prove Theorem 10.5.

It follows from Theorem 10.5 that a sequence (a_n) is unbounded if for every positive number B there exists a natural number n for which $|a_n| > B$.

Example 10.6. We show the Fibonacci sequence f from Example 9.6 is unbounded. Since all the terms of f are positive, it suffices to show that for every positive number B there is a term of f larger than B.

Assume $B > 0$ and use the Archimedean property to obtain a natural number N such that $N > B + 1$; then $N - 1$ is also a natural number and is greater than B. Making use of the results from Theorem 9.8 and Exercise 3.13, we have

$$f_{2N} \geq (\sqrt{2})^{2N-2} = 2^{N-1} > N - 1 > B,$$

so we have found a term of the Fibonacci sequence that is greater than the given positive number B. Thus, f is unbounded.

Among the bounded sequences of real numbers are all the convergent sequences. This follows because any neighborhood of a convergent sequence's limit must contain all but finitely many terms of the sequence.

Theorem 10.7. *Every convergent sequence is bounded.*

Proof. Suppose the sequence (a_n) converges to p. Then there exists an index N for which $|a_n - p| < 1$, that is, $a_n \in (p - 1, p + 1)$, whenever $n \geq N$. Let

$$B = \max\{|a_1|, |a_2|, \ldots, |a_{N-1}|, |p - 1|, |p + 1|\}$$

and note that $B > 0$. From the way in which B is defined, it is clear that $|a_n| \leq B$ when $n < N$. So suppose $n \geq N$. If $a_n \geq 0$, then

$$0 \leq |a_n| = a_n < p + 1 = |p + 1| \leq B.$$

If $a_n < 0$, then

$$0 < |a_n| = -a_n < -(p - 1) = |p - 1| \leq B.$$

Thus, $|a_n| \leq B$ for all natural numbers n, so we may conclude that (a_n) is bounded. □

The contrapositive of Theorem 10.7 provides the following criterion for divergence of a sequence.

Theorem 10.8. *Every unbounded sequence diverges.*

Example 10.9. As they are unbounded, the sequence of natural numbers $(1, 2, 3, 4, \ldots)$ and the Fibonacci sequence $(1, 1, 2, 3, 5, 8, 13, \ldots)$ both diverge.

Exercise 10.6. Show that the sequence $(n \cdot (-1)^n)$ must diverge by showing it is unbounded. (In Exercise 9.18, you used a different approach to show this same sequence diverges.)

Even though a convergent sequence must be bounded, a bounded sequence need not converge. For instance, in Example 9.17, we demonstrated that the bounded alternating sequence $(-1, 1, -1, 1, \dots)$ diverges.

Exercise 10.7. Suppose the sequence (a_n) converges to p. Prove that the set $\{a_n \mid n \in \mathbb{N}\} \cup \{p\}$ is compact.

We have shown that an unbounded sequence must diverge. In particular, we say that a sequence (a_n)

(1) **diverges to** ∞, denoted $a_n \to \infty$ or $\lim a_n = \infty$, if for every positive number B, there exists an index N such that when $n \geq N$, it follows that $a_n \geq B$ (i. e., (a_n) becomes arbitrarily large from some index onward);

(2) **diverges to** $-\infty$, denoted $a_n \to -\infty$ or $\lim a_n = -\infty$, if for every positive number B, there exists an index N such that when $n \geq N$, it follows that $a_n \leq -B$ (i. e., (a_n) becomes arbitrarily small, in the negative sense, from some index onward).

Keep in mind that the symbols ∞ and $-\infty$ do not represent real numbers; rather, they are being employed only as notational conveniences. Also, the index N generally depends on the value of B, so this is another situation where we could write N_B in place of N to make this dependence more explicit.

Example 10.6 (Continued). Our earlier work in this example shows that the Fibonacci sequence f diverges to ∞. Thus, we may write $f_n \to \infty$ or $\lim f_n = \infty$.

Exercise 10.8. Use the definition of divergence to ∞ to show that the sequence (n^2) diverges to ∞.

Exercise 10.9. Use the definition of divergence to $-\infty$ to show that $-\sqrt{n} \to -\infty$.

Exercise 10.10. Show that if (a_n) is a sequence of nonzero real numbers that diverges to ∞, then the sequence $(\frac{1}{a_n})$ converges to 0.

Tails of sequences and convergence

A **tail** of a sequence is the sequence formed by taking all terms of the original sequence from some particular index onward. In other words, a **tail** of the sequence (a_n) is a sequence (b_n) for which $b_n = a_{N+n}$ for some nonnegative integer N and every natural number n.

Example 10.10. Among the tails of the sequence $(1, \frac{1}{2}, \frac{1}{3}, \frac{1}{4}, \dots)$ whose nth term is $\frac{1}{n}$ are:

(1) the tail $(\frac{1}{2}, \frac{1}{3}, \frac{1}{4}, \ldots)$ that begins with the second term of the original sequence;
(2) the tail $(\frac{1}{100}, \frac{1}{101}, \frac{1}{102}, \ldots)$ that begins with the one-hundredth term of the original sequence; and,
(3) the original sequence $(1, \frac{1}{2}, \frac{1}{3}, \frac{1}{4}, \ldots)$ itself.

Note that if $a = (1, \frac{1}{2}, \frac{1}{3}, \frac{1}{4}, \ldots)$, then the tail $b = (\frac{1}{100}, \frac{1}{101}, \frac{1}{102}, \ldots)$ of a is defined so that $b_n = a_{n+99}$, which means

$$b_1 = a_{100} = \frac{1}{100},$$
$$b_2 = a_{101} = \frac{1}{101},$$
$$b_3 = a_{102} = \frac{1}{102},$$

and so on.

Exercise 10.11. Identify three different tails of the Fibonacci sequence.

Whether a sequence converges is completely determined by the behavior of the sequence "way out" beyond any finite number of the sequence's initial terms. In other words, convergence of a sequence is really about the behavior of the tails of the sequence.

Theorem 10.11. *Given a sequence (a_n) and a real number p, the following are equivalent:*
(i) *the sequence (a_n) converges to p;*
(ii) *some tail of (a_n) converges to p;*
(iii) *every tail of (a_n) converges to p.*

Proof. (i) ⇒ (ii) Assume (a_n) converges to p. As the sequence (a_n) is a tail of itself, it automatically follows that some tail of (a_n) converges to p.

(ii) ⇒ (iii) Suppose some tail (b_n) of (a_n) converges to p. Then there exists a nonnegative integer N for which $b_n = a_{N+n}$ for every natural number n. Now consider any tail (c_n) of (a_n). Then there exists a nonnegative integer K for which $c_n = a_{K+n}$ for every natural number n. We show (c_n) converges to p.

Suppose $\varepsilon > 0$. As $b_n \to p$, there exists an index J such that $|b_n - p| < \varepsilon$ when $n \geq J$. As $b_n = a_{N+n}$, this means that $|a_n - p| < \varepsilon$ when $n \geq N + J$. Now, when $n \geq N + J$, as it follows that $K + n \geq K + N + J \geq N + J$, we may then conclude that $|c_n - p| = |a_{K+n} - p| < \varepsilon$, so by definition $c_n \to p$.

(iii) ⇒ (i) Assume that every tail of (a_n) converges to p. As the sequence (a_n) is a tail of itself, it automatically follows that (a_n) converges to p. □

The significance of Theorem 10.11 cannot be overemphasized: The convergence or divergence of a sequence has nothing to do with the first finitely many terms of the sequence. It is impossible to determine whether a sequence converges by looking only at some finite number of the sequence's initial terms.

Exercise 10.12. The first one thousand terms of the sequence a are unknown, but it is known that $a_n = \frac{(n+1)^2}{(2n-2)(3n-4)}$ when $n > 1000$.

(a) Must the sequence a converge? If so, to what limit? If not, why not?

(b) Must some tail of the sequence a converge? If so, to what limit? If not, why not?

(c) Must every tail of the sequence a converge? If so, can different tails of the sequence have different limits?

Arithmetic properties of convergent sequences

We can form new sequences by adding, subtracting, multiplying, or dividing the corresponding terms of given sequences (being sure to avoid division by zero, of course). That is, given sequences a and b, we define

(1) the **sum** $a + b$ to be the sequence whose nth term is $a_n + b_n$;

(2) the **difference** $a - b$ to be the sequence whose nth term is $a_n - b_n$;

(3) the **product** $a \cdot b$ to be the sequence whose nth term is $a_n \cdot b_n$;

(4) and the **quotient** $\frac{a}{b}$ to be the sequence whose nth term is $\frac{a_n}{b_n}$, provided that $b_n \neq 0$ for all n.

Example 10.12. Given the sequences $a = (\frac{1}{n+1})$ and $b = (\frac{1}{n})$, we have

$$a + b = \left(\frac{1}{n+1} + \frac{1}{n}\right) = \left(\frac{2n+1}{n^2+n}\right),$$
$$a - b = \left(\frac{1}{n+1} - \frac{1}{n}\right) = \left(\frac{-1}{n^2+n}\right),$$
$$a \cdot b = \left(\frac{1}{n+1} \cdot \frac{1}{n}\right) = \left(\frac{1}{n^2+n}\right),$$

and

$$\frac{a}{b} = \left(\frac{\frac{1}{n+1}}{\frac{1}{n}}\right) = \left(\frac{n}{n+1}\right).$$

The limits of sums, differences, products, and quotients of convergent sequences can be determined in the expected manner. After presenting the proofs, we discuss the preliminary analysis that went into their development.

Theorem 10.13. *Suppose the sequence $a = (a_n)$ converges to p and the sequence $b = (b_n)$ converges to q.*

(1) *The sequence $a + b = (a_n + b_n)$ converges to $p + q$.*

(2) *The sequence $a - b = (a_n - b_n)$ converges to $p - q$.*

(3) *The sequence* $a \cdot b = (a_n \cdot b_n)$ *converges to* $p \cdot q$.
(4) *The sequence* $\frac{a}{b} = (\frac{a_n}{b_n})$ *converges to* $\frac{p}{q}$, *as long as* $q \neq 0$ *and* $b_n \neq 0$ *for all* n.

Proof. Assume $a_n \to p$ and $b_n \to q$.

First, we prove (1). Assume $\varepsilon > 0$. As $a_n \to p$, there exists an index N_1 such that when $n \geq N_1$, we have $|a_n - p| < \frac{\varepsilon}{2}$. Also, as $b_n \to q$, there exists an index N_2 such that when $n \geq N_2$, we have $|b_n - q| < \frac{\varepsilon}{2}$. Let $N = \max\{N_1, N_2\}$ and suppose $n \geq N$. Then

$$\big|(a_n + b_n) - (p + q)\big| = \big|(a_n - p) + (b_n - q)\big| \leq |a_n - p| + |b_n - q| < \frac{\varepsilon}{2} + \frac{\varepsilon}{2} = \varepsilon.$$

As ε is an arbitrary positive number, we may conclude that $(a_n + b_n)$ converges to $p + q$.

We now establish (3). Assume $\varepsilon > 0$. Then there exist indices N_1, N_2, N_3, and N_4 such that

(a) when $n \geq N_1$, it follows that $|a_n - p| < \sqrt{\frac{\varepsilon}{3}}$;
(b) when $n \geq N_2$, it follows that $|b_n - q| < \sqrt{\frac{\varepsilon}{3}}$;
(c) when $n \geq N_3$, it follows that $|a_n - p| < \frac{\varepsilon}{3(|q|+1)}$;
(d) when $n \geq N_4$, it follows that $|b_n - q| < \frac{\varepsilon}{3(|p|+1)}$.

Let $N = \max\{N_1, N_2, N_3, N_4\}$ and suppose $n \geq N$. Then

$$\begin{aligned}
|a_n \cdot b_n - p \cdot q| &= \big|[(a_n - p) + p] \cdot [(b_n - q) + q] - pq\big| \\
&= \big|(a_n - p)(b_n - q) + (a_n - p)q + p(b_n - q)\big| \\
&\leq |a_n - p| \cdot |b_n - q| + |a_n - p| \cdot |q| + |p| \cdot |b_n - q| \\
&< \sqrt{\frac{\varepsilon}{3}} \cdot \sqrt{\frac{\varepsilon}{3}} + \frac{\varepsilon}{3(|q| + 1)} \cdot |q| + |p| \cdot \frac{\varepsilon}{3(|p| + 1)} \\
&= \frac{\varepsilon}{3} + \frac{|q|}{|q| + 1} \cdot \frac{\varepsilon}{3} + \frac{|p|}{|p| + 1} \cdot \frac{\varepsilon}{3} \\
&< \frac{\varepsilon}{3} + \frac{\varepsilon}{3} + \frac{\varepsilon}{3} \\
&= \varepsilon.
\end{aligned}$$

As ε is an arbitrary positive number, we may conclude that $(a \cdot b_n)$ converges to $p \cdot q$.

Next we prove (2). As the constant sequence (-1) converges to -1, and we have just proven that the product of convergent sequences converges to the product of the sequences' limits, we may conclude that the sequence $((-1) \cdot b_n)$ converges to $(-1) \cdot q = -q$. Then, as we have already proved that the sum of convergent sequences converges to the sum of the sequences' limits, we may conclude that the sequence $(a_n + (-1) \cdot b_n)$ converges to $p + (-q) = p - q$. Since $a_n - b_n = a_n + (-1) \cdot b_n$, we have therefore shown that $(a_n - b_n)$ converges to $p - q$.

Finally, we prove (4). Here, along with the assumptions that $a_n \to p$ and $b_n \to q$, we further assume that $q \neq 0$ and $b_n \neq 0$ for all n. Note that as $\frac{a_n}{b_n} = a_n \cdot \frac{1}{b_n}$, and as we have already shown that the product of convergent sequences converges to the product of the sequences' limits, we need only prove that $\frac{1}{b_n} \to \frac{1}{q}$ and the desired result, that $(\frac{a_n}{b_n})$ converges to $\frac{p}{q}$, immediately follows.

Note that as $b_n \to q$ and $q \neq 0$, there exists an index N_1 such that $|b_n| \geq \frac{1}{2}|q|$ whenever $n \geq N_1$. Now let ε be an arbitrary positive number. As $b_n \to q$, there exists an index N_2 for which $N_2 \geq N_1$ and $|b_n - q| < \frac{1}{2}\varepsilon|q|^2$ whenever $n \geq N_2$. Then for all n such that $n \geq N_2$, it follows that

$$\left|\frac{1}{b_n} - \frac{1}{q}\right| = \left|\frac{q - b_n}{qb_n}\right| = \frac{1}{|q|} \cdot \frac{1}{|b_n|} \cdot |b_n - q| < \frac{1}{|q|} \cdot \frac{2}{|q|} \cdot \frac{1}{2}\varepsilon|q|^2 = \varepsilon.$$

Thus, we may conclude that $\frac{1}{b_n} \to \frac{1}{q}$. □

As promised, we now discuss the development of the arguments employed to establish Theorem 10.13. In proving (1), that the limit of a sum of convergent sequences is the sum of the limits of these sequences, we have to think about how the distance

$$\left|(a_n + b_n) - (p + q)\right|$$

between the terms $a_n + b_n$ of the sum sequence and the number $p + q$ is related to the distances

$$|a_n - p|$$

and

$$|b_n - q|$$

between the terms of the sequences (a_n) and (b_n) being added and their respective limits, p and q. Rewriting $|(a_n + b_n) - (p + q)|$ as

$$\left|(a_n - p) + (b_n - q)\right|$$

and then applying the triangle inequality to get

$$\left|(a_n - p) + (b_n - q)\right| \leq |a_n - p| + |b_n - q|$$

gives us the relationship we are looking for. We then realize that if we cut the given positive number ε in half, we can make the terms of both of the convergent sequences (a_n) and (b_n) stay a distance less than $\frac{\varepsilon}{2}$ from their respective limits, as long as we move out far enough in these sequences, which then gives us

$$|a_n - p| + |b_n - q| < \frac{\varepsilon}{2} + \frac{\varepsilon}{2} = \varepsilon,$$

which is precisely what we need.

The argument for (3), that the limit of a product of convergent sequences is the product of their limits, is more involved. Here we need to relate

$$|a_n \cdot b_n - p \cdot q|$$

to the quantities

$$|a_n - p|$$

and

$$|b_n - q|,$$

which we do by rewriting a_n as $(a_n - p) + p$ and b_n as $(b_n - q) + q$, so that $|a_n b_n - pq|$ becomes

$$\left|[(a_n - p) + p] \cdot [(b_n - q) + q] - pq\right|,$$

helping to establish a relationship between the quantity $|a_n b_n - pq|$ we hope to "make small" and the quantities $|a_n - p|$ and $|b_n - q|$ we know we can "make small." Multiplying out $[(a_n - p) + p] \cdot [(b_n - q) + q]$ while keeping intact the expressions $(a_n - p)$ and $(b_n - q)$ yields

$$(a_n - p)(b_n - q) + (a_n - p)q + p(b_n - q) + pq,$$

so that $|[(a_n - p) + p] \cdot [(b_n - q) + q] - pq|$ then becomes

$$\left|(a_n - p)(b_n - q) + (a_n - p)q + p(b_n - q)\right|.$$

Next, the triangle inequality tells us that

$$\begin{aligned} &\left|(a_n - p)(b_n - q) + (a_n - p)q + p(b_n - q)\right| \\ &\quad \le \left|(a_n - p)(b_n - q)\right| + \left|(a_n - p)q\right| + \left|p(b_n - q)\right|, \end{aligned}$$

which motivates us to separate the given positive ε into three equal-sized pieces $\frac{\varepsilon}{3}$ and try to determine how to make each of the three terms

$$\left|(a_n - p)(b_n - q)\right| \quad \text{and} \quad \left|(a_n - p)q\right| \quad \text{and} \quad \left|p(b_n - q)\right|$$

on the right side of this inequality less than $\frac{\varepsilon}{3}$.

(1) As $|(a_n - p)(b_n - q)| = |a_n - p||b_n - q|$, we see that $|(a_n - p)(b_n - q)| < \frac{\varepsilon}{3}$ if we can make $|a_n - p| < \sqrt{\frac{\varepsilon}{3}}$ and $|b_n - q| < \sqrt{\frac{\varepsilon}{3}}$.

(2) As $|(a_n-p)q| = |a_n-p||q|$, it at first looks like we can get $|(a_n-p)q| < \frac{\varepsilon}{3}$ if we are able to make $|a_n-p| < \frac{\varepsilon}{3|q|}$, but as it is possible that $q = 0$, we instead make $|a_n-p| < \frac{\varepsilon}{3(|q|+1)}$ so as to avoid the possibility of division by zero (the larger denominator also makes the resulting fraction smaller).
(3) Similarly, as $|p(b_n-q)| = |p||b_n-q|$, we see that we can get $|p(b_n-q)| < \frac{\varepsilon}{3}$ by making $|b_n-q| < \frac{\varepsilon}{3(|p|+1)}$.

As indicated in the proof itself, the natural numbers N_1, N_2, N_3, and N_4 represent index values beyond which the respective conditions we just identified must hold. Then, for indices at least as large as the maximum of these four numbers, all the conditions hold simultaneously and our calculations in the proof show that the plan we have outlined here does indeed succeed.

Finally, let us describe the thinking behind our approach to showing $\frac{1}{b_n} \to \frac{1}{q}$ when it is known that $b_n \to q$ (with q and all b_n nonzero), needed to establish (4) concerning the limit of the quotient of convergent sequences. We require that the quantity

$$\left|\frac{1}{b_n} - \frac{1}{q}\right|$$

be small when the quantity

$$|b_n - q|$$

is appropriately small. But as

$$\left|\frac{1}{b_n} - \frac{1}{q}\right| = \left|\frac{q-b_n}{qb_n}\right| = \frac{1}{|q|} \cdot \frac{1}{|b_n|} \cdot |b_n - q|,$$

we realize that not only must we make $|b_n - q|$ small, but we also have to make $\frac{1}{|b_n|}$ suitably small as well. This is achievable because, as we know the limit q of (b_n) is not zero, the terms of (b_n) from some index onward must all lie in the neighborhood of q having radius $\frac{1}{2}|q|$, a neighborhood that does not include 0 or, in fact, any numbers less than distance $\frac{1}{2}|q|$ from 0. In our proof, therefore, we are able to choose an index N_1 so that when $n \geq N_1$ we have $|b_n| \geq \frac{1}{2}|q|$. It would then follow that

$$\frac{1}{|q|} \cdot \frac{1}{|b_n|} \leq \frac{2}{|q|^2},$$

so in order to get

$$\frac{1}{|q|} \cdot \frac{1}{|b_n|} \cdot |b_n - q| \leq \frac{2}{|q|^2} \cdot |b_n - q| < \varepsilon,$$

we need to choose an index N_2 at least as large as N_1 so that when $n \geq N_1$, we have

$$|b_n - q| < \frac{|q|^2}{2} \cdot \varepsilon,$$

which is possible as $\frac{|q|^2}{2} \cdot \varepsilon > 0$ and $b_n \to q$.

Example 10.14. As we have already shown that the sequence $(\frac{1}{n})$ converges to 0, we can use Theorem 10.13 to conclude that
(1) the sequence $(5 + \frac{1}{n})$ converges to $5 + 0 = 5$;
(2) the sequence $(\frac{5}{n}) = (5 \cdot \frac{1}{n})$ converges to $5 \cdot 0 = 0$; and,
(3) the sequence $(\frac{3-n}{1+2n}) = (\frac{3 \cdot \frac{1}{n} - 1}{\frac{1}{n} + 2})$ converges to $\frac{3 \cdot 0 - 1}{0+2} = -\frac{1}{2}$.

Exercise 10.13. Use the relevant part of Theorem 10.13 to explain why the sequence $(\frac{1}{n+1} - \frac{1}{n})$ converges to 0. (You used the definition of sequential convergence to verify this limit in Exercise 9.12(d).)

Exercise 10.14. Suppose (a_n) and (b_n) are sequences, and both of the sequences (a_n) and $(a_n + b_n)$ converge. Must the sequence (b_n) also converge? Defend your conclusion.

Exercise 10.15. Prove that if (a_n) is a sequence of nonnegative numbers and $a_n \to p$, then $\sqrt{a_n} \to \sqrt{p}$.

The squeeze theorem for sequences

When each term of a sequence lies between the corresponding terms of two other sequences, each of which converges to the same limit, the "trapped" sequence must also converge to this limiting value. In the proof, we use the characterization of sequential convergence via neighborhoods given in Theorem 9.15.

Theorem 10.15 (The squeeze theorem). *If (a_n), (x_n), and (y_n) are sequences such that $x_n \leq a_n \leq y_n$ for every natural number n, and for which both $x_n \to p$ and $y_n \to p$, then it follows that $a_n \to p$.*

Proof. Assume (a_n), (x_n), and (y_n) are sequences with $x_n \leq a_n \leq y_n$ for all natural numbers n, and that $x_n \to p$ and $y_n \to p$. Consider any positive number ε. Then there exist indices N_1 and N_2 for which $x_n \in (p - \varepsilon, p + \varepsilon)$ when $n \geq N_1$ and $y_n \in (p - \varepsilon, p + \varepsilon)$ when $n \geq N_2$. Let $N = \max\{N_1, N_2\}$ and suppose $n \geq N$. Then

$$p - \varepsilon < x_n \leq a_n \leq y_n < p + \varepsilon,$$

so that $a_n \in (p - \varepsilon, p + \varepsilon)$. Thus, $a_n \to p$. □

Example 10.16. As $0 < \frac{1}{2^n} < \frac{1}{n}$ for every natural number n (see Exercise 3.13), and both the constant zero sequence (0) and the sequence $(\frac{1}{n})$ of reciprocals of the natural numbers converge to 0, the squeeze theorem permits us to conclude that the sequence $(\frac{1}{2^n})$ also converges to 0.

Exercise 10.16. Use the squeeze theorem to show that the sequence $(\frac{(-1)^{n+1}}{4n+1})$ converges to 0.

Exercise 10.17. Suppose $a_n \to p$ and $b_n \to q$, and $a_n \leq b_n$ for every natural number n. Show that $p \leq q$.

Exercise 10.18. Use the result of Exercise 10.17 to deduce each of the following.
(a) If $a_n \to p$ and $a_n \geq 0$ for all natural numbers n, then $p \geq 0$.
(b) If $a_n \to p$ and $u \leq a_n \leq v$ for all natural numbers n, then $u \leq p \leq v$.

Exercise 10.19. Show the sequence $(\sqrt[n]{n})$ converges to 1. *Hint*: Let $s_n = \sqrt[n]{n}-1$. Rewrite this equation and then apply Theorem 3.23, the binomial theorem. Consider in particular the term involving s_n^2 in order to obtain an inequality involving s_n. Then apply the squeeze theorem.

Exercise 10.20. Show that if $a > 0$, then the sequence $(\sqrt[n]{a})$ converges to 1. *Hint*: Consider separately the cases $a = 1$, $a > 1$, and $0 < a < 1$. For $a > 1$, use a strategy similar to that described in the hint for Exercise 10.19. For $0 < a < 1$, note that $\frac{1}{a} > 1$.

Monotone sequences

A sequence (a_n) is said to be
(1) **increasing** if $a_{n+1} > a_n$ for all natural numbers n (i. e., the terms are always becoming larger);
(2) **decreasing** if $a_{n+1} < a_n$ for all natural numbers n (i. e., the terms are always becoming smaller);
(3) **nondecreasing** if $a_{n+1} \geq a_n$ for all natural numbers n (the terms never become smaller);
(4) **nonincreasing** if $a_{n+1} \leq a_n$ for all natural numbers n (i. e., the terms never become larger).

Note that an increasing sequence is nondecreasing, and a decreasing sequence is nonincreasing. A sequence is called **monotone** if it is either nondecreasing or nonincreasing.

Example 10.17. The sequence $(1, \frac{1}{2}, \frac{1}{3}, \frac{1}{4}, \dots)$ of reciprocals of the natural numbers is decreasing, hence, is monotone.

The Fibonacci sequence $(1, 1, 2, 3, 5, 8, 13, \dots)$ from Examples 9.6 and 10.6 is nondecreasing, hence, monotone, but is not increasing as its first two terms are equal.

The alternating sequence $(-1, 1, -1, 1, \ldots)$ is neither nondecreasing nor nonincreasing, hence, is not monotone.

Exercise 10.21. Determine whether the given sequence is monotone. If it is monotone, tell whether it is any of the following: increasing, decreasing, nonincreasing, nondecreasing.

(a) $(\frac{2-n}{n+3})$;

(b) $(\frac{(-1)^n}{n})$;

(c) the constant sequence $(4, 4, 4, \ldots)$;

(d) the sequence x defined so that $x_n = \begin{cases} \frac{1}{2}(n+1), & \text{if } n \text{ is odd,} \\ \frac{1}{2}n, & \text{if } n \text{ is even.} \end{cases}$

Among the convergent sequences are the bounded monotone sequences. As we would anticipate, a bounded nondecreasing sequence converges to the least upper bound of its terms, while a bounded nonincreasing sequence converges to the greatest lower bound of its terms.

Theorem 10.18 (The monotone convergence theorem). *A monotone sequence converges if and only if it is bounded. More specifically, we have the following results.*

(1) *If (a_n) is a bounded nondecreasing sequence, then $a_n \to p$, where $p = \sup\{a_n \mid n \in \mathbb{N}\}$.*

(2) *If (a_n) is a bounded nonincreasing sequence, then $a_n \to p$, where $p = \inf\{a_n \mid n \in \mathbb{N}\}$.*

Proof. A monotone sequence that converges must be bounded, since according to Theorem 10.7 any convergent sequence is bounded. A monotone sequence is either nondecreasing or nonincreasing, so to complete the proof we need to show (1) and (2). We provide an argument for (1); the proof of (2) is assigned as Exercise 10.23.

Suppose (a_n) is a bounded nondecreasing sequence. Thus, $a_1 \le a_2 \le a_3 \le \ldots$ and the set $A = \{a_n \mid n \in \mathbb{N}\}$ has an upper bound, hence, a supremum; let $p = \sup(A)$. To show $a_n \to p$, we begin with an arbitrary positive number ε. As p is the least upper bound of A, the number $p - \varepsilon$ is not an upper bound of A, so there exists a natural number N such that $a_N > p - \varepsilon$. Now as (a_n) is nondecreasing and p is an upper bound of A, it follows that for all n where $n \ge N$, we have

$$p - \varepsilon < a_N \le a_n \le p < p + \varepsilon,$$

from which we may conclude that $a_n \to p$. □

Example 10.19. We can use the monotone convergence theorem to show that the sequence $(\frac{1}{\sqrt[3]{n}})$ converges to 0. First, note that as

$$\frac{1}{\sqrt[3]{1}} > \frac{1}{\sqrt[3]{2}} > \frac{1}{\sqrt[3]{3}} > \cdots,$$

we see that $(\frac{1}{\sqrt[3]{n}})$ is a decreasing sequence. Since the range $A = \{\frac{1}{\sqrt[3]{n}} \mid n \in \mathbb{N}\}$ of this sequence is bounded below by 0, the monotone convergence theorem tells us that the sequence converges to the greatest lower bound of A. We show this greatest lower bound is 0, so it then follows that $\frac{1}{\sqrt[3]{n}} \to 0$.

To see that $\inf(A) = 0$, having already observed that 0 is a lower bound of A, we need only show that no positive number is a lower bound of A. So consider any positive number ε. By the Archimedean Property, there exists a natural number N such that $\frac{1}{N} < \varepsilon^3$, from which it follows that $\frac{1}{\sqrt[3]{n}} < \varepsilon$. Thus, ε is not a lower bound of A and we have $\inf(A) = 0$.

Exercise 10.22. Give an example of a decreasing sequence of real numbers that converges to 5. Then give an example of an increasing sequence of real numbers that converges to 5.

Exercise 10.23. Prove (2) of Theorem 10.18.

Exercise 10.24. Show that an increasing sequence that is not bounded above must diverge to ∞.

Exercise 10.25. Show that the sequence $(\sqrt{n+1} - \sqrt{n})$ must converge.

Example 10.20. We show that if $0 < r < 1$, then the sequence

$$(r^n) = (r, r^2, r^3, r^4, \dots)$$

converges to 0. Note first that this sequence is decreasing and bounded below by 0; thus, by the monotone convergence theorem, (r^n) converges to p, the greatest lower bound of $A = \{r^n \mid n \in \mathbb{N}\}$.

Suppose to the contrary that $p > 0$. Then $\varepsilon = \frac{p(1-r)}{r}$ is positive, so as $r^n \to p$, there is a natural number N for which $r^N < p + \varepsilon$. It follows that

$$r^{N+1} < r(p+\varepsilon) = rp + r\left(\frac{p(1-r)}{r}\right) = p,$$

a contradiction to the fact that p is a lower bound for A. Hence, $p = 0$.

Exercise 10.26. Prove that if $r > 1$, then the sequence $(r^n) = (r, r^2, r^3, r^4, \dots)$ diverges to ∞.

Exercise 10.27. Assume $a_n > 0$ for every natural number n and $\lim \frac{a_{n+1}}{a_n} = L$ for some (necessarily nonnegative) real number L.

(a) Prove that if $L < 1$, then the sequence (a_n) converges to 0.
(b) Prove that if $L > 1$, then the sequence (a_n) diverges to ∞.
(c) Give examples showing that if $L = 1$, it is possible for the sequence (a_n) to converge or diverge.

Exercise 10.28. A sequence is **eventually monotone** if one of its tails is monotone. Explain why a sequence that is bounded and eventually monotone must converge.

Exercise 10.29. The Bakhshali manuscript, an ancient Indian mathematical text dating to before the year 1000, contains the following method for approximating square roots.

Suppose $p > 0$ and recursively define the sequence a so that

$$a_1 = 1$$

and

$$a_{n+1} = \frac{a_n^2 + p}{2a_n}$$

for every natural number n.

(a) Show the sequence a converges. *Hint*: First show $a_n^2 \geq p$ when $n \geq 2$. Then use this result to show that $a_{n+1} \leq a_n$ when $n \geq 2$. Finally, apply the result of Exercise 10.28.
(b) Show that $\lim a_n = \sqrt{p}$. *Hint*: Apply appropriate arithmetic properties of convergent sequences to the recurrence $a_{n+1} = \frac{a_n^2+p}{2a_n}$.
(c) Show that when $n \geq 2$ we have $0 \leq a_n - \sqrt{p} \leq \frac{a_n^2-p}{a_n}$. This inequality permits $\sqrt{p}$ to be approximated with any desired degree of accuracy.
(d) Approximate $\sqrt{2}$ accurately to four decimal places.

Whenever a nonempty subset of $\mathbb{R}$ is bounded above (respectively, below) we can create a sequence in the subset that converges to the least upper bound (respectively, greatest lower bound) of the subset.

Theorem 10.21. *Let A be a nonempty subset of $\mathbb{R}$.*
(1) *If A is bounded above, then there exists a nondecreasing sequence (a_n) in A such that $a_n \to \sup(A)$.*
(2) *If A is bounded below, then there exists a nonincreasing sequence (a_n) in A such that $a_n \to \inf(A)$.*

Exercise 10.30. Prove (1) of Theorem 10.21.

Exercise 10.31. The density of the rational numbers and the irrational numbers (see Theorem 5.17) in $\mathbb{R}$ permits us, given any real number p, to find sequences (a_n) in $\mathbb{Q}$ and (b_n) in $\mathbb{R} - \mathbb{Q}$, none of whose terms are equal to p, such that $a_n \to p$ and $b_n \to p$. Prove that this is indeed the case.

Exercise 10.32. Define the sequence u so that $u_n = (\frac{1}{n} + 1)^n$.

(a) Use Theorem 3.23, the binomial theorem, to show that

$$u_n = 1 + 1 + \sum_{k=2}^{n} \frac{1}{k!} \cdot \left(1 - \frac{1}{n}\right) \cdot \left(1 - \frac{2}{n}\right) \cdot \dots \cdot \left(1 - \frac{k-1}{n}\right)$$

and

$$u_{n+1} = 1 + 1 + \sum_{k=2}^{n} \frac{1}{k!} \cdot \left(1 - \frac{1}{n+1}\right) \cdot \left(1 - \frac{2}{n+1}\right) \cdot \dots \cdot \left(1 - \frac{k-1}{n+1}\right) + \frac{1}{(n+1)^{n+1}}.$$

(b) Use (a) to show that $u_n < u_{n+1}$ for every natural number n; hence, the sequence u is increasing.

(c) From (a) it follows that $2 < u_n$ for every n. Use (a) to show that $u_n < 3$ for every n. Thus, the sequence u is bounded. *Hint*: Make use of the results from Exercise 3.17 and Theorem 3.21(6).

As the sequence u is increasing and bounded, we may conclude, using the monotone convergence theorem, that u converges to the least upper bound of its range. Moreover, Exercise 10.18(b) permits us to conclude that this limit is between 2 and 3. The number to which the sequence u converges can be taken as the definition for **Euler's number** e. The terms of the sequence u provide approximations to the numerical value of e, with u_n producing a better approximation as n becomes larger. For example,

$$u_{1000} = \left(\frac{1}{1000} + 1\right)^{1000} \approx 2.717,$$

which gives an approximation to e that is accurate to two decimal places.

An introduction to infinite series

Imagine walking across a room in such a way that you walk halfway across in your first move and half the remaining distance in each successive move. Intuitively, this suggests that

$$\frac{1}{2} + \frac{1}{4} + \frac{1}{8} + \cdots = 1$$

since the sequence

$$\left(\frac{1}{2}, \frac{1}{2} + \frac{1}{4}, \frac{1}{2} + \frac{1}{4} + \frac{1}{8}, \frac{1}{2} + \frac{1}{4} + \frac{1}{8} + \frac{1}{16}, \dots\right) = (0.5, 0.75, 0.875, 0.9375, \dots)$$

of "partial sums" appears to converge to 1.

Given a real sequence (a_n), the **(infinite) series**

$$a_1 + a_2 + a_3 + \cdots = \sum_{n=1}^{\infty} a_n$$

with summands $a_1, a_2, a_3, \ldots$ is defined to be the sequence

$$(a_1, a_1 + a_2, a_1 + a_2 + a_3, \ldots)$$

whose nth term is

$$a_1 + a_2 + \cdots + a_n = \sum_{k=1}^{n} a_k,$$

which is also referred to as the n**th partial sum** of the series. Thus, by definition, the series $\sum_{n=1}^{\infty} a_n$ is the sequence $(\sum_{k=1}^{n} a_k)$ consisting of its partial sums.

If the sequence of partial sums of the series $\sum_{n=1}^{\infty} a_n$ converges to a real number S, we say the series **converges**, we refer to S as the **sum** of the series, and we write $\sum_{n=1}^{\infty} a_n = S$. If the sequence of partial sums diverges, we say the series itself **diverges**.

Note that when

$$\sum_{n=1}^{\infty} a_n = a_1 + a_2 + a_3 + \cdots = S$$

for some number S, we have essentially used the notion of sequential convergence to perform an addition that incorporates infinitely many summands.

Example 10.22. The series

$$\sum_{n=1}^{\infty} \frac{1}{2^n}$$

is the sequence

$$\left(\frac{1}{2}, \frac{1}{2} + \frac{1}{4}, \frac{1}{2} + \frac{1}{4} + \frac{1}{8}, \ldots\right),$$

where $\frac{1}{2}$ is the first partial sum of the series, $\frac{1}{2} + \frac{1}{4}$ is the second partial sum of the series, $\frac{1}{2} + \frac{1}{4} + \frac{1}{8}$ is the third partial sum of the series, and so on. Calculating these partial sums reveals that

$$\left(\frac{1}{2}, \frac{1}{2} + \frac{1}{4}, \frac{1}{2} + \frac{1}{4} + \frac{1}{8}, \ldots\right) = \left(\frac{1}{2}, \frac{3}{4}, \frac{7}{8}, \ldots\right) = \left(\frac{2^n - 1}{2^n}\right).$$

Because the sequence $(\frac{2^n-1}{2^n})$ of partial sums of the series converges to 1, we can say that the series itself converges to 1 and has sum 1, which permits us to write

$$\sum_{n=1}^{\infty} \frac{1}{2^n} = 1.$$

Example 10.23. The series

$$\sum_{n=1}^{\infty} \frac{n+2}{n+1}$$

is the sequence

$$\left(\frac{3}{2}, \frac{3}{2} + \frac{4}{3}, \frac{3}{2} + \frac{4}{3} + \frac{5}{4}, \dots\right).$$

Observe that as $\frac{k+2}{k+1} > 1$ for every natural number k, it follows that

$$\sum_{k=1}^{n} \frac{k+2}{k+1} > n,$$

that is, the nth partial sum of the series is greater than n. Therefore, the sequence of partial sums diverges to ∞, meaning the series itself diverges to ∞.

Exercise 10.33. Find the first four partial sums of the given series.
(a) $\sum_{n=1}^{\infty} n$;
(b) $\sum_{n=0}^{\infty} (\frac{1}{3})^n$;
(c) $\sum_{n=1}^{\infty} \frac{(-1)^n}{n}$.

Convergent infinite series behave in predictable ways with respect to sums, differences, and constant multiples.

Theorem 10.24. *Suppose the series $\sum_{n=1}^{\infty} a_n$ and $\sum_{n=1}^{\infty} b_n$ both converge, with $\sum_{n=1}^{\infty} a_n = S$ and $\sum_{n=1}^{\infty} b_n = T$. Then each of the series $\sum_{n=1}^{\infty}(a_n + b_n)$ and $\sum_{n=1}^{\infty}(a_n - b_n)$ also converges, with*

$$\sum_{n=1}^{\infty}(a_n + b_n) = \sum_{n=1}^{\infty} a_n + \sum_{n=1}^{\infty} b_n = S + T$$

and

$$\sum_{n=1}^{\infty}(a_n - b_n) = \sum_{n=1}^{\infty} a_n - \sum_{n=1}^{\infty} b_n = S - T.$$

Proof. As $\sum_{n=1}^{\infty} a_n = S$ and $\sum_{n=1}^{\infty} b_n = T$, the sequence $(\sum_{k=1}^{n} a_k)$ converges to S and the sequence $(\sum_{k=1}^{n} b_k)$ converges to T. So, as $\sum_{k=1}^{n}(a_k \pm b_k) = \sum_{k=1}^{n} a_k \pm \sum_{k=1}^{n} b_k$, it follows, using arithmetic properties of convergent sequences, that the sequence $(\sum_{k=1}^{n}(a_k + b_k))$ converges to $S+T$ and the sequence $(\sum_{k=1}^{n}(a_k - b_k))$ converges to $S-T$. Thus, by definition, the sum of the series $\sum_{n=1}^{\infty}(a_n + b_n)$ is $S+T$ and the sum of the series $\sum_{n=1}^{\infty}(a_n - b_n)$ is $S-T$, meaning $\sum_{n=1}^{\infty}(a_n \pm b_n) = \sum_{n=1}^{\infty} a_n \pm \sum_{n=1}^{\infty} b_n = S \pm T$. □

Theorem 10.25. *Suppose the series $\sum_{n=1}^{\infty} a_n$ converges, with $\sum_{n=1}^{\infty} a_n = S$. Then, for any real number constant c, the series $\sum_{n=1}^{\infty} ca_n$ converges, with*

$$\sum_{n=1}^{\infty} ca_n = c\sum_{n=1}^{\infty} a_n = cS.$$

Exercise 10.34. Prove Theorem 10.25.

We shall conduct a more comprehensive investigation of infinite series beginning in Chapter 22.

11 Subsequences

Taking infinitely many terms of a sequence in such a way as to maintain the order in which these terms appeared, by index value, in the original sequence produces what is called a *subsequence* of the sequence. Even if a sequence does not converge, it may contain a subsequence that converges and, in fact, may contain a variety of subsequences that converge to different limits. Our study of subsequences in this chapter also permits us to prove the important Cauchy criterion for convergence of a sequence.

Subsequences of a sequence

We obtain what is referred to as a *subsequence* of a sequence by taking infinitely many terms of the sequence and keeping them in the same relative order as they appeared in the original sequence according to their index values. That is, the formation of a subsequence allows for the deletion of terms in the original sequence as long as infinitely many terms are retained and kept in the same relative order.

Example 11.1. Each of the following sequences is a subsequence of the sequence $(1, \frac{1}{2}, \frac{1}{3}, \frac{1}{4}, \ldots)$ of reciprocals of the natural numbers:
(1) the tail $(\frac{1}{3}, \frac{1}{4}, \frac{1}{5}, \ldots)$ (in fact, any tail of a sequence is a subsequence of that sequence);
(2) the sequence $(1, \frac{1}{3}, \frac{1}{5}, \ldots)$ obtained by retaining only the odd-indexed terms of the original sequence (or, equivalently, by deleting all of the even-indexed terms);
(3) the sequence $(\frac{1}{1}, \frac{1}{2\cdot 1}, \frac{1}{3\cdot 2\cdot 1}, \frac{1}{4\cdot 3\cdot 2\cdot 1}, \frac{1}{5\cdot 4\cdot 3\cdot 2\cdot 1}, \ldots) = (\frac{1}{n!})$.

More formally, if (a_n) is a sequence and $k_1 < k_2 < k_3 < \cdots$ forms an increasing sequence of natural numbers, then the sequence

$$(a_{k_1}, a_{k_2}, a_{k_3}, \ldots)$$

is called a **subsequence** of (a_n). This subsequence can also be represented as (a_{k_n}).

Note also that, according to this definition, a sequence is always considered to be a subsequence of itself.

Example 11.1 (Continued). Let a be the sequence

$$(a_1, a_2, a_3, \ldots) = \left(1, \frac{1}{2}, \frac{1}{3}, \frac{1}{4}, \ldots\right).$$

Then
(1) the subsequence $(\frac{1}{3}, \frac{1}{4}, \frac{1}{5}, \ldots)$ of a is $(a_3, a_4, a_5, \ldots)$,
(2) the subsequence $(1, \frac{1}{3}, \frac{1}{5}, \ldots)$ of a is $(a_1, a_3, a_5, \ldots)$, and
(3) the subsequence $(\frac{1}{1}, \frac{1}{2\cdot 1}, \frac{1}{3\cdot 2\cdot 1}, \frac{1}{4\cdot 3\cdot 2\cdot 1}, \frac{1}{5\cdot 4\cdot 3\cdot 2\cdot 1}, \ldots)$ of a is $(a_1, a_2, a_6, a_{24}, a_{120}, \ldots)$.

https://doi.org/10.1515/9783111429564-011

For any subsequence (a_{k_n}) of a sequence (a_n), we may observe that k is actually just an increasing function from $\mathbb{N}$ into $\mathbb{N}$ that picks out the indices of the terms that are included in the subsequence.

Example 11.1 (Continued). For the subsequence

$$(a_3, a_4, a_5, \dots) = \left(\frac{1}{3}, \frac{1}{4}, \frac{1}{5}, \dots\right)$$

of a, the function $k : \mathbb{N} \to \mathbb{N}$ is defined so that $k(n) = k_n = n + 2$, whereas for the subsequence

$$(a_1, a_3, a_5, \dots) = \left(1, \frac{1}{3}, \frac{1}{5}, \dots\right)$$

of a, the function $k : \mathbb{N} \to \mathbb{N}$ is defined so that $k(n) = k_n = 2n - 1$.

Exercise 11.1. Identify the function k used to create the subsequence $(\frac{1}{1}, \frac{1}{2\cdot1}, \frac{1}{3\cdot2\cdot1}, \frac{1}{4\cdot3\cdot2\cdot1}, \frac{1}{5\cdot4\cdot3\cdot2\cdot1}, \dots)$ of the sequence $(1, \frac{1}{2}, \frac{1}{3}, \frac{1}{4}, \dots)$.

Based on the formal definition, there are some things we are not allowed to do when creating a subsequence.

Example 11.1 (Continued). When forming a subsequence of a sequence, we are not permitted to rearrange the terms of the original sequence. For example,

$$\left(\frac{1}{2}, 1, \frac{1}{3}, \frac{1}{4}, \dots\right)$$

is not a subsequence of $(1, \frac{1}{2}, \frac{1}{3}, \frac{1}{4}, \dots)$.

Neither are we allowed to insert values into a sequence that were not present to begin with. For instance,

$$\left(0, 1, 0, \frac{1}{2}, 0, \frac{1}{3}, \dots\right)$$

is not a subsequence of $(1, \frac{1}{2}, \frac{1}{3}, \frac{1}{4}, \dots)$.

Also, we are not allowed to insert additional repeats of values appearing in a sequence when forming a subsequence. For instance,

$$\left(1, 1, \frac{1}{2}, \frac{1}{2}, \frac{1}{3}, \frac{1}{3}, \dots\right)$$

is not a subsequence of $(1, \frac{1}{2}, \frac{1}{3}, \frac{1}{4}, \dots)$.

Exercise 11.2. Which of the following sequences might be subsequences of the Fibonacci sequence from Example 9.6?
(a) $(2, 3, 5, 8, 13, \dots)$;
(b) $(2, 1, 3, 1, 5, 8, 13, \dots)$;
(c) $(-1, -1, -2, -3, -5, -8, -13, \dots)$;
(d) $(2, 8, 34, 144 \dots)$;
(e) $(2, 3, 4, 5, 8, 13, \dots)$;
(f) $(1, 1, 3, 5, 13, 21, \dots)$.

When a sequence converges, so do all its subsequences, and to the same limit.

Theorem 11.2. *The sequence (a_n) converges to the real number p if and only if every subsequence of (a_n) converges to p.*

Proof. ($\Rightarrow$) Suppose $a_n \to p$, consider any subsequence (a_{k_n}) of (a_n), and let ε be any positive number. As $a_n \to p$, there is an index N such that when $n \geq N$, we have $|a_n - p| < \varepsilon$. But as (k_n) is an increasing sequence in $\mathbb{N}$, it follows that when $n \geq N$, we must have $k_n \geq k_N \geq N$, so that $|a_{k_n} - p| < \varepsilon$. Thus, $a_{k_n} \to p$.

($\Leftarrow$) Assume that every subsequence of (a_n) converges to p. Then, as (a_n) is a subsequence of itself, it follows that (a_n) converges to p. □

One important consequence of this theorem is that a sequence necessarily diverges if it possesses two subsequences that converge to different limits.

Corollary 11.3. *If a sequence has two subsequences that converge to different limits, then the sequence diverges.*

Example 11.4. As the alternating sequence

$$a = (-1, 1, -1, 1, \dots) = ((-1)^n)$$

has a subsequence

$$(a_{2n-1}) = (-1, -1, -1, \dots)$$

that converges to -1, as well as a subsequence

$$(a_{2n}) = (1, 1, 1, \dots)$$

that converges to 1, we may conclude that the sequence a itself diverges.

Exercise 11.3. Define the sequence a so that

$$a_n = \begin{cases} 1 - 2^{-(n+1)/2}, & \text{if } n \text{ is odd;} \\ \frac{1}{n}, & \text{if } n \text{ is even.} \end{cases}$$

Show that this sequence diverges.

Exercise 11.4. For each natural number n, suppose $I_n = [a_n, b_n]$ is a closed bounded interval. Also suppose that $\bigcap_{n=1}^{\infty} I_n = \{p\}$ for some real number p.
(a) Show that if the intervals form a nested sequence with $I_1 \supseteq I_2 \supseteq I_3 \supseteq \cdots$, then $a_n \to p$ and $b_n \to p$.
(b) Give an example to show that if the intervals are not nested as in (a), then it is possible that neither of the sequences (a_n) or (b_n) converges to p.

Exercise 11.5. Let (a_n) be a sequence for which both of the subsequences (a_{2n-1}) and (a_{2n}) converge to the same number p. Show that (a_n) converges to p.

Recall that the Bolzano–Weierstrass theorem, Theorem 7.5, tells us that every bounded infinite subset of $\mathbb{R}$ has a limit point. The original formulation of this theorem states that a bounded sequence of real numbers must have a convergent subsequence. These two versions of the Bolzano–Weierstrass theorem are in fact equivalent.

Theorem 11.5 (The Bolzano–Weierstrass theorem). *Every bounded sequence of real numbers has a convergent subsequence.*

Proof. Consider any bounded sequence (a_n). If $A = \{a_n \mid n \in \mathbb{N}\}$ is finite, then there exists a real number p such that $a_n = p$ for infinitely many n, thus making the constant sequence (p) a convergent subsequence of (a_n). On the other hand, if A is infinite, our earlier version of the Bolzano–Weierstrass theorem tells us, since A is also bounded, that A has a limit point p. Hence, for each natural number n, there exists a natural number k_n such that $a_{k_n} \in (p - \frac{1}{n}, p + \frac{1}{n})$ and such that $k_1 < k_2 < k_3 < \cdots$. It follows that (a_{k_n}) is a subsequence of (a_n) and (a_{k_n}) converges to p (see Exercise 11.6). □

Exercise 11.6. Provide the details showing that the subsequence (a_{k_n}) created in the proof of Theorem 11.5 actually does converge to p.

Exercise 11.7. Give an example of an unbounded sequence that has a convergent subsequence. Does the existence of such a sequence contradict the Bolzano–Weierstrass theorem for sequences?

Exercise 11.8. We used Theorem 7.5 to prove Theorem 11.5. Show that Theorem 11.5 can be used to prove Theorem 7.5. Thus, the two theorems are equivalent.

Exercise 11.9. Prove that a subset A of $\mathbb{R}$ is compact if and only if every sequence in A has a subsequence that converges to a point of A.

Exercise 11.10. Prove that if (a_n) is a sequence and p is a limit point of $\{a_n \mid n \in \mathbb{N}\}$, then (a_n) has a subsequence that converges to p.

Exercise 11.11. Show that if the sequence (a_n) has a subsequence that converges to p and p is not a limit point of $\{a_n \mid n \in \mathbb{N}\}$, then $a_n = p$ for infinitely many n.

Exercise 11.12. Show that if (a_n) is a bounded sequence of distinct real numbers and $\{a_n \mid n \in \mathbb{N}\}$ has exactly one limit point p, then (a_n) converges to p.

Limit superior and limit inferior of a sequence

When a sequence diverges but is bounded, the Bolzano–Weierstrass theorem guarantees the sequence has a convergent subsequence. Actually, the theorem guarantees such a sequence must have at least two convergent subsequences that converge to different limits.

Theorem 11.6. *Every bounded divergent sequence of real numbers has at least two convergent subsequences having different limits.*

Proof. Assume the sequence (a_n) is bounded and does not converge. By the Bolzano–Weierstrass theorem, there is a subsequence (a_{k_n}) of (a_n) for which $a_{k_n} \to p$ for some real number p. But since (a_n) does not converge to p, there is a positive number ε and an increasing sequence (j_n) of natural numbers such that for every n we have $a_{j_n} \notin (p-\varepsilon, p+\varepsilon)$. The sequence (a_{j_n}) is bounded, being a subsequence of the bounded sequence (a_n), and thus by the Bolzano–Weierstrass theorem, has a convergent subsequence (b_n), which is itself a subsequence of (a_n). But (b_n) cannot converge to p as none of its terms are in the neighborhood $(p-\varepsilon, p+\varepsilon)$ of p. Hence, (b_n) converges to a number q different from p. Therefore, we have found two convergent subsequences of (a_n), the subsequence (a_{k_n}) which converges to p and the subsequence (b_n) which converges to q, where $p \neq q$. □

Example 11.4 (Continued). The alternating sequence $a = (-1, 1, -1, 1, \dots)$ is bounded and divergent, but has a subsequence that converges to -1 and another subsequence that converges to 1.

A real number is called a **subsequential limit** of a sequence if the sequence has a subsequence that converges to the specified number.

Example 11.4 (Continued). As the alternating sequence $a = (-1, 1, -1, 1, \ldots)$ has subsequences converging to -1 and 1, each of the numbers -1 and 1 is a subsequential limit of a.

Exercise 11.13. Explain why a convergent sequence has precisely one subsequential limit.

Just as we say a sequence is bounded if its range is bounded, we also say a sequence is **bounded above** if its range is bounded above (i. e., has an upper bound) and is **bounded below** if its range is bounded below (i. e., has a lower bound). In analyzing the convergent subsequences of a bounded divergent sequence, it is convenient to make use of the concepts of *limit superior* and *limit inferior* of a sequence.

Given a sequence (a_n) that is bounded above, the **limit superior** of (a_n), denoted $\limsup a_n$, is defined so that

$$\limsup a_n = \lim s_n,$$

where s is the sequence whose nth term is $s_n = \sup\{a_k \mid k \geq n\}$, assuming the sequence s converges. In Exercise 11.16 below, you show the sequence s is nonincreasing, which means that if s does not converge it is because s diverges to $-\infty$, in which case we may write

$$\limsup a_n = -\infty.$$

Given a sequence (a_n) that is bounded below, the **limit inferior** of (a_n), denoted $\liminf a_n$, is defined so that

$$\liminf a_n = \lim i_n,$$

where i is the sequence whose nth term is $i_n = \inf\{a_k \mid k \geq n\}$, assuming the sequence i converges. In Exercise 11.16 below you show the sequence i is nondecreasing, which means that if i does not converge it is because i diverges to ∞, in which case we may write

$$\liminf a_n = \infty.$$

Thus, the limit superior of a sequence is the limiting value of the suprema of tails of the sequence, while the limit inferior is the limiting value of the infima of tails of the sequence. The uniqueness of sequential limits guarantees a sequence cannot have more than one limit superior or more than one limit inferior.

When a sequence (a_n) is not bounded above, we may write

$$\limsup a_n = \infty$$

and when (a_n) is not bounded below, we may write

$$\liminf a_n = -\infty.$$

Keep in mind that the use of the symbols ∞ and $-\infty$ does not mean that a limit superior or a limit inferior exists. These are notational conveniences only and help to convey in what sense a limit superior or limit inferior does not exist.

Example 11.4 (Continued). For the alternating sequence

$$a = (-1, 1, -1, 1, \ldots) = ((-1)^n),$$

observe that for every n we have

$$\sup\{(-1)^k \mid k \geq n\} = 1$$

and

$$\inf\{(-1)^k \mid k \geq n\} = -1.$$

Therefore, the sequence s for which $s_n = \sup\{(-1)^k \mid k \geq n\}$ is the constant sequence $(1, 1, 1, \ldots)$, so that

$$\limsup a_n = \lim s_n = 1,$$

and the sequence i for which $i_n = \inf\{a_k \mid k \geq n\}$ is the constant sequence $(-1, -1, -1, \ldots)$ so that

$$\liminf a_n = \lim i_n = -1.$$

Example 11.7. Define the sequence a so that

$$a_n = \frac{1}{n} + r_n,$$

where r_n is the remainder 0, 1, or 2 that results when the division algorithm, Theorem 3.7, is applied to divide n by 3. Note that for all natural numbers i, j, and k, we have

$$0 < \frac{1}{i} < \frac{1}{j} + 1 < \frac{1}{k} + 2 \leq 3,$$

so both the limit superior and limit inferior of a exist. We now show $\limsup a_n = 2$.

If we let $b_n = \frac{1}{n} + 2$, it follows using the inequality above that $a_n \leq b_n$ for each natural number n. Hence, taking $s_n = \sup\{a_k \mid k \geq n\}$, we have

$$s_n \leq \sup\{b_k \mid k \geq n\} = \frac{1}{n} + 2,$$

from which we may conclude that

$$\limsup a_n = \lim s_n \leq \lim\left(\frac{1}{n} + 2\right) = 2.$$

Now consider any natural number n. We know that $r_n \in \{0, 1, 2\}$. If $r_n = 0$, then $r_{n+2} = 2$, so that $a_{n+2} = \frac{1}{n+2} + 2 > 2$. If $r_n = 1$, then $r_{n+1} = 2$, so that $a_{n+1} = \frac{1}{n+1} + 2 > 2$. If $r_n = 2$, then $a_n = \frac{1}{n} + 2 > 2$. Thus, for any n, there exists k for which $k \geq n$ and $a_k > 2$. Therefore, for any n, we have

$$s_n = \sup\{a_k \mid k \geq n\} \geq 2,$$

from which it follows that

$$\limsup a_n = \lim s_n \geq 2.$$

As we have shown that $\limsup a_n \leq 2$ and $\limsup a_n \geq 2$, we may now conclude that $\limsup a_n = 2$.

Example 11.8. The sequence (n^2) is not bounded above, so $\limsup n^2 = \infty$. This sequence is bounded below by 1, but since $i_n = \inf\{k^2 \mid k \geq n\} = n^2$, the sequence (i_n) diverges to ∞, and so $\liminf n^2 = \infty$.

The sequence $(-n^2)$ is not bounded below, so we have $\liminf -n^2 = -\infty$. This sequence is bounded above by -1, but since $s_n = \sup\{-k^2 \mid k \geq n\} = -n^2$, the sequence (s_n) diverges to $-\infty$, and so $\limsup -n^2 = -\infty$.

The sequence

$$((-1)^n n^2) = (-1, 4, -9, 16, -25, 36, \ldots)$$

is not bounded above, so $\limsup(-1)^n n^2 = \infty$, and is also not bounded below, so $\liminf(-1)^n n^2 = -\infty$.

Exercise 11.14. Define the sequence (a_n) so that $a_n = \frac{1-(-1)^n n}{2n+1}$. Find the limits superior and inferior of (a_n), justifying your conclusions.

Exercise 11.15. Prove that $\liminf a_n = 0$ for the sequence (a_n) of Example 11.7.

Exercise 11.16. Prove each of the following.

(a) If the sequence (a_n) is bounded above, then the sequence s whose nth term is $s_n = \sup\{a_k \mid k \geq n\}$ is nonincreasing.

(b) If the sequence (a_n) is bounded below, then the sequence i whose nth term is $i_n = \inf\{a_k \mid k \geq n\}$ is nondecreasing.

The following theorem includes two conditions that the limit superior always satisfies and that, taken together, provide an equivalent characterization of the concept.

Theorem 11.9. *A real number p is the limit superior of a sequence (a_n) that is bounded above if and only if both of the following conditions hold:*
(a) *whenever $r > p$, there exists an index N such that $a_n < r$ for all $n \geq N$; and,*
(b) *whenever $r < p$ and n is a natural number, there exists an index k_n such that $k_n \geq n$ and $a_{k_n} > r$.*

Assuming p is the limit superior of a sequence, condition (a) says that the sequence's terms are *eventually less than* any number larger than p, while condition (b) says the terms are *frequently greater than* any number smaller than p. When the terms of a sequence are always eventually less than every number larger than a specific number p and also frequently greater than every number smaller than this number p, then p is the limit superior of the sequence.

Example 11.7 (Continued). We have found that $\limsup a_n = 2$ for the sequence a defined so that $a_n = \frac{1}{n} + r_n$, where r_n is the remainder obtained by dividing n by 3. We now verify that both of the conditions (a) and (b) given in Theorem 11.9 hold.

For condition (a), consider any number r greater than 2. By the Archimedean property, there exists a natural number N for which $\frac{1}{N} < r - 2$. Consider any natural number n for which $n \geq N$. There are three cases to consider depending on the value of the remainder r_n. If $r_n = 0$, then

$$a_n = \frac{1}{n} \leq 1 < 2 < r.$$

If $r_n = 1$, then

$$a_n = \frac{1}{n} + 1 \leq 2 < r.$$

If $r_n = 2$, then

$$a_n = \frac{1}{n} + 2 \leq \frac{1}{N} + 2 < (r - 2) + 2 = r,$$

since $n \geq N$. Thus, whenever $n \geq N$, it follows that $a_n < r$, meaning the sequence a is eventually less than r whenever $r > 2$.

For condition (b), consider any number r less than 2 and any natural number n. Then $3n + 2 \geq n$ and

$$a_{3n+2} = \frac{1}{3n + 2} + 2 > 2 > r.$$

Thus, the sequence a is frequently greater than r whenever $r < 2$.

Exercise 11.17. Use Theorem 11.9 to establish that $\limsup a_n = 1$ for the alternating sequence $a = (-1, 1, -1, 1, \ldots)$. (We already established this result in Example 11.4 using the definition of limit superior.)

Proof of Theorem 11.9. ($\Rightarrow$) Assume $\limsup a_n = p$ and, for each n, let $s_n = \sup\{a_k \mid k \geq n\}$.

First, suppose to the contrary that condition (a) does not hold. Then there exists both a number r for which $r > p$ and a subsequence (a_{j_n}) of (a_n) such that $a_{j_n} \geq r$ for every n. Then, as $j_n \geq n$ for each n, we have $s_n \geq a_{j_n} \geq r$, which implies that

$$\limsup a_n = \lim s_n \geq r,$$

a contradiction since $\limsup a_n = p$ and $r > p$.

Now suppose to the contrary that condition (b) does not hold. Then there exists a number r for which $r < p$ and $a_n \leq r$ for all n at least as large as some index N. Thus, whenever $n \geq N$, it follows that

$$s_n = \sup\{a_k \mid k \geq n\} \leq r,$$

so that

$$\limsup a_n = \lim s_n \leq r,$$

a contradiction since $\limsup a_n = p$ and $r < p$.

($\Leftarrow$) Assume both of the conditions (a) and (b) hold for a sequence (a_n) that is bounded above and a real number p.

Suppose $r > p$. Then from condition (a), there exists an index N such that $a_n < r$ for all $n \geq N$. Hence, when $n \geq N$ we have

$$s_n = \sup\{a_k \mid k \geq n\} \leq r,$$

from which we may conclude that

$$\limsup a_n = \lim s_n \leq r.$$

Now suppose $r < p$. Then from condition (b), for each natural number n, there exists a natural number k_n such that $k_n \geq n$ and $a_{k_n} > r$. Hence,

$$s_n = \sup\{a_k \mid k \geq n\} \geq a_{k_n} > r,$$

from which we may conclude that

$$\limsup a_n = \lim s_n \geq r.$$

Having shown that $\limsup a_n$ is less than or equal to every number larger than p, and also greater than or equal to every number smaller than p, it follows that $\limsup a_n = p$. □

The next theorem is the counterpart to Theorem 11.9 for the limit inferior. The listed conditions say that the sequence (a_n) is (a) *eventually greater than* any number less than p and (b) *frequently less than* any number greater than p.

Theorem 11.10. *A real number p is the limit inferior of a sequence (a_n) that is bounded below if and only if both of the following conditions hold:*
(a) *whenever $r < p$, there exists an index N such that $a_n > r$ for all $n \geq N$; and,*
(b) *whenever $r > p$ and n is a natural number, there exists an index k_n such that $k_n \geq n$ and $a_{k_n} < r$.*

Exercise 11.18. In Exercise 11.15 you verified that $\liminf a_n = 0$ for the sequence (a_n) of Example 11.7. Now verify both conditions (a) and (b) of Theorem 11.10 are satisfied by $p = 0$.

Exercise 11.19. Prove Theorem 11.10.

It is precisely when the limit superior and limit inferior of a sequence are identical that the sequence converges, the limit of the sequence being the common value of the limits superior and inferior.

Theorem 11.11. *Given a bounded sequence (a_n) and a real number p, the following are equivalent:*
(i) $\lim a_n = p$;
(ii) $\limsup a_n = p = \liminf a_n$.

Example 11.12. As the sequence $(\frac{1}{n})$ converges to 0, it follows that $\limsup \frac{1}{n} = 0$ and $\liminf \frac{1}{n} = 0$.

Exercise 11.20. Prove Theorem 11.11.

The limit superior and limit inferior reveal, respectively, the largest and smallest limiting values a subsequence of a sequence achieves. It is this fact that makes the limits superior and inferior useful in analyzing the convergent subsequences of a bounded divergent sequence.

Theorem 11.13. *When the limit superior of a sequence exists, it is the greatest subsequential limit of the sequence.*

When the limit inferior of a sequence exists, it is the least subsequential limit of the sequence.

Example 11.7 (Continued). For the sequence a defined so that $a_n = \frac{1}{n} + r_n$, where r_n is the remainder obtained by dividing n by 3, we showed that $\limsup a_n = 2$ and you showed, in Exercise 11.15, that $\liminf a_n = 0$. Hence, according to Theorem 11.13, there is a subsequence of a that converges to 2, there is a subsequence of a that converges to 0, and the limit of any convergent subsequence of a is neither larger than 2 nor smaller than 0.

A sequence can have subsequential limits that lie between its limit inferior and limit superior.

Exercise 11.21. For the sequence a from Example 11.7, identify a subsequential limit that lies between the limit inferior and the limit superior.

Exercise 11.22. Prove Theorem 11.13.

When a sequence has two convergent subsequences, each of which can be obtained from the original sequence by deleting the terms of the other subsequence, the limits of these subsequences are the only subsequential limits of the sequence.

Theorem 11.14. *Let (a_{j_n}) and (a_{k_n}) be subsequences of the sequence (a_n), and take $J = \{j_n \mid n \in \mathbb{N}\}$ and $K = \{k_n \mid n \in \mathbb{N}\}$. Suppose also that $J \cup K = \mathbb{N}$ and $J \cap K = \emptyset$.*

If p and q are real numbers for which $a_{j_n} \to p$ and $a_{k_n} \to q$, then p and q are the only subsequential limits of (a_n).

Example 11.4 (Continued). The odd-indexed terms of the alternating sequence $a = (-1, 1, -1, 1, \ldots)$ form the subsequence $(-1, -1, -1, \ldots)$ and the even-indexed terms form the subsequence $(1, 1, 1, \ldots)$. As each subsequence can be obtained from a by deleting the terms of the other subsequence, the limits of these subsequences, -1 and 1, are the only subsequential limits of the sequence a.

Exercise 11.23. Prove Theorem 11.14.

Theorem 11.14, or a variation of it, is useful when a sequence is essentially formed by weaving together several other sequences.

Exercise 11.24. Consider again the sequence a of Example 11.7. Apply a variation of Theorem 11.14 to identify all the subsequential limits of a. Then use your conclusion to identify the limits superior and inferior of a.

The Cauchy criterion for convergence of a sequence

If we want to know whether or not a given sequence converges, the definition of convergence is not always of much help. It may be used to eliminate a real number as a limiting value for the sequence or to verify a (given or conjectured) limiting value. But what do we do if we have no idea whether the sequence of interest converges or diverges? The so-called *Cauchy criterion* for sequential convergence is useful in many such situations.

The great French mathematician Augustin-Louis Cauchy (1789–1857) can be credited with founding modern mathematical analysis. Among his many contributions to the subject is the notion of what is nowadays called a *Cauchy sequence*. A real sequence (a_n) is a **Cauchy sequence** if for every positive number ε, there is an index N such that when both $m \geq N$ and $n \geq N$, it follows that $|a_m - a_n| < \varepsilon$. In other words, a Cauchy sequence is a sequence for which the terms eventually become arbitrarily close to one another, meaning they stay less than any pre-assigned positive distance from each other once we are sufficiently far out in the sequence. Note that the value of the index N can, and usually will, vary with the value of the positive number ε.

Example 11.15. We prove that $(\frac{5}{\sqrt[3]{n}})$ is a Cauchy sequence. To do so, we begin with an arbitrary positive number ε and must use it to find an index N beyond which all the terms of $(\frac{5}{\sqrt[3]{n}})$ are less than ε from each other. That is, we need to find N so that

$$\left| \frac{5}{\sqrt[3]{m}} - \frac{5}{\sqrt[3]{n}} \right| < \varepsilon$$

for all indices m and n that are greater than or equal to N. Since it is necessarily the case that either $n \geq m$ or $m \geq n$, there is no harm in assuming that $n \geq m$. With $n \geq m$ it follows that $\frac{5}{\sqrt[3]{m}} \geq \frac{5}{\sqrt[3]{n}}$, so

$$\left| \frac{5}{\sqrt[3]{m}} - \frac{5}{\sqrt[3]{n}} \right| = \frac{5}{\sqrt[3]{m}} - \frac{5}{\sqrt[3]{n}},$$

and since $\frac{5}{\sqrt[3]{n}} > 0$, we also have

$$\frac{5}{\sqrt[3]{m}} - \frac{5}{\sqrt[3]{n}} < \frac{5}{\sqrt[3]{m}}.$$

Thus, to make the inequality $|\frac{5}{\sqrt[3]{m}} - \frac{5}{\sqrt[3]{n}}| < \varepsilon$ true, it suffices to make $\frac{5}{\sqrt[3]{m}} < \varepsilon$ for suitably large m. In the following write-up of the proof, note how we use the Archimedean property to achieve this goal.

Proof. We are using the definition of Cauchy sequence to prove that $(\frac{5}{\sqrt[3]{n}})$ is a Cauchy sequence. Suppose $\varepsilon > 0$. Then $(\frac{\varepsilon}{5})^3 > 0$ and by the Archimedean property there exists a natural number N such that $\frac{1}{N} < (\frac{\varepsilon}{5})^3$. From this inequality it follows that $\frac{5}{\sqrt[3]{N}} < \varepsilon$. Now suppose $m \geq N$ and $n \geq N$, and assume without loss of generality that $n \geq m$. Then

$$\left|\frac{5}{\sqrt[3]{m}} - \frac{5}{\sqrt[3]{n}}\right| = \frac{5}{\sqrt[3]{m}} - \frac{5}{\sqrt[3]{n}} < \frac{5}{\sqrt[3]{m}} \leq \frac{5}{\sqrt[3]{N}} < \varepsilon.$$

□

Exercise 11.25. Use the definition of Cauchy sequence to prove the given sequence is Cauchy.
(a) $(\frac{1-n}{n})$
(b) $(\frac{1}{n^2})$

Exercise 11.26. Give an example of a bounded sequence that is not a Cauchy sequence.

Exercise 11.27. Use the definition of Cauchy sequence to prove that if (a_n) and (b_n) are Cauchy sequences, then $(a_n + b_n)$ is also a Cauchy sequence.

It turns out that every convergent sequence of real numbers is Cauchy and, more significantly, every Cauchy sequence converges.

Theorem 11.16 (The Cauchy criterion for sequential convergence). *A sequence of real numbers converges if and only if it is a Cauchy sequence.*

Proof. ($\Rightarrow$) This part of the proof is assigned as Exercise 11.28(a).

($\Leftarrow$) Consider any Cauchy sequence (a_n). It is straightforward to show that (a_n) is bounded (assigned as Exercise 11.28(b)). Hence, by the Bolzano–Weierstrass theorem, (a_n) has a convergent subsequence (a_{k_n}). Let p be the limit of (a_{k_n}).

Consider any positive number ε. Then $\frac{\varepsilon}{2} > 0$, and since (a_n) is a Cauchy sequence, there is an index N_1 such that when $m \geq N_1$ and $n \geq N_1$, it follows that $|a_m - a_n| < \frac{\varepsilon}{2}$. Then, as the subsequence (a_{k_n}) of (a_n) converges to p, there is an index N_2 such that $N_2 \geq N_1$ and $|a_{k_{N_2}} - p| < \frac{\varepsilon}{2}$. Thus, whenever $n \geq N_1$, we have

$$|a_n - p| = |a_n - a_{k_{N_2}} + a_{k_{N_2}} - p| \leq |a_n - a_{k_{N_2}}| + |a_{k_{N_2}} - p| < \frac{\varepsilon}{2} + \frac{\varepsilon}{2} = \varepsilon.$$

Thus, $a_n \to p$. □

In the argument just above that a Cauchy sequence converges, we first used the given positive number ε to move out far enough in the sequence, specifically to index N_1, so that all the terms from this index onward are "close to each other," specifically, less than $\frac{\varepsilon}{2}$ from each other. Then, since a Cauchy sequence is bounded, we were able to use the Bolzano–Weierstrass theorem to obtain a convergent subsequence of the Cauchy sequence, thereby permitting us to move even further out in the sequence, specifically to index k_{N_2}, to obtain a term, specifically $a_{k_{N_2}}$, of the subsequence that is "close to its limit," specifically, less than $\frac{\varepsilon}{2}$ from its limit. The triangle inequality is then employed to guarantee all terms of the Cauchy sequence from index N_1 onward are less than the given ε from the limit of the subsequence.

Exercise 11.28. Complete the proof of Theorem 11.16 by showing each of the following.
(a) If $a_n \to p$, then (a_n) is a Cauchy sequence.
(b) If (a_n) is a Cauchy sequence, then (a_n) is bounded.

According to the Cauchy criterion, if we discover that a sequence is not a Cauchy sequence, the sequence must diverge.

Example 11.17. For each natural number n, let

$$s_n = \sum_{k=1}^{n} \frac{1}{k} = 1 + \frac{1}{2} + \frac{1}{3} + \cdots + \frac{1}{n}.$$

The sequence (s_n) is the sequence

$$\left(1, 1 + \frac{1}{2}, 1 + \frac{1}{2} + \frac{1}{3}, 1 + \frac{1}{2} + \frac{1}{3} + \frac{1}{4}, \ldots\right)$$

of partial sums of the infinite series $\sum_{n=1}^{\infty} \frac{1}{n}$, called the **harmonic series.** We can show the harmonic series diverges by showing that it is not a Cauchy sequence. Take $\varepsilon = \frac{1}{2}$ and note that, for every n,

$$|s_{2n} - s_n| = s_{2n} - s_n = \frac{1}{n+1} + \frac{1}{n+2} + \cdots + \frac{1}{2n} > n \cdot \frac{1}{2n} = \frac{1}{2} = \varepsilon,$$

so that (s_n) is not Cauchy. Thus, the harmonic series diverges.

Exercise 11.29. Let m be a fixed natural number. Use induction to prove that for every natural number n, we have $\frac{1}{2^m} > \frac{1}{2^{m+1}} + \frac{1}{2^{m+2}} + \cdots + \frac{1}{2^{m+n}}$.

Exercise 11.30. For each natural number n, let $a_n = \sum_{k=1}^{n} \frac{1}{2^k} = \frac{1}{2} + \frac{1}{4} + \cdots + \frac{1}{2^n}$. Prove that the sequence (a_n) is a Cauchy sequence. What can we then conclude? *Hint*: Make use of the result from Exercise 11.29.

12 Limits of functions

Our interest shifts now, and for much of the remainder of the book, to the analysis of functions having real number inputs and outputs. Therefore, unless stated otherwise, we assume the domain and the range of a function are both subsets of $\mathbb{R}$.

In the last three chapters, we explored the notion of *sequential limits* by asking whether the terms of a sequence we were examining tended to become arbitrarily close to some particular number, though perhaps never actually equal to this number, as we allowed the index values of the terms to become ever larger without bound. In this chapter, we perform a similar study of *function limits*, asking whether the outputs of a function we are examining are becoming arbitrarily close to some particular number, though again perhaps never actually equal to this number, as we allow the inputs to become arbitrarily close to, but not equal to, some specified number.

The limit of a function: graphical and numerical perspectives

First, let us remind ourselves of how we probably thought about function limits back in our study of calculus. Consider the function f defined so that

$$f(x) = \begin{cases} 3x + 1, & \text{if } x \neq 2; \\ 4, & \text{if } x = 2; \end{cases}$$

and whose graph is displayed in Figure 12.1.

Our attention is almost immediately drawn to the behavior of this function at and near $x = 2$, either because of the disruption in the function's graph at this point or because the function has been defined "differently" at the input 2 than for all other inputs. For the input 2, the function f has been defined to have value/output 4; that is,

$$f(2) = 4.$$

But the behavior of the function f is quite different for inputs very close to, but not equal to, 2. Looking at the graph of f, it appears that the function values $f(x)$ become arbitrarily close to 7, though they are never equal to 7, as the inputs x become arbitrarily close to, but distinct from, 2.

Not only does graphical evidence suggest that

$$\lim_{x \to 2} f(x) = 7,$$

even though $f(2) = 4$, numerical evidence as provided in Table 12.1 suggests the same limiting value of 7.

https://doi.org/10.1515/9783111429564-012

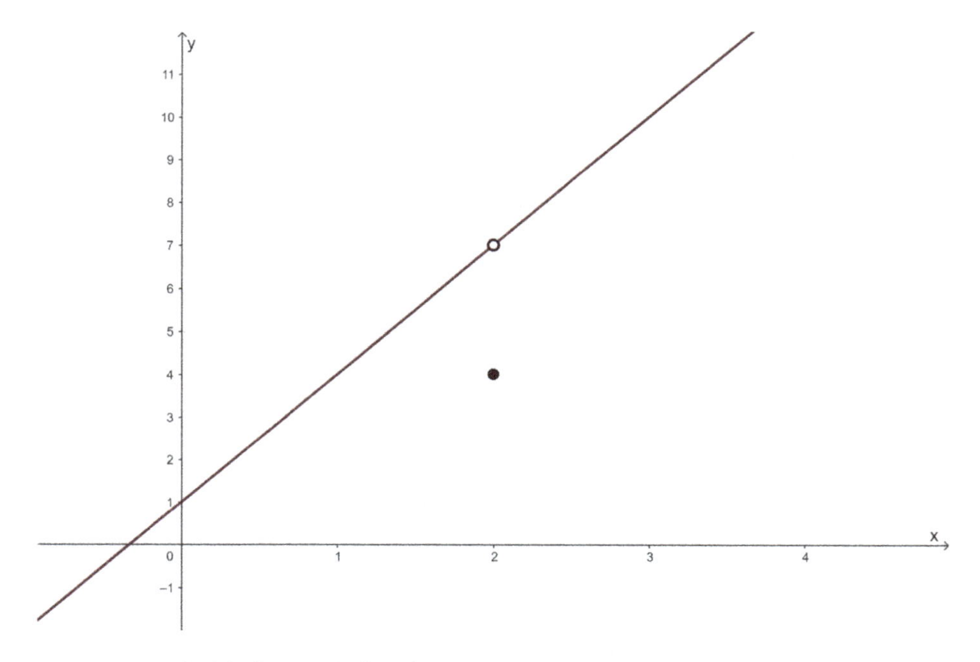

Figure 12.1: Graph of the function f, where $f(2) = 4$ and $f(x) = 3x + 1$ if $x \neq 2$.

Table 12.1: Values of function f near the input 2.

x	1.9	1.99	1.999	→	2	←	2.001	2.01	2.1
$f(x)$	6.7	6.97	6.997	→	?	←	7.003	7.03	7.3

Exercise 12.1. For the function f discussed above, $\lim_{x\to 2} f(x) \neq f(2)$. It can, of course, happen that a function's limit as the inputs approach a particular number is the same as the function's output at this number. Graph the function g defined so that

$$g(x) = 3x + 1$$

and use the graph to convince yourself that $\lim_{x\to 2} g(x) = g(2)$.

Exercise 12.2. The limit of a function as the inputs approach a particular number can exist even if the function is undefined at this number. Make a sketch of the graph of the function h defined so that

$$h(x) = \frac{3x^2 - 5x - 2}{x - 2}$$

and use the graph to convince yourself that $\lim_{x\to 3} h(x) = 7$ even though h is not defined at $x = 2$.

The limit of a function: a precise mathematical definition

Our goal here is to make mathematically precise, in a way that does not depend upon pictures or tables of numerical values, the notion, symbolized as

$$\lim_{x\to p} f(x) = L,$$

that the outputs of a function f approach arbitrarily closely, but perhaps do not ever equal, a certain number L as the inputs approach arbitrarily closely, but are not chosen equal to, a certain number p.

Rather than deal with the most general situation, we instead focus on a specific example, and generalize from it. We saw, via graphical and numerical evidence, that the limit as the inputs approach 2 of the function f defined by

$$f(x) = \begin{cases} 3x+1, & \text{if } x \neq 2; \\ 4, & \text{if } x = 2; \end{cases}$$

appears to be 7. What we mean is that as the inputs approach (i. e., get arbitrarily close to) 2, without being equal to 2, the corresponding outputs approach (i. e., get arbitrarily close to) 7, without perhaps ever being equal to 7. That is, the outputs should stay less than any pre-assigned positive distance from 7 as long as the inputs are chosen close enough to, but not equal to, 2.

Example 12.1. How close to 2 must we choose the inputs to the function f for which

$$f(x) = \begin{cases} 3x+1, & \text{if } x \neq 2; \\ 4, & \text{if } x = 2; \end{cases}$$

without choosing 2 itself, so that the corresponding outputs are less than 0.6 units from 7?

To require that an input x yield an output $f(x)$ that is less than 0.6 units from 7 means that

$$f(x) \in (7 - 0.6, 7 + 0.6),$$

that is,

$$7 - 0.6 < f(x) < 7 + 0.6.$$

Moreover, as we also require that the input x not be 2 itself, we know that $f(x) = 3x + 1$, which permits us to rewrite the above inequality as

$$6.4 < 3x + 1 < 7.6.$$

Solving for x yields the equivalent inequality

$$1.8 < x < 2.2,$$

which can be expressed as

$$2 - 0.2 < x < 2 + 0.2.$$

Since the algebraic steps applied here are reversible, it is now clear that for every input x which is less than 0.2 units from 2, and different from 2 itself, the corresponding output $f(x)$ is less than 0.6 units from 7.

Figure 12.2 visualizes this scenario by showing that, with the exception of the point for which $x = 2$, the part of the graph of f with points having x-coordinates between 1.8 and 2.2 (between the dashed vertical lines) has y-coordinates between 6.4 and 7.6 (between the dashed horizontal lines).

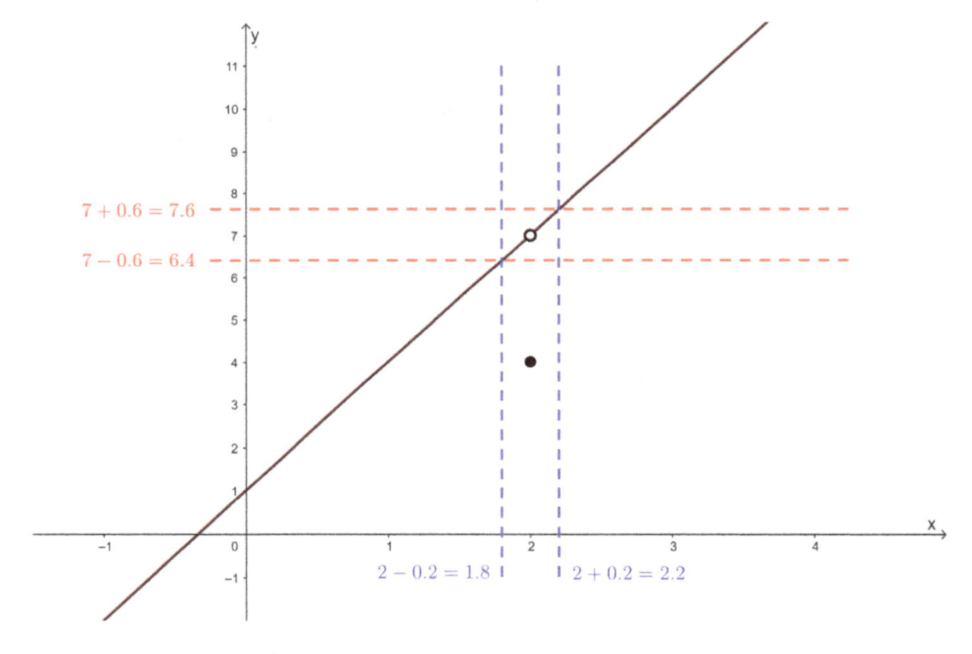

Figure 12.2: Inputs x to the function f of Example 12.1 must be less than 0.2 units from 2, and different from 2, in order for the corresponding outputs $f(x)$ to be less than 0.6 units from 7.

Exercise 12.3. How close to 2 must we choose the inputs to the function f from Example 12.1, without choosing 2 itself, so that the corresponding outputs are less than $\frac{1}{1000}$ units from 7?

We refer to the pre-assigned positive distance from the limiting value that the outputs must stay less than as the *output tolerance* and denote it by ε. We have just seen, in Example 12.1 and Exercise 12.3, that, for the function f, where

$$f(x) = \begin{cases} 3x + 1, & \text{if } x \neq 2; \\ 4, & \text{if } x = 2; \end{cases}$$

we must choose the inputs less than 0.2 units from 2 in order to guarantee the corresponding outputs stay less than $\varepsilon = 0.6$ units from 7, and less than $\frac{1}{3000}$ units from 2 in order to guarantee the corresponding outputs stay less than $\varepsilon = \frac{1}{1000}$ units from 7, and that, with either output tolerance, we must not choose the input 2 itself.

Note that we have produced an *input tolerance* for each of two specified output tolerances. However, in order for the limit of the outputs to be 7 as the inputs approach 2, we must be able to generate such a positive input tolerance δ (*delta*) for each possible positive output tolerance ε. Since it is not feasible to work with every possible positive output tolerance individually, we try to find a positive input tolerance that corresponds to an arbitrary positive output tolerance ε.

Example 12.1 (Continued). In terms of an arbitrary positive output tolerance ε, how close to 2 must the inputs to the function f defined by

$$f(x) = \begin{cases} 3x + 1, & \text{if } x \neq 2; \\ 4, & \text{if } x = 2; \end{cases}$$

be chosen, without choosing 2 itself, so that the corresponding outputs are less than ε units from 7?

The analysis of the general situation parallels that performed earlier when we worked with the particular output tolerance $\varepsilon = 0.6$. For our arbitrary positive number ε, we require that

$$f(x) \in (7 - \varepsilon, 7 + \varepsilon),$$

that is,

$$7 - \varepsilon < f(x) < 7 + \varepsilon$$

for inputs x sufficiently close to, but not equal to, 2. Hence, as $x \neq 2$ we may substitute $3x + 1$ for $f(x)$ to obtain

$$7 - \varepsilon < 3x + 1 < 7 + \varepsilon.$$

Solving for x yields

$$2 - \frac{\varepsilon}{3} < x < 2 + \frac{\varepsilon}{3}.$$

Thus, as the algebra here can be reversed, whenever an input x is less than $\frac{\varepsilon}{3}$ units from, and different from, 2, the corresponding output $f(x)$ is less than ε units from 7.

In other words, for each possible positive output tolerance ε that may be specified, we may take $\delta = \frac{\varepsilon}{3}$ as the associated positive input tolerance (Figure 12.3). Note that this conclusion agrees with those reached before: For $\varepsilon = 0.6$ we found $\delta = 0.2 = \frac{0.6}{3} = \frac{\varepsilon}{3}$, and for $\varepsilon = \frac{1}{1000}$ you found, in Exercise 12.3, that $\delta = \frac{1}{3000} = \frac{\frac{1}{1000}}{3} = \frac{\varepsilon}{3}$.

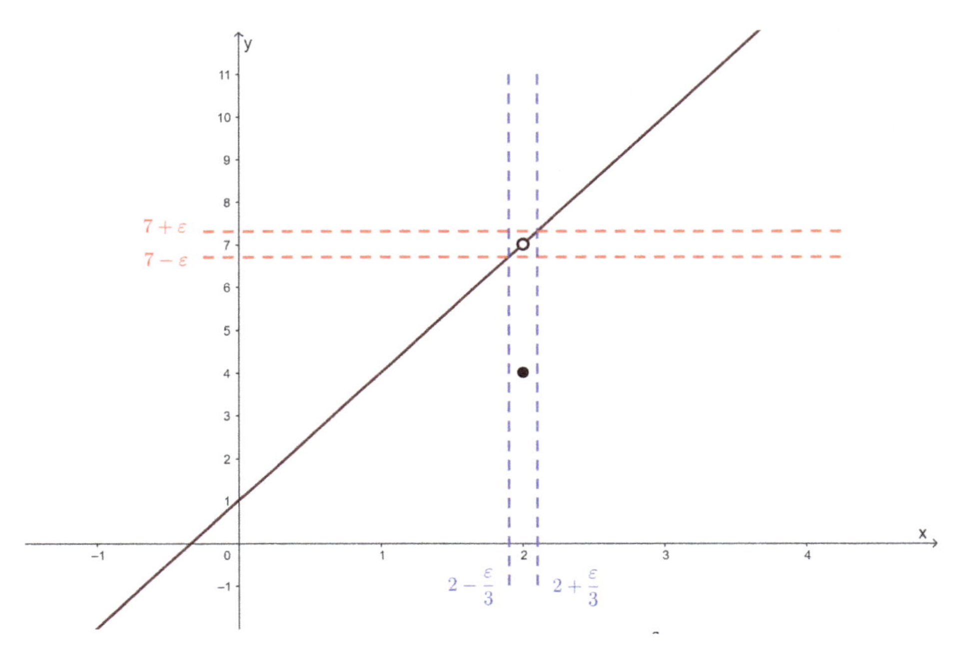

Figure 12.3: Inputs x to the function f of Example 12.1 must be less than $\frac{\varepsilon}{3}$ units from 2, and different from 2, in order for the corresponding outputs $f(x)$ to be less than ε units from 7.

It is worth noting that once a positive input tolerance δ has been found that works for a positive output tolerance ε, any positive number that is smaller than this δ could be used in its place. For instance, as we found $\delta = 0.2$ could serve as the input tolerance associated with the output tolerance $\varepsilon = 0.6$, we could replace δ with any positive number less than 0.2 if we wanted to (and, in general, we could replace $\delta = \frac{\varepsilon}{3}$ with any positive number less than $\frac{\varepsilon}{3}$, for example, $\frac{\varepsilon}{5}$, if convenient).

We are now in a position to generalize what we have learned from the specific limiting situation we have considered. We intend

$$\lim_{x \to p} f(x) = L$$

to mean that for any positive output tolerance ε, there is a corresponding positive input tolerance δ such that whenever an input x different from p is chosen so that its distance from p is less than the input tolerance δ, then it automatically follows that the distance from the corresponding output $f(x)$ to L is less than the output tolerance ε. Translating into precise mathematical language yields the following formal definition.

Definition 12.2 (The $\varepsilon\delta$-definition of limit of a function). Given a limit point p of the domain of a function f, a real number L is a **limit of the function f at p**, denoted

$$\lim_{x \to p} f(x) = L,$$

if for every positive number ε, there exists a positive number δ such that whenever x is in the domain of f and $0 < |x - p| < \delta$, it follows that $|f(x) - L| < \varepsilon$.

Bear in mind that the value of δ in this definition usually varies with the value of ε, so there are times where we may write δ_ε to emphasize this dependence.

The statement $\lim_{x \to p} f(x) = L$ is commonly read as

$$f(x) \textbf{ approaches } L \textbf{ as } x \textbf{ approaches } p,$$

and can also be expressed by writing

$$f(x) \to L \textbf{ as } x \to p.$$

The reason that we have assumed p to be a limit point of the domain of f is so that it is possible to approach p arbitrarily closely via points in the domain of f. That is, the notion that we want x to approach p really only makes sense if there are points of the domain of f distinct from, but arbitrarily close to, p. This is typically an implicit assumption anytime we make reference to a limit of a function.

As we almost always do, we are also interpreting the absolute value of a difference as the distance between the quantities being subtracted. Thus, $|x - p|$ is the distance between x and p, while $|f(x) - L|$ is the distance between $f(x)$ and L. As the absolute value of a number is never negative, we may note that $0 < |x - p|$ is really just another way to convey that $x \neq p$.

Hopefully, the process we have undertaken to develop a rigorous definition for function limits is helping you to really think carefully about this concept and come to a fuller understanding of it. We are now able to use our precise formulation of function limit to not only verify limiting values for specific functions, but to mathematically justify various properties of limits and, in Chapter 15, to define the important notion of *derivative* of a function.

It was Augustin-Louis Cauchy who, in his 1821 book *Cour d'Analyse*, first put forth the rigorous $\varepsilon\delta$ definition of function limit and used it to prove many theorems concerning limits. Cauchy also rigorously defined sequential convergence, continuity of a function, and differentiability of a function. His work has had a profound influence on modern mathematics, especially analysis.

Example 12.1 (Continued). We can use the $\varepsilon\delta$-definition of limit to prove that, for the function f, where

$$f(x) = \begin{cases} 3x+1, & \text{if } x \neq 2; \\ 4, & \text{if } x = 2; \end{cases}$$

we have

$$\lim_{x \to 2} f(x) = 7.$$

Our write-up makes use of what we learned in the preliminary analysis performed earlier.

Proof. Consider any positive number ε. Let $\delta = \frac{\varepsilon}{3}$, which is positive, and assume $0 < |x-2| < \delta$. Then

$$|f(x) - 7| = |(3x+1) - 7| = |3x - 6| = 3|x-2| < 3 \cdot \delta = 3 \cdot \frac{\varepsilon}{3} = \varepsilon.$$ □

Exercise 12.4. In the proof given above in Example 12.1, we chose $\delta = \frac{\varepsilon}{3}$. Explain why we could have chosen $\delta = \frac{\varepsilon}{10}$ instead. Then explain why we could not choose $\delta = \frac{\varepsilon}{2}$.

Exercise 12.5. Use the $\varepsilon\delta$-definition of limit to verify each of the following.
(a) $\lim_{x \to 8} (\frac{x}{5} - 1) = \frac{3}{5}$.
(b) $\lim_{x \to 2} f(x) = 5$, where $f(x) = \begin{cases} 4x - 3, & \text{if } x < 2; \\ 6x - 7, & \text{if } x > 2. \end{cases}$

We now apply the $\varepsilon\delta$-definition of function limit to formally establish a couple of limiting values that are intuitively apparent.

Theorem 12.3 (Two essential limits).
(1) *If c is a real number constant, then* $\lim_{x \to p} c = c$.
(2) $\lim_{x \to p} x = p$.

Proof. (1) Assume c is a real number constant and define $f : \mathbb{R} \to \mathbb{R}$ so that $f(x) = c$ for all x in $\mathbb{R}$. Assume $\varepsilon > 0$, take $\delta = 1$, and suppose $0 < |x - p| < \delta$. Then

$$|f(x) - c| = |c - c| = 0 < \varepsilon.$$

(2) Define $f : \mathbb{R} \to \mathbb{R}$ so that $f(x) = x$ for all x in $\mathbb{R}$. Assume $\varepsilon > 0$, take $\delta = \varepsilon$, and suppose $0 < |x - p| < \delta$. Then

$$|f(x) - p| = |x - p| < \delta = \varepsilon.$$ □

Exercise 12.6. Is there anything special about the choice of 1 as the value of δ in the proof of Theorem 12.3(1)? In this particular case does the value of δ depend on the value of ε?

Because

$$0 < |x - p| < \delta \quad \text{if and only if} \quad x \in (p - \delta, p + \delta) - \{p\}$$

and

$$|f(x) - L| < \varepsilon \quad \text{if and only if} \quad f(x) \in (L - \varepsilon, L + \varepsilon),$$

we immediately obtain the following characterization of the notion of function limit by means of neighborhoods of a point.

Theorem 12.4 (Neighborhood characterization of limit of a function). *Given a limit point p of the domain of a function f, the following are equivalent:*
(i) $\lim_{x \to p} f(x) = L$;
(ii) *for every positive number ε, there exists a positive number δ such that whenever x is in the domain of f and $x \in (p - \delta, p + \delta) - \{p\}$, it follows that $f(x) \in (L - \varepsilon, L + \varepsilon)$.*

Condition (ii) of Theorem 12.4 is often viewed as saying that for every neighborhood of L, there exists a corresponding neighborhood of p for which all points in this neighborhood of p that are also in the domain of f, except perhaps p itself, are mapped by f to points in the originally specified neighborhood of L.

Example 12.1 (Continued). We can rewrite our proof that $\lim_{x \to 2} f(x) = 7$ using the neighborhood characterization of function limit as follows.

Proof. Consider any positive number ε. Let $\delta = \frac{\varepsilon}{3}$, which is positive, and assume $x \in (2 - \frac{\varepsilon}{3}, 2 + \frac{\varepsilon}{3}) - \{2\}$. Then $2 - \frac{\varepsilon}{3} < x < 2 + \frac{\varepsilon}{3}$ and $x \neq 2$. It follows that

$$3\left(2 - \frac{\varepsilon}{3}\right) + 1 < 3x + 1 < 3\left(2 + \frac{\varepsilon}{3}\right) + 1$$

so that, since we know $x \neq 2$,

$$7 - \varepsilon < f(x) < 7 + \varepsilon,$$

Thus, we may conclude that $f(x) \in (7 - \varepsilon, 7 + \varepsilon)$. □

Conceptually this version of the proof is really the same as the original version; it is just written a bit differently, incorporating neighborhoods to measure distance rather than measuring distance via absolute value. Note that in the version presented here, we have shown that given any neighborhood $(7-\varepsilon, 7+\varepsilon)$ of 7, there is a corresponding neighborhood $(2-\frac{\varepsilon}{3}, 2+\frac{\varepsilon}{3})$ of 2 for which all points in this neighborhood of 2, except 2 itself (in the circumstance where $\varepsilon \leq 3$), are mapped by f into $(7-\varepsilon, 7+\varepsilon)$.

Exercise 12.7. Write a proof using Theorem 12.4 that verifies $\lim_{x\to 8}(\frac{x}{5}-1) = \frac{3}{5}$. Compare with the proof you wrote in Exercise 12.5(a) verifying this same limit.

Exercise 12.8. Prove that if $\lim_{x\to p} f(x) = L$ for some positive real number L, then there exists a neighborhood of p for which $f(x) > 0$ for all x in the domain of f that are in this neighborhood of p, except perhaps p itself.

It is not possible for a function to have multiple limits as the inputs approach a specified real number. That is, if we discover that L and M are both limits of the function f at p, then it must follow that $L = M$.

Theorem 12.5 (Uniqueness of function limits). *A function cannot have more than one limit as the inputs approach a particular real number.*

Exercise 12.9. Suitably modify the proof we gave for Theorem 9.14 concerning the uniqueness of sequential limits to prove Theorem 12.5.

Further useful strategies for verifying a limit via the $\varepsilon\delta$-definition

When applying the $\varepsilon\delta$-definition of limit to verify $\lim_{x\to p} f(x) = L$, we begin with an arbitrary positive number ε and attempt to make $|f(x) - L|$ smaller than this ε, for all x different from p for which $|x-p|$ is less than a suitably chosen positive number δ. In many such situations, it is not immediately apparent how the hypothesis that $|x-p| < \delta$ yields the desired conclusion that $|f(x) - L| < \varepsilon$. One strategy that may be helpful is to rewrite $f(x) - L$ so that it is expressed in terms of $x - p$ by replacing x with $(x-p)+p$.

Example 12.6. Suppose we want to demonstrate that

$$\lim_{x\to 2} x^2 = 4.$$

Based on the definition of limit, we know we need to examine

$$|x^2 - 4|.$$

If not prompted to do otherwise, many of us would probably rewrite this expression as follows:

$$|x^2 - 4| = |(x+2)(x-2)| = |x+2||x-2|.$$

In doing so, we are pleased to see the appearance of $|x-2|$, but probably left wondering what to do about $|x+2|$.

While rewriting $|x^2-4|$ in this way can be used to establish the desired limit, we instead employ what we believe to be a simpler approach by replacing x in the expression $|x^2-4|$ with $(x-2)+2$ and, after doing so, always treating the expression $x-2$ as an inseparable entity. Observe that

$$\begin{aligned} |x^2-4| &= \left|[(x-2)+2]^2-4\right| \\ &= \left|[(x-2)^2+4(x-2)+4]-4\right| \\ &= \left|(x-2)^2+4(x-2)\right| \\ &\le \left|(x-2)^2\right| + \left|4(x-2)\right| \\ &= |x-2|^2+4|x-2|, \end{aligned}$$

where we have applied the triangle inequality, and where the final expression, $|x-2|^2+4|x-2|$, is written in terms of $|x-2|$.

It is now possible to determine how to define the input tolerance δ in terms of a given output tolerance ε so that the assumption

$$0<|x-2|<\delta$$

leads to

$$|x-2|^2+4|x-2|<\varepsilon.$$

This last inequality can be achieved if both terms on the left side,

$$|x-2|^2$$

and

$$4|x-2|,$$

can each be made less than $\frac{\varepsilon}{2}$. The inequality

$$|x-2|^2<\frac{\varepsilon}{2}$$

can be achieved if we require that

$$|x-2|<\sqrt{\frac{\varepsilon}{2}}$$

and the inequality

$$4|x - 2| < \frac{\varepsilon}{2}$$

can be achieved if we require that

$$|x - 2| < \frac{\varepsilon}{8}.$$

Thus, if we choose δ to be the minimum of $\sqrt{\frac{\varepsilon}{2}}$ and $\frac{\varepsilon}{8}$, we are then able to achieve the desired inequality

$$|x - 2|^2 + 4|x - 2| < \varepsilon$$

for all x satisfying the condition

$$0 < |x - 2| < \delta.$$

Our write-up of the proof that $\lim_{x\to 2} x^2 = 4$ incorporates all that we have learned here.

Proof. Assume $\varepsilon > 0$ and let $\delta = \min\{\sqrt{\frac{\varepsilon}{2}}, \frac{\varepsilon}{8}\}$, which we may observe is positive. Suppose $0 < |x - 2| < \delta$. Then

$$\begin{aligned}
|x^2 - 4| &= \left|[(x - 2) + 2]^2 - 4\right| \\
&= \left|[(x - 2)^2 + 4(x - 2) + 4] - 4\right| \\
&= \left|(x - 2)^2 + 4(x - 2)\right| \\
&\leq \left|(x - 2)^2\right| + \left|4(x - 2)\right| \\
&= |x - 2|^2 + 4|x - 2| \\
&< \delta^2 + 4\delta \\
&\leq \left(\sqrt{\frac{\varepsilon}{2}}\right)^2 + 4 \cdot \frac{\varepsilon}{8} \\
&= \frac{\varepsilon}{2} + \frac{\varepsilon}{2} \\
&= \varepsilon.
\end{aligned}$$

□

In verifying certain limits, we may need to carefully analyze whether, and if so precisely how, we can make "small" a quantity that happens to arise in our up front planning of the proof.

Example 12.7. To show

$$\lim_{x\to -1} \frac{3x}{2x + 1} = 3$$

we first perform the following preliminary analysis:

$$\left|\frac{3x}{2x+1}-3\right| = \left|\frac{3x}{2x+1}-\frac{3(2x+1)}{2x+1}\right| = \left|\frac{-3x-3}{2x+1}\right| = \frac{3|x+1|}{|2x+1|} = 3\cdot\frac{1}{|2x+1|}\cdot|x-(-1)|.$$

We recognize that as $x \to -1$ we are able to make $|x-(-1)|$ arbitrarily small. Our concern is with the factor

$$\frac{1}{|2x+1|},$$

which we also need to make "small." Of course, this quantity is small when its reciprocal

$$|2x+1|$$

is, relatively speaking, "large." Thus, we are led to investigate how we might get the quantity $|2x+1|$ to be larger than some fixed positive number for all x within some neighborhood of the number -1 that x is approaching.

As $2x+1$ is negative for x sufficiently close to -1, this means we want to be sure x stays close enough to -1 so that $2x+1$ is not equal to 0. Noting that $2x+1 = 0$ when $x = -0.5$, and -0.5 is 0.5 units away from -1, we try constraining x to be less than 0.4 units away from -1, that is, we take x within the neighborhood

$$(-1-0.4, -1+0.4) = (-1.4, -0.6)$$

of -1 (our reasoned guess at this point is that any choice of a neighborhood radius less than 0.5 might work).

Observe that

$$-1.4 < x < -0.6$$

implies that

$$2(-1.4)+1 < 2x+1 < 2(-0.6)+1,$$

that is,

$$-1.8 < 2x+1 < -0.2.$$

As $2x+1$ is negative for the specified values of x, we have

$$|2x+1| = -(2x+1),$$

and as we have determined that $2x+1 < -0.2$, it follows that

$$-(2x+1) > -(-0.2),$$

that is,

$$|2x + 1| > 0.2.$$

Hence, when we eventually choose the input tolerance δ associated with a given output tolerance ε, we must be sure to choose it no larger than 0.4, as we have demonstrated that for every x in $(-1 - 0.4, -1 + 0.4)$, it follows that $|2x + 1|$ is "large" in the sense of being larger than 0.2, which then makes $\frac{1}{|2x+1|}$ "small" in the sense of being smaller than $\frac{1}{0.2} = 5$.

This preliminary analysis is taken into account in the following proof that $\lim_{x\to -1} \frac{3x}{2x+1} = 3$.

Proof. Let ε be any positive number. Observe that if x is in the 0.4-nhood of -1, that is, if $-1.4 < x < -0.6$, it follows that $-1.8 < 2x + 1 < -0.2$ so that

$$|2x + 1| = -(2x + 1) > 0.2.$$

Let $\delta = \min\{0.4, \frac{\varepsilon}{15}\}$, which is positive, and assume $0 < |x - (-1)| < \delta$. Then

$$\begin{aligned}
\left|\frac{3x}{2x+1} - 3\right| &= \left|\frac{3x}{2x+1} - \frac{3(2x+1)}{2x+1}\right| \\
&= \left|\frac{-3x-3}{2x+1}\right| \\
&= \frac{3|x+1|}{|2x+1|} \\
&= 3 \cdot \frac{1}{|2x+1|} \cdot |x - (-1)| \\
&< 3 \cdot \frac{1}{0.2} \cdot \frac{\varepsilon}{15} \\
&= \varepsilon.
\end{aligned}$$

□

Exercise 12.10. Use the $\varepsilon\delta$-definition of limit to verify each of the following.

(a) $\lim_{x\to 3}(x^2 + 2x) = 15$.

(b) $\lim_{x\to 2}(2x^3 - 3x - 2) = 8$.

(c) $\lim_{x\to -4} \frac{x+2}{x+5} = -2$.

(d) $\lim_{x\to 9} \sqrt{x} = 3$.

A sequential characterization of function limits

Given a sequence (x_n) in the domain of a function f, we can form an associated sequence $(f(x_n))$, whose terms are the function values of the corresponding terms of (x_n).

Example 12.8. Consider the function $f : \mathbb{R} \to \mathbb{R}$ defined by $f(x) = x^2$ and the sequence $(x_n) = (\frac{1}{n})$. Then the sequence $(f(x_n))$ is

$$(f(x_n)) = \left(f\left(\frac{1}{n}\right)\right) = \left(\frac{1}{n^2}\right) = \left(1, \frac{1}{4}, \frac{1}{9}, \dots\right).$$

Exercise 12.11. Define the function $f : \mathbb{R} \to \mathbb{R}$ so that $f(x) = 3x - 5$. Identify the first three terms of the specified sequence.
(a) $(f(\frac{2n}{3}))$
(b) $(f(\frac{1}{3n^3}))$
(c) $(f((-1)^n))$

It is possible to use sequences to verify function limits. Moreover, such an approach is sometimes easier to employ than the $\varepsilon\delta$-definition of limit.

Theorem 12.9 (Sequential characterization of function limits). *The following are equivalent:*
(i) $\lim_{x\to p} f(x) = L$;
(ii) *for every sequence* (x_n) *in the domain of* f*, all of whose terms are different from* p*, if* $x_n \to p$*, it follows that* $f(x_n) \to L$.

Proof. (i) $\Rightarrow$ (ii) Suppose $\lim_{x\to p} f(x) = L$ and consider any sequence (x_n) in the domain of f whose terms are all different from p and for which $x_n \to p$. We must show $f(x_n) \to L$. To do so we begin with an arbitrary positive number ε. As $\lim_{x\to p} f(x) = L$, there exists a positive number δ such that whenever x is in the domain of f and $0 < |x - p| < \delta$, it follows that $|f(x) - L| < \varepsilon$. Then, as $x_n \to p$, there exists an index N such that when $n \geq N$ we have $|x_n - p| < \delta$. So, as it is also the case that each term of the sequence (x_n) is different from p, it follows that $0 < |x_n - p| < \delta$ when $n \geq N$. Hence, we may conclude that $|f(x_n) - L| < \varepsilon$ when $n \geq N$. Therefore, we have shown that $f(x_n) \to L$.

(ii) $\Rightarrow$ (i) We apply contraposition. Suppose that $\lim_{x\to p} f(x) \neq L$. Then there exists a positive number ε such that for each natural number n, there exists a number x_n in the domain of f for which $0 < |x_n - p| < \frac{1}{n}$ and $|f(x_n) - L| \geq \varepsilon$. The result is a sequence (x_n) in the domain of f, no term of which is p, for which $x_n \to p$ and $f(x_n) \nrightarrow L$. (Justification of these conclusions is left to Exercise 12.12.) □

Exercise 12.12. Provide justification of the conclusions listed in the final sentence of the argument for (ii) $\Rightarrow$ (i) in the proof of Theorem 12.9.

Example 12.6 (Continued). We have already used the $\varepsilon\delta$-definition of function limit to verify that

$$\lim_{x\to 2} x^2 = 4,$$

but we can use the sequential characterization of function limits to confirm this result much more easily.

Consider any sequence (x_n) none of whose terms is 2 and for which $x_n \to 2$; to show $\lim_{x\to 2} x^2 = 4$, we need only show that $x_n^2 \to 4$. But this conclusion follows immediately as the limit of the product of convergent sequences is the product of the limits of these sequences, and the sequence (x_n^2) is the product of the convergent sequence (x_n) with itself.

Function limits need not exist, and the sequential characterization of limits is often useful in explaining why a certain limit does not exist.

Example 12.10. We analytically develop the cosine and sine functions in Chapter 28, but in a very few examples prior, we make use of some results involving these functions. For instance, in this example, we use the facts that $\cos(2n\pi) = 1$ and $\cos(\pi + 2n\pi) = -1$, where n is any integer. These facts are established geometrically in Appendix D.

Consider the function $f : \mathbb{R} - \{0\} \to \mathbb{R}$ defined so that

$$f(x) = \cos\left(\frac{1}{x}\right).$$

Define sequences (a_n) and (b_n) so that

$$a_n = \frac{1}{2n\pi}$$

and

$$b_n = \frac{1}{\pi + 2n\pi},$$

and note that none of the terms of these sequences is equal to 0. As $\frac{1}{n} \to 0$ it follows that

$$a_n = \frac{1}{2n\pi} \to \frac{1}{2\pi} \cdot 0 = 0.$$

It is a straightforward application of the squeeze theorem for sequences, Theorem 10.15, to then show that $b_n \to 0$ (assigned as Exercise 12.13). However, as for all n, we have

$$f(a_n) = \cos(2n\pi) = 1$$

and

$$f(b_n) = \cos(\pi + 2n\pi) = -1,$$

we see that $f(a_n) \to 1$ while $f(b_n) \to -1$.

Now the sequential characterization of limit of a function tells us that if $\lim_{x\to 0} f(x) = L$, then every time we encounter a sequence (x_n) whose terms lie in the domain of f, for which no term of (x_n) is equal to 0, and for which (x_n) converges to 0, we would be able to conclude that the sequence $(f(x_n))$ converges to L. We have found two such sequences, (a_n) and (b_n), but with $(f(a_n))$ and $(f(b_n))$ converging to different numbers, 1 and −1. Thus, by the uniqueness of limits, it is impossible for $\lim_{x\to 0} f(x)$ to exist.

Exercise 12.13. Use Theorem 10.15, the squeeze theorem for sequences, to prove that the sequence (b_n) given in Example 12.10 converges to 0.

Exercise 12.14. Use an online or handheld graphing calculator to make a graph of the function f from Example 12.10. Try "zooming in" on the portion of the graph near the origin. How does the graph suggest that $\lim_{x\to 0} f(x)$ does not exist?

Exercise 12.15. Use the sequential characterization of function limit to verify each statement.

(a) $\lim_{x\to 3}(x^2 + 2x) = 15$.

(b) $\lim_{x\to -4} \frac{x+2}{x+5} = -2$.

(c) $\lim_{x\to 0} \frac{1}{x}$ does not exist.

Exercise 12.16. Consider the function $g : \mathbb{R} \to \mathbb{R}$ defined so that

$$g(x) = \begin{cases} 2x, & \text{if } x \text{ is rational;} \\ \frac{x}{2}, & \text{if } x \text{ is irrational.} \end{cases}$$

(a) Prove that $\lim_{x\to 0} g(x) = 0$.

(b) In Exercise 10.31, you showed that, given any real number p, there exist sequences (a_n) in $\mathbb{Q}$ and (b_n) in $\mathbb{R} - \mathbb{Q}$, none of whose terms are equal to p, such that $a_n \to p$ and $b_n \to p$. Use this information, along with the sequential characterization of limit, to prove that, for the function g defined above, if $p \neq 0$, then $\lim_{x\to p} g(x)$ does not exist.

Exercise 12.17. Prove that if $\lim_{x\to p} f(x) = L$, then $\lim_{x\to p} |f(x)| = |L|$. Then give an example to show that it is possible that $\lim_{x\to p} f(x)$ does not exist even if $\lim_{x\to p} |f(x)|$ does exist.

Exercise 12.18. Show that $\lim_{x\to p} f(x) = 0$ if and only if $\lim_{x\to p} |f(x)| = 0$.

Limits that do not exist because of unboundedness

One particular way in which a function limit does not exist results from the function values becoming unbounded as the inputs approach a particular number.

Example 12.11. Figure 12.4 displays the graph of the function g where $g(x) = \frac{1}{x^2}$.

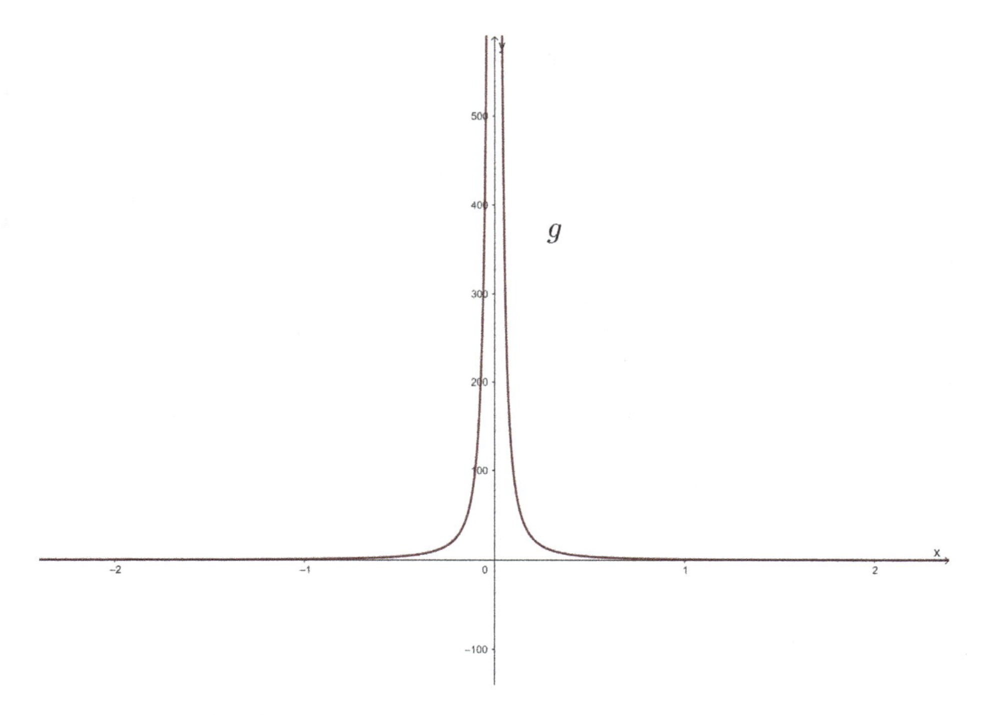

Figure 12.4: Graph of the function from Example 12.11.

The graph reveals that the values of g are growing without bound as the inputs are chosen ever closer to 0.

We can use the sequential characterization of limits to prove that $\lim_{x\to 0} g(x)$ does not exist. Consider the sequence $(\frac{1}{n})$, which converges to 0, but none of whose terms are 0. Note that the sequence $(g(\frac{1}{n})) = (\frac{1}{(\frac{1}{n})^2}) = (n^2)$, however, diverges to ∞, hence, does not converge. Thus, it is impossible for g to have a limit at 0.

When the values of a function f become larger than any positive number as the inputs approach p, we say that $f(x)$ **diverges to** ∞ **as** x **approaches** p, and we may write

$$\lim_{x\to p} f(x) = \infty$$

(you are asked to make this somewhat informal definition more precise in Exercise 12.19). Similarly, if the values of f become smaller than any negative number as the inputs approach p, we say that $f(x)$ **diverges to** $-\infty$ **as** x **approaches** p, and we may write

$$\lim_{x\to p} f(x) = -\infty.$$

Keep in mind, though, that these kinds of statements are simply notational conveniences. The symbols ∞ and $-\infty$ do not designate real numbers, and so-called *infinite limits* are really not limits at all, but simply provide some specificity as to why a limit does not exist.

When the values of a function become unbounded as the inputs approach p we also say the line $x = p$ is a **vertical asymptote** of the graph of the function.

Example 12.11 (Continued). Since the values of $g(x) = \frac{1}{x^2}$ become arbitrarily large and positive as x approaches 0, we may write $\lim_{x\to 0} g(x) = \infty$. Thus, as is evident from the graph displayed earlier, the line $x = 0$ is a vertical asymptote of the graph of g.

Exercise 12.19. Consult Definition 12.2, the $\varepsilon\delta$-definition of function limit, and the definition given in Chapter 10 for divergence of a sequence to ∞, in order to formulate a precise definition for $\lim_{x\to p} f(x) = \infty$.

Exercise 12.20. Suppose that f and g are both real-valued functions with domain $\mathbb{R} - \{p\}$, that $g(x) \geq f(x)$ for all x in $\mathbb{R} - \{p\}$, and that $\lim_{x\to p} f(x) = \infty$. Explain why it must also be true that $\lim_{x\to p} g(x) = \infty$.

Given a subset A of the domain of a function f, we say f is **bounded on** A if there is a positive number B for which $|f(x)| \leq B$ for all x in A. A function is called **bounded** if it is bounded on its entire domain, that is, if its range is a bounded subset of $\mathbb{R}$.

Exercise 12.21. Prove that if $\lim_{x\to p} f(x) = 0$ and the function g is bounded on some neighborhood of p, then $\lim_{x\to p} f(x)g(x) = 0$.

Limits that do not exist because of differing behavior from the left and right

Sometimes a limit does not exist because the behavior a function exhibits as the inputs approach a particular number through values less than the number (i. e., from the left) is different from the behavior the function exhibits as the inputs approach the same number through values greater than the number (i. e., from the right).

Example 12.12. The graph of the function h defined by $h(x) = \frac{x^4+|x|}{x}$ is displayed in Figure 12.5.

The graph suggests that $\lim_{x\to 0} h(x)$ does not exist, because the function values are approaching -1 as the inputs approach 0 from below 0, but are approaching 1 as the inputs approach 0 from above 0.

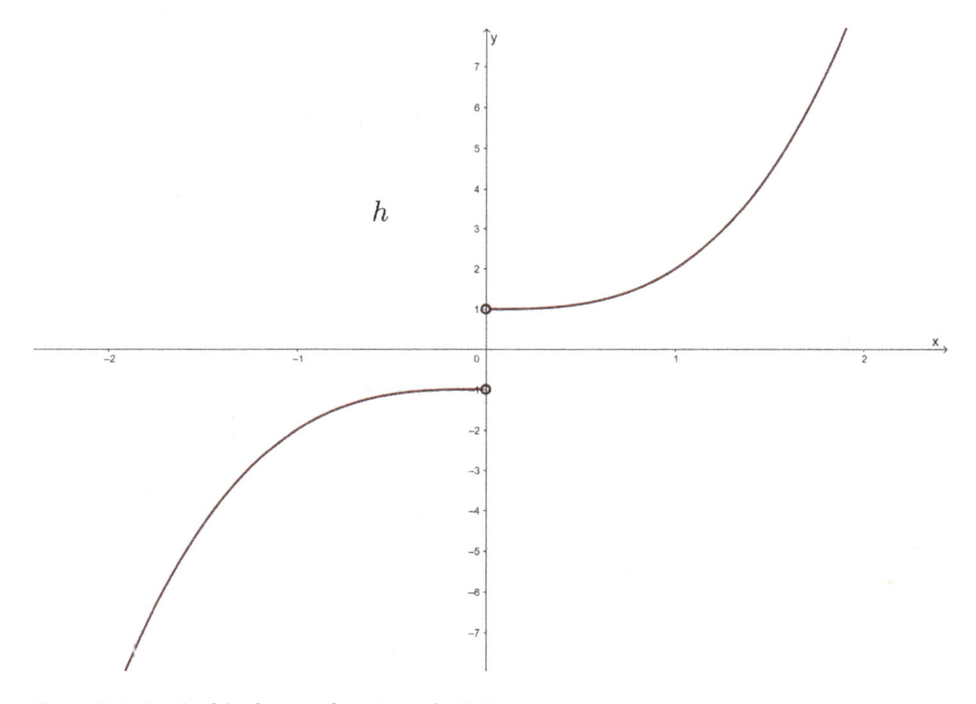

Figure 12.5: Graph of the function from Example 12.12.

To facilitate discussion of the limiting behavior of a function when the inputs approach a particular number from only the left or only the right, we introduce the notion of *one-sided limits*.

Assuming that every neighborhood of p includes points of the domain of a function f that are less than p, a real number L is a **left-hand limit of f at p**, denoted $\lim_{x\to p^-} f(x) = L$, if for every positive number ε, there exists a positive number δ such that whenever x is in the domain of f and $p - \delta < x < p$, it follows that $|f(x) - L| < \varepsilon$.

Assuming that every neighborhood of p includes points of the domain of a function f that are greater than p, a real number L is a **right-hand limit of f at p**, denoted $\lim_{x\to p^+} f(x) = L$, if for every positive number ε, there exists a positive number δ such that whenever x is in the domain of f and $p < x < p + \delta$, it follows that $|f(x) - L| < \varepsilon$.

Example 12.12 (Continued). Note that

$$\lim_{x\to 0^-} \frac{x^4 + |x|}{x} = -1$$

as, assuming $\varepsilon > 0$, if we take $\delta = \sqrt[3]{\varepsilon}$, then whenever $0 - \delta = -\sqrt[3]{\varepsilon} < x < 0$, it follows that

$$\left|\frac{x^4 + |x|}{x} - (-1)\right| = |x^3 - 1 + 1| = -x^3 < \left(\sqrt[3]{\varepsilon}\right)^3 = \varepsilon.$$

Similarly,

$$\lim_{x\to 0^+} \frac{x^4 + |x|}{x} = 1,$$

as, assuming $\varepsilon > 0$, if we again take $\delta = \sqrt[3]{\varepsilon}$, then whenever $0 < x < \sqrt[3]{\varepsilon} = 0 + \delta$, it follows that

$$\left| \frac{x^4 + |x|}{x} - 1 \right| = \left| x^3 + 1 - 1 \right| = x^3 < \left(\sqrt[3]{\varepsilon}\right)^3 = \varepsilon.$$

If it is possible for the inputs of a function to approach a number p from one side only, and this particular one-sided limit does exist, it is also the function's limit at p.

Exercise 12.22. The principal square root function $y = \sqrt{x}$ is undefined for negative inputs, so it is impossible for inputs to this function to approach 0 from the left.
(a) Use the definition of right-hand limit to show that $\lim_{x\to 0^+} \sqrt{x} = 0$.
(b) Use the definition of function limit to show that $\lim_{x\to 0} \sqrt{x} = 0$.

Exercise 12.23. Use Theorem 12.9, the sequential characterization of function limits, to prove that $\lim_{x\to 0} h(x)$ does not exist, where h is the function defined in Example 12.12.

It is apparent that one-sided limits are related to (two-sided) limits, as described by the following theorem.

Theorem 12.13. *Assuming* $\lim_{x\to p^-} f(x)$ *and* $\lim_{x\to p^+} f(x)$ *both exist, the following are equivalent:*
(i) $\lim_{x\to p} f(x) = L$;
(ii) $\lim_{x\to p^-} f(x) = \lim_{x\to p^+} f(x) = L$.

Exercise 12.24. Use the definitions of left-hand and right-hand limits to show that $\lim_{x\to 0^-} |x| = 0$ and $\lim_{x\to 0^+} |x| = 0$. What does Theorem 12.13 then allow us to conclude?

Many of the results that hold for function limits have natural counterparts for one-sided limits. For instance, when it exists a one-sided limit is unique, there are sequential characterizations for one-sided limits, and so forth.

Exercise 12.25. Formulate a sequential characterization for $\lim_{x\to p^-} f(x) = L$ and then prove your characterization is equivalent to the $\varepsilon\delta$-definition of left-hand limit.

Exercise 12.26. Formulate definitions for each of the following statements: $\lim_{x\to p^-} f(x) = \infty$; $\lim_{x\to p^-} f(x) = -\infty$; $\lim_{x\to p^+} f(x) = \infty$; and $\lim_{x\to p^+} f(x) = -\infty$. Then use them to determine $\lim_{x\to 0^-} \frac{1}{x}$ and $\lim_{x\to 0^+} \frac{1}{x}$.

Arithmetic properties of limits

Given real-valued functions f and g, we define functions called the **sum** $f + g$, the **difference** $f - g$, the **product** fg, and the **quotient** $\frac{f}{g}$ so that, for any x in the intersection of the domains of f and g,

$$(f + g)(x) = f(x) + g(x),$$
$$(f - g)(x) = f(x) - g(x),$$
$$(fg)(x) = f(x) \cdot g(x),$$

and

$$\left(\frac{f}{g}\right)(x) = \frac{f(x)}{g(x)},$$

with the additional stipulation that $g(x) \neq 0$ when forming $\frac{f}{g}$.

Also, as a special case of the formation of a product function, given a real number c and a real-valued function f, we define a function called the **constant multiple** cf so that, for any x in the domain of f,

$$(cf)(x) = c \cdot f(x).$$

Example 12.14. Consider the functions $f : \mathbb{R} - \{6\} \to \mathbb{R}$ and $g : [1, \infty) \to \mathbb{R}$ defined by

$$f(x) = \frac{1}{x - 6}$$

and

$$g(x) = \sqrt{x - 1}.$$

Then for each x in the intersection $[1, \infty) - \{6\}$ of the domains of f and g, we have

$$(f + g)(x) = \frac{1}{x - 6} + \sqrt{x - 1},$$
$$(f - g)(x) = \frac{1}{x - 6} - \sqrt{x - 1},$$

and

$$(fg)(x) = \frac{\sqrt{x - 1}}{x - 6}.$$

Also, for each x in $(1, \infty) - \{6\}$, the set of points in the intersection of the domains of f and g for which $\sqrt{x - 1} \neq 0$, we have

$$\left(\frac{f}{g}\right)(x) = \frac{1}{(x-6)\sqrt{x-1}}.$$

Additionally, for each x in the domain $[1, \infty)$ of g, we have

$$(-2g)(x) = -2\sqrt{x-1}.$$

Limits of functions, provided they exist, behave in the expected ways with respect to the arithmetic operations of addition, subtraction, multiplication, and division.

Theorem 12.15. *Let p be a limit point of the intersection of the domains of the functions f and g, and suppose $\lim_{x\to p} f(x) = L$ and $\lim_{x\to p} g(x) = M$.*
(1) $\lim_{x\to p}(f(x) + g(x)) = L + M$.
(2) $\lim_{x\to p}(f(x) - g(x)) = L - M$.
(3) $\lim_{x\to p}(f(x) \cdot g(x)) = L \cdot M$.
(4) $\lim_{x\to p} \frac{f(x)}{g(x)} = \frac{L}{M}$, *provided $M \neq 0$ and $g(x) \neq 0$ for all x in the domain of g lying within some neighborhood of p, except perhaps for $x = p$.*
(5) $\lim_{x\to p} cf(x) = c \lim_{x\to p} f(x)$, *for any real number c.*

The results stated in this theorem can be proved directly from the $\varepsilon\delta$-definition of limit, Definition 12.2, or using the sequential characterization of limits, Theorem 12.9. We establish (1) using both approaches.

Proof of Theorem 12.15(1) *using the $\varepsilon\delta$-definition of limit.* Consider any positive number ε. As $\lim_{x\to p} f(x) = L$, there exists a positive number δ_1 such that whenever x is in the domain of f and $0 < |x - p| < \delta_1$, it follows that

$$|f(x) - L| < \frac{\varepsilon}{2}.$$

Similarly, as $\lim_{x\to p} g(x) = M$, there exists a positive number δ_2 such that whenever x is in the domain of g and $0 < |x - p| < \delta_2$, it follows that

$$|g(x) - M| < \frac{\varepsilon}{2}.$$

Now let $\delta = \min\{\delta_1, \delta_2\}$, which is positive, and consider any x in the intersection of the domains of f and g for which $0 < |x - p| < \delta$. It follows that x is in the domains of both f and g, and that $0 < |x - p| < \delta_1$ and $0 < |x - p| < \delta_2$, from which we may conclude that

$$\begin{aligned} |(f(x) + g(x)) - (L + M)| &= |(f(x) - L) + (g(x) - M)| \\ &\leq |f(x) - L| + |g(x) - M| \\ &< \frac{\varepsilon}{2} + \frac{\varepsilon}{2} \\ &= \varepsilon. \end{aligned}$$

Thus, by definition, $\lim_{x\to p}(f(x) + g(x)) = L + M$. □

Proof of Theorem 12.15(1) *using sequences.* Consider any sequence (x_n) in the intersection of the domains of f and g for which no term is equal to p; further assume that $x_n \to p$. As $\lim_{x\to p} f(x) = L$, it follows that $f(x_n) \to L$. Similarly, as $\lim_{x\to p} g(x) = M$, it follows that $g(x_n) \to M$. Then, as the sum of convergent sequences converges to the sum of the sequences' limits, we may conclude that the sequence

$$((f+g)(x_n)) = (f(x_n) + g(x_n))$$

converges to $L + M$. Hence, according to the sequential characterization of limit of a function, we must have

$$\lim_{x\to p}(f(x) + g(x)) = L + M. \qquad \square$$

Exercise 12.27. Use the sequential characterization of limit of a function to prove Theorem 12.15(3).

Example 12.16. Applying the arithmetic properties of limits, we have

$$\begin{aligned}
\lim_{x\to 2} \frac{x+5}{x^2 - 7x + 4} &= \frac{\lim_{x\to 2}(x+5)}{\lim_{x\to 2}(x^2 - 7x + 4)} \\
&= \frac{\lim_{x\to 2} x + \lim_{x\to 2} 5}{\lim_{x\to 2} x^2 - \lim_{x\to 2} 7x + \lim_{x\to 2} 4} \\
&= \frac{2+5}{(\lim_{x\to 2} x)^2 - (\lim_{x\to 2} 7)(\lim_{x\to 2} x) + 4} \\
&= \frac{7}{2^2 - 7\cdot 2 + 4} \\
&= -\frac{7}{6}.
\end{aligned}$$

Recall that a function $f : \mathbb{R} \to \mathbb{R}$ is called a **polynomial function** if

$$f(x) = a_n x^n + a_{n-1}x^{n-1} + \cdots + a_2 x^2 + a_1 x + a_0,$$

for some nonnegative integer n and some real numbers $a_0, a_1, a_2, \ldots, a_n$. The arithmetic properties of limits tell us that limits of polynomial functions can be found by "direct substitution," that is, by evaluating the polynomial function at the number the inputs are approaching.

Theorem 12.17. *If f is a polynomial function, then* $\lim_{x\to p} f(x) = f(p)$.

Example 12.18. We can evaluate

$$\lim_{x\to -1}(x^3 - 2x + 4)$$

by simply evaluating $f(-1)$, where $f(x) = x^3 - 2x + 4$, since the function f is a polynomial function. Thus,

$$\lim_{x \to -1} (x^3 - 2x + 4) = f(-1) = (-1)^3 - 2(-1) + 4 = 5.$$

A **rational function** is a quotient of polynomial functions. The arithmetic properties of limits permit us to evaluate the limit of a rational function via direct substitution, provided that the limit is being taken at a point of the function's domain.

Theorem 12.19. *If f is a rational function and p is in the domain of f, then* $\lim_{x \to p} f(x) = f(p)$.

Example 12.16 (Continued). Note that the function f, where

$$f(x) = \frac{x + 5}{x^2 - 7x + 4}$$

is a rational function. Earlier we determined that

$$\lim_{x \to 2} f(x) = -\frac{7}{6}.$$

Note, however, that the arithmetic properties of limits essentially allow us to obtain this limit by evaluating f at the input 2. That is,

$$\lim_{x \to 2} f(x) = f(2) = \frac{2 + 5}{2^2 - 7 \cdot 2 + 4} = -\frac{7}{6}.$$

Exercise 12.28. Evaluate the following limits using the arithmetic properties of limits and/or the results established above for finding limits of polynomial and rational functions.

(a) $\lim_{x \to 10} (\frac{x^2 - x}{x - 4} + 5x)$

(b) $\lim_{x \to -1} (x^{101} - \frac{1}{x^{100}})$

Exercise 12.29. Evaluate $\lim_{x \to 0^-} \frac{x}{|x|}$ and $\lim_{x \to 0^+} \frac{x}{|x|}$. What does Theorem 12.13 then allow us to conclude?

Exercise 12.30. Suppose both $\lim_{x \to p} f(x)$ and $\lim_{x \to p} (f(x) - g(x))$ exist. Must $\lim_{x \to p} g(x)$ exist? Defend your conclusion.

Exercise 12.31. Suppose $\lim_{x \to p} f(x)$ exists, but $\lim_{x \to p} g(x)$ does not exist. Is it possible that $\lim_{x \to p} (f(x) + g(x))$ exists? Defend your conclusion.

The squeeze theorem

When the values of a function lie between the corresponding values of two other functions that both have the same limit at a particular point, the limit at this point of the function which is “squeezed” between these other two functions also exists and must be equal to the common limit of the functions doing the “squeezing.”

Theorem 12.20 (The squeeze theorem). *If f, g, and h are functions with $g(x) \leq f(x) \leq h(x)$ for all x in some neighborhood of p, except perhaps at p itself, and if $\lim_{x\to p} g(x) = \lim_{x\to p} h(x) = L$, then $\lim_{x\to p} f(x) = L$.*

The proof of the squeeze theorem is assigned as Exercise 12.33.

Example 12.21. We can use the squeeze theorem to evaluate

$$\lim_{x\to 0} x\cos\left(\frac{1}{x}\right).$$

Note that as

$$-1 \leq \cos\left(\frac{1}{x}\right) \leq 1$$

for all nonzero real numbers x, it follows that

$$-|x| \leq x\cos\left(\frac{1}{x}\right) \leq |x|.$$

As you have already shown in Exercise 12.24 that $\lim_{x\to 0} |x| = 0$, from which it follows that $\lim_{x\to 0} -|x| = 0$ as well, the squeeze theorem permits us to conclude that

$$\lim_{x\to 0} x\cos\left(\frac{1}{x}\right) = 0.$$

Exercise 12.32. Use an online or handheld graphing calculator to examine the graph of the function g where $g(x) = x\cos(\frac{1}{x})$. Does the graph suggest the same conclusion that we reached in Example 12.21?

Exercise 12.33. Prove Theorem 12.20, the squeeze theorem.

Exercise 12.34. Prove that if $\lim_{x\to p} f(x) = L$ and $a \leq f(x) \leq b$ for all x in some neighborhood of p, except perhaps at p itself, then $a \leq L \leq b$.

The fundamental property of limits

The limit of an arithmetic combination of functions exhibits an **indeterminate form** when no conclusion as to the existence or value of the limit can be reached solely on the basis of the corresponding limits of the functions being combined.

Theorem 12.22. *Given any real numbers p and c, there exist functions f and g for which* $\lim_{x\to p} f(x) = 0$ *and* $\lim_{x\to p} g(x) = 0$, *and for which* $\lim_{x\to p} \frac{f(x)}{g(x)} = c$.

Moreover, given any real number p, there exist functions f and g for which $\lim_{x\to p} f(x) = 0$ *and* $\lim_{x\to p} g(x) = 0$, *and for which* $\lim_{x\to p} \frac{f(x)}{g(x)}$ *is undefined.*

In other words, in the context of function limits, $\frac{0}{0}$ *is an indeterminate form.*

Proof. Let p and c be any real numbers. Observe that $\lim_{x\to p} c(x-p) = 0$ and $\lim_{x\to p}(x-p) = 0$, while

$$\lim_{x\to p} \frac{c(x-p)}{(x-p)} = c,$$

as, given any positive number ε, if we take $\delta = 1$ and consider any number x such that $x \neq p$, it follows that

$$\left|\frac{c(x-p)}{x-p} - c\right| = |c-c| = 0 < \varepsilon.$$

Now let p be any real number. Observe that $\lim_{x\to p}(x-p) = 0$ and $\lim_{x\to p}(x-p)^2 = 0$, while

$$\lim_{x\to p} \frac{x-p}{(x-p)^2}$$

does not exist because the values of $\frac{x-p}{(x-p)^2}$ become unbounded as x approaches p (see Exercise 12.35). □

Exercise 12.35. Verify that $\lim_{x\to p} \frac{x-p}{(x-p)^2}$ does not exist because of unboundedness.

Thus, when we encounter an indeterminate form while attempting to evaluate a limit, we can conclude nothing without further analysis: The limit may or may not exist, and if it exists, it may have any real number value. To evaluate such limits, we often make use of the following theorem, which is an immediate consequence of the definition of function limit.

Theorem 12.23 (The fundamental property of limits). *If f and g are functions for which* $f(x) = g(x)$ *for all x in some neighborhood of p, except perhaps at p itself, and if* $\lim_{x\to p} g(x)$ *exists, then* $\lim_{x\to p} f(x)$ *exists and* $\lim_{x\to p} f(x) = \lim_{x\to p} g(x)$.

We have chosen to label this theorem as the *fundamental property* of limits because it makes explicit both of the following attributes of function limits.

(1) The value, or lack thereof, of a function at a number p that the function's inputs are approaching really has no bearing at all on the existence or value of the limit of the function at p.
(2) Functions which agree with one another for all inputs "near," but different from, a number p must have the same limiting behavior at p, even if they do not agree at p itself.

Example 12.24. The limit

$$\lim_{x\to -2} \frac{(x+2)^2}{x^2+3x+2}$$

is indeterminate of form $\frac{0}{0}$. We can use the fundamental property of limits to evaluate this limit by recognizing that the function f defined by

$$f(x) = \frac{(x+2)^2}{x^2+3x+2}$$

agrees with (i. e., takes the same values as) the function g defined by

$$g(x) = \frac{x+2}{x+1}$$

in the neighborhood, say, $(-2-1, -2+1) = (-3, -1)$, of -2, except at -2 itself.

To see that this is the case, note that

$$f(x) = \frac{(x+2)^2}{x^2+3x+2} = \frac{(x+2)(x+2)}{(x+2)(x+1)} = \frac{x+2}{x+1} = g(x)$$

for all real numbers x other than -2 and -1. We do not care about the fact that f is undefined at -2, whereas $g(-2) = \frac{-2+2}{2+1} = 0$, because the limit is being taken as x *approaches* -2, which conceptually does not involve what either function, f or g, is doing at -2. Furthermore, the only other number where f and g "disagree" is -1, which is "far" from -2 in the sense that there are neighborhoods, for example $(-3, -1)$, of -2 which do not include -1. This is the reasoning behind the "factor and cancel" method of evaluating limits used in elementary calculus courses:

$$\lim_{x\to -2} \frac{(x+2)^2}{x^2+3x+2} = \lim_{x\to -2} \frac{(x+2)(x+2)}{(x+2)(x+1)} = \lim_{x\to -2} \frac{x+2}{x+1} = 0.$$

Exercise 12.36. Evaluate $\lim_{x\to 1} \frac{x^3-x}{x^2+2x-3}$.

Exercise 12.37. Prove Theorem 12.23, the fundamental property of limits.

Exercise 12.38. Suppose $\lim_{x\to p} f(x) = L$ for some real number L and that $\lim_{x\to p} g(x) = \infty$.

(a) Show that if $L > 0$, then $\lim_{x\to p} f(x) \cdot g(x) = \infty$.

(b) Give examples to show that if $L = 0$, it is possible for $\lim_{x\to p} f(x) \cdot g(x) = \infty$, but also possible for $\lim_{x\to p} f(x) \cdot g(x) = c$, where c can be any real number. What does this make $0 \cdot \infty$?

Limits at infinity

There are situations in which a function's values approach a particular real number as the inputs become arbitrarily large or arbitrarily small (in the negative sense).

Example 12.25. Shown in Figure 12.6 is the graph of the reciprocal function f, where $f(x) = \frac{1}{x}$.

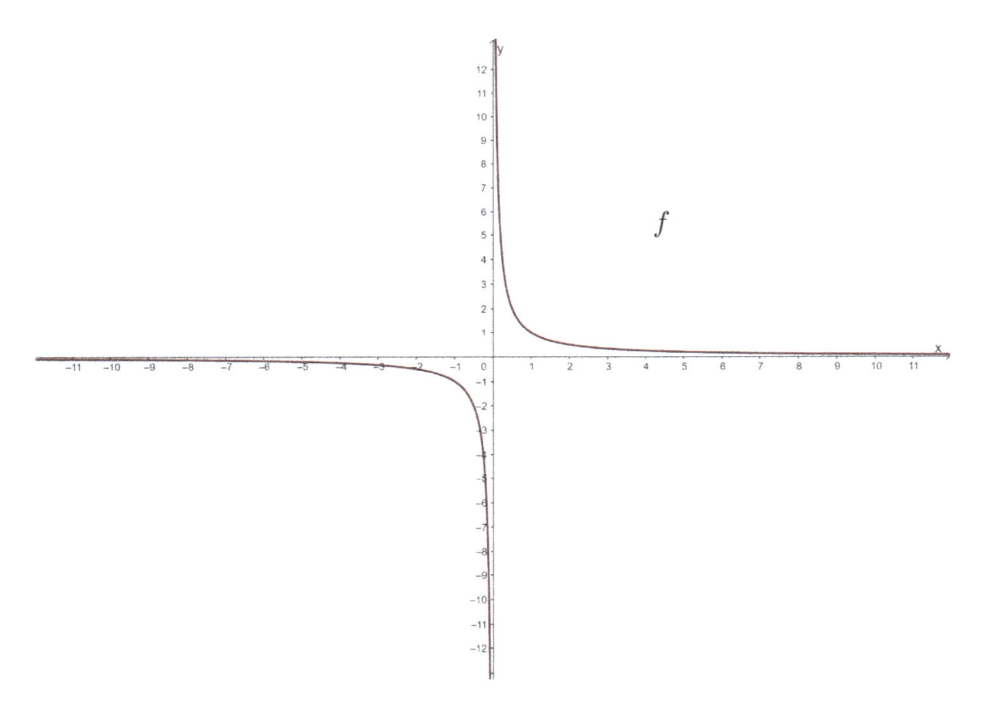

Figure 12.6: Graph of the function from Example 12.25.

Observe that as the inputs increase without bound (i. e., "tend toward ∞"), it appears that the corresponding outputs become ever closer to 0; similar behavior is exhibited as the inputs decrease without bound (i. e., "tend toward $-\infty$").

To accommodate such situations, we define so-called *limits at infinity*.

A number L is called a **limit of the function** f **at** ∞, denoted $\lim_{x\to\infty} f(x) = L$, if for every positive number ε, there exists a positive number ω such that whenever x is in the domain of f and $x > \omega$, it follows that $|f(x) - L| < \varepsilon$.

A number L is called a **limit of the function** f **at** $-\infty$, denoted $\lim_{x\to-\infty} f(x) = L$, if for every positive number ε, there exists a positive number ω such that whenever x is in the domain of f and $x < -\omega$, it follows that $|f(x) - L| < \varepsilon$.

In these definitions, ω is the lowercase Greek letter *omega*. Note that the value of ω tends to vary with the value of ε (we anticipate having to choose a larger value of ω as ε becomes smaller).

When either $\lim_{x\to\infty} f(x) = L$ or $\lim_{x\to-\infty} f(x) = L$, we call the line $y = L$ a **horizontal asymptote** of the graph of f.

Example 12.25 (Continued). The graph displayed in Figure 12.6 suggests that $\lim_{x\to\infty} \frac{1}{x} = 0$ and $\lim_{x\to-\infty} \frac{1}{x} = 0$. We verify the former, leaving the latter as an exercise.

Suppose $\varepsilon > 0$; then $\omega = \frac{1}{\varepsilon} > 0$. Now whenever $x > \omega$, we have

$$\left|\frac{1}{x} - 0\right| = \frac{1}{x} < \frac{1}{\omega} = \varepsilon.$$

Hence, $\lim_{x\to\infty} \frac{1}{x} = 0$. Thus, as is suggested graphically, the line $y = 0$ is a horizontal asymptote of the graph of the reciprocal function.

Exercise 12.39. Show that $\lim_{x\to-\infty} \frac{1}{x} = 0$.

Exercise 12.40. Suppose we have a function $f : [1, \infty) \to \mathbb{R}$ for which $\lim_{x\to\infty} f(x) = L$. Define the sequence y so that $y_n = f(n)$. Prove that $y_n \to L$.

---

13 An introduction to continuous functions

In high school algebra, a *continuous* function is one whose graph has no breaks, jumps, holes, and so on, so that the graph can be “drawn without lifting our pencil off the paper.” Continuity of a function imparts a level of predictability for the values of the function that permits us to prove a number of results that are not always true more generally. To explore these results, we must first move beyond a purely intuitive pictorial perspective and develop an analytic definition of continuity of a function.

Continuity of a function: a precise mathematical definition

As an aid to understanding the analytic definition of continuity we shall present, consider the graph, displayed in Figure 13.1, of a generic function f that appears to be continuous at, among other inputs, the real number p.

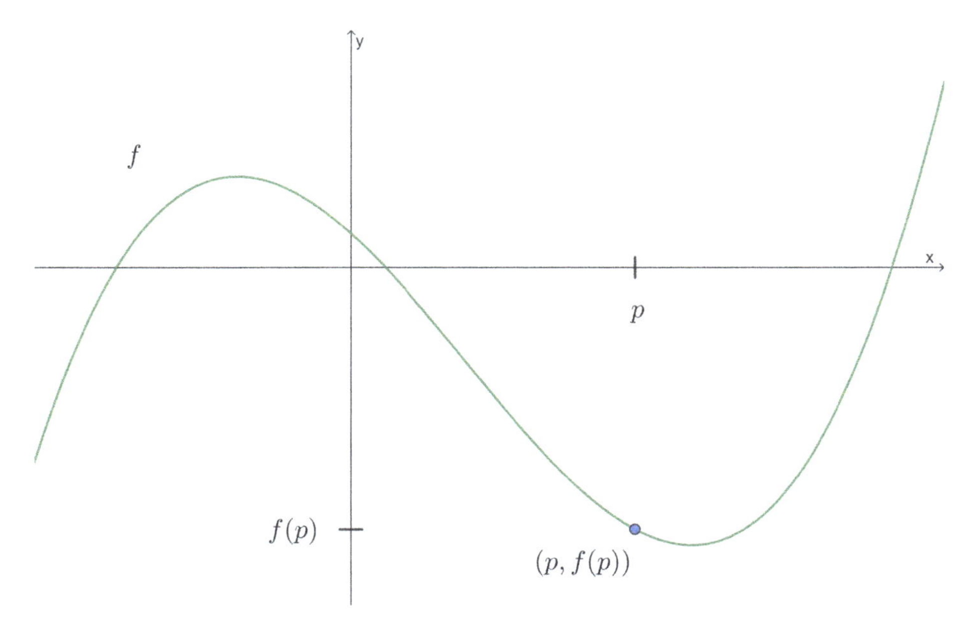

Figure 13.1: Graph of a function f that appears to be continuous at p.

Visually, the graph is uninterrupted at the point $(p, f(p))$, which suggests the outputs of f stay “as close as we want” to $f(p)$ for all inputs to f “sufficiently close” to p. The following definition makes this idea precise.

Definition 13.1 (The $\varepsilon\delta$-definition of continuity of a function). The function f is **continuous at the point** p of its domain if for every positive number ε, there exists a positive

https://doi.org/10.1515/9783111429564-013

number δ such that whenever x is in the domain of f and $|x - p| < \delta$, it follows that $|f(x) - f(p)| < \varepsilon$.

As used in this definition, and as was also the case for the $\varepsilon\delta$-definition of *function limit* presented in Chapter 12, the positive number ε is viewed as an *output tolerance*, the positive number δ as an *input tolerance*, and different values of ε may (in fact, usually do) give rise to different values of δ.

The fact that the output tolerance ε is allowed to be any positive number, no matter how small, accommodates our expectation that for a function f that is continuous at p, we can make the outputs "as close as we want" to $f(p)$. The existence of the corresponding positive input tolerance δ describes what is meant by inputs being "sufficiently close" or "close enough" to p for the corresponding outputs to be "close to $f(p)$." Thus, in words, continuity of a function f at an input p means that for each positive output tolerance there is a corresponding positive input tolerance with the property that all inputs closer to p than the input tolerance are mapped by f to outputs that are closer to $f(p)$ than the originally specified output tolerance.

Figure 13.2 provides a visualization in which a specific input tolerance δ corresponding to a specific output tolerance ε is displayed. Note how the part of the graph of f with x-coordinates (inputs) between $p - \delta$ and $p + \delta$ (between the dashed vertical lines) has y-coordinates (outputs) between $f(p) - \varepsilon$ and $f(p) + \varepsilon$ (between the dashed horizontal lines).

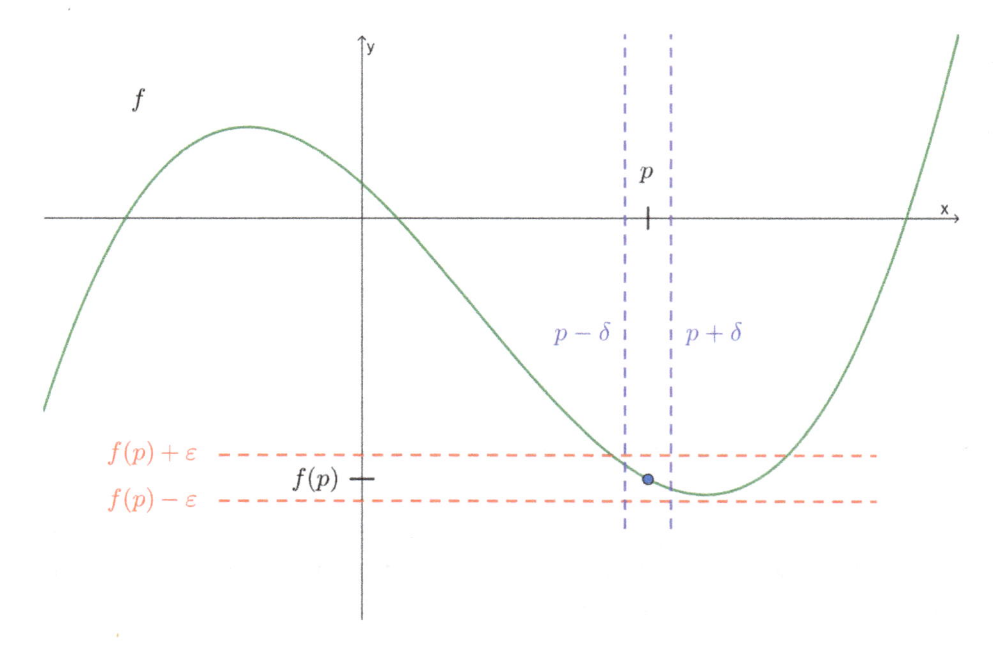

Figure 13.2: Inputs x that are less than δ units from p are mapped to outputs $f(x)$ that are less than ε units from $f(p)$.

Example 13.2. The graph of the function $f : \mathbb{R} \to \mathbb{R}$, where $f(x) = 3x + 1$ is displayed in Figure 13.3.

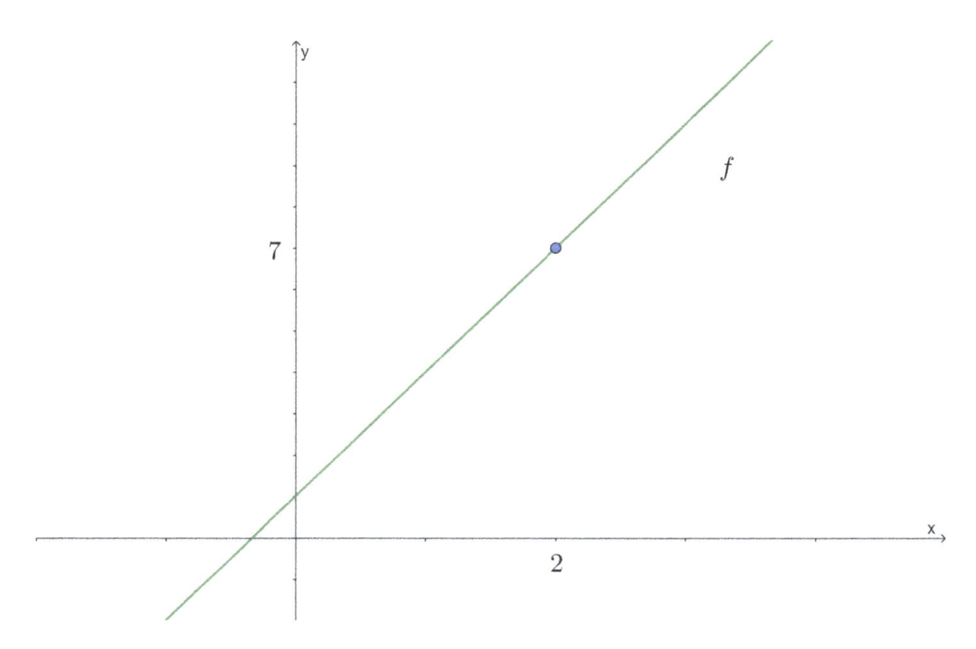

Figure 13.3: Graph of the function from Example 13.2.

The graph suggests that f is continuous at all of its inputs, in particular, the input 2. According to the $\varepsilon\delta$-definition of continuity, then, for any specified positive output tolerance, say $\varepsilon = 0.6$, there should exist a corresponding positive input tolerance δ so that whenever $|x - 2| < \delta$, it follows that $|f(x) - f(2)| < \varepsilon$, that is, $|(3x + 1) - 7| < 0.6$. We can find a suitable value for δ by observing that

$$\left|(3x + 1) - 7\right| = |3x - 6| = 3|x - 2|,$$

which tells us the desired inequality

$$\left|(3x + 1) - 7\right| < 0.6$$

is equivalent to

$$3|x - 2| < 0.6,$$

that is,

$$|x - 2| < 0.2.$$

Hence, we may take $\delta = 0.2$ as our input tolerance corresponding to the original output tolerance $\varepsilon = 0.6$. Thus, the outputs $f(x)$ of this function are always less than $\varepsilon = 0.6$ units from $f(2) = 7$ when the inputs x are chosen so that they are less than $\delta = 0.2$ units from 2.

Exercise 13.1. For the function $f : \mathbb{R} \to \mathbb{R}$ with $f(x) = 3x + 1$, find the numerical value of a positive input tolerance δ so that whenever the distance $|x-2|$ between an input x and the input 2 is less than δ, the distance $|f(x) - f(2)|$ between the corresponding outputs $f(x)$ and $f(3)$ is less than the output tolerance $\varepsilon = 0.0036$.

We have demonstrated how we could find a numerical value for the input tolerance δ when a numerical value for the output tolerance ε has been specified. But in order to satisfy the definition of continuity, we must be able to generate a positive value for δ no matter what positive number has been specified for ε. So, in general, to establish continuity, we must show how to find δ from an anonymous ε. Finding δ usually begins with a careful analysis of the condition $|f(x) - f(p)| < \varepsilon$ that we ultimately hope to achieve, and this analysis is usually based on the way in which the particular function f has been defined.

Example 13.2 (Continued). Suppose now that ε is an arbitrary positive number. We are looking to determine a corresponding positive number δ for which $|x - 2| < \delta$ implies $|f(x) - f(2)| < \varepsilon$.

To discover a candidate for δ we examine what our desired conclusion $|f(x)-f(2)| < \varepsilon$ tells us about an input x, much the same as we did in the special case above where ε was assigned the particular value 0.6. Specifically, we observe that

$$\left|f(x) - f(2)\right| = \left|(3x + 1) - 7\right| = |3x - 6| = 3|x - 2|,$$

which means the inequality

$$\left|f(x) - f(2)\right| < \varepsilon$$

is equivalent to the inequality

$$|x - 2| < \frac{\varepsilon}{3}.$$

Thus, for any specified output tolerance ε, we are conjecturing that a corresponding input tolerance δ can be obtained by dividing the output tolerance by 3. For instance, if the output tolerance is $\varepsilon = 0.6$, we should be able to take $\delta = \frac{\varepsilon}{3} = \frac{0.6}{3} = 0.2$ (which agrees with our earlier conclusion), and if the output tolerance is $\varepsilon = 0.0036$, as in Exercise 13.1, we could take $\delta = \frac{\varepsilon}{3} = \frac{0.0036}{3} = 0.0012$.

Here is a write-up of the proof that f is continuous at 2, a write-up based on the above preliminary analysis which suggests taking $\delta = \frac{\varepsilon}{3}$.

Proof. Consider any positive number ε and let $\delta = \frac{\varepsilon}{3}$, which is also a positive number. Suppose $|x - 2| < \delta$. Then

$$|f(x) - f(2)| = |(3x+1) - 7| = |3x - 6| = 3|x - 2| < 3\delta = 3 \cdot \frac{\varepsilon}{3} = \varepsilon. \qquad \square$$

Because

$$|x - p| < \delta \quad \text{if and only if} \quad x \in (p - \delta, p + \delta)$$

and

$$|f(x) - f(p)| < \varepsilon \quad \text{if and only if} \quad f(x) \in (f(p) - \varepsilon, f(p) + \varepsilon),$$

we immediately obtain the following characterization of continuity of a function by means of neighborhoods.

Theorem 13.3 (Neighborhood characterization of continuity). *Given a function f and a point p in the domain of f, the following are equivalent:*
(i) *f is continuous at p;*
(ii) *for every positive number ε, there exists a positive number δ such that whenever x is in the domain of f and $x \in (p - \delta, p + \delta)$, it follows that $f(x) \in (f(p) - \varepsilon, f(p) + \varepsilon)$.*

Condition (ii) of the theorem is often viewed as saying that for any neighborhood of $f(p)$, there is a corresponding neighborhood of p for which all points in the domain of f and in this neighborhood of p are mapped by f to points in the originally specified neighborhood of $f(p)$.

One other detail to note in relation to the $\varepsilon\delta$-definition of continuity is that when there does exist an input tolerance δ corresponding to a specified output tolerance ε, this input tolerance is far from unique, as any smaller positive number could also serve as the value of δ.

Exercise 13.2. In establishing the continuity of the function f of Example 13.2 at the input 2, we found that for any given positive output tolerance ε, the number $\delta = \frac{\varepsilon}{3}$ could serve as the corresponding input tolerance.
(a) Note that $\frac{\varepsilon}{4} < \frac{\varepsilon}{3}$. Show that our proof that f is continuous at 2 still works if we take $\delta = \frac{\varepsilon}{4}$ instead of $\delta = \frac{\varepsilon}{3}$.
(b) Explain why our proof works if any positive number less than or equal to $\frac{\varepsilon}{3}$ is taken as the value of δ.

Exercise 13.3. Prove that the function f defined so that $f(x) = \frac{1}{x}$ is continuous at 5.

Exercise 13.4. Prove that the function g defined so that

$$g(x) = \begin{cases} (x-2)^2, & \text{if } x \leq 3; \\ \frac{3}{x}, & \text{if } x > 3; \end{cases}$$

is continuous at 3.

A function f is **continuous on a subset** A of its domain provided that it is continuous at every point of A. When a function f is continuous on its entire domain, we often simply say that f is **continuous.** In such circumstances, it is sometimes possible to write a proof of the continuity of the function at all of its inputs by leaving the input at which continuity is established anonymous.

Exercise 13.5. The function f, where $f(x) = 3x + 1$ of Example 13.2, appears to be continuous at all real numbers, not just the real number 2. Show that our proof that f is continuous at 2 can be generalized to establish that f is continuous at any real number p if we replace 2 with an anonymous real number p.

A function is **constant** (or a **constant function**) if it takes the same real number value at all members of its domain (every input produces the same output). A function is **constant on a subset** A **of its domain** if it takes the same real number value at all members of A.

Exercise 13.6. Prove that every constant function on $\mathbb{R}$ is continuous. That is, given any real number c, prove that the function $f : \mathbb{R} \to \mathbb{R}$, defined by $f(x) = c$, is continuous at every real number p.

Exercise 13.7. The function $f : \mathbb{R} \to \mathbb{R}$ defined by $f(x) = x$ is called the **identity function** on $\mathbb{R}$. Prove that this function is continuous.

Exercise 13.8. Prove that the function $f : \mathbb{R} \to \mathbb{R}$ defined so that $f(x) = x^2$ is continuous.

Example 13.4. Here we show the principal square root function, that is, the function $f : [0, \infty) \to \mathbb{R}$ defined by $f(x) = \sqrt{x}$, is continuous.

Proof. First, we show f is continuous at 0. Assume $\varepsilon > 0$ and let $\delta = \varepsilon^2$, which is also positive. Consider any x in $[0, \infty)$, the domain of f, for which $|x - 0| < \delta$; it follows that $0 \leq x < \delta = \varepsilon^2$. Hence, $0 \leq \sqrt{x} < \varepsilon$, which means

$$|f(x) - f(0)| = |\sqrt{x} - \sqrt{0}| = \sqrt{x} < \varepsilon.$$

Thus, f is continuous at 0.

Now we show f is continuous on $(0, \infty)$. Consider any positive real number p and assume $\varepsilon > 0$. Let $\delta = \varepsilon\sqrt{p}$, which is positive, and assume $x \geq 0$ and $|x - p| < \delta$. It follows that

$$\begin{aligned}
|f(x) - f(p)| &= |\sqrt{x} - \sqrt{p}| \\
&= \left|(\sqrt{x} - \sqrt{p}) \cdot \frac{\sqrt{x} + \sqrt{p}}{\sqrt{x} + \sqrt{p}}\right| \\
&= \frac{1}{\sqrt{x} + \sqrt{p}}|x - p| \\
&\leq \frac{1}{\sqrt{p}}|x - p| \\
&< \frac{1}{\sqrt{p}}\delta \\
&= \frac{1}{\sqrt{p}}(\varepsilon\sqrt{p}) \\
&= \varepsilon.
\end{aligned}$$

Thus, f is continuous at p.

As we have been able to show that the principal square root function is continuous at every point of its domain $[0, \infty)$, we may conclude that this function is continuous. □

Exercise 13.9. Prove that the absolute value function is continuous on $\mathbb{R}$.

Continuity at a limit point and continuity at an isolated point

If p is a point of the domain of a function f and is also a limit point of the domain, then the $\varepsilon\delta$-definitions of limit and continuity tell us that f is continuous at p if and only if $\lim_{x \to p} f(x) = f(p)$ (assigned as Exercise 13.10).

Exercise 13.10. Verify that if p is both a point of the domain of a function f and a limit point of the domain, then f is continuous at p if and only if $\lim_{x \to p} f(x) = f(p)$.

A number in a subset A of $\mathbb{R}$ that is not a limit point of A is called an **isolated point** of A. If p is an isolated point of the domain of a function f, then the $\varepsilon\delta$-definition of continuity tells us that f is necessarily continuous in p, even though $\lim_{x \to p} f(x)$ does not exist.

Example 13.5. Define the function $f : \mathbb{Z} \to \mathbb{Z}$ so that $f(n)$ is the unique remainder in the set $\{0, 1, 2, 3, 4\}$ obtained when the integer n is divided by 5 (see Theorem 3.7, the division algorithm). For instance, $f(17) = 2$ because the remainder when 17 is divided by 5 is 2.

The set $\mathbb{Z}$ of all integers, the domain of f, has no limit points, so $\lim_{n \to p} f(n)$ does not exist for any real number p since no real number can be approached arbitrarily closely from within the set of integers.

Thus, every point of the domain $\mathbb{Z}$ of f is an isolated point of $\mathbb{Z}$, and we claim that as a result, f is continuous on its domain. To see why, consider any integer k and any positive number ε, and choose $\delta = 1$. Assume that n is in the domain of f, that is, assume n is an integer, and that $|n-k| < 1$. It follows that $n = k$, so that $|f(n)-f(k)| = |f(k)-f(k)| = 0 < \varepsilon$. Thus, by definition, f is continuous at k.

Essentially, the same argument given in Example 13.5 may be applied to show that any function is continuous at an isolated point of its domain.

Analytically verifying a discontinuity

When a function f is not continuous at a real number p, we say that f is **discontinuous** at p and the point p is then referred to as a **discontinuity** of f.

Example 13.6. The graph of the function $f : \mathbb{R} \to \mathbb{R}$ defined by

$$f(x) = \begin{cases} 3x + 1, & \text{if } x \neq 2; \\ 4, & \text{if } x = 2 \end{cases}$$

suggests that f is discontinuous at the input 2 (Figure 13.4).

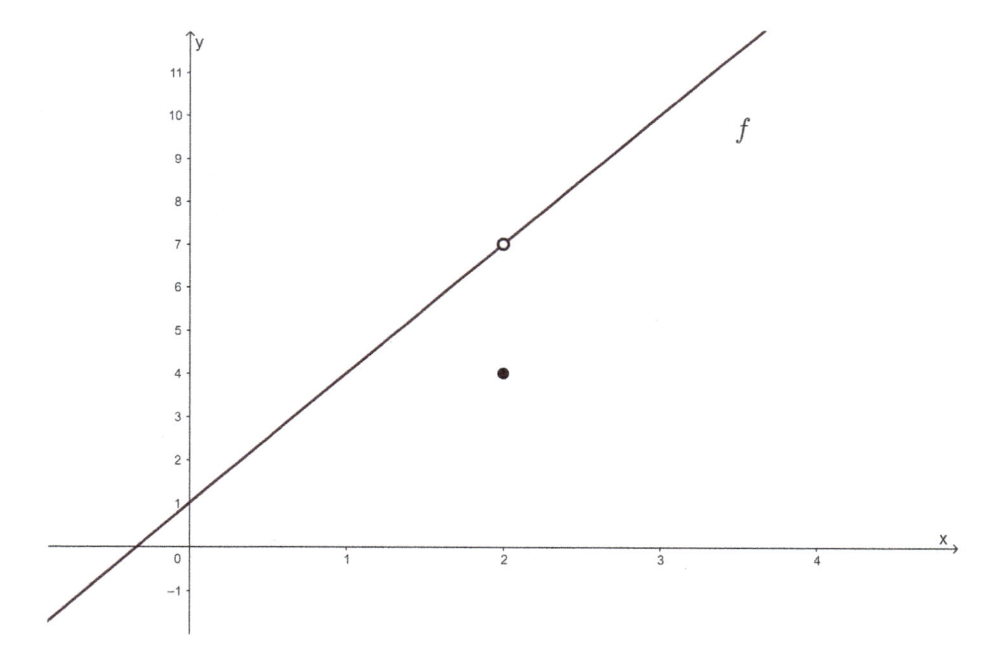

Figure 13.4: Graph of the function from Example 13.6.

Specifically, there is a hole in the graph at the point $(2, 7)$, and it appears that inputs close to but different from 2 are not taken by f to outputs close to $f(2) = 4$; in fact, they appear to be taken close to 7.

The following theorem, obtained by negating Definition 13.1, provides an analytic characterization of what it means for a function to be discontinuous at a particular input.

Theorem 13.7 (Verifying a discontinuity). *Given a function f and a point p in the domain of f, the following are equivalent:*
(i) *f is discontinuous at p;*
(ii) *There exists a positive number ε such that, for each positive number δ, there exists a number x in the domain of f for which $|x - p| < \delta$ and $|f(x) - f(p)| \geq \varepsilon$.*

Example 13.6 (Continued). Graphically, we have already observed that inputs close to but different from 2 are not mapped by the function f to outputs close to $f(2) = 4$, suggesting that f is discontinuous at 2. To analytically verify this is the case, we apply the discontinuity criterion of Theorem 13.7.

Specifically, we claim that if we take the output tolerance to be $\varepsilon = 3$, then for every choice of a positive input tolerance δ we can find an input x for which $|x - 2| < \delta$ and $|f(x) - f(2)| \geq 3$. In other words, for any prescribed measure of closeness of inputs to the specific input 2, we can find an input this close to 2 for which the corresponding output's distance from $f(2)$ is at least 3, hence, not close to $f(2)$. Here is our argument.

Proof. Take $\varepsilon = 3$ and consider any positive number δ. By the density property there exists a number x for which $2 < x < 2 + \delta$. Then $|x - 2| < \delta$ and $f(x) = 3x + 1$. Now as $2 < x$, it follows that $3 \cdot 2 + 1 < 3x + 1$, in other words, $7 < f(x)$, which implies

$$|f(x) - f(2)| = |f(x) - 4| = f(x) - 4 > 7 - 4 = 3 = \varepsilon.$$

Hence, according to our analytic criterion for establishing a discontinuity, the function $f : \mathbb{R} \to \mathbb{R}$, where

$$f(x) = \begin{cases} 3x + 1, & \text{if } x \neq 2; \\ 4, & \text{if } x = 2; \end{cases}$$

is discontinuous at 2. □

Exercise 13.11. In establishing the function f in Example 13.6 is discontinuous at 2, note that the chosen input x depends on the value of δ (this is typically the case when establishing a discontinuity).
(a) Let $\delta = 0.01$ and find a numerical value for x so that $|x - 2| < 0.01$ and $|f(x) - f(2)| \geq 3$.
(b) Find a numerical value of $\delta > 0$ for which your choice of a numerical value of x in (a) results in $|x - 2| \geq \delta$.
(c) Find a different numerical value for x so that, for your choice of a numerical value of δ in (b), both $|x - 2| < \delta$ and $|f(x) - f(2)| \geq 3$.

Exercise 13.12. Consider again the function f from Example 13.6.
(a) Verify that if we take the output tolerance to be $\varepsilon = 6$ and the input tolerance to be $\delta = 1$, then whenever $|x - 2| < \delta$ it follows that $|f(x) - f(2)| < \varepsilon$.
(b) Explain why (a) does not mean that f is continuous at 2.

Exercise 13.13. Consider the function $h : \mathbb{R} \to \mathbb{R}$ defined so that

$$h(x) = \begin{cases} 1 + ((x-1)^2 + 0.001), & \text{if } x \geq 1; \\ 1 - ((x-1)^2 + 0.001), & \text{if } x < 1. \end{cases}$$

(a) Use an online or handheld graphing calculator to graph h within the viewing window for which $0 \leq x \leq 2$ and $0 \leq y \leq 2$. Does it appear from the picture you have generated that h is continuous at the input 1?
(b) Repeat (a), but using a small enough viewing window to graphically detect the discontinuity of h at 1.
(c) Prove that h is discontinuous at 1.

Characterizing continuity using sequences

Just as we were able to express the notion of limit of a function using sequences, we can express the notion of continuity of a function by means of sequences.

Theorem 13.8 (Sequential characterization of continuity). *Given a function f and a point p in the domain of f, the following are equivalent:*
(i) *f is continuous at p;*
(ii) *for any sequence (x_n) in the domain of f, if $x_n \to p$, then $f(x_n) \to f(p)$.*

This theorem is sometimes expressed by saying that continuous functions preserve convergent sequences, and if a function preserves all sequences that converge to a specified point, the function must be continuous at that point.

Exercise 13.14. Prove Theorem 13.8.

Example 13.9. Here we use the sequential characterization of continuity to show that the squaring function is continuous.

Proof. Define $f : \mathbb{R} \to \mathbb{R}$ so that $f(x) = x^2$. Let p be any real number and consider any sequence (x_n) that converges to p; we need only show that $f(x_n) \to f(p)$ in order to deduce that f is continuous at p. As the limit of a product of convergent sequences is the product of the sequences' limits, it follows from the assumption that $x_n \to p$ that

$$x_n \cdot x_n \to p \cdot p,$$

that is, $x_n^2 \to p^2$, which, as f is the squaring function, means $f(x_n) \to f(p)$. Because p is an arbitrary real number, we may conclude that the squaring function is continuous on $\mathbb{R}$. □

Negating (ii) in Theorem 13.8 yields the following method for using sequences to establish a function is discontinuous at a point.

Theorem 13.10 (Sequential characterization of discontinuity). *Given a function f and a point p in the domain of f, the following are equivalent:*
(i) *f is discontinuous at p;*
(ii) *there exists a sequence (x_n) in the domain of f for which $x_n \to p$ and $f(x_n) \nrightarrow f(p)$.*

Exercise 13.15. Use Theorem 13.10 and Exercise 10.31 to show that the function $f : \mathbb{R} \to \mathbb{R}$ defined so that

$$f(x) = \begin{cases} 1, & \text{if } x \text{ is rational;} \\ 0, & \text{if } x \text{ is irrational;} \end{cases}$$

is not continuous at any real number, in other words, is everywhere discontinuous.

Exercise 13.16. Give an example of a function $f : \mathbb{R} \to \mathbb{R}$ that is discontinuous at every real number, but for which the function $|f|$ is continuous at every real number.

Exercise 13.17. Define the function $g : [0,1] \to \mathbb{R}$ so that when x is an irrational number, $g(x) = 0$, and when x is a rational number, $g(x) = \frac{1}{n}$, where n is the denominator of the fraction for which $x = \frac{m}{n}$, where m and n are integers having greatest common divisor 1 (so the fraction is in "lowest terms"). Prove that g is discontinuous at every rational number in $[0,1]$, but is continuous at every irrational number in $[0,1]$.

Arithmetic properties of continuous functions

As we would probably expect, continuous functions behave well with respect to the arithmetic operations of addition, subtraction, multiplication, and division.

Theorem 13.11. *Suppose the functions f and g are both continuous at p.*
(1) *Each of the sum $f + g$, difference $f - g$, and product $f \cdot g$ is continuous at p.*
(2) *The quotient $\frac{f}{g}$ is continuous at p as long as $g(x) \neq 0$ for all x in the domain of g that lie in some neighborhood of p.*
(3) *If c is any real number, the constant multiple cf is continuous at p.*

It is a straightforward exercise to establish the results of this theorem using the sequential characterization of continuity, together with appropriate properties of convergent sequences (see, for instance, Exercise 13.18 below).

Corollary 13.12. *Every polynomial function is continuous on $\mathbb{R}$ and every rational function is continuous on its domain.*

Example 13.13. As the function f, where $f(x) = 1000x^{97} - 4x + 12$, is a polynomial function, it is continuous on $\mathbb{R}$.

The reciprocal function g, where $g(x) = \frac{1}{x}$, being a rational function, is continuous on its domain; that is, g is continuous at all nonzero real numbers.

Exercise 13.18. Use the sequential characterization of continuity, Theorem 13.8, to prove that if the functions f and g are both continuous at p, so is the function $f + g$.

Exercise 13.19. What is it that allows us to immediately conclude that

$$|x_n| \to |p|,$$
$$x_n^{100} \to p^{100},$$

and

$$(x_n - 7) \to (p - 7)$$

if we know the sequence (x_n) converges to p?

Exercise 13.20. Suppose the real-valued functions f and g are both continuous on $\mathbb{R}$ and that $f(x) = g(x)$ for all rational numbers x. Must $f(x) = g(x)$ for all real numbers x? Prove your conclusion.

Continuity of composite functions

The composition of continuous functions, provided it is defined, is continuous.

Theorem 13.14 (The composition of continuous functions is continuous). *If the function f is continuous at p and the function g is continuous at $f(p)$, then $g \circ f$ is continuous at p.*

Example 13.15. Consider the functions f and g, where $f(x) = x^3 + x$ and $g(x) = \sqrt{x}$. The function f is continuous as it is a polynomial function. In Example 13.4, we proved that the function g is continuous. Thus, the composite function $g \circ f$, where

$$(g \circ f)(x) = g(f(x)) = g(x^3 + x) = \sqrt{x^3 + x}$$

is continuous on its domain, the set of all nonnegative real numbers.

Exercise 13.21. Use the sequential characterization of continuity, Theorem 13.8, to prove Theorem 13.14.

Exercise 13.22. Explain how we know the function $f : \mathbb{R} \to \mathbb{R}$ defined so that $f(x) = x - \sqrt{x^2 + 1}$ is continuous on $\mathbb{R}$.

Types of discontinuities

Suppose p is both a point of discontinuity of the function f and a limit point of the domain of f. This discontinuity p is called **removable** if $\lim_{x \to p} f(x)$ exists and **nonremovable** otherwise.

Example 13.16. Consider once again the function $f : \mathbb{R} \to \mathbb{R}$ defined by

$$f(x) = \begin{cases} 3x + 1, & \text{if } x \neq 2; \\ 4, & \text{if } x = 2. \end{cases}$$

In Example 13.6, we showed this function is discontinuous at 2. Back in Example 12.1, we proved that $\lim_{x \to 2} f(x) = 7$. Thus, the discontinuity 3 is removable. The terminology "removable" is evoked by the graph of f (see Figure 13.4), which shows a hole at the point $(2, 7)$, which could be "repaired" by changing $f(2)$ from the value 4 to the value 7 (i. e., just "fill in" the hole).

Certain nonremovable discontinuities can be further classified. When both $\lim_{x \to p^-} f(x)$ and $\lim_{x \to p^+} f(x)$ exist, but are unequal, the nonremovable discontinuity p is called a **jump discontinuity**.

Example 13.17. In Example 12.12, we showed that

$$\lim_{x \to 0^-} \frac{x^4 + |x|}{x} = -1 \quad \text{and} \quad \lim_{x \to 0^+} \frac{x^4 + |x|}{x} = 1,$$

so we may conclude that the function $h : \mathbb{R} - \{0\} \to \mathbb{R}$ defined by

$$h(x) = \frac{x^4 + |x|}{x}$$

has a jump discontinuity at 0. The graph of h, displayed back in Figure 12.5, shows the jump in values that occurs at 0.

When f is not bounded on any neighborhood of p, the nonremovable discontinuity p is called an **infinite discontinuity**.

Example 13.18. Consider the function $g : \mathbb{R} - \{0\} \to \mathbb{R}$, where

$$g(x) = \frac{1}{x^2}.$$

Back in Example 12.11, we showed that $\lim_{x\to 0} g(x) = \infty$. Since the values of g become unbounded as the inputs approach 0, we conclude that g has an infinite discontinuity at 0. Note that infinite discontinuities correspond to vertical asymptotes (see Figure 12.4).

A nonremovable discontinuity need not be either a jump discontinuity or an infinite discontinuity. A function can exhibit a variety of other behaviors at a point of discontinuity.

Example 13.19. Consider the function $f : \mathbb{R} - \{0\} \to \mathbb{R}$ defined so that

$$f(x) = \cos\left(\frac{1}{x}\right).$$

Our work in Example 12.10 shows that $\lim_{x\to 0^+} f(x)$ does not exist. However, as the cosine function never takes a value greater than 1 or less than -1, we see that the function f is bounded on every neighborhood of 0. Thus, the discontinuity at 0 is nonremovable, but neither a jump nor an infinite discontinuity. Instead, it can be described as an *oscillatory discontinuity*, as the graph of f exhibits ever more rapid oscillation of the values of f between -1 and 1 as the inputs approach 0 (Figure 13.5).

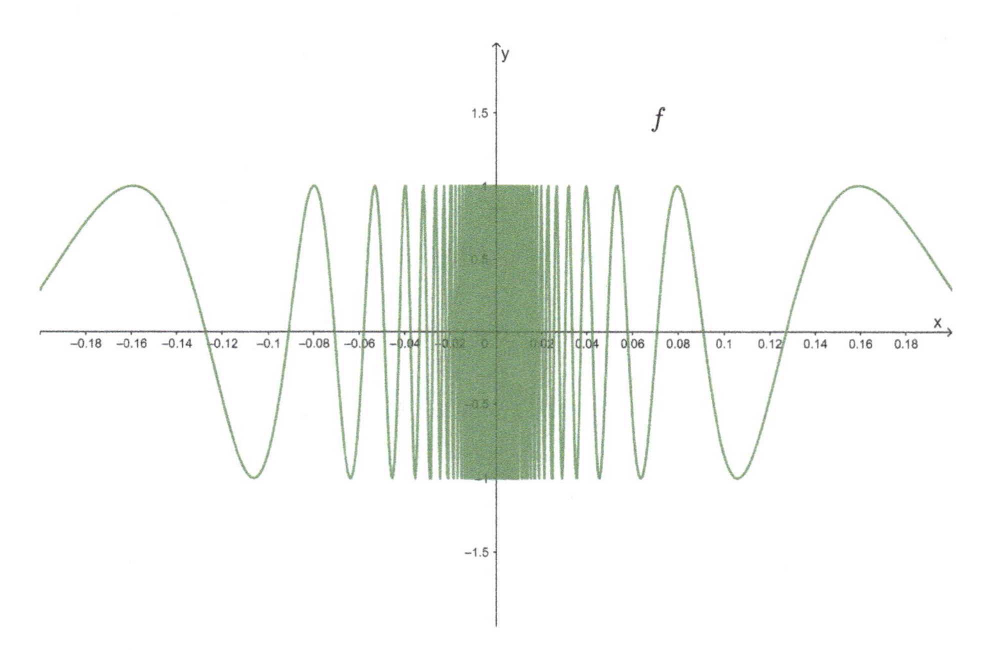

Figure 13.5: Graph of the function from Example 13.19.

Exercise 13.23. Suppose the function f is defined so that $f(x) = \frac{x^2-2x-15}{x+3}$ for all real numbers x other than -3. Is it possible to define f at -3 so that f is continuous at -3? If so, give the numerical value of $f(-3)$. If not, explain why not.

Exercise 13.24. Show that the function h from Exercise 13.13 has a jump discontinuity at 1.

14 More about continuity

In this chapter, we prove the two major theorems concerning continuous functions, the *intermediate value theorem* and the *extreme value theorem*. We also provide several characterizations of functions that are continuous on their entire domains, show that the inverse of a strictly monotone continuous function is also strictly monotone and continuous, and introduce the important notion of *uniform continuity*.

The intermediate value theorem

The intermediate value theorem makes precise our intuitive sense that the outputs of a continuous function defined on an interval cannot "skip over" any real numbers lying between two specified outputs. The common description of continuous functions as those whose graphs can be drawn without lifting pencil from paper is an informal way of expressing the intermediate value theorem.

Theorem 14.1 (The intermediate value theorem). *A continuous real-valued function defined on an interval achieves as values all numbers lying between any two of its values.*

That is, if I is an interval, the function $f : I \to \mathbb{R}$ is continuous, the numbers a and b are in I with $a < b$, and r is a number between $f(a)$ and $f(b)$, then there exists a number c for which $a < c < b$ and $f(c) = r$.

Before we prove the intermediate value theorem, we illustrate how it can be applied and provide some evidence of its significance.

Example 14.2. Consider the function $f : [0.5, 5] \to \mathbb{R}$, where $f(x) = \frac{1}{x}$, which is continuous on the interval $[0.5, 5]$ as it is a rational function. The intermediate value theorem tells us that f achieves each real number between $f(0.5) = \frac{1}{0.5} = 2$ and $f(5) = \frac{1}{5} = 0.2$ as an actual output. For instance, as 1.5 lies between 0.2 and 2, there must exist a number c for which $0.5 < c < 5$ and $f(c) = \frac{1}{c} = 1.5$ (Figure 14.1). Solving this equation for c yields $c = \frac{2}{3}$. Thus, $f(\frac{2}{3}) = 1.5$, where certainly $0.5 < \frac{2}{3} < 5$.

Basic geometry also makes the conclusion we have reached via the intermediate value theorem believable: Since the points $(0.5, 2)$ and $(5, 0.2)$ are joined by an unbroken (i. e., "continuous") curve, it must follow that there is a point $(c, 1.5)$ lying on this curve for some number c such that $0.5 < c < 5$.

Exercise 14.1. Consider the continuous function $f : [-1, 3] \to \mathbb{R}$ defined so that $f(x) = x^2$. As all of the hypotheses of the intermediate value theorem are satisfied by f, we should be able to find a number c for which $-1 < c < 3$ and $f(c) = r$ for any specified number r between $f(-1) = 1$ and $f(3) = 9$. Find such a value for c if we take $r = 6$.

---

https://doi.org/10.1515/9783111429564-014

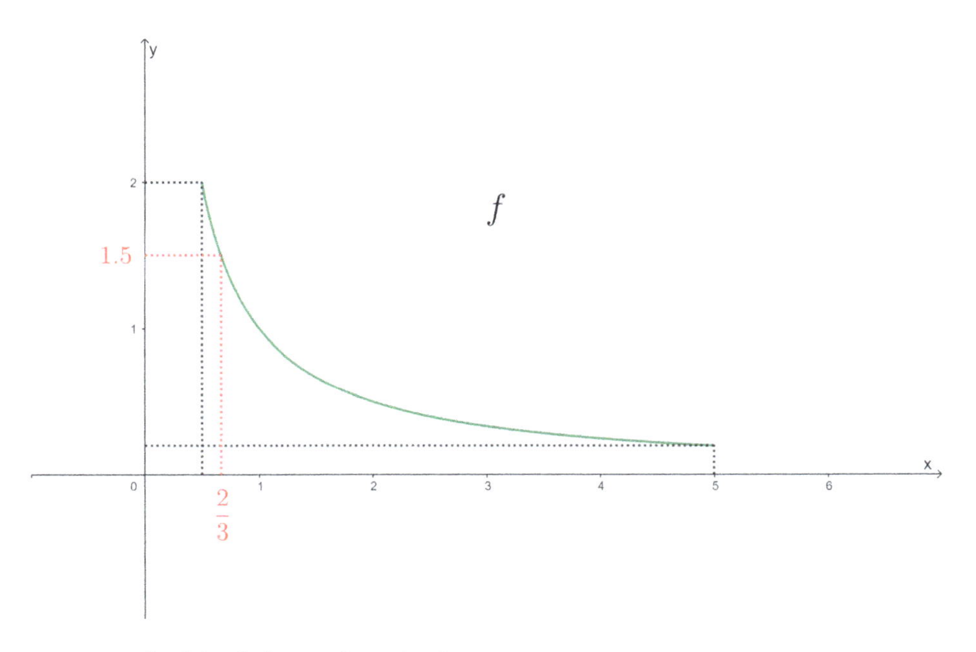

Figure 14.1: Visualizing the intermediate value theorem.

Exercise 14.2. If all of the hypotheses of the intermediate value theorem are not satisfied, the conclusion of the theorem may not hold. Consider the function $g : [-1, 3] \to \mathbb{R}$ defined so that

$$g(x) = \begin{cases} x^2, & \text{if } x \neq 2; \\ -4, & \text{if } x = 2. \end{cases}$$

Note that the number 4 lies between $g(-1) = 1$ and $g(3) = 9$. Show, however, that there is no number c for which $-1 < c < 3$ and $g(c) = 4$. Which part of the hypotheses of the intermediate value theorem is not satisfied?

We used the intermediate value theorem at the beginning of Chapter 4 to motivate the importance of including irrational numbers in the real number system. In the next example, we present again the specific scenario we considered.

Example 14.3. The function $f : \mathbb{R} \to \mathbb{R}$ defined so that $f(x) = x^2 - 2$ is continuous, being a polynomial function. As $f(1) = -1 < 0$ and $f(2) = 2 > 0$, the intermediate value theorem allows us to conclude that there is a real number c for which $1 < c < 2$ and $f(c) = c^2 - 2 = 0$, that is, $c^2 = 2$. Since we proved in Theorem 3.28 that there is no rational number c such that $c^2 = 2$, the intermediate value theorem therefore demonstrates the existence of an irrational number, namely, $c = \sqrt{2}$, within the set of real numbers.

In Theorem 4.21, we generalized the argument employed in Example 14.3 to show how the intermediate value theorem guarantees the existence of $\sqrt[n]{a}$ for any positive real number a and any natural number n. Arguments similar to that given in the proof of Theorem 3.28 can be used to show that when a is a natural number and $\sqrt[n]{a}$ is not, $\sqrt[n]{a}$ is irrational.

The proof of the intermediate value theorem uses the nested intervals property and the monotone convergence theorem, both of which rely on the completeness axiom for their own proofs. So, although the completeness axiom is never mentioned explicitly in our proof of the intermediate value theorem, it is pivotal to the argument.

A diagram illustrating the construction that appears within the proof is provided in Figure 14.2.

Proof of Theorem 14.1, *the intermediate value theorem.* Assume I is an interval, that the function $f : I \to \mathbb{R}$ is continuous, that a and b are numbers in I for which $a < b$, and that r is a number for which $f(a) < r < f(b)$; the case where $f(b) < r < f(a)$ can be handled similarly.

Let $p_1 = a$ and $q_1 = b$, and define $m_1 = \frac{1}{2}(p_1 + q_1)$, which is the midpoint of the interval $[p_1, q_1]$. If $f(m_1) \leq r$, define $p_2 = m_1$ and $q_2 = q_1$; if $f(m_1) > r$, define $p_2 = p_1$ and $q_2 = m_1$. We continue to recursively define sequences (p_n) and (q_n) so that if $f(m_n) \leq r$, we take $p_{n+1} = m_n$ and $q_{n+1} = q_n$, but if $f(m_n) > r$, we take $p_{n+1} = p_n$ and $q_{n+1} = m_n$, where for each n, we are taking $m_n = \frac{1}{2}(p_n + q_n)$, the midpoint of $[p_n, q_n]$.

Note that for every n, we have

$$a \leq p_n \leq p_{n+1} < q_{n+1} \leq q_n \leq b,$$

as well as $f(p_n) \leq r$ and $f(q_n) > r$. Furthermore,

$$q_n - p_n = \frac{b-a}{2^{n-1}} \to 0.$$

Hence, by the nested intervals property, Theorem 6.7,

$$\bigcap_{n=1}^{\infty} [p_n, q_n] = \{c\}$$

for some real number c in $[a, b]$. According to the monotone convergence theorem, Theorem 10.18, each of the sequences (p_n) and (q_n) converges to c, so by the continuity of f it follows using Theorem 13.8 that $f(p_n) \to f(c)$ and $f(q_n) \to f(c)$. Now, as $f(p_n) \leq r$ for all n, it follows that $f(c) \leq r$. However, as $f(q_n) > r$ for all n, it also follows that $f(c) \geq r$. Thus, we may conclude that $f(c) = r$. Finally, as $f(a) < r < f(b)$, it is not possible that $c = a$ or $c = b$, so it follows, using our earlier conclusion that c is in $[a, b]$, that $a < c < b$. □

Figure 14.2 illustrates the beginning of the construction performed in the proof for the value of r shown. Also included is the location of the number c that would be determined within the proof. Make sure you agree with the specifications of p_2, q_2, p_3, and q_3,

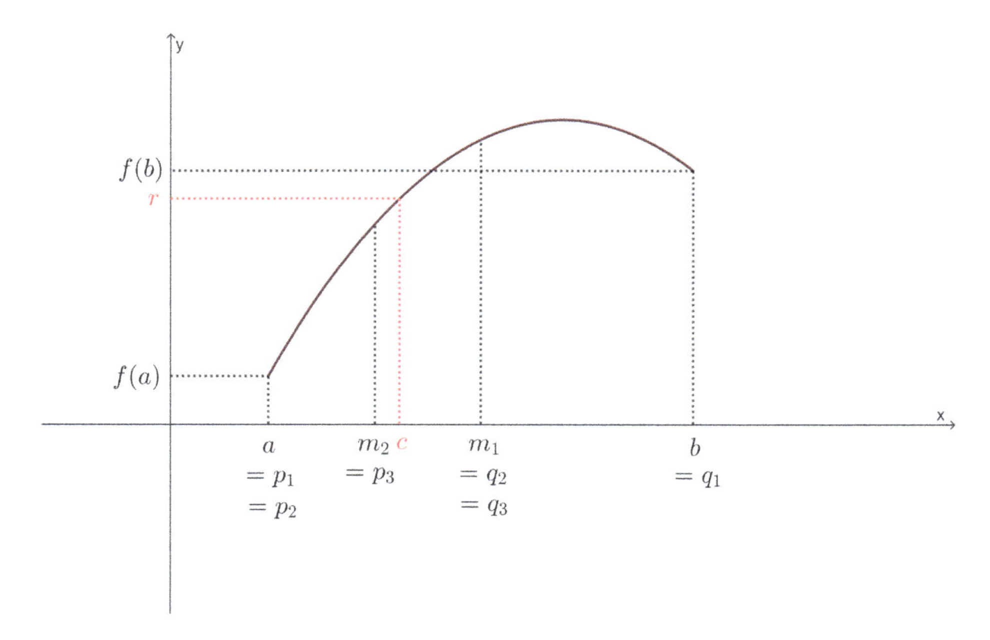

Figure 14.2: Visualizing the construction employed in the proof of the intermediate value theorem.

which are determined by the relevant values of f, as indicated by its graph, when they are compared to the number r.

Exercise 14.3. Suppose $a < b$. Which continuous functions $f : [a, b] \to \mathbb{R}$ take only rational values?

Exercise 14.4. A function $f : I \to \mathbb{R}$, where I is an interval in $\mathbb{R}$, is said to have the **intermediate value property** if, whenever the numbers a and b are in I with $a < b$, and r is a number between $f(a)$ and $f(b)$, then there exists a number c for which $a < c < b$ and $f(c) = r$. Show that the function $g : [0, \infty) \to \mathbb{R}$ defined so that

$$g(x) = \begin{cases} \cos(\frac{1}{x}), & \text{if } x > 0; \\ 1, & \text{if } x = 0; \end{cases}$$

has the intermediate value property on its domain $[0, \infty)$ even though g is discontinuous at 0. You will need to use some basic facts about the cosine function, including its continuity, which you can find in Appendix D.

Exercise 14.5. Show that if $f : [0, 1] \to [0, 1]$ is continuous, then $f(p) = p$ for some p in $[0, 1]$.

Images and preimages

If f is a function and A is a subset of the domain of f, we define the **image of A under f** to be the set $f[A] = \{f(x) \mid x \in A\}$, the set of all outputs produced when f is applied to all the elements of A. Note that if D is the domain of a function f, then $f[D]$ is the function's range.

Example 14.4. For the function $f : \mathbb{R} \to \mathbb{R}$, where $f(x) = |x|$, we have

$$f[\{-2, 1\}] = \{2, 1\}$$

and

$$f[(-2, 1)] = [0, 2).$$

Since $\mathbb{R}$ is the domain of f we may observe that $f[\mathbb{R}] = [0, \infty)$ is the range of f.

If f is a function and A is a subset of the codomain of f, we define the **preimage of A under f** to be the set $f^{-1}[A] = \{x \mid f(x) \in A\}$, the set of all inputs to f that produce outputs in A.

Example 14.4 (Continued). Observe that for the absolute value function f it follows that

$$f^{-1}[\{2, -1\}] = \{2, -2\},$$
$$f^{-1}[[0, 3)] = (-3, 3),$$

and

$$f^{-1}[\mathbb{R}^+] = \mathbb{R} - \{0\}.$$

Exercise 14.6. For the function $f : \mathbb{R} \to \mathbb{R}$ defined by $f(x) = x^2$, determine each of the following.
(a) $f[\{-2, 2\}]$
(b) $f[[-3, 2)]$
(c) $f^{-1}[(4, 9]]$
(d) $f^{-1}[\mathbb{R}^-]$

The use of square brackets in the notations $f[A]$ and $f^{-1}[A]$ for images and preimages helps to distinguish them from values $f(a)$ and $f^{-1}(a)$ of a function and inverse function (introduced later in this chapter), where round brackets are employed.

The image of an interval under a continuous function is always an interval. This makes sense intuitively as the intervals in $\mathbb{R}$ are precisely the connected subsets of $\mathbb{R}$ (Theorem 8.3), and we would not expect a continuous function to break apart a connected set.

Theorem 14.5 (Continuous functions preserve intervals). *If the interval I is a subset of the domain of a continuous function f, then $f[I]$ is also an interval.*

Exercise 14.7. Use the intermediate value theorem and the definition of interval to prove Theorem 14.5.

Continuity from a global perspective

In Chapter 7, we defined the distance between a real number p and a subset A of $\mathbb{R}$ to be

$$\inf\{|p - a| \mid a \in A\},$$

the greatest lower bound of the distances between p and the elements of A. When the distance between p and A is zero, it is natural to view the number p as being close to the set A. The only way this distance can be zero is if either p is in A or there are points of A that are arbitrarily close to p, which would make p a limit point of A. Thus, it is precisely the numbers in the set

$$\{x \mid x \text{ is in } A \text{ or } x \text{ is a limit point of } A\},$$

called the **closure** of A and denoted by $\overline{A}$, that would be reasonably construed as being close to A.

Example 14.6. The closure of a set is formed by taking the union of the set with the set of all its limit points.

In Example 7.3, we showed that 0 is the only limit point of the set $\{\frac{1}{n} \mid n \in \mathbb{N}\}$ of reciprocals of the natural numbers. Thus, $\overline{\{\frac{1}{n} \mid n \in \mathbb{N}\}} = \{\frac{1}{n} \mid n \in \mathbb{N}\} \cup \{0\}$.

Between Example 7.1 and Exercise 7.1, we concluded that all the numbers in $[0, 1]$ are limit points of $(0, 1)$, so $\overline{(0, 1)} = (0, 1) \cup [0, 1] = [0, 1]$.

Exercise 14.8. Find the closure of the given set.

(a) $[-1, 4)$
(b) $\{-1, 0, 1\}$
(c) $\mathbb{Q}$

Exercise 14.9. Suppose A is a subset of $\mathbb{R}$. Prove each of the following.

(a) The set A is closed if and only if $A = \overline{A}$.
(b) A real number p is in $\overline{A}$ if and only if every neighborhood of p includes at least one element of A.
(c) The set $\overline{A}$ is closed.

Exercise 14.10. Show that if B is a subset of $\mathbb{R}$ and $A \subseteq B$, then $\overline{A} \subseteq \overline{B}$.

The following theorem provides several conditions that are equivalent to a function being continuous on its entire domain. In particular, condition (ii) may be interpreted as saying that a continuous function always maps inputs close to a set to outputs that are close to the image of the set, which is precisely what we would expect a continuous function to do (the preservation of closeness is what ensures the function's graph is not disrupted). Condition (iii) characterizes continuity via closed sets and condition (iv) does the same via open sets.

Theorem 14.7. *Given a function $f : A \to \mathbb{R}$, where A is a subset of $\mathbb{R}$, the following are equivalent:*
(i) *f is continuous;*
(ii) *for every subset S of A, the set $f[\overline{S} \cap A]$ is a subset of $\overline{f[S]}$;*
(iii) *for every closed set E, there exists a closed set F such that $f^{-1}[E] = F \cap A$;*
(iv) *for every open set G, there exists an open set H such that $f^{-1}[G] = H \cap A$.*

Example 14.8. We illustrate the application of Theorem 14.7 by using condition (iv) to prove that every nonconstant linear function is continuous on $\mathbb{R}$. Note that when the domain of a function f is $\mathbb{R}$, condition (iv) reduces to the requirement that for every open set G, the set $f^{-1}[G]$ is open.

Proof. Suppose f is the linear function defined on $\mathbb{R}$ so that $f(x) = mx + b$ for some constants m and b with $m \neq 0$. Observe that, when $m > 0$,

$$s < x < t \quad \text{if and only if} \quad ms + b < mx + b < mt + b,$$

and when $m < 0$,

$$s < x < t \quad \text{if and only if} \quad mt + b < mx + b < ms + b.$$

Hence, for each open interval (s, t), it follows that $f^{-1}[(s, t)] = (u, v)$ for some open interval (u, v).

Now consider any open set G. From Theorem 7.14, we know that $G = \bigcup_{i \in I}(s_i, t_i)$, where each (s_i, t_i) is an open interval. It follows that

$$f^{-1}[G] = f^{-1}\left[\bigcup_{i \in I}(s_i, t_i)\right] = \bigcup_{i \in I} f^{-1}[(s_i, t_i)] = \bigcup_{i \in I}(u_i, v_i),$$

for some open intervals (u_i, v_i). As open intervals are open sets, and as any union of open sets is an open set (Exercise 7.13), it follows that $f^{-1}[G]$ is open. Thus, by condition (iv) of Theorem 14.7, we may conclude that f is continuous on $\mathbb{R}$. □

Exercise 14.11. Use condition (iv) of Theorem 14.7 to prove that the function f defined so that $f(x) = \frac{1}{x}$ is continuous on $(0, \infty)$.

Exercise 14.12. Verify Theorem 14.7 by proving each of the implications (i) ⇒ (ii), (ii) ⇒ (iii), (iii) ⇒ (iv), and (iv) ⇒ (i).

Monotone functions and inverse functions

Let A be a subset of the domain of a function f. Then f is
(1) **increasing on** A if whenever x_1 and x_2 are in A with $x_1 < x_2$, it follows that $f(x_1) < f(x_2)$;
(2) **decreasing on** A if whenever x_1 and x_2 are in A with $x_1 < x_2$, it follows that $f(x_1) > f(x_2)$;
(3) **nondecreasing on** A if whenever x_1 and x_2 are in A with $x_1 < x_2$, it follows that $f(x_1) \le f(x_2)$;
(4) **nonincreasing on** A if whenever x_1 and x_2 are in A with $x_1 < x_2$, it follows that $f(x_1) \ge f(x_2)$;
(5) **monotone on** A if f is either nondecreasing or nondecreasing on A;
(6) **strictly monotone on** A if f is either increasing or decreasing on A.

In your study of calculus, you learned how to find the intervals on which a function is increasing and those on which it is decreasing, and then used this information to locate local extreme values of the function. We consider these matters in Chapter 16, after we introduce the derivative.

Example 14.9. Consider the function $f : \mathbb{R} \to \mathbb{R}$, where $f(x) = x^2$, whose graph is displayed in Figure 14.3.

Observe that f is not monotone or strictly monotone on $\mathbb{R}$. For instance, $-2 < 0 < 2$, but as $0 < 4$ we see that $f(-2) > f(0)$, whereas $f(0) < f(2)$.

However, f is increasing on $[0, \infty)$ and is decreasing on $(-\infty, 0]$.

Example 14.10. The **sign function** is the function $\operatorname{sgn} : \mathbb{R} \to \mathbb{R}$ defined so that

$$\operatorname{sgn}(x) = \begin{cases} -1, & \text{if } x < 0; \\ 0, & \text{if } x = 0; \\ 1, & \text{if } x > 0. \end{cases}$$

Not to be confused with the trigonometric function *sine*, the sign function essentially indicates whether the input is negative, zero, or positive. As the graph of this function indicates (Figure 14.4), sgn is nondecreasing on $\mathbb{R}$, hence, monotone on $\mathbb{R}$. However, sgn is not increasing on $\mathbb{R}$, as for example $\operatorname{sgn}(-3) = \operatorname{sgn}(-2) = -1$, so we may then conclude that sgn is not strictly monotone on $\mathbb{R}$.

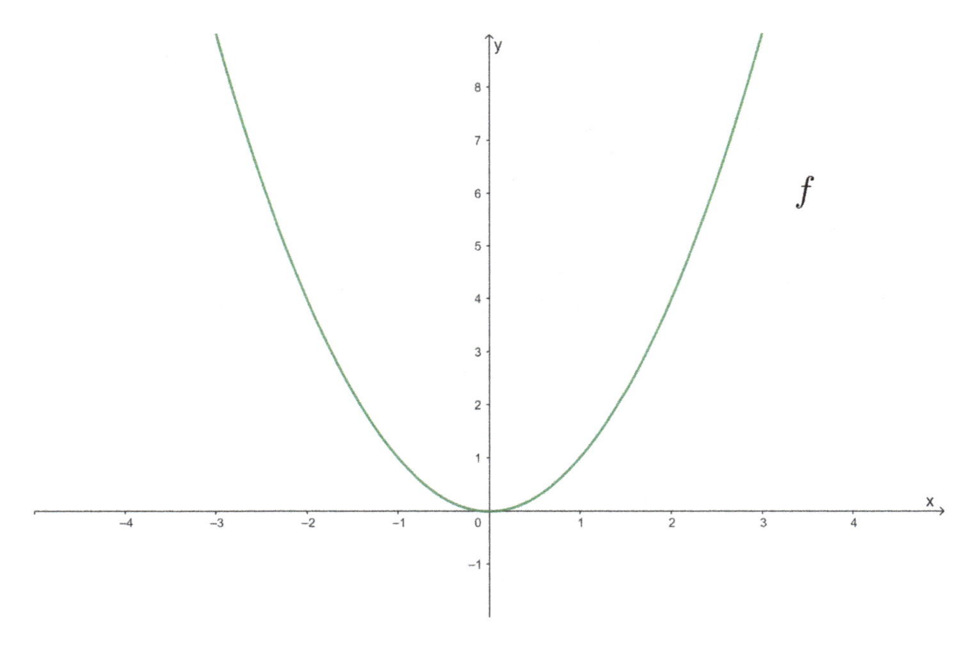

Figure 14.3: Graph of the function from Example 14.9.

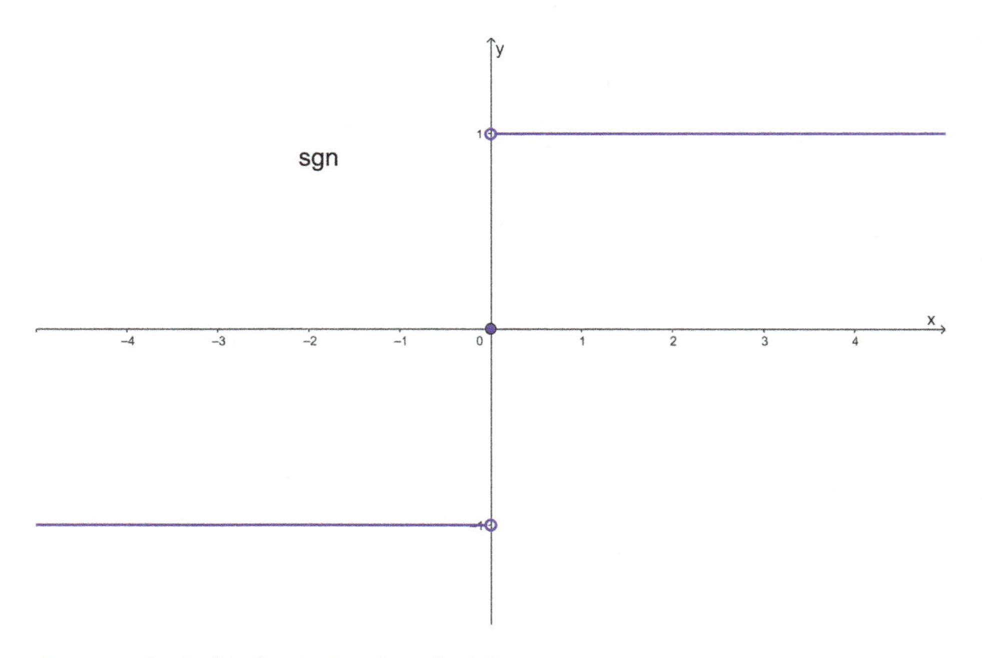

Figure 14.4: Graph of the function from Example 14.10.

Exercise 14.13. Let A be a subset of $\mathbb{R}$ and suppose f, g, and h are real-valued functions all having domain A. Assume also that f and g are nondecreasing on A, while h is increasing on A.
(a) Prove that $f + g$ is nondecreasing on A.
(b) Prove that $f + h$ is increasing on A.

Exercise 14.14. Give an example to show that if the functions f and g are increasing on $\mathbb{R}$, it need not follow that fg is increasing on $\mathbb{R}$.

Exercise 14.15. Show that if the function f is strictly monotone on an interval (a, b) and continuous at a, then f is strictly monotone on the interval $[a, b)$. One can similarly show that if f is strictly monotone on (a, b) and continuous at b, then f is strictly monotone on $(a, b]$.

Recall that a function is **one-to-one** if different inputs to the function always yield different outputs.

Exercise 14.16. Let I be an interval in $\mathbb{R}$. Prove that if the function $f : I \to \mathbb{R}$ is strictly monotone on I, then f is one-to-one.

Given a one-to-one function $f : X \to Y$, the function $f^{-1} : f[X] \to X$, called the **inverse** of f, is defined for each y in the range $f[X]$ of f so that $f^{-1}(y) = x$, where x is the unique member of X for which $f(x) = y$. The inverse of a one-to-one function reverses the function's input-output process, and is the set of all the ordered pairs obtained by interchanging the coordinates of the ordered pairs comprising the original function.

Example 14.11. It is readily checked that the function $f : \mathbb{R} \to \mathbb{R}$ defined so that $f(x) = x + 2$ is one-to-one (actually a bijection). The inverse of f is the function $f^{-1} : \mathbb{R} \to \mathbb{R}$ defined by $f^{-1}(x) = x - 2$. These conclusions are sensible as subtracting 2 and adding 2 are processes that "undo" one another.

Exercise 14.17. For the particular functions f and f^{-1} given in Example 14.11, verify that $(a, b) \in f$ if and only if $(b, a) \in f^{-1}$.

Exercise 14.18. Consider the function $g : \mathbb{R} - \{-1\} \to \mathbb{R}$ defined by $g(x) = 5 - \frac{2}{x+1}$.
(a) Show the function g is one-to-one.
(b) Find the domain of g^{-1}.
(c) Find a formula for $g^{-1}(x)$.

Exercise 14.19. Prove that the inverse of a one-to-one function is also one-to-one.

Among the one-to-one functions defined on an interval of real numbers are those that are strictly monotone (see Exercise 14.16). The inverse of a strictly monotone continuous function defined on an interval turns out to itself be both strictly monotone and continuous.

Theorem 14.12 (Inverses of strictly monotone continuous functions). *If I is an interval in $\mathbb{R}$ and the function $f : I \to \mathbb{R}$ is strictly monotone and continuous, then the function f^{-1} is also strictly monotone and continuous.*

Proof. Suppose I is an interval in $\mathbb{R}$ and the function $f : I \to \mathbb{R}$ is strictly monotone and continuous. It follows, because a continuous function preserves intervals (Theorem 14.5), that $f[I]$, which is the domain of f^{-1}, is an interval. The fact that f^{-1} is strictly monotone (in fact, increasing if f is increasing, and decreasing if f is decreasing) can be established by contradiction and is left as Exercise 14.20. Therefore, to complete the proof, we show that f^{-1} is continuous on $f[I]$. We do so under the assumption that f is increasing on I (the argument in the case where f is decreasing on I is similar).

Suppose to the contrary that f^{-1} is not continuous at a point q in $f[I]$. As $q \in f[I]$ there exists a point p in I such that $f(p) = q$. The assumption that f^{-1} is not continuous at q implies there is a sequence (y_n) in $f[I]$ for which $y_n \to q$, but for which $f^{-1}(y_n) \nrightarrow f^{-1}(q) = p$. For each natural number n, let $x_n = f^{-1}(y_n)$, which is in I. Observe that $x_n \nrightarrow p$. Hence, there exists a positive number ε and a subsequence (x_{k_n}) of (x_n) such that $x_{k_n} \notin (p - \varepsilon, p + \varepsilon)$ for every n. It follows that either $x_{k_n} \le p - \varepsilon$ for infinitely many n or else $x_{k_n} \ge p + \varepsilon$ for infinitely many n. Assume, without loss of generality, the latter. Thus, there is a subsequence (x_{j_n}) of (x_n) with $x_{j_n} \ge p + \varepsilon$ for every n.

Now since p is in I and each x_{j_n} is in I, the fact that I is an interval permits us to conclude that $p + \varepsilon$ is in I. Furthermore, as f is increasing, we may conclude that $f(p) < f(p + \varepsilon) \le f(x_{j_n})$ for every n. Hence, none of the terms of the sequence $(f(x_{j_n}))$ are in the neighborhood of $f(p)$ with radius $f(p+\varepsilon) - f(p)$, which implies $(f(x_{j_n}))$ does not converge to $f(p) = q$. This produces a contradiction, since $(f(x_{j_n}))$ is a subsequence of (y_n), and (y_n) converges to q. □

Example 14.13. The function $f : \mathbb{R} \to \mathbb{R}$ defined by $f(x) = x^3$ is continuous and increasing on $\mathbb{R}$. Thus, using Theorem 14.12, we may conclude that its inverse $f^{-1} : \mathbb{R} \to \mathbb{R}$, defined by $f^{-1}(x) = \sqrt[3]{x}$, is continuous (and increasing) on $\mathbb{R}$ (Figure 14.5).

Exercise 14.20. Complete the proof of Theorem 14.12 by using contradiction to show that f^{-1} is increasing if f is increasing (one could similarly show that f^{-1} is decreasing if f is decreasing).

Exercise 14.21. Give an example of a function $f : \mathbb{R} \to \mathbb{R}$ that is strictly monotone on $\mathbb{R}$ but discontinuous at 0.

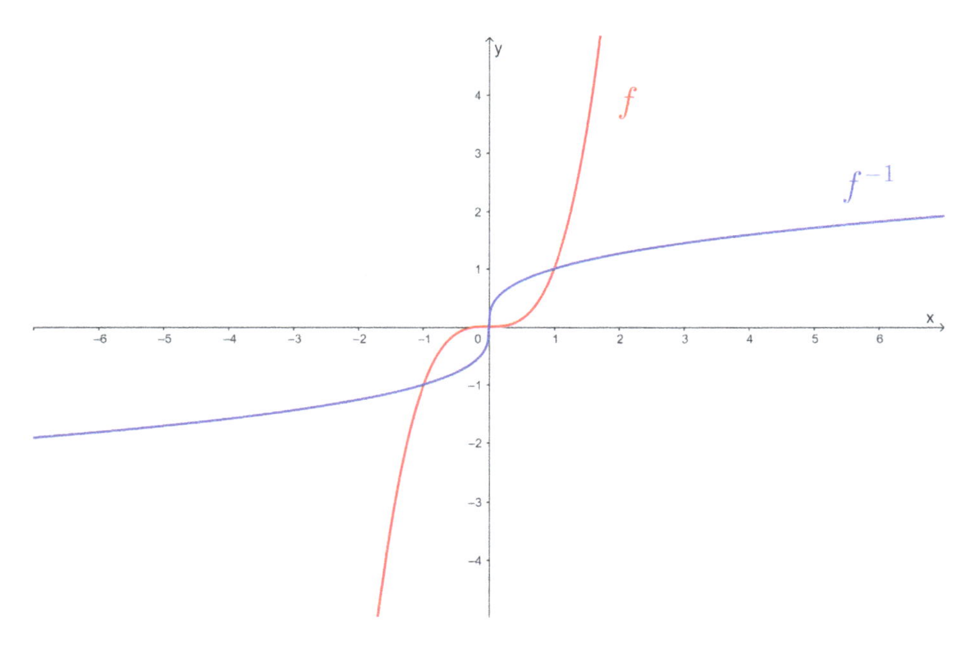

Figure 14.5: Graphs of the functions from Example 14.13.

The extreme value theorem

A function f **achieves a maximum value on a set** A contained in its domain if the set $f[A]$ of outputs of f produced from inputs in A has a maximum value and **achieves a minimum value on** A if $f[A]$ has a minimum value.

From what we have learned about suprema and infima, it is clear that a function does not have to achieve a maximum or minimum on a set. However, the so-called extreme value theorem provides us with sufficient conditions for a function to do so. This theorem is often used to guarantee that certain optimization problems in calculus can be solved.

Theorem 14.14 (The extreme value theorem). *A continuous real-valued function defined on a closed bounded interval in* $\mathbb{R}$ *achieves both a maximum value and a minimum value on this interval.*

That is, if the function $f : [a,b] \to \mathbb{R}$ *is continuous, then there exist numbers* m *and* M *in* $[a,b]$ *such that* $f(m) \leq f(x) \leq f(M)$ *for all* x *in* $[a,b]$.

Proof. Suppose $f : [a,b] \to \mathbb{R}$ is continuous. First, we show that f achieves a maximum value. To do so, we begin by showing the range R of f is bounded above. Otherwise, for every natural number n there exists a number x_n in $[a,b]$ such that $f(x_n) > n$. By Theorem 11.5, the Bolzano–Weierstrass theorem, as the sequence (x_n) is bounded, there exists a subsequence (x_{k_n}) of (x_n) that converges to a point p, and as each x_n is in the closed set $[a,b]$, it follows, using Theorem 10.3, that p is also in $[a,b]$. Then, as f

is continuous, we may use Theorem 13.8 to conclude that $f(x_{k_n}) \to f(p)$. Now, by the Archimedean property, there exists a natural number N for which $N > f(p)$. It follows, using our earlier hypothesis that $f(x_n) > n$ for every n, that whenever $n \geq N$, we must have $f(x_{k_n}) > k_n \geq n \geq N > f(p)$, which contradicts the convergence of $(f(x_{k_n}))$ to $f(p)$ (*do you see why?*).

Now, as the range R of f is bounded above and nonempty, it has a supremum s. We show that there is a number M in $[a, b]$ such that $f(M) = s$. For each natural number n, choose t_n in $[a, b]$ so that $s - \frac{1}{n} < f(t_n) \leq s$ (*can you explain why this is possible?*). Then, by the squeeze theorem for sequences, Theorem 10.15, we may conclude that $f(t_n) \to s$. Once again applying the Bolzano–Weierstrass theorem and using the fact that each t_n is in the closed set $[a, b]$, we obtain a subsequence (t_{j_n}) of (t_n) that converges to a number M in $[a, b]$, and since f is continuous at M, we may then conclude that $f(t_{j_n}) \to f(M)$. But, since $(f(t_{j_n}))$ is a subsequence of the sequence $(f(t_n))$, which itself converges to s, it then follows that $f(M) = s$, which is what we wanted to show.

To show that f achieves a minimum value, observe that as f is continuous on $[a, b]$, so is $-f$ (*do you see why?*). From what we have just proved it follows that $-f$ achieves a maximum on $[a, b]$. A point in $[a, b]$ that yields a maximum value for $-f$ yields a minimum value for f (*can you provide details?*). □

Exercise 14.22. Within the proof of the extreme value theorem, there are four places where we parenthetically asked you to think about why a certain conclusion can be reached. Provide explicit justification for each of these conclusions.

Example 14.15. Consider the function $f : [-1, 3] \to \mathbb{R}$ defined so that $f(x) = x^2$. The domain $[-1, 3]$ of f is a closed bounded interval and f is continuous on this domain. Thus, all of the hypotheses of the extreme value theorem are satisfied by f, so f must achieve maximum and minimum values on $[-1, 3]$.

As f is increasing on $[0, 3]$ and is decreasing on $[-1, 0]$, we can determine that the maximum value of f is 9, which is achieved at the input 3, while the minimum value of f is 0, which is achieved at the input 0. These conclusions are consistent with what we see by looking at the graph of f (Figure 14.6).

Exercise 14.23. Consider the function $g : [3, 5] \to \mathbb{R}$ defined by $g(x) = \frac{x}{x-2}$.

(a) Verify that the hypotheses of the extreme value theorem are satisfied by g.

(b) What does the extreme value theorem permit us to conclude about g?

(c) Find the maximum and minimum values that g achieves on its domain, as well as the inputs that produce these extreme values.

If the hypotheses of the extreme value theorem are not satisfied, the conclusion of this theorem may not hold.

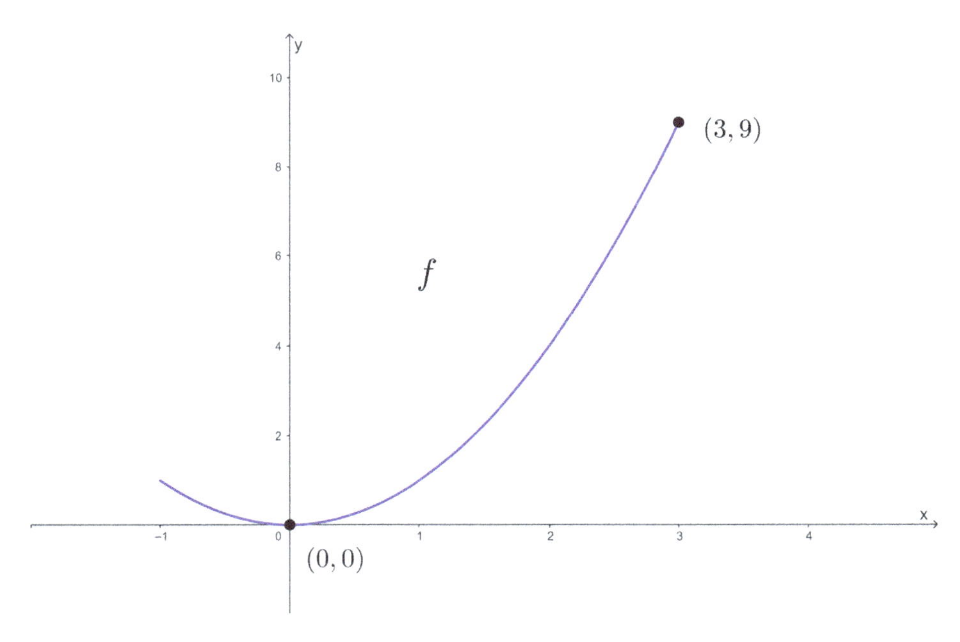

Figure 14.6: Graph of the function from Example 14.15.

Exercise 14.24. Explain why the function described does not achieve the specified extreme value and what part of the hypotheses of the extreme value theorem is not satisfied.

(a) The function $h : [-1, 3) \to \mathbb{R}$ defined so that $h(x) = x^2$ does not achieve a maximum value on $[-1, 3)$.

(b) The function $i : [-1, \infty) \to \mathbb{R}$ defined so that $i(x) = x^2$ does not achieve a maximum value on $[-1, \infty)$.

(c) The function $j : [-1, 3] \to \mathbb{R}$ defined so that

$$j(x) = \begin{cases} x^2, & \text{if } x \neq 0; \\ 1, & \text{if } x = 0; \end{cases}$$

does not achieve a minimum value on $[-1, 3]$.

Exercise 14.25. Prove that the image of a closed bounded interval under a continuous function is a closed bounded interval. That is, prove that if $[a, b]$ is a subset of the domain of a continuous function f, then $f[[a, b]] = [c, d]$ for some real numbers c and d with $c \leq d$. *Hint*: Consult Theorems 14.5 and 14.14.

Exercise 14.26. Prove that if A is a compact subset of $\mathbb{R}$ and the function $f : A \to \mathbb{R}$ is continuous, then $f[A]$ is compact.

Exercise 14.27. Prove a more general version of the extreme value theorem, Theorem 14.14, in which the closed bounded interval $[a, b]$ is replaced by an arbitrary nonempty compact subset A of $\mathbb{R}$.

Uniform continuity

When a function $f : A \to \mathbb{R}$ is continuous, we know that given a point p in A and a positive number ε, we are able to generate a positive number δ such that

$$|f(x) - f(p)| < \varepsilon,$$

whenever x is in A and $|x - p| < \delta$. Note carefully that the number δ may depend not only on ε but also on p. That is, a value of δ that works for a specified ε and specified point p may not work for the same value of ε if the point p is changed.

Example 14.16. Consider the function $f : (0,\infty) \to \mathbb{R}$ defined by $f(x) = \frac{1}{x^2}$. As f is a rational function it is continuous on its domain. Let $\varepsilon = 0.2$ and $\delta = 0.5$.

We first consider the point $p = 2$. Observe that if $|x - p| < \delta$, that is, $|x - 2| < 0.5$, we have

$$1.5 = 2 - 0.5 < x < 2 + 0.5 = 2.5,$$

from which it follows that

$$\frac{1}{2.5^2} < \frac{1}{x^2} < \frac{1}{1.5^2}.$$

As $\frac{1}{2.5^2} = 0.16$ and $\frac{1}{1.5^2} \approx 0.\overline{4}$, we see that

$$f(2) - \varepsilon = \frac{1}{4} - 0.2 = 0.05 < 0.16 < \frac{1}{x^2} < 0.\overline{4} < 0.45 = \frac{1}{4} + 0.2 = f(2) + \varepsilon,$$

meaning

$$|f(x) - f(2)| = \left|\frac{1}{x^2} - \frac{1}{4}\right| < 0.2 = \varepsilon.$$

That is, all inputs to f that are less than $\delta = 0.5$ units from 2 are mapped to outputs that are less than $\varepsilon = 0.2$ units from $f(2) = \frac{1}{4}$. We might also say that for $p = 2$, a permissible input tolerance δ for the output tolerance $\varepsilon = 0.2$ is $\delta = 0.5$.

Now consider the point $q = 0.7$ and take $x = 1$. Observe that

$$|x - q| = |1 - 0.7| = 0.3 < 0.5 = \delta,$$

but as $\frac{1}{0.7^2} \approx 2.0408 < 2.041$,

$$|f(x) - f(q)| = |f(1) - f(0.7)| = \left|1 - \frac{1}{0.7^2}\right| > 1.041 > 0.2 = \varepsilon.$$

So the input $x = 1$, though less than $\delta = 0.5$ units from $q = 0.7$, is mapped to an output $f(1) = 1$ that is more than $\varepsilon = 0.2$ units from $f(q) = f(0.7) \approx 2.0408$. In other words, for the point $q = 0.7$, the number $\delta = 0.5$ is not a permissible input tolerance for the output tolerance $\varepsilon = 0.2$.

Even though the function f is continuous at both $p = 2$ and $q = 0.7$, we see that a value of δ that works for $\varepsilon = 0.2$ and the point $p = 2$, does not work for $\varepsilon = 0.2$ when the point is changed to $q = 0.7$. This situation is visualized in Figure 14.7, where $\delta_1 = \delta = 0.5$ and δ_2 is about as large an input tolerance that will work for $\varepsilon = 0.2$ and $q = 0.7$.

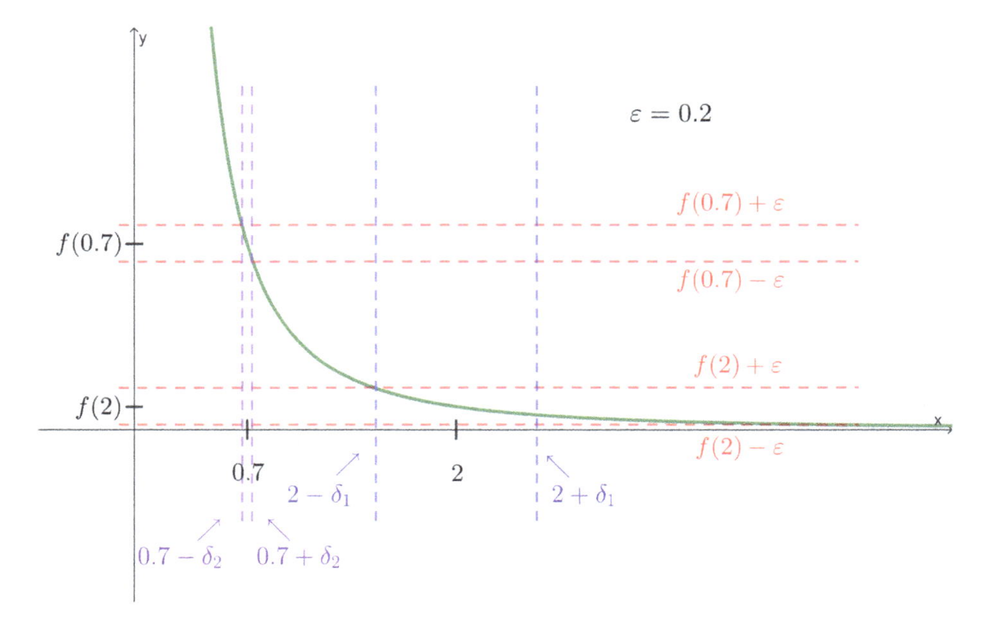

Figure 14.7: Visualizing how the input tolerance can be different for the same output tolerance at two different points of continuity.

The value of δ_2 is much smaller than $\delta_1 = 0.5$ because the graph of f is considerably steeper in the vicinity of $q = 0.7$ than it is near $p = 2$.

There are circumstances, however, in which a function f is continuous on a subset A of $\mathbb{R}$ and the same value of δ can be employed for a fixed choice of ε regardless of what point is selected from A. In such a circumstance, f is said to be *uniformly continuous* on A.

Definition 14.17 (Definition of uniform continuity). The function f is **uniformly continuous on a subset** A of its domain if for every positive number ε, there exists a positive number δ such that whenever x and y are in A and $|x - y| < \delta$, it follows that $|f(x) - f(y)| < \varepsilon$.

A comparison of Definitions 14.17 and 13.1 makes it clear that a function that is uniformly continuous on a set must be continuous on that set.

Example 14.16 (Continued). We prove the continuous function $f : (0, \infty) \to \mathbb{R}$ defined by $f(x) = \frac{1}{x^2}$ is uniformly continuous on any interval $[a, \infty)$, where $a > 0$.

Proof. Consider any positive number ε and take $\delta = \frac{a^3\varepsilon}{2}$, which we may observe is positive. Without loss of generality, assume that $a \le x \le y$, with $|x - y| = y - x < \delta$. Then

$$|f(x) - f(y)| = \left|\frac{1}{x^2} - \frac{1}{y^2}\right| = \frac{|y^2 - x^2|}{x^2y^2} = \frac{x+y}{x^2y^2}|x - y| \le \frac{2y}{x^2y^2} \cdot \delta = \frac{2}{x^2y} \cdot \frac{a^3\varepsilon}{2} < \frac{2}{a^3} \cdot \frac{a^3\varepsilon}{2} = \varepsilon.$$

Thus, by definition, f is uniformly continuous on $[a, \infty)$ for any positive number a. □

Exercise 14.28. Use the definition of uniform continuity to prove the function $f : \mathbb{R} \to \mathbb{R}$, where $f(x) = x^2$ is uniformly continuous on $[0, 4]$.

Condition (ii) of the following theorem, obtained by negating Definition 14.17, or condition (iii), are typically used to show a function is not uniformly continuous on a set.

Theorem 14.18. *Given a function $f : A \to \mathbb{R}$, the following are equivalent:*

(i) *f is not uniformly continuous on A;*

(ii) *there exists a positive number ε such that, for every positive number δ, there exist x and y in A for which both $|x - y| < \delta$ and $|f(x) - f(y)| \ge \varepsilon$;*

(iii) *there exists a positive number ε, along with sequences (x_n) and (y_n) in A, such that for every natural number n, both $|x_n - y_n| < \frac{1}{n}$ and $|f(x_n) - f(y_n)| \ge \varepsilon$.*

Exercise 14.29. Prove Theorem 14.18.

We now show that continuity on a set does not guarantee uniform continuity on that set.

Example 14.16 (Continued). We prove, using condition (ii) of Theorem 14.18, that the function $f : (0, \infty) \to \mathbb{R}$ defined by $f(x) = \frac{1}{x^2}$ is not uniformly continuous on the interval $(0, \infty)$, even though f is continuous on $(0, \infty)$.

Proof. Let $\varepsilon = 1$ and consider any positive number δ. Then by the Archimedean property there exists a natural number N such that $\frac{1}{N} < \delta$. Take $x = \frac{1}{N}$ and $y = \frac{1}{N^2+1}$, which are both positive numbers, hence, in the interval $(0, \infty)$. It follows that

$$|x - y| = \frac{1}{N} - \frac{1}{N^2 + 1} < \frac{1}{N} < \delta,$$

but

$$|f(x) - f(y)| = \left|\frac{1}{x^2} - \frac{1}{y^2}\right| = \left|N^2 - (N^2+1)^2\right| = N^4 + N^2 + 1 > 1 = \varepsilon.$$

So, by definition, f is not uniformly continuous on $(0, \infty)$. □

Note that when we proved f is uniformly continuous on $[a, \infty)$, we took $\delta = \frac{a^3\varepsilon}{2}$, which tends toward 0 as $a \to 0$. This is suggestive of the lack of uniform continuity of f on $(0, \infty)$. It is also suggested by the slope of the graph of f becoming unbounded as the inputs to f approach 0 along the x-axis.

Exercise 14.30. Prove the function $f : \mathbb{R} \to \mathbb{R}$ where $f(x) = x^2$ is not uniformly continuous on $[0, \infty)$.

The following theorem provides a useful sufficient condition for establishing uniform continuity.

Theorem 14.19. *If A is a subset of the domain of a function f and there exists a positive number C such that*

$$|f(x) - f(y)| \le C|x - y|$$

for all x and y in A, then f is uniformly continuous on A.

Example 14.16 (Continued). Within our proof that the function $f : (0, \infty) \to \mathbb{R}$ defined by $f(x) = \frac{1}{x^2}$ is uniformly continuous on $[a, \infty)$, we actually showed that $|f(x) - f(y)| < \frac{2}{a^3} \cdot |x - y|$ for all positive numbers x and y. Since $\frac{2}{a^3}$ is a positive constant, it would then automatically follow, by Theorem 14.19, that f is uniformly continuous on $[a, \infty)$.

By rewriting the inequality

$$|f(x) - f(y)| \le C|x - y|$$

appearing in Theorem 14.19 as

$$\frac{|f(x) - f(y)|}{|x - y|} \le C,$$

we see that the theorem is really saying that uniform continuity follows from knowing the slope of the graph of f is bounded.

Exercise 14.31. Show that Theorem 14.19 can be applied to establish that the function $f : \mathbb{R} \to \mathbb{R}$, where $f(x) = x^2$ is uniformly continuous on $[0, 4]$.

Exercise 14.32. Prove Theorem 14.19.

It is not always possible to use Theorem 14.19 to establish uniform continuity (see Exercise 14.34). The next, very important, theorem tells us that continuous functions are automatically uniformly continuous on closed bounded intervals.

Theorem 14.20. *A function that is continuous on a closed bounded interval is uniformly continuous on that interval.*

That is, if the function $f : [a, b] \to \mathbb{R}$ is continuous, then f is uniformly continuous on $[a, b]$.

Proof. Assume $f : [a, b] \to \mathbb{R}$ is continuous. If the desired conclusion does not hold, then there exists a positive number ε and sequences (x_n) and (y_n) in $[a, b]$ with the property that for every natural number n, we have both $|x_n - y_n| < \frac{1}{n}$ and $|f(x_n) - f(y_n)| \geq \varepsilon$. Since (x_n) is bounded, the Bolzano–Weierstrass theorem, Theorem 11.5, tells us (x_n) has a convergent subsequence (x_{k_n}), and since $[a, b]$ is closed, by Theorem 10.3, the limit p of (x_{k_n}) is in $[a, b]$. It follows that the subsequence (y_{k_n}) of (y_n) must also converge to p, since

$$|y_{k_n} - p| = |y_{k_n} - x_{k_n} + x_{k_n} - p| \leq |y_{k_n} - x_{k_n}| + |x_{k_n} - p|$$

(*do you see why?*). Then, since f is continuous at p, it follows that $f(x_{k_n}) \to f(p)$ and $f(y_{k_n}) \to f(p)$, which contradicts the fact that $|f(x_n) - f(y_n)| \geq \varepsilon$ for all n (*do you see why?*). □

One important application of uniform continuity, and Theorem 14.20 in particular, is its use in proving that a continuous function on a closed bounded interval $[a, b]$ is integrable on $[a, b]$; we establish this result in Chapter 19.

Exercise 14.33. Within the proof of Theorem 14.20, there are two places where we parenthetically asked you to think about why a certain conclusion can be reached. Provide explicit justification for each of these conclusions.

Exercise 14.34. Suppose $b > 0$ and consider the continuous function $f : [0, b] \to \mathbb{R}$ defined by $f(x) = \sqrt[3]{x}$.

(a) Explain why f is uniformly continuous on $[0, b]$.

(b) Show that there is no positive constant C for which $\sqrt[3]{x} \leq Cx$ for all x in $[0, b]$. Then explain why this shows that Theorem 14.19 cannot be used to establish the uniform continuity of f on $[0, b]$.

Exercise 14.35. Prove a more general version of Theorem 14.20 in which the closed bounded interval $[a, b]$ is replaced by an arbitrary nonempty compact subset A of $\mathbb{R}$.

15 The derivative

The notion of *derivative* of a function arose from (ultimately successful) attempts by Isaac Newton (1643–1727) and Gottfried Leibniz (1646–1716) to solve problems involving the motion of objects. In our study of calculus we learned that the derivative may be interpreted geometrically as the slope of the tangent line to the graph of a function and, in context, as the function's instantaneous rate of change. This chapter forms a bridge from the often relatively intuitive experiences with the derivative found in calculus courses to a more rigorous treatment of the concept, and includes proofs of the rules learned in calculus for computing derivatives efficiently.

Motivation for the derivative

Before introducing a precise definition of the derivative of a function, we consider a familiar scenario from calculus in order to help recall some of the motivation behind this concept.

Example 15.1. An object is dropped from a height of 100 feet above the ground. Its height after t seconds is given by $h(t) = 100 - 16t^2$ for $0 \leq t \leq 2.5$. We determine the object's instantaneous velocity 2 seconds after it begins to fall.

Consider a time t that is just a "bit later" than the time 2 seconds of interest. The elapsed time from time 2 to time t is

$$t - 2$$

and the change in height from time 2 to time t is

$$h(t) - h(2).$$

Thus, the average velocity of the object during the time interval beginning at time 2 seconds and ending at time t seconds is

$$\frac{h(t) - h(2)}{t - 2}.$$

Now recall that the instantaneous velocity of the object can be obtained as the limit of this average velocity, the limit being taken as the slightly later time t approaches the time 2 at which we want to find the instantaneous velocity. In other words, the instantaneous velocity at time 2 seconds can be represented as

$$\lim_{t \to 2} \frac{h(t) - h(2)}{t - 2},$$

https://doi.org/10.1515/9783111429564-015

where we have allowed a bidirectional approach of t toward 2 as, in this situation, we can see that the value of the limit is unaffected by whether t approaches 2 from above or from below:

$$\begin{aligned}\lim_{t\to 2}\frac{h(t)-h(2)}{t-2} &= \lim_{t\to 2}\frac{(100-16t^2)-36}{t-2}\\ &= \lim_{t\to 2}\frac{-16t^2+64}{t-2}\\ &= \lim_{t\to 2}\frac{-16(t+2)(t-2)}{t-2}\\ &= \lim_{t\to 2}-16(t+2)\\ &= -64 \quad \text{feet per second.}\end{aligned}$$

Of course, we interpret the negative sign as telling us the object's direction of motion is downward. That is, exactly 2 seconds after it is dropped, the object is *falling* at the rate of 64 feet per second.

The definition of the derivative

We continue to assume, unless otherwise indicated, that functions have real number inputs and outputs.

Definition 15.2 (The definition of the derivative of a function). Let f be a function and let p be a point that is both in the domain of f and also a limit point of the domain of f. The **derivative** $f'(p)$ of f at p is defined so that

$$f'(p) = \lim_{x\to p}\frac{f(x)-f(p)}{x-p},$$

provided this limit exists.

When the derivative of a function f exists at p, we say f is **differentiable** at p. We may view f' as a function having as its domain the set of all inputs to f at which f is differentiable. The process of obtaining the derivative of a function is known as **differentiation.**

Example 15.1 (Continued). We have determined that the function h is differentiable at 2 with $h'(2) = -64$.

The requirement in the definition of $f'(p)$ that p be a limit point of the domain of f means that any time we assume $f'(p)$ exists, we are assuming there are infinitely many points of the domain of f arbitrarily close to p. This assumption is not explicitly written into the theorems we state and prove, and in practice differentiability of f at p usually means that there is a neighborhood of p in which f is defined (and we assume this is the case when it is needed).

Often, we can find a general formula for the derivative of a function by applying the definition of the derivative to an arbitrary point x at which the function is differentiable.

Example 15.3. We use the definition of derivative to determine the derivative f' for the function f where $f(x) = \sqrt{x}$. Consider an arbitrary positive real number x. Then

$$\begin{aligned} f'(x) &= \lim_{t\to x} \frac{f(t)-f(x)}{t-x} = \lim_{t\to x} \frac{\sqrt{t}-\sqrt{x}}{t-x} = \lim_{t\to x} \frac{\sqrt{t}-\sqrt{x}}{(\sqrt{t}+\sqrt{x})(\sqrt{t}-\sqrt{x})} \\ &= \lim_{t\to x} \frac{1}{\sqrt{t}+\sqrt{x}} = \frac{1}{2\sqrt{x}}. \end{aligned}$$

Hence, we may conclude that f is differentiable at all positive real numbers x with

$$f'(x) = \frac{1}{2\sqrt{x}}.$$

Exercise 15.1. Use the definition of the derivative to find the derivative of the given function.
(a) $f(x) = x^3$
(b) $g(x) = \frac{1}{x}$

Exercise 15.2 (Derivative of a constant). Let c be any real number and consider the constant function f defined so that $f(x) = c$. Use the definition of the derivative to show that $f'(x) = 0$.

Exercise 15.3 (Derivative of a linear function). Let m and b be any fixed real numbers. Consider the linear function f, where $f(x) = mx + b$. Use the definition of the derivative to show that $f'(x) = m$.

Exercise 15.4. Consider the function f defined so that

$$f(x) = \begin{cases} x + x^2, & \text{if } x \text{ is rational;} \\ x - x^3, & \text{if } x \text{is irrational.} \end{cases}$$

Show that f is differentiable at 0 by using the definitions of derivative and function limit to evaluate $f'(0)$.

While we usually denote the derivative of a function f by f', if $y = f(x)$, then $f'(x)$ may also be denoted by $\frac{dy}{dx}$, by $\frac{df}{dx}$, and by $\frac{d}{dx}[f(x)]$.

Interpreting the derivative in context: instantaneous rate of change

In the study of real analysis, our interest in the derivative is primarily theoretical, but for additional motivation we recall from calculus that the value $f'(p)$ of the derivative of a function f represents *the instantaneous rate of change of the outputs of f with respect to the inputs at the particular input p.*

Example 15.1 (Continued). We found that $h'(2) = -64$ and, in context, this told us that the instantaneous rate of change of the height of the falling object two seconds after it is dropped is 64 feet per second in the downward direction (physicists would say this is the object's *velocity* at this moment).

Example 15.4. Suppose the annual revenue $R(w)$, in thousands of dollars, of a company is a function of the number w of workers it employs.

If $R'(500) = 3$, then when the company employs 500 workers, the company's revenue is increasing at the rate of $3000 per employee.

If $R'(700) = -4.5$, then when the company employs 700 workers, the company's revenue is decreasing at the rate of $4500 per employee.

Exercise 15.5. Suppose that $P(t)$ is the population of a city t years from now. Interpret each statement in context.

(a) $P'(4) = 100$.

(b) $P'(5) = -200$.

(c) $P'(6) = 0$.

Exercise 15.6. What does $g'(30)$ represent and in what units is it measured if $g(x)$ represents

(a) the amount of money in a bank account x months after it was opened?

(b) the number of gallons of gas a car's engine has consumed after x miles?

Local linearization and the approximation of function values

Assuming the function f is differentiable at p, as we continue to zoom in on the point $(p,f(p))$, the graph of f looks more and more like a straight line. Intuitively, this line is the limit of the so-called *secant lines* passing through the points $(p,f(p))$ and $(x,f(x))$ on the graph of f, where $(p,f(p))$ remains fixed and $(x,f(x))$ varies in such a way that x approaches p (see Figure 15.1).

Therefore, the slope of this line is the limiting value of the slope

$$\frac{f(x)-f(p)}{x-p} = \frac{\text{output change}}{\text{input change}} = \frac{\text{rise}}{\text{run}}$$

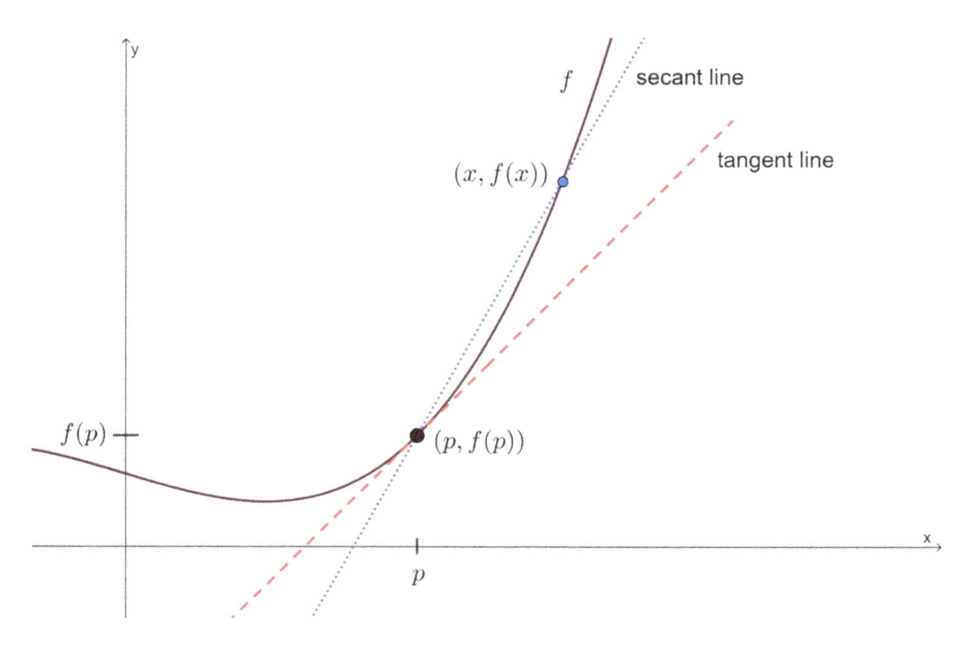

Figure 15.1: Secant line and tangent line to the graph of a function.

of the secant line, the limit being taken as x approaches p, which means the slope is given by

$$f'(p) = \lim_{x \to p} \frac{f(x) - f(p)}{x - p}.$$

We refer to this line as the **tangent line** to the graph of f at the point $(p, f(p))$; by definition, it is the line passing through the point $(p, f(p))$ having slope $f'(p)$ (again, see Figure 15.1). The function L whose graph is this tangent line is called the **local linearization** of f at p; you are asked to verify that $L(x) = f(p) + f'(p)(x - p)$ in Exercise 15.7.

Exercise 15.7. Show that the local linearization L of a function f that is differentiable at p can be defined so that $L(x) = f(p) + f'(p)(x - p)$.

Figure 15.2 displays the graph of a function f along with the graph of its local linearization (tangent line) L at the input p.

The local linearization L may be viewed as the best linear approximation to the function f in the vicinity of the input p, in the sense that when we zoom in on the point $(p, f(p))$, the graph of f more and more closely coincides with the graph of L (the tangent line). Thus, for inputs x to f sufficiently close to p, the local linearization L yields values $L(x)$ close to the values $f(x)$ of the function f itself. Of course, as we move further

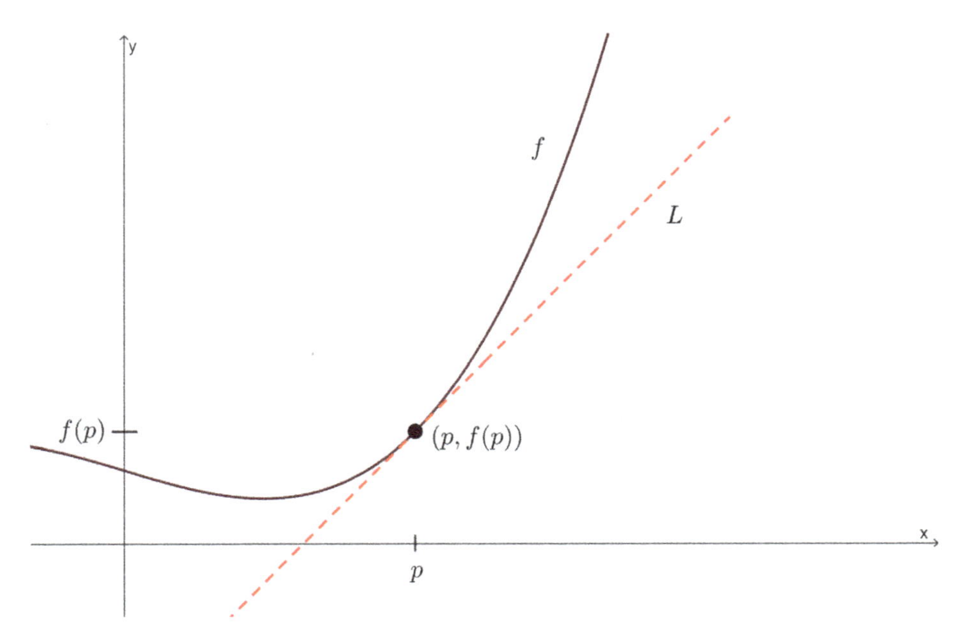

Figure 15.2: Graphs of function f and its local linearization L at input p.

away from $(p, f(p))$, the graph of f and the tangent line may separate from one another, resulting in less accurate (or perhaps very inaccurate) approximations to the values of f.

Example 15.5. We find the local linearization L of the principal square root function f defined by $f(x) = \sqrt{x}$ at the input 4. Using our work from Example 15.3, we see that the slope of L is

$$f'(4) = \frac{1}{2\sqrt{4}} = \frac{1}{4}.$$

Thus, as $f(4) = \sqrt{4} = 2$, the local linearization is

$$L(x) = f(4) + f'(4)(x-4) = 2 + \frac{1}{4}(x-4)$$

or

$$L(x) = \frac{1}{4}x + 1.$$

The graphs of f and L are both shown in Figure 15.3.

Observe that for values of x quite close to 4, the graphs are virtually indistinguishable, though once x is far enough from 4 the graphs begin to separate from one another.

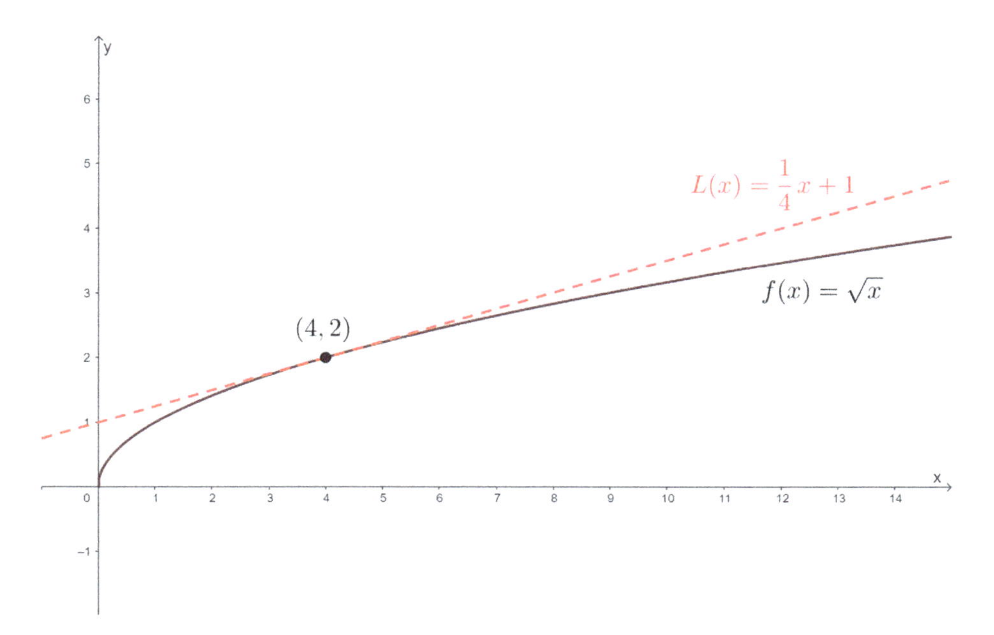

Figure 15.3: Graphs of function f and local linearization L from Example 15.5.

We can use L to approximate $\sqrt{5}$ as

$$\sqrt{5} = f(5) \approx L(5) = \frac{1}{4} \cdot 5 + 1 = 2.25,$$

which is fairly accurate as, to three decimal places, $\sqrt{5}$ is 2.236. On the other hand, if we try to use L to approximate $\sqrt{16}$ we get

$$\sqrt{16} = f(16) \approx L(16) = \frac{1}{4} \cdot 16 + 1 = 5,$$

which, as $\sqrt{16} = 4$, would probably not be viewed as providing a good approximation (essentially, the input 16 is not close enough to the input 4 at which the local linearization was constructed).

Exercise 15.8. Let $f(x) = x^3$ and note that, from your work in Exercise 15.1(a) above, $f'(x) = 3x^2$. Use this information to determine the equation of the tangent line to the graph of f at the point $(2, 8)$. Then use an online or handheld graphing calculator to graph both f and the tangent line you found in the vicinity of the point $(2, 8)$; both graphs should pass through this point and be almost identical near this point.

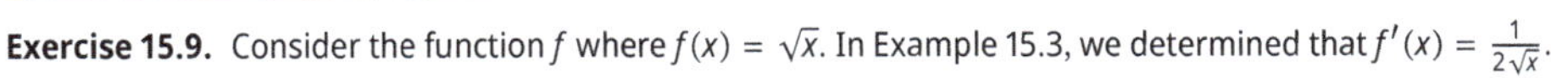

Exercise 15.9. Consider the function f where $f(x) = \sqrt{x}$. In Example 15.3, we determined that $f'(x) = \frac{1}{2\sqrt{x}}$.

(a) Find the local linearization of f at 9 and use it to approximate the value of $\sqrt{8}$. Compare your answer to that provided by a calculator.

(b) Use an online or handheld graphing calculator to display the graphs of f and its local linearization at the input 9. Zoom in on the graph of f around the point $(9, 3)$. As you zoom in, does it appear that the graphs of f and its local linearization at 9 are becoming virtually indistinguishable?

Situations in which the derivative does not exist

In order for a function's derivative to exist at a certain input, the function must be continuous at that input.

Theorem 15.6. *If the function f is differentiable at p, then f is continuous at p.*

Proof. Suppose f is differentiable at p. It follows that

$$\begin{aligned}\lim_{x\to p}(f(x)-f(p)) &= \lim_{x\to p}\left(\frac{f(x)-f(p)}{x-p}\cdot(x-p)\right)\\ &= \lim_{x\to p}\frac{f(x)-f(p)}{x-p}\cdot\lim_{x\to p}(x-p)\\ &= f'(p)\cdot 0\\ &= 0.\end{aligned}$$

Therefore, $\lim_{x\to p} f(x) = f(p)$, so we may conclude that f is continuous at p. □

Thus, a lack of continuity automatically implies the derivative does not exist. However, just because a function is continuous at a point does not mean it must be differentiable at the point.

Example 15.7. The absolute value function is continuous at all real numbers (see Exercise 13.9), in particular, at 0. However, it is not differentiable at 0 as

$$\lim_{x\to 0^+}\frac{|x|-|0|}{x-0} = \lim_{x\to 0^+}\frac{x}{x} = \lim_{x\to 0^+} 1 = 1,$$

whereas

$$\lim_{x\to 0^-}\frac{|x|-|0|}{x-0} = \lim_{x\to 0^-}\frac{-x}{x} = \lim_{x\to 0^-} -1 = -1,$$

which together imply that $\lim_{x\to 0}\frac{|x|-|0|}{x-0}$ does not exist. As depicted in Figure 15.4, this lack of differentiability at 0 is a consequence of the sharp bend in the graph of the absolute value function at the point $(0, 0)$.

Example 15.8. The square root function is continuous at 0, but as

$$\lim_{x\to 0}\frac{\sqrt{x}-\sqrt{0}}{x-0} = \lim_{x\to 0}\frac{\sqrt{x}}{x} = \lim_{x\to 0}\frac{1}{\sqrt{x}} = \infty,$$

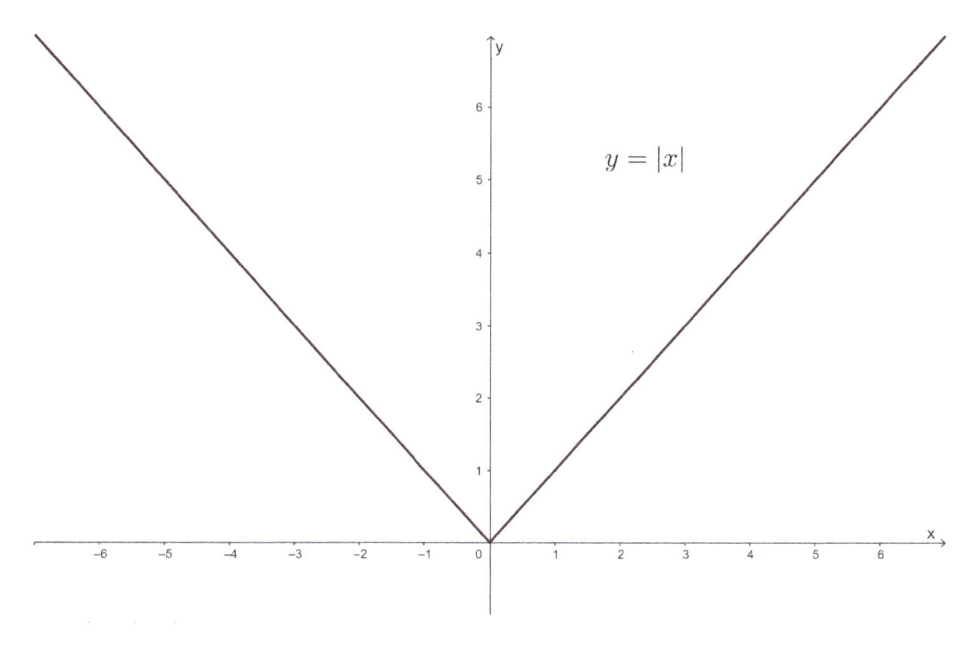

Figure 15.4: Graph of the absolute value function from Example 15.7.

the square root function is not differentiable at 0. Geometrically, the nonexistence of the derivative at 0 is the result of a vertical tangent line (namely, the y-axis) to the graph of the square root function at the point $(0, 0)$ (see Figure 15.3).

Exercise 15.10. For the given function, use the definition of the derivative to verify that the function is not differentiable at 0. Then use the graph of the function to provide a geometric reason as to why the function is not differentiable at 0.

(a) the sign function sgn where $\operatorname{sgn}(x) = \begin{cases} -1, & \text{if } x < 0; \\ 0, & \text{if } x = 0; \\ 1, & \text{if } x > 0; \end{cases}$

(b) the function g where $g(x) = \begin{cases} 2x + 1, & \text{if } x \leq 0; \\ 1 - x, & \text{if } x > 0; \end{cases}$

(c) the function h where $h(x) = \sqrt[5]{x}$.

Higher-order derivatives

The derivative of the derivative of a function f is called the **second derivative** of f and is denoted by f''. Other so-called "higher-order" derivatives are defined in the analogous manner via recursion. For instance, the **third derivative** f''' of f is the derivative of the second derivative f'' of f. For any natural number n, the nth derivative of f may be denoted by $f^{(n)}$, and this notation is especially employed when $n \geq 4$. It is also convenient to sometimes use the notation $f^{(0)}$ for the function f itself.

Example 15.9. As velocity is the derivative of position and acceleration is the derivative of velocity, it follows that acceleration is the second derivative of position.

Exercise 15.11. A ball is thrown into the air from a bridge. The height $h(t)$, in feet, of the ball above the ground t seconds after it is released is given by

$$h(t) = -16t^2 + 30t + 65,$$

for those times t while the ball is in the air.

(a) Use the definition of the derivative to find the velocity of the ball at time t.

(b) Use the definition of the derivative to find the acceleration of the ball at time t.

Arithmetic properties of the derivative

The following properties of derivatives are learned in a first calculus course.

Theorem 15.10 (Sum, difference, product, and quotient rules for derivatives). *Suppose the functions f and g are both differentiable at x.*

(1) *The sum $f + g$ is differentiable at x and*

$$(f + g)'(x) = f'(x) + g'(x).$$

(2) *The difference $f - g$ is differentiable at x and*

$$(f - g)'(x) = f'(x) - g'(x).$$

(3) *The product $f \cdot g$ is differentiable at x and*

$$(f \cdot g)'(x) = f(x) \cdot g'(x) + f'(x) \cdot g(x).$$

(4) *Provided $g(x) \neq 0$, the quotient $\frac{f}{g}$ is differentiable at x and*

$$\left(\frac{f}{g}\right)'(x) = \frac{f'(x) \cdot g(x) - f(x) \cdot g'(x)}{(g(x))^2}.$$

Proof. The proofs of (1) and (2) are left as exercises. We also leave the proof of (4) as an exercise that involves (3) and the chain rule (see Exercise 15.20). Here we demonstrate the product rule (3) under the (necessary) hypothesis that $f'(x)$ and $g'(x)$ both exist. Observe that

$$
\begin{aligned}
(f \cdot g)'(x) &= \lim_{t\to x} \frac{f(t)\cdot g(t) - f(x)\cdot g(x)}{t-x} \\
&= \lim_{t\to x} \frac{f(t)\cdot g(t) - f(t)\cdot g(x) + f(t)\cdot g(x) - f(x)\cdot g(x)}{t-x} \\
&= \lim_{t\to x} \frac{f(t)\cdot (g(t) - g(x)) + (f(t) - f(x))\cdot g(x)}{t-x} \\
&= \lim_{t\to x} f(t)\cdot \lim_{t\to x} \frac{g(t)-g(x)}{t-x} + \lim_{t\to x} \frac{f(t)-f(x)}{t-x}\cdot g(x) \\
&= f(x)\cdot g'(x) + f'(x)\cdot g(x),
\end{aligned}
$$

where $\lim_{t\to x} f(t) = f(x)$ because the differentiability of f at x implies f is continuous at x. □

Exercise 15.12. Prove (1) of Theorem 15.10.

Exercise 15.13 (Constant multiple rule for derivatives). Let c be any real number and suppose that the function f is differentiable at x. Use the product rule (3) of Theorem 15.10 to show that

$$(cf)'(x) = c \cdot f'(x),$$

which demonstrates the differentiability of the constant multiple function cf at x.

Exercise 15.14. Define functions f and g on $\mathbb{R}$ so that $f(x) = 1$ and $g(x) = x$. Show that $(f(x) \cdot g(x))' \neq f'(x) \cdot g'(x)$. Thus, in general, the derivative of a product is not the product of the derivatives of the factors.

Parallel to Σ-notation for sums, there is **Π-notation** or **product notation** for products. For example, we may write

$$6! = 6\cdot 5\cdot 4\cdot 3\cdot 2\cdot 1 = \prod_{i=1}^{6} i$$

and

$$(x-4)(x-9)(x-16) = \prod_{k=2}^{4} (x-k^2).$$

Theorem 15.11 (Extended product rule for derivatives). *If all of the functions $f_1, f_2, \ldots, f_n$ are differentiable at x, then so is their product $f_1 f_2 \cdots f_n$, with*

$$(f_1 f_2 \cdots f_n)'(x) = \sum_{k=1}^{n} \left(\frac{f_k'(x)}{f_k(x)} \prod_{i=1}^{n} f_i(x) \right).$$

Exercise 15.15. Use induction to establish Theorem 15.11. (Note that the base step for the induction is already established as (3) from Theorem 15.10.)

Theorem 15.12 (Power rule for derivatives). *Let r be any real number and define f so that $f(x) = x^r$. Then for each x at which f is differentiable, $f'(x) = rx^{r-1}$.*

The proof of Theorem 15.12 is carried out in stages. It is easily accomplished for positive integer powers by induction (Exercise 15.16). The chain rule helps us to establish the result for negative integer powers (Exercise 15.21). The proof for powers that are reciprocals of positive integers is an application of the relationship between the derivatives of a function and its inverse (Example 15.16). The proof for rational powers is assigned as Exercise 15.23. The proof for irrational powers can be done easily once we have introduced the natural logarithmic function and some of its fundamental properties (see Exercises 20.23 and 20.24).

Exercise 15.16. Use induction to show that for any positive integer n, if $f(x) = x^n$, then $f'(x) = nx^{n-1}$.

Exercise 15.17. Determine the points at which the function f given by $f(x) = x \cdot |x|$ is differentiable and then find the derivative.

Exercise 15.18. Consider the function g defined so that $g(x) = x^2 \cdot |x|$.
(a) Find $g'(x)$ for arbitrary real x.
(b) Find $g''(x)$ for arbitrary real x.
(c) Find $g'''(x)$ for arbitrary nonzero real x.
(d) Show that $g'''(0)$ does not exist.

The chain rule

To provide motivation for the differentiation procedure used for composite functions, suppose your car gets 20 miles per gallon when driven at 60 miles per hour. Then the number of gallons of gas the car is consuming each hour can be computed as

$$\left(\frac{1}{20}\,\frac{\text{gallon}}{\text{mile}}\right)\cdot\left(60\,\frac{\text{miles}}{\text{hour}}\right) = 3\,\frac{\text{gallons}}{\text{hour}}.$$

Letting x be the time (in hours) you have been driving, u the distance traveled (in miles), and y the amount of gasoline consumed (in gallons), we have observed that

$$\frac{dy}{du} \cdot \frac{du}{dx} = \frac{dy}{dx}$$

or, introducing function notation with $u = f(x)$ and $y = g(u)$, hence, $y = (g \circ f)(x)$, that

$$(g \circ f)'(x) = g'(u) \cdot f'(x) = g'(f(x)) \cdot f'(x).$$

This familiar result from elementary calculus is the chain rule for computing the derivatives of composite functions.

Example 15.13. According to the chain rule,

$$\frac{d}{dx}[\sqrt{5x+1}] = \frac{d}{dx}[(5x+1)^{1/2}] = \frac{1}{2}(5x+1)^{-1/2} \cdot 5 = \frac{5}{2\sqrt{5x+1}}.$$

Theorem 15.14 (The chain rule). *If the function f is differentiable at x and the function g is differentiable at $f(x)$, then the function $g \circ f$ is differentiable at x, with $(g \circ f)'(x) = g'(f(x)) \cdot f'(x)$.*

Proof. Assume that f is differentiable at x and g is differentiable at $f(x)$. Let $y = f(x)$ and define the function h so that

$$h(s) = \begin{cases} \frac{g(s)-g(y)}{s-y}, & \text{if } s \neq y; \\ g'(y), & \text{if } s = y. \end{cases}$$

Observe that h is continuous at y since

$$h(y) = g'(y) = \lim_{s \to y} \frac{g(s) - g(y)}{s - y} = \lim_{s \to y} h(s).$$

Also, as differentiability implies continuity, f is continuous at x. Thus, the composition $h \circ f$ is continuous at x, so that

$$\lim_{t \to x} h(f(t)) = h(f(x)) = h(y).$$

Now, for each s, note that

$$g(s) - g(y) = h(s) \cdot (s - y),$$

which implies that for every t,

$$g(f(t)) - g(f(x)) = h(f(t)) \cdot (f(t) - f(x)).$$

Therefore,

$$
\begin{aligned}
(g \circ f)'(x) &= \lim_{t \to x} \frac{g(f(t)) - g(f(x))}{t - x} \\
&= \lim_{t \to x} \frac{h(f(t)) \cdot (f(t) - f(x))}{t - x} \\
&= \lim_{t \to x} h(f(t)) \cdot \lim_{t \to x} \frac{f(t) - f(x)}{t - x} \\
&= h(y) \cdot f'(x) \\
&= g'(y) \cdot f'(x) \\
&= g'(f(x)) \cdot f'(x).
\end{aligned}
$$

□

Exercise 15.19. Find a more direct route to proving the chain rule under the assumption that there is a neighborhood of x in which $f(t) \neq f(x)$ for all $t \neq x$.

Exercise 15.20. In Exercise 15.1(b), you showed that

$$\frac{d}{dx}\left[\frac{1}{x}\right] = -\frac{1}{x^2}.$$

Use this fact, along with the product rule and the chain rule, to prove the quotient rule for derivatives stated in (4) of Theorem 15.10.

Exercise 15.21. The power rule for derivatives is stated in Theorem 15.12; you proved this result for positive integer powers in Exercise 15.16. You also proved the quotient rule for derivatives in Exercise 15.20. Use these results to establish the power rule for negative integer exponents.

Derivatives of inverse functions

Recall that the inverse of a function simply interchanges the roles of input and output. Because the derivative may be interpreted geometrically as representing a slope, and the interchange of input and output in the slope formula would result in a reciprocal being taken, the following relationship between the derivative of a function and the derivative of the function's inverse can be anticipated.

Theorem 15.15 (Derivatives of inverse functions). *Suppose the function f is strictly monotone (i. e., increasing or decreasing) and continuous on an interval I that includes x. If f is differentiable at x and $f'(x) \neq 0$, then f^{-1} is differentiable at $f(x)$ and*

$$(f^{-1})'(f(x)) = \frac{1}{f'(x)}.$$

Proof. Under the stated hypotheses, we have

$$(f^{-1})'(f(x)) = \lim_{s\to f(x)} \frac{f^{-1}(s) - f^{-1}(f(x))}{s - f(x)} = \lim_{f^{-1}(s)\to x} \frac{f^{-1}(s) - x}{s - f(x)} = \lim_{t\to x} \frac{t - x}{f(t) - f(x)} = \frac{1}{f'(x)},$$

where we are letting $t = f^{-1}(s)$ so that $s = f(t)$, and making use of the fact that as f is strictly monotone and continuous at x, according to Theorem 14.12, we must have f^{-1} continuous at $f(x)$. □

Exercise 15.22. The function $f : [0,\infty) \to [0,\infty)$ defined so that $f(x) = x^2$ is increasing and continuous on its domain, as is its inverse $f^{-1} : [0,\infty) \to [0,\infty)$, where $f^{-1}(x) = \sqrt{x}$.

(a) Use Theorem 15.15 and the fact that $f'(2) = 4$ to show that $(f^{-1})'(4) = \frac{1}{4}$.

(b) Use an online or handheld graphing calculator to obtain the graphs of f and f^{-1}, along with the tangent lines to these graphs at the respective points $(2,4)$ and $(4,2)$, where you use the same scale on both coordinate axes. Geometrically, do the tangent lines appear to have reciprocal slopes?

Example 15.16. We can now show that the power rule for computing derivatives applies to powers that are reciprocals of positive integers.

Consider any positive integer n and let f be the function defined by $f(x) = x^n$, defined on $[0,\infty)$ if n is even and on $\mathbb{R}$ if n is odd. From Exercise 10.4(a), we know that f is increasing on $[0,\infty)$; if n is odd a similar argument may be used to extend to $\mathbb{R}$ the conclusion that f is increasing. Thus, f is continuous and strictly monotone on its domain.

The inverse of f is the function f^{-1} defined so that $f^{-1}(x) = \sqrt[n]{x} = x^{1/n}$ for all real x when n is odd, and non-negative real x when n is even. We already know that $f'(x) = nx^{n-1}$, so from Theorem 15.15 we may conclude that

$$(f^{-1})'(f(x)) = \frac{1}{f'(x)} = \frac{1}{nx^{n-1}} = \frac{1}{n}x^{1-n}.$$

If we now let $y = f(x) = x^n$, it follows that $x = y^{1/n}$ so that

$$(f^{-1})'(y) = \frac{1}{n}(y^{1/n})^{1-n} = \frac{1}{n}y^{\frac{1-n}{n}} = \frac{1}{n}y^{\frac{1}{n}-1},$$

that is, $(f^{-1})'(x) = \frac{1}{n}x^{\frac{1}{n}-1}$, which is the desired result.

Exercise 15.23. Use the fact that the power rule for derivatives has already been established for integer powers and for powers that are reciprocals of positive integers to prove the power rule in the case where the power is an arbitrary rational number.

---

16 The mean value theorem

Applications of derivatives abound throughout the sciences, engineering, and business. The derivative also provides a means for locating the extreme values (maxima and minima) of a function when they exist, for approximating the values of important types of functions such as trigonometric and logarithmic, and for evaluating limits involving certain indeterminate forms. The linchpin for all these activities is the mean value theorem, which directly relates values of a function to values of the function's derivative.

Local extrema

The number $f(c)$ is a
(1) **local maximum** of the function f if there is a neighborhood of c on which $f(c)$ is the maximum value of f;
(2) **local minimum** of the function f if there is a neighborhood of c on which $f(c)$ is the minimum value of f.

A **local extremum** is a value of a function that is either a local maximum or a local minimum. Local extrema correspond to "turning points" on the graph of a function, with local maxima corresponding to relative high points and local minima to relative low points.

Example 16.1. Consider the function f, whose graph is displayed in Figure 16.1.

It appears that f has a local maximum of approximately 1.64 at (or at least very near) the input -0.68, because the graph suggests $f(-0.68) \approx 1.64$ is the largest value taken by f for some positive distance either side of -0.68, that is, within some neighborhood of -0.68.

Since the graph of f reaches points vertically higher than 1.64 if we look far enough to the left and far enough to the right, we see that f takes larger values than 1.64 once the input to f is chosen sufficiently smaller or larger than -0.68. Thus, although 1.64 is a local maximum, it is not the overall maximum value of f.

The graph also suggests that f has a local minimum of approximately 1.43 at $x \approx -1.27$ and another local minimum of approximately 0.52 at $x \approx 0.45$.

A number c in the domain of a function f is called a **critical number** of f if $f'(c)$ is zero or is undefined. The significance of critical numbers is revealed by the following theorem.

Theorem 16.2. *Let f be a function that is defined within some neighborhood of c. If $f(c)$ is a local extremum of f and f is differentiable at c, then $f'(c) = 0$. That is, local extrema of a function can occur only at critical numbers.*

https://doi.org/10.1515/9783111429564-016

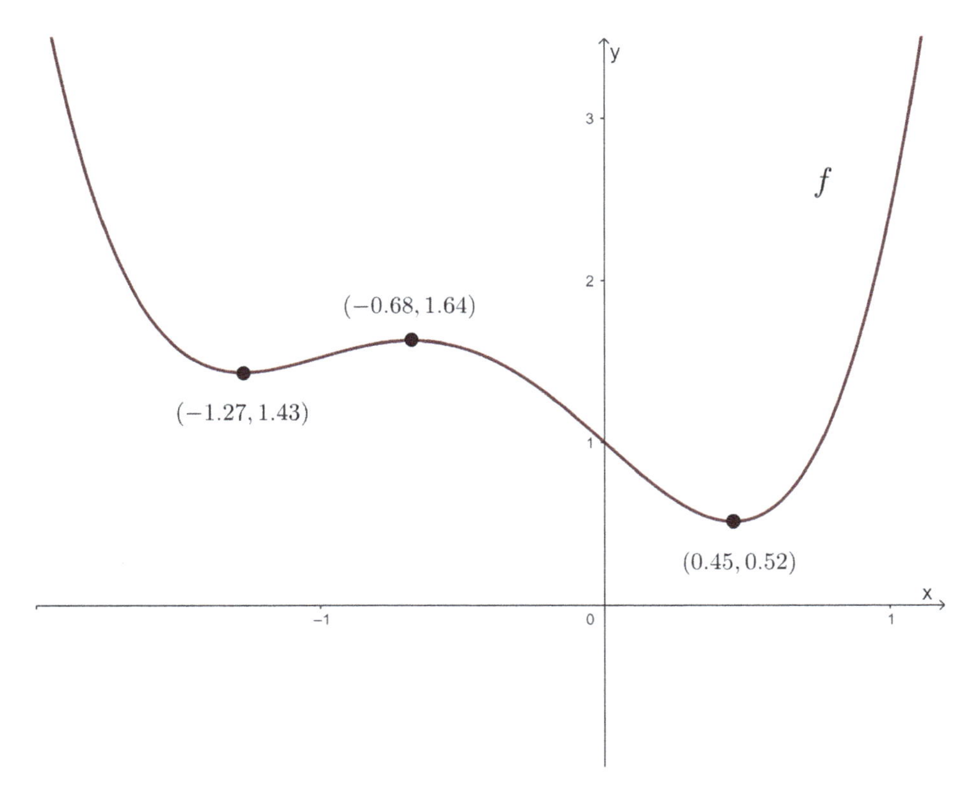

Figure 16.1: Graph of the function from Example 16.1.

Proof. Suppose $f(c)$ is a local extremum of f and f is differentiable at c. Suppose to the contrary that $f'(c) \neq 0$. Then either $f'(c) > 0$ or $f'(c) < 0$. We show the former assumption produces a contradiction, leaving the reader to show the latter also produces a contradiction.

So suppose $f'(c) = \lim_{x \to c} \frac{f(x)-f(c)}{x-c} > 0$. Then, using the result of Exercise 12.8, there exists a positive number δ such that when $0 < |x - c| < \delta$, it follows that

$$\frac{f(x) - f(c)}{x - c} > 0.$$

Hence, for x in $(c, c + \delta)$, we have $f(x) > f(c)$, while for x in $(c - \delta, c)$, we have $f(x) < f(c)$. Hence, there can be no neighborhood of c on which $f(c)$ is either the maximum or minimum value of f, contradicting the hypothesis that $f(c)$ is a local extremum of f. □

Example 16.3. Consider the function f defined so that

$$f(x) = 5x^4 - 4x^3.$$

It is easily verified that f is differentiable on $\mathbb{R}$ with

$$f'(x) = 20x^3 - 12x^2.$$

Thus, the only possible inputs to f at which local extrema can occur are the critical numbers obtained by solving $f'(x) = 0$. The solutions to this equation are $x = 0$ and $x = \frac{3}{5}$. Thus, the critical numbers of f are 0 and $\frac{3}{5}$, and these numbers are the only inputs that have the potential to produce local extreme values for f.

Example 16.4. Consider the function g defined so that

$$g(x) = \sqrt[3]{(x^2 - 4)^2}$$

and observe that

$$g'(x) = \frac{4x}{3\sqrt[3]{x^2 - 4}}.$$

Thus, g is differentiable for all real numbers except 2 and −2. As g' is undefined at ±2, but g itself is defined at ±2 (note, in fact, that the domain of g is $\mathbb{R}$), the numbers 2 and −2 are critical numbers of g. The only other critical number of g is 0, as $g'(0) = 0$ and no other points in the domain of g produce a derivative value of 0. Thus, the only candidates for the locations of local extreme values of the function g are −2, 0, and 2.

Exercise 16.1. Find the critical numbers of each function.

(a) the function f defined by $f(x) = 2x^3 + 3x^2 - 36x + 3$

(b) the function g defined by $g(x) = 4x + \frac{1}{x}$

(c) the function h defined by $h(x) = 4 - \sqrt[3]{(x-2)^2}$

(d) the function j defined by $j(x) = \frac{2(x^2-9)}{x^2-4}$

Finding global extrema

According to the extreme value theorem, a continuous function f defined on a closed bounded interval $[a, b]$ achieves both a maximum and a minimum value on $[a, b]$. Based on the theory we have developed thus far, these **global extreme values** (i. e., **global extrema**) of f on $[a, b]$ can occur only at one of the endpoints a or b of the interval, or at a critical number of f lying between a and b.

Example 16.5. The hypotheses of the extreme value theorem are satisfied by the polynomial function f defined by

$$f(x) = 4x^3 + 9x^2 - 12x + 1$$

on the interval $[-1, 2]$. Thus, f achieves global extreme values on $[-1, 2]$. The only possible inputs at which these extrema can occur are the endpoints -1 and 2 of the interval, along with any critical numbers of f lying within the interval. As f is differentiable on $\mathbb{R}$, with

$$f'(x) = 12x^2 + 18x - 12,$$

the only critical numbers f possesses are the solutions to the equation

$$12x^2 + 18x - 12 = 0,$$

which are -2 and $\frac{1}{2}$. Of these critical numbers, only $\frac{1}{2}$ lies between the endpoints of $[-1, 2]$. Thus, we have determined that the only candidates for the locations of the extreme values of f on $[-1, 2]$ are the inputs -1, $\frac{1}{2}$, and 2. Observing that $f(-1) = 18$, $f(\frac{1}{2}) = -\frac{9}{4}$, and $f(2) = 45$, we may then conclude that on $[-1, 2]$, the function f achieves its maximum value of 45 at the right endpoint 2 of the interval and its minimum value $-\frac{9}{4}$ at the critical number $\frac{1}{2}$, which lies between the endpoints of the interval.

Exercise 16.2. Find the global extreme values, and their locations, of the function f given in Exercise 16.1(a) on the given interval.
(a) $[-4, 0]$
(b) $[0, 4]$
(c) $[-4, 4]$
(d) $[0, 1.5]$

The mean value theorem

The mean value theorem is one of the most important theorems in calculus and analysis. It relates values of a function's derivative to values taken by the function itself. In context, the mean value theorem may also be viewed as relating a function's instantaneous rate of change to its average rate of change. Because of its importance in laying the theoretical foundation for much of the rest of the calculus, the mean value theorem is sometimes referred to as the fundamental theorem of *differential* calculus. We shall eventually see that the mean value theorem plays a pivotal role in the proof of the fundamental theorem of (integral) calculus.

We begin by providing some motivation by asking the question, does an object traveling at an average speed of 50 miles per hour (mph) ever have to achieve an instantaneous speed of 50 mph? Under the most general circumstances, no. For example, an object that maintains a constant speed of 25 mph for one hour and then a constant speed of 75 mph for another hour would have an average speed of 50 mph over the two-hour time period, but the object would never achieve an instantaneous speed of 50 mph at any specific time during the two hours.

In many situations, though, for example driving a car, it would be reasonable to assume that an object cannot abruptly change its speed, meaning the object's position function would be differentiable. In such situations, the mean value theorem tells us that there is at least one moment in time when the speed of the object at that time is equal to the object's average (mean) speed over the entire trip.

Before dealing with the mean value theorem in a general setting, we consider a special case.

Lemma 16.6 (Rolle's theorem). *If the function f is continuous on the interval $[a, b]$ and differentiable on the interval (a, b), with $f(a) = f(b)$, then there exists a number c in (a, b) such that $f'(c) = 0$.*

Proof. Suppose the function f is continuous on $[a, b]$ and differentiable on (a, b), and that $f(a) = f(b)$. If f is constant on $[a, b]$, the result holds as $f'(c) = 0$ for every c in (a, b). Otherwise, f takes a value different from that taken at the endpoints of $[a, b]$ at some interior point of this interval. If this value is greater than $f(a)$, it follows that the maximum of f on $[a, b]$, which must exist according to the extreme value theorem, occurs at a point c in the interior (a, b) of $[a, b]$; this maximum is also a local maximum, and since f is differentiable at c, we may conclude, via Theorem 16.2, that $f'(c) = 0$. A similar argument can be employed if f takes a value smaller than $f(a)$. □

Michel Rolle (1652–1719) proved this result for polynomial functions in 1691. Cauchy established it more generally as a corollary to the mean value theorem.

Exercise 16.3. What does Rolle's theorem say in the context in which $f(t)$ represents the position of an object moving along the number line at time t?

The mean value theorem tells us that, under the right circumstances, a function's average rate of change on some input interval must be achieved instantaneously at some specific input in the interval. It was Cauchy who in 1823 first proved the mean value theorem.

Theorem 16.7 (The mean value theorem). *If the function f is continuous on the interval $[a, b]$ and differentiable on the interval (a, b), then there exists a number c in (a, b) such that*

$$f'(c) = \frac{f(b) - f(a)}{b - a}.$$

Exercise 16.4. Suppose a car is traveling along a highway and passes under a bridge at time $t = 0$. Let $d(t)$ be the car's distance from the bridge, in miles, as measured along the highway, t hours after passing under the bridge.

(a) Mathematically, what would we need to know about the function d in order to apply the mean value theorem to d on the input interval $[0.5, 5]$?

(b) Assuming the requirements you cited in (a) are met, what would the mean value theorem let us conclude, in context, if $d(0.5) = 40$ and $d(5) = 355$?

(c) Based on the information that has been provided, is there any evidence to suggest the car exceeded the posted speed limit of 75 mph at any time since it passed under the bridge?

Exercise 16.5. Prove the mean value theorem by applying Rolle's theorem to the function $f - g$ that is the difference between the given function f and the linear function g, whose graph passes through the points $(a, f(a))$ and $(b, f(b))$.

Example 16.8. Consider the function f defined by $f(x) = x^3$. As f is differentiable on $\mathbb{R}$, with derivative $f'(x) = 3x^2$, we see that f is differentiable, and hence, continuous, on the interval $[1, 3]$. So all of the hypotheses necessary for applying the mean value theorem to f on $[1, 3]$ are satisfied. Note that the average rate of change of f on the interval $[1, 3]$ is

$$\frac{f(3) - f(1)}{3 - 1} = \frac{3^3 - 1^3}{3 - 1} = \frac{27 - 1}{2} = 13.$$

Therefore, the mean value theorem allows us to conclude that there is an input c between 1 and 3 for which the instantaneous rate of change $f'(c)$ of f at c is equal to 13. Solving

$$f'(c) = 3c^2 = 13,$$

yields

$$c = \pm\sqrt{\frac{13}{3}}.$$

As we are looking for c to be between 1 and 3, we discard $c = -\sqrt{\frac{13}{3}}$ and conclude that $c = \sqrt{\frac{13}{3}} \approx 2.08$ is an input to the cubing function that lies between the endpoints of $[1, 3]$ and at which the instantaneous rate of change is the same value, 13, as the average rate of change on the entire interval.

Exercise 16.6. The conclusion of the mean value theorem can be interpreted geometrically as stating that, under the necessary hypotheses, the slopes of two lines must be equal. What are these two lines? What does the fact that these two lines have equal slopes tell us about the lines?

Exercise 16.7. Draw a picture illustrating the conclusions from Exercise 16.6 using the situation from Example 16.8.

Exercise 16.8. Suppose the function f is differentiable on $\mathbb{R}^+$, with $f(4) = 5$ and $f(10) = 1$. Indicate precisely how the mean value theorem can be applied and what conclusion we can reach using it.

Exercise 16.9. The mean value theorem can be used to establish quite a number of useful inequalities. Use it to show that $\sqrt{x+1} \leq \frac{1}{2}x + 1$ when $x > -1$. *Hint*: Note that equality holds if $x = 0$. Consider separately the cases where $x > 0$ and $-1 < x < 0$.

Exercise 16.10. Suppose the functions f and g are differentiable on $[0, \infty)$, with $f(0) = g(0)$ and $f'(x) \leq g'(x)$ for all x in $[0, \infty)$. Show that $f(x) \leq g(x)$ for all x in $[0, \infty)$. *Hint*: Apply the mean value theorem to the function $g - f$.

Exercise 16.11. Let f be a function with domain $[a, b]$ and let p be in (a, b). Suppose f is continuous on $[a, b]$ and differentiable on $(a, b) - \{p\}$. Use the mean value theorem to show that if $\lim_{x\to p} f'(x) = L$, then f is differentiable at p and $f'(p) = L$.

Determining intervals on which a function is monotone

Observe that the conclusion

$$f'(c) = \frac{f(b) - f(a)}{b - a}$$

one can reach when the mean value theorem is applicable relates values $f(a)$ and $f(b)$ of a function f to a value $f'(c)$ of the function's derivative f'. Thus, it is always reasonable to consider the use of the mean value theorem when we want to prove a result that involves the relationship between the values of a function and the values of the function's derivative.

To illustrate, we now establish how the derivative of a function can be used to determine the intervals on which the function is monotone.

Theorem 16.9 (The monotonicity theorem). *Suppose the function f is differentiable on an interval I.*

(1) *The function f is nondecreasing on I if and only if $f'(x) \geq 0$ for all x in I.*
(2) *The function f is nonincreasing on I if and only if $f'(x) \leq 0$ for all x in I.*
(3) *If $f'(x) > 0$ for all x in I, then f is increasing on I. If f is increasing on I, then $f'(x) \geq 0$ for all x in I, but it need not be the case that $f'(x) > 0$ for all x in I.*
(4) *If $f'(x) < 0$ for all x in I, then f is decreasing on I. If f is decreasing on I, then $f'(x) \leq 0$ for all x in I, but it need not be the case that $f'(x) < 0$ for all x in I.*

Proof. (1) First, assume f is nondecreasing on I and consider any x in I. Note that if $t \in I$ and $t < x$, we have $f(t) \leq f(x)$, from which it follows that $\frac{f(t)-f(x)}{t-x} \geq 0$. Also note that if

$t \in I$ and $t > x$ we must have $f(t) \geq f(x)$, from which it again follows that $\frac{f(t)-f(x)}{t-x} \geq 0$. Thus, using Exercise 12.34, we may conclude that

$$f'(x) = \lim_{t \to x} \frac{f(t) - f(x)}{t - x} \geq 0.$$

As x is an arbitrary member of I, we may therefore conclude that $f'(x) \geq 0$ for all x in I.

Now suppose $f'(x) \geq 0$ for all x in I, and consider any a and b in I such that $a < b$. By the mean value theorem, there exists c in (a, b) such that $f'(c) = \frac{f(b)-f(a)}{b-a}$ (*do you agree that the mean value theorem may be legitimately applied here?*). Since $a < b$ and, by hypothesis, $f'(c) \geq 0$, it follows that $f(a) \leq f(b)$. Thus, f is nondecreasing.

(2) This proof is similar to that of (1) and is left as an exercise.

(3) The argument demonstrating that if $f'(x) > 0$ for all x in I, then f is increasing on I, is similar to the argument presented in the proof of (1) that if $f'(x) \geq 0$ for all x in I, then f is nondecreasing on I.

Note that if f is increasing on I, then f is nondecreasing on I, so it then follows from (1) that $f'(x) \geq 0$ for all x in I.

The function f where $f(x) = x^3$ provides an example of a function that is increasing on the interval $\mathbb{R}$, but for which the derivative f' is not always positive; you are asked to provide the details in Exercise 16.14.

(4) This proof is similar to that of (3). □

Exercise 16.12. Prove that if f is differentiable on an open interval I that includes c and $f'(c) < 0$, then there exists a positive number δ such that whenever $c < x < c + \delta$, it follows that $f(x) < f(c)$.

Example 16.3 (Continued). Consider again the function f, for which

$$f(x) = 5x^4 - 4x^3$$

and

$$f'(x) = 20x^3 - 12x^2.$$

Note that $f'(x) = 4x^2(5x - 3)$, so it follows that
(1) $f'(x) > 0$ if and only if $x > \frac{3}{5}$, and,
(2) $f'(x) < 0$ if and only if either $x < 0$ or $0 < x < \frac{3}{5}$.

Thus, we may use the monotonicity theorem to conclude that f is increasing on the interval $(\frac{3}{5}, \infty)$, and is decreasing on the intervals $(-\infty, 0)$ and $(0, \frac{3}{5})$. These conclusions are borne out by the graph of f, depicted in Figure 16.2.

In fact, since we may also observe that $f(x) > 0$ for $x < 0$, that $f(0) = 0$, and that $f(x) < 0$ for $0 < x < \frac{3}{5}$, we may further conclude that f is actually decreasing on $(-\infty, \frac{3}{5})$.

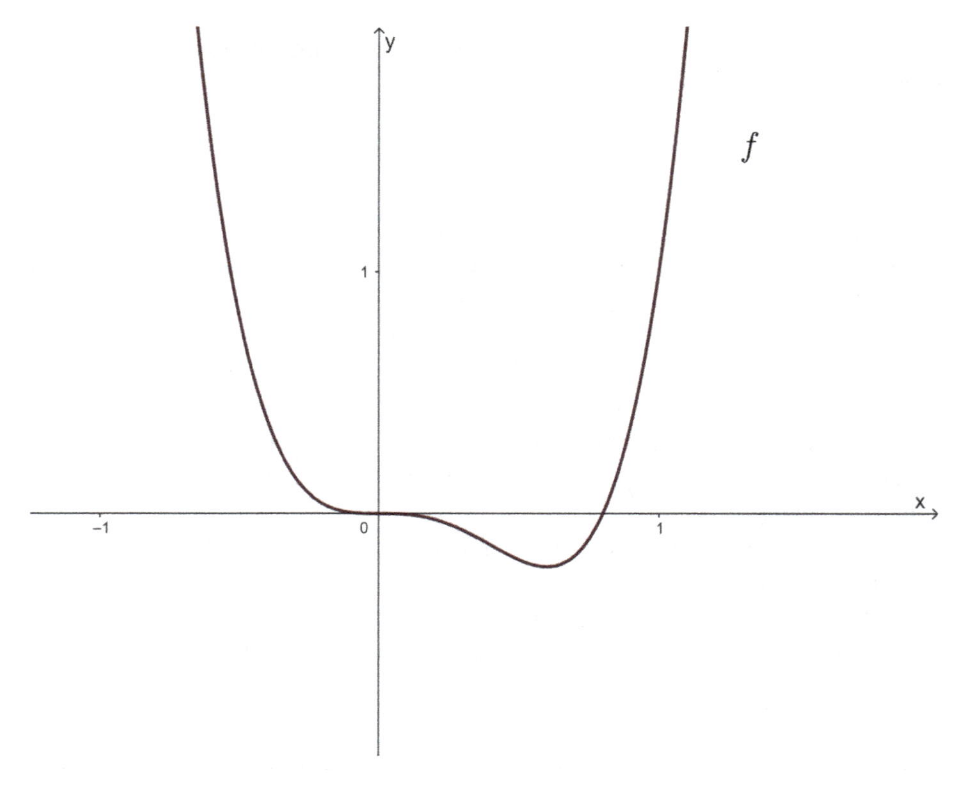

Figure 16.2: Graph of the function from Example 16.3.

Example 16.4 (Continued). Consider again the function g for which

$$g(x) = \sqrt[3]{(x^2 - 4)^2}$$

and

$$g'(x) = \frac{4x}{3\sqrt[3]{x^2 - 4}}.$$

Note that the numerator of $g'(x)$ is positive when $x > 0$ and is negative when $x < 0$. Also, the denominator of $g'(x)$ is

(1) positive precisely when $x^2 > 4$, that is, when either $x < -2$ or $x > 2$; and,
(2) negative precisely when $x^2 < 4$, that is, when $-2 < x < 2$.

Thus, we may conclude that $g'(x)$ is negative on the intervals $(-\infty, -2)$ and $(0, 2)$, and positive on the intervals $(-2, 0)$ and $(2, \infty)$. Hence, g is decreasing on the intervals $(-\infty, -2)$ and $(0, 2)$, but is increasing on the intervals $(-2, 0)$ and $(2, \infty)$. The graph of g, displayed in Figure 16.3, is consistent with these conclusions.

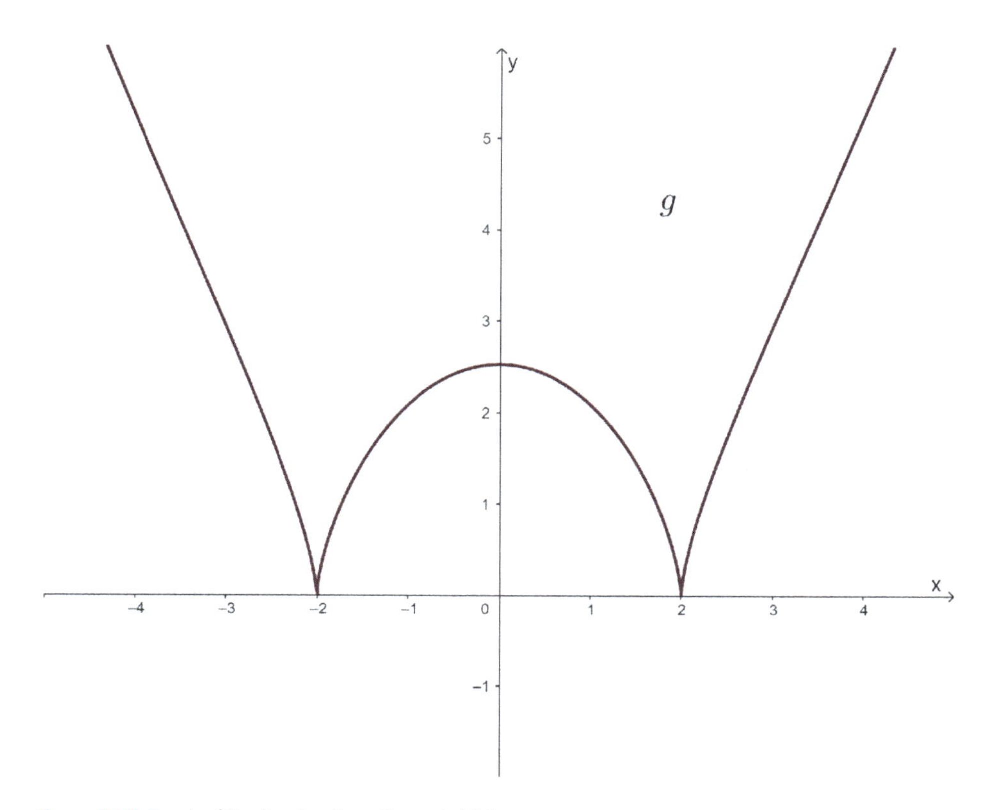

Figure 16.3: Graph of the function from Example 16.4.

In applying the monotonicity theorem in Examples 16.3 and 16.4, we have taken care to determine those intervals on which the derivative is positive and those intervals on which the derivative is negative before declaring these intervals to be, respectively, the ones on which the function itself is increasing and the ones on which it is decreasing.

In calculus courses, it is usually stated that a function's derivative can only change sign, from positive to negative or from negative to positive, at a critical number or a point of discontinuity. While this is true, it requires validation as a derivative need not be continuous. That is, while differentiability of a function f forces f to be continuous, it does not mean f' must be continuous.

Example 16.10. Define the function $f : \mathbb{R} \to \mathbb{R}$ so that

$$f(x) = \begin{cases} x^2 \cos(\frac{1}{x}), & \text{if } x \neq 0; \\ 0, & \text{if } x = 0. \end{cases}$$

Observe that

$$f'(x) = \begin{cases} 2x\cos(\frac{1}{x}) + \sin(\frac{1}{x}), & \text{if } x \neq 0; \\ 0, & \text{if } x = 0; \end{cases}$$

where for $x \neq 0$, we have made use of the product rule and the chain rule, and for $x = 0$, we have calculated

$$f'(0) = \lim_{x\to 0} \frac{f(x) - f(0)}{x - 0} = \lim_{x\to 0} \frac{x^2\cos(\frac{1}{x}) - 0}{x} = \lim_{x\to 0} x\cos\left(\frac{1}{x}\right) = 0,$$

using the result of Exercise 12.21. Thus, f is differentiable on $\mathbb{R}$. However, f' is not continuous at 0 since

$$\lim_{x\to 0} f'(x) = \lim_{x\to 0}\left(2x\cos\left(\frac{1}{x}\right) + \sin\left(\frac{1}{x}\right)\right)$$

does not exist, as $\lim_{x\to 0} 2x\cos(\frac{1}{x}) = 0$, whereas $\lim_{x\to 0}\sin(\frac{1}{x})$ does not exist by essentially the same reasoning used in Example 12.10 to show that $\lim_{x\to 0}\cos(\frac{1}{x})$ does not exist.

Therefore, while f is differentiable at 0, its derivative f' is not continuous at 0.

A function can change sign at a point of discontinuity, so as derivatives need not be continuous, it is conceivable that a function's derivative could change sign at a point of the domain of the function that is not a critical number of the function. It turns out this cannot happen because, although not necessarily continuous, the derivative of a function that is differentiable on an interval possesses the intermediate value property. This result is credited to the French mathematician Jean Gaston Darboux (1842–1917).

Theorem 16.11 (Darboux's theorem). *If the function f is differentiable on an open interval that includes both a and b, where $a < b$, and if k is any number between $f'(a)$ and $f'(b)$, then there exists a number c for which $a < c < b$ and $f'(c) = k$.*

Proof. Assume the function f is differentiable on an open interval I that includes both a and b, where $a < b$, and that $f'(a) < k < f'(b)$; the case where $f'(b) < k < f'(a)$ can be handled similarly.

Define the function $g : [a, b] \to \mathbb{R}$ so that $g(x) = f(x) - kx$. Note that g is differentiable on $[a, b]$ with derivative $g'(x) = f'(x) - k$. Differentiability of g on $[a, b]$ implies continuity of g on $[a, b]$, so by the extreme value theorem, the function g achieves a minimum value on $[a, b]$.

As $f'(a) < k$, we have $g'(a) = f'(a) - k < 0$, so by Exercise 16.12, there exists a positive number δ_1 such that whenever $a < x < a + \delta_1$, it follows that $g(x) < g(a)$. In a similar manner, as $f'(b) > k$, we have $g'(b) = f'(b) - k > 0$, so there exists a positive number δ_2 such that whenever $b - \delta_2 < x < b$, it follows that $g(x) < g(b)$. From these conclusions, we deduce that the minimum value of g on $[a, b]$ does not occur at either endpoint, a or b, hence, occurs at a number c between a and b. Therefore, $g(c)$ is not only a global minimum, but also a local minimum, so as g is differentiable at c, we may conclude via Theorem 16.2, that $g'(c) = f'(c) - k = 0$, from which it follows that $f'(c) = k$. □

It is Darboux's theorem that permits us to use a function's critical numbers and points of discontinuity to create intervals on which to test the sign of the function's derivative, secure that a sign change cannot occur at any other point.

Example 16.3 (Continued). We determined that the function f defined by

$$f(x) = 5x^4 - 4x^3$$

has critical numbers 0 and $\frac{3}{5}$, and no points of discontinuity. Darboux's theorem assures us that f' cannot change sign at any points other than 0 and $\frac{3}{5}$.

Thus, since f' must maintain the same sign throughout the interval $(-\infty, 0)$, it is possible to determine this sign by evaluating $f'(x) = 20x^3 - 12x^2$ at a representative test point in the interval. For instance, if we select -1, we see that

$$f'(-1) = 20(-1)^3 - 12(-1)^2 = -32,$$

and we may conclude that f' is negative throughout the interval $(-\infty, 0)$.

The sign of f' on each of the intervals $(0, \frac{3}{5})$ and $(\frac{3}{5}, \infty)$ can be determined similarly through the selection of a test point in each of them.

Exercise 16.13. For each function from Exercise 16.1, determine the intervals on which it is increasing and the intervals on which it is decreasing.

Exercise 16.14. Show that the function $f(x) = x^3$ is increasing on $\mathbb{R}$, but that its derivative f' is not always positive.

Exercise 16.15. Define the function f so that $f(x) = |x|$. In Example 15.7, we showed that f is not differentiable at 0. Find $f'(x)$ for $x \neq 0$. Based on your conclusion, do we have a contradiction to Darboux's theorem?

Classifying local extrema

The following result is an immediate consequence of the monotonicity theorem and establishes criteria that enable us to distinguish when a critical number yields a local maximum and when it yields a local minimum.

Theorem 16.12 (The first derivative test). *Suppose $a < c < b$, that the function f is differentiable on the interval (a, b) except perhaps at c, and that f is continuous at c.*

(1) *If $f'(x) > 0$ for all x in (a, c) and $f'(x) < 0$ for all x in (c, b), then $f(c)$ is a local maximum of f.*

(2) *If $f'(x) < 0$ for all x in (a, c) and $f'(x) > 0$ for all x in (c, b), then $f(c)$ is a local minimum of f.*

Example 16.3 (Continued). We have determined that for the function f where $f(x) = 5x^4 - 4x^3$, the derivative f' is negative on the intervals $(-\infty, 0)$ and $(0, \frac{3}{5})$, and positive on the interval $(\frac{3}{5}, \infty)$. Thus, the first derivative test tells us that since f' changes sign from negative to positive at $x = \frac{3}{5}$, we may conclude that $f(\frac{3}{5}) = -\frac{27}{125}$ is a local minimum of f. Moreover, since f' does not change sign at $x = 0$, this critical number yields neither a local minimum nor a local maximum. Note that these conclusions are consistent with the graph of f displayed in Figure 16.2.

The previous example illustrates that, though it is true that local extrema can occur only at critical numbers, a critical number of a function need not yield a local extremum.

Example 16.4 (Continued). We have already concluded that for the function g, where $g(x) = \sqrt[3]{(x^2 - 4)^2}$, the derivative g' is negative on the intervals $(-\infty, -2)$ and $(0, 2)$, but is positive on the intervals $(-2, 0)$ and $(2, \infty)$. We may therefore apply the first derivative test to conclude that g achieves a local minimum of 0 at both $x = -2$ and $x = 2$, and a local maximum of $2\sqrt[3]{2}$ at $x = 0$.

Exercise 16.16. For each function from Exercise 16.1, use your conclusions in Exercise 16.13 along with the first derivative test to classify each critical number of the function as the location of a local maximum, the location of a local minimum, or the location of neither.

The second derivative of a function can sometimes be used to determine whether a critical number yields a local maximum or a local minimum.

Theorem 16.13 (The second derivative test). *Suppose the second derivative f'' of f exists on an open interval containing c and that $f'(c) = 0$.*
(1) *If $f''(c) < 0$, then $f(c)$ is a local maximum of f.*
(2) *If $f''(c) > 0$, then $f(c)$ is a local minimum of f.*

Proof. (1) Suppose $f''(c) < 0$. Then there exists a positive number δ such that $\frac{f'(x)-f'(c)}{x-c} < 0$ whenever $0 < |x - c| < \delta$. But as $f'(c) = 0$, this implies that $f'(x) > 0$ for all x in $(c - \delta, c)$ and $f'(x) < 0$ for all x in $(c, c + \delta)$. Then, by the first derivative test, $f(c)$ is a local maximum of f.

(2) This proof is similar to that of (1) and is, therefore, left to the reader. □

Exercise 16.17. Provide details indicating why the number δ in the proof of (1) of Theorem 16.13 must exist.

Example 16.3 (Continued). Once again consider the function f defined so that $f(x) = 5x^4 - 4x^3$. Previously we used the first derivative test to classify the critical number $\frac{3}{5}$ of f as the location of a local minimum. As $f'(\frac{3}{5}) = 0$, we can use the second derivative test to reach the same conclusion. Note that

$$f''(x) = 60x^2 - 24x$$

so that $f''(\frac{3}{5}) = \frac{36}{5} > 0$, hence, by the second derivative test, the function f has a local minimum at $\frac{3}{5}$.

While it is always possible to apply the first derivative test to determine whether a critical number of a function yields a local maximum, a local minimum, or neither, the second derivative test is not always applicable

Example 16.4 (Continued). Consider once again the function g, where $g(x) = \sqrt[3]{(x^2-4)^2}$. We found its critical numbers to be 0, −2, and 2, but note that the only critical number to which the second derivative test may be applied is 0. This is because the first derivative of g, given by

$$g'(x) = \frac{4x}{3\sqrt[3]{x^2-4}},$$

is undefined at ±2, and the first derivative must be 0 at a critical number in order to apply the second derivative test. The fact that the first derivative is undefined at ±2 also automatically implies the second derivative g'' is undefined at ±2.

Exercise 16.18. Explain why, if $f'(c)$ is undefined, it must follow that all higher-order derivatives of the function f are also undefined at c.

Example 6.3 (Continued). The critical number 0 of the function f where $f(x) = 5x^4 - 4x^3$ cannot be classified using the second derivative test as, even though $f'(0) = 0$, the fact that $f''(0) = 60 \cdot 0^2 - 24 \cdot 0 = 0$ means the test is not applicable. In this case, we determined earlier that the critical number 0 yields neither a local minimum nor a local maximum, but it would be wrong to conclude this is because $f''(0) = 0$ (see Exercise 16.19).

Exercise 16.19. If $f'(c) = f''(c) = 0$, note that the second derivative test is not applicable. In this circumstance, the first derivative test should be used to determine whether the critical number c yields a local maximum, a local minimum, or neither.

(a) Show that the function f defined by $f(x) = x^3$ has neither a local maximum nor a local minimum at 0, and that $f'(0) = f''(0) = 0$.

(b) Show that the function g defined by $g(x) = x^4$ has a local minimum at 0 and that $g'(0) = g''(0) = 0$.

(c) Show that the function h defined by $h(x) = -x^4$ has a local maximum at 0 and that $h'(0) = h''(0) = 0$.

Exercise 16.20. The graph of a function f that is differentiable on an interval is said to be **concave up** on the interval if f' is increasing on the interval and is said to be **concave down** on an interval if f' is decreasing on the interval. Provide an intuitive explanation of the second derivative test using the notion of concavity.

Exercise 16.21. For each function from Exercise 16.1, where possible use the second derivative test to classify each critical number of the function as the location of a local maximum or the location of a local minimum. If the second derivative test is not applicable, state that this is so.

The constant difference theorem

When a function is differentiable on an open interval, not only is it the case that if the function is constant its derivative must be zero, but the converse is also true.

Theorem 16.14. *Let f be a function that is differentiable on an open interval I. The function f is constant on I if and only if $f'(x) = 0$ for all x in I.*

Proof. ($\Rightarrow$) You established this result in Exercise 15.2.

($\Leftarrow$) Suppose $f'(x) = 0$ for all x in I. Consider any a and b in I for which $a < b$; we now show that $f(a) = f(b)$. As f is differentiable, hence continuous, on $[a, b]$, the mean value theorem tells us there exists a number c such that $a < c < b$ and $f'(c) = \frac{f(b)-f(a)}{b-a}$. But as $f'(x) = 0$ for all x in I, we have $f'(c) = 0$, from which it follows that $f(b) - f(a) = 0$, that is, $f(a) = f(b)$. □

Theorem 16.14 implies, as we state in the next theorem, that functions having identical derivatives on an open interval can differ by only a constant. This observation itself then implies that functions having identical derivatives on an open interval that also agree in value at even one input in that interval must agree in value at all inputs in the interval. We have therefore acquired an additional method for proving that two functions are equal.

Theorem 16.15 (The constant difference theorem). *Let f and g be functions that are both differentiable on an open interval I.*

The functions f and g differ by a constant on I, that is, there exists a real number C for which $f(x) - g(x) = C$ for all x in I, if and only if $f'(x) = g'(x)$ for all x in I.

Furthermore, the functions f and g are equal on I, that is, $f(x) = g(x)$ for all x in I, if and only if both $f'(x) = g'(x)$ for all x in I and $f(p) = g(p)$ for some p in I.

Exercise 16.22. Use Theorem 16.14 to prove Theorem 16.15.

Differential equations

Some of the elementary functions commonly employed in mathematics and the sciences transcend the fundamental algebraic operations of addition and multiplication in the sense that they require the use of limiting processes for their formal definitions. Such functions are known as **transcendental functions** and include the trigonometric, inverse trigonometric, logarithmic, and exponential functions.

It is specifically via the limiting process involved in the notion of derivative that we can analytically define the trigonometric functions *cosine* and *sine* as solutions to certain so-called differential equations (in Chapter 20 we define the *natural logarithmic* and *natural exponential functions* using a somewhat different, but related, approach). A **differential equation** is an equation involving derivatives of one or more dependent (output) variables with respect to one or more independent (input) variables. We shall only be interested in differential equations containing a single dependent variable, typically represented by y, and a single independent variable, typically represented by x.

Example 16.16. The equation

$$y' = 2x$$

is a differential equation containing the single dependent variable y and the single independent variable x. So is the equation

$$y'' + y = 0,$$

under the assumption that y must be a function of the single variable x.

A **solution** to a differential equation incorporating a single dependent variable y and a single independent variable x is a function that when substituted into the equation makes the equation true.

Example 16.16 (Continued). The function

$$y = x^2$$

is a solution to the differential equation

$$y' = 2x$$

as

$$(x^2)' = 2x$$

is a true statement for every real number x. Note that there are (many) other solutions to the differential equation $y' = 2x$, for example, $y = x^2 + 10$.

The **general solution** of a differential equation is an equation that may contain one or more arbitrary (real number) constants that describes all the solutions to the differential equation. A **particular solution** is obtained from the general solution by assigning specific numerical values to the arbitrary constants.

Example 16.16 (Continued). The general solution to

$$y' = 2x$$

is

$$y = x^2 + C,$$

where C is an arbitrary constant, as any two functions differentiable on $\mathbb{R}$ with the same derivative must differ by a constant according to the constant difference theorem. If we assign the numerical value 0 to C, we obtain the particular solution

$$y = x^2.$$

On the other hand, when C is given the value 10 we obtain the particular solution

$$y = x^2 + 10.$$

Given a differential equation incorporating the dependent variable y and the independent variable x, an **initial condition** is a requirement that y or one of its derivatives must have a specified numerical value at a specified numerical value of x. A differential equation together with one or more initial conditions is called an **initial value problem.** A **solution** to an initial value problem is a function that is a solution to the differential equation that also satisfies all specified initial conditions.

Example 16.16 (Continued). The initial value problem

$$\begin{cases} y' = 2x, \\ y(3) = 5 \end{cases}$$

is comprised of the differential equation

$$y' = 2x$$

along with the initial condition

$$y(3) = 5,$$

which specifies that $y = 5$ when $x = 3$. We have already seen that the general solution to this differential equation is

$$y = x^2 + C,$$

where C is an arbitrary constant. As a solution to the initial value problem must satisfy the initial condition $y(3) = 5$ as well, the value of C must make the equation

$$5 = 3^2 + C$$

true, which can only happen if $C = -4$. Thus, there is only one solution to the initial value problem, namely,

$$y = x^2 - 4.$$

Exercise 16.23. Show that both of the functions

$$y = \frac{5}{24}x^4 - \frac{11}{6}x + \frac{13}{8}$$

and

$$y = \frac{5}{24}x^4 - \frac{1}{2}x^2 - \frac{5}{6}x + \frac{9}{8}$$

are solutions to the initial value problem

$$\begin{cases} y''' = 5x, \\ y(1) = 0, \\ y'(1) = -1. \end{cases}$$

The cosine and sine functions

Given a nonempty subset A of $\mathbb{R}$, the **constant zero function** on A is the function having domain A for which the value of the function is 0 at every member of A. The following result concerning the constant zero function on $\mathbb{R}$ initiates an analytic process for defining the trigonometric functions *cosine* and *sine* as solutions to certain initial value problems.

Theorem 16.17. *The unique solution to the initial value problem*

$$\begin{cases} y'' + y = 0, \\ y(0) = 0, \\ y'(0) = 0 \end{cases} \tag{1}$$

is the constant zero function $y = 0$ *on* $\mathbb{R}$.

Proof. It is straightforward to verify that $y = 0$ is a solution to (1). Now suppose that f is any solution to (1) and define the function $g : \mathbb{R} \to \mathbb{R}$ so that

$$g(x) = (f(x))^2 + (f'(x))^2.$$

Then, for every real number x, we have

$$g'(x) = 2f(x)f'(x) + 2f'(x)f''(x) = 2f'(x)(f(x) + f''(x)) = 2f'(x) \cdot 0 = 0,$$

where we have used our assumption that f is a solution to the differential equation $y'' + y = 0$. Thus, we may use Theorem 16.14 to conclude that g must be a constant function, and as

$$g(0) = (f(0))^2 + (f'(0))^2 = 0^2 + 0^2 = 0$$

because f must satisfy the initial conditions $y(0) = 0$ and $y'(0) = 0$ stated in (1), we conclude that g is the constant zero function on $\mathbb{R}$. Thus, for every real number x, we have

$$0 = g(x) = (f(x))^2 + (f'(x))^2,$$

which implies that $f(x) = 0$ for every real number x, and we may conclude that f is the constant zero function on $\mathbb{R}$. □

We now reveal the initial value problems that enable us to analytically define the cosine and sine functions.

Theorem 16.18 (Initial value problems used to define the cosine and sine functions). *The initial value problem*

$$\begin{cases} y'' + y = 0, \\ y(0) = 1, \\ y'(0) = 0 \end{cases} \tag{2}$$

*has a unique solution defined on $\mathbb{R}$, which is called the **cosine function**, abbreviated **cos**. The **sine function**, abbreviated **sin**, is then defined on $\mathbb{R}$ so that*

$$\sin(x) = -\cos'(x).$$

It follows that the sine function is the unique solution to the initial value problem

$$\begin{cases} y'' + y = 0, \\ y(0) = 0, \\ y'(0) = 1 \end{cases} \tag{3}$$

and that

$$\cos(0) = 1 \quad \text{and} \quad \sin(0) = 0.$$

Proof. The existence of a solution to the initial value problem (2) is demonstrated in Chapter 28 after we explore the notion of a *series of functions*. To see that there cannot be more than one solution to (2), suppose that, on the contrary, there are two different solutions f and g. Then it can be shown (details left as part of Exercise 16.24) that $f - g$ is a solution to the initial value problem (1) stated in Theorem 16.17 that is different from the constant zero function, a contradiction.

That the sine function, as defined, is a solution to (3) is also left to Exercise 16.24.

As the cosine function is a solution to (2), it satisfies the initial condition $y(0) = 1$, and we therefore have $\cos(0) = 1$. Similarly, as the sine function is a solution to (3), it satisfies the initial condition $y(0) = 0$ and we have $\sin(0) = 0$. □

Exercise 16.24. Provide the details omitted from the proof of Theorem 16.18 that $f - g$ is a solution to the initial value problem (1) from Theorem 16.17. Then show that the sine function, as defined in Theorem 16.18, is the unique solution to the initial value problem (3) stated in that theorem.

Theorem 16.19 (Derivatives of cosine and sine). *The cosine and sine functions are differentiable on* $\mathbb{R}$ *with* $\cos'(x) = -\sin(x)$ *and* $\sin'(x) = \cos(x)$.

Proof. Because the cosine and sine functions are solutions to the differential equation $y'' + y = 0$, their second derivatives exist on $\mathbb{R}$, which implies their first derivatives exist on $\mathbb{R}$.

By definition, $\sin(x) = -\cos'(x)$. Solving this equation for $\cos'(x)$ yields $\cos'(x) = -\sin(x)$.

Differentiating on both sides of $\sin(x) = -\cos'(x)$ and recalling that the cosine function must satisfy the differential equation $y'' + y = 0$, we see that

$$\sin'(x) = -\cos''(x) = \cos(x).$$

□

Because differentiability implies continuity, it is the case that both the cosine and sine functions are continuous.

Exercise 16.25. Find a formula for $f'(x)$ if f is defined by $f(x) = \frac{\sin(\cos(x))}{\sqrt{x}}$.

It is now possible to derive some familiar trigonometric identities, but in a manner different from what would be done in a high school trigonometry class. Recall that $\cos^2(x)$ and $\sin^2(x)$ are abbreviations for, respectively, $(\cos(x))^2$ and $(\sin(x))^2$.

Theorem 16.20 (Some fundamental trigonometric identities).

(1) Cosine is an even function*: For every real number x, we have*

$$\cos(-x) = \cos(x).$$

(2) Sine is an odd function*: For every real number x, we have*

$$\sin(-x) = -\sin(x).$$

(3) Pythagorean identity: *For every real number x, we have*

$$\cos^2(x) + \sin^2(x) = 1.$$

(4) Sum formula for cosine: *For all real numbers x and y, we have*

$$\cos(x + y) = \cos(x)\cos(y) - \sin(x)\sin(y).$$

(5) Sum formula for sine: *For all real numbers x and y, we have*

$$\sin(x + y) = \sin(x)\cos(y) + \cos(x)\sin(y).$$

Proof. The proofs of (1) and (3) are exercises. The proof of (2) is similar to that of (1). To establish (4) and (5), fix the real number y and define the function $f : \mathbb{R} \to \mathbb{R}$ so that

$$f(x) = \sin(x + y) - \sin(x)\cos(y) - \cos(x)\sin(y).$$

Observe that

$$f'(x) = \cos(x + y) - \cos(x)\cos(y) + \sin(x)\sin(y)$$

so that

$$f''(x) = -\sin(x + y) + \sin(x)\cos(y) + \cos(x)\sin(y).$$

It follows that f is a solution to the initial value problem (1) stated in Theorem 16.17 and which we have shown has only constant zero solution. Hence,

$$0 = f(x) = \sin(x + y) - \sin(x)\cos(y) - \cos(x)\sin(y)$$

from which it follows that

$$\sin(x + y) = \sin(x)\cos(y) + \cos(x)\sin(y).$$

Also, as f is a constant function, its derivative f' is the constant zero function. So we now have

$$0 = f'(x) = \cos(x + y) - \cos(x)\cos(y) + \sin(x)\sin(y)$$

and it follows that

$$\cos(x + y) = \cos(x)\cos(y) - \sin(x)\sin(y). \qquad \square$$

Exercise 16.26. Prove Theorem 16.20(1). *Hint*: Define the function $f : \mathbb{R} \to \mathbb{R}$ so that $f(x) = \cos(-x)$ and show that f is a solution to the initial value problem (2) from Theorem 16.18, which has already been shown to have the cosine function as its only solution.

Exercise 16.27. Prove Theorem 16.20(3). *Hint*: Define the function $f : \mathbb{R} \to \mathbb{R}$ so that $f(x) = \cos^2(x) + \sin^2(x)$. Show that $f'(x) = 0$ for every x in $\mathbb{R}$ and that $f(0) = 1$.

It follows from the Pythagorean identity $\cos^2(x) + \sin^2(x) = 1$ that

$$-1 \le \cos(x) \le 1$$

and

$$-1 \le \sin(x) \le 1.$$

The continuity of the cosine and sine functions then tells us each has range $[-1, 1]$.

Theorem 16.21. *The range of the cosine function is* $[-1, 1]$, *as is the range of the sine function.*

Exercise 16.28. Use the mean value theorem to prove that for all real numbers x and y, we have

$$\left|\sin(x) - \sin(y)\right| \le |x - y|.$$

L'Hôpital's rule

In Chapter 12, we discussed the idea that the limit of an arithmetic combination of functions could exhibit what is called an **indeterminate form**, meaning a form from which no conclusion as to the existence or value of the limit can be reached solely on the basis of the limits of the functions being combined. The specific indeterminate form $\frac{0}{0}$ was considered in Theorem 12.22, and it is worth noting that this form is exhibited by the limit

$$\lim_{x \to p} \frac{f(x) - f(p)}{x - p}$$

that defines the derivative $f'(p)$. In your study of calculus you encountered other indeterminate forms such as $\frac{\pm\infty}{\pm\infty}$ and $0 \cdot \infty$.

The evaluation of limits involving indeterminate forms is often facilitated through the application of *L'Hôpital's rule*, a result that, under certain circumstances, relates the limiting behavior of the ratio of two functions to that of the ratio of the functions' derivatives. L'Hôpital's rule was introduced by Johann Bernoulli in 1694, popularized in the first calculus textbook, *Infinitesimal Calculus with Applications to Curved Lines*, written by Guillaume de L'Hôpital in 1696, and rigorously proved by Cauchy.

Theorem 16.22 (L'Hôpital's rule). *Assume the functions f and g are differentiable on the interval (a, b), and that $g'(x) \neq 0$ for all x in (a, b). If either*
(1) *both* $\lim_{x\to a^+} f(x) = 0$ *and* $\lim_{x\to a^+} g(x) = 0$,

or

(2) $\lim_{x\to a^+} g(x) = \pm\infty$,

then

$$\lim_{x\to a^+} \frac{f(x)}{g(x)} = \lim_{x\to a^+} \frac{f'(x)}{g'(x)},$$

assuming the latter limit exists or is infinite.

Some comments about this theorem are in order. First, observe that L'Hôpital's rule tells us that
(1) when both $f(x)$ and $g(x)$ tend toward 0 as $x \to a^+$, then $\lim_{x\to a^+} \frac{f(x)}{g(x)}$ depends on the *relative rates* at which $f(x)$ and $g(x)$ are heading toward 0;
(2) and when both $|f(x)|$ and $|g(x)|$ grow without bound as $x \to a^+$, then $\lim_{x\to a^+} \frac{f(x)}{g(x)}$ depends on the *relative rates* at which $|f(x)|$ and $|g(x)|$ are growing.

Second, when we state that $\lim_{x\to a^+} \frac{f'(x)}{g'(x)}$ may be infinite, we mean that if $\lim_{x\to a^+} \frac{f'(x)}{g'(x)} = \infty$, it must follow that $\lim_{x\to a^+} \frac{f(x)}{g(x)} = \infty$, and if $\lim_{x\to a^+} \frac{f'(x)}{g'(x)} = -\infty$, it must follow that $\lim_{x\to a^+} \frac{f(x)}{g(x)} = -\infty$, assuming all the hypotheses of the theorem are met.

Third, note that when evaluating a limit of the form

$$\lim_{x\to a^+} \frac{f(x)}{g(x)},$$

where it is known that $\lim_{x\to a^+} g(x) = \pm\infty$ and $\lim_{x\to a^+} f(x) = L$ for some real number L, it follows immediately that $\lim_{x\to a^+} \frac{f(x)}{g(x)} = \lim_{x\to a^+} f(x) \cdot \frac{1}{g(x)} = 0$. Thus, from a practical perspective, the application of L'Hôpital's rule when condition (2) holds typically involves $\lim_{x\to a^+} f(x) = \pm\infty$ as well as $\lim_{x\to a^+} g(x) = \pm\infty$. As we shall see, though, the conclusion that $\lim_{x\to a^+} \frac{f(x)}{g(x)} = \lim_{x\to a^+} \frac{f'(x)}{g'(x)}$ can be reached under the assumption that $\lim_{x\to a^+} g(x) = \pm\infty$ without any regard to whether $\lim_{x\to a^+} f(x)$ exists or has a particular value.

Finally, provided the differentiability assumptions concerning f and g stated in the hypothesis are suitably modified, a similar conclusion can be reached if $\lim_{x\to a^+}$ is replaced by any of

$$\lim_{x\to a^-} \quad \text{or} \quad \lim_{x\to a} \quad \text{or} \quad \lim_{x\to\infty} \quad \text{or} \quad \lim_{x\to-\infty}.$$

In order to prove L'Hôpital's rule, we need the following generalization of the mean value theorem, which may be interpreted as telling us that, under the stated hypotheses, the ratio of the average rates of change of two functions over an interval must be achieved at some specific input by the ratio of the functions' instantaneous rates of change.

Theorem 16.23 (The Cauchy mean value theorem). *If the functions f and g are both continuous on the interval $[a,b]$ and both differentiable on the interval (a,b), and if $g'(x) \neq 0$ for all x in (a,b), then there exists a number c in (a,b) such that*

$$\frac{f'(c)}{g'(c)} = \frac{f(b)-f(a)}{g(b)-g(a)}.$$

Proof. Suppose the functions f and g are continuous on $[a,b]$ and differentiable on (a,b), and that $g'(x) \neq 0$ for all x in (a,b). Note that the mean value theorem is applicable to g on $[a,b]$, so the hypothesis that $g'(x) \neq 0$ for all x in (a,b) implies that $g(a) \neq g(b)$. Thus, we may define the function $h : [a,b] \to \mathbb{R}$ so that

$$h(x) = f(x) - f(a) - \frac{f(b)-f(a)}{g(b)-g(a)}(g(x)-g(a)),$$

observing that h is continuous on $[a,b]$ and differentiable on (a,b), and also that $h(a) = 0 = h(b)$. Thus, according to Rolle's theorem, there exists a number c in (a,b) such that

$$h'(c) = f'(c) - \frac{f(b)-f(a)}{g(b)-g(a)} \cdot g'(c) = 0,$$

from which it follows that

$$\frac{f'(c)}{g'(c)} = \frac{f(b)-f(a)}{g(b)-g(a)}.$$

□

Exercise 16.29. Show that Theorem 16.23 is a generalization of Theorem 16.7, the mean value theorem. *Hint*: Take $g(x) = x$.

Exercise 16.30. The Cauchy mean value theorem may be interpreted geometrically by considering the curve in the xy-plane defined parametrically by the function $p : [a,b] \to \mathbb{R} \times \mathbb{R}$, where

$$p(t) = \big(g(t), f(t)\big).$$

(a) Find the slope of the line passing through the points $p(a)$ and $p(b)$.
(b) Write down an expression for the slope of the tangent line to the graph of p at the point $(g(t), f(t))$.
(c) What does the conclusion of the Cauchy mean value theorem tell us in light of (a) and (b)? Draw a picture that suggests this conclusion.

We now use the Cauchy mean value theorem to establish L'Hôpital's rule.

Proof of Theorem 16.22, L'Hôpital's rule. First, we show that if both $\lim_{x\to a^+} f(x) = 0$ and $\lim_{x\to a^+} g(x) = 0$, and $\lim_{x\to a^+} \frac{f'(x)}{g'(x)} = L$ for some real number L, then it must follow that $\lim_{x\to a^+} \frac{f(x)}{g(x)} = L$; you consider the parallel scenario in which $\lim_{x\to a^+} \frac{f'(x)}{g'(x)} = \infty$ in Exercise 16.31.

Define functions $F : [a,b) \to \mathbb{R}$ and $G : [a,b) \to \mathbb{R}$ so that

$$F(x) = \begin{cases} f(x), & \text{if } a < x < b; \\ 0, & \text{if } x = a; \end{cases}$$

and

$$G(x) = \begin{cases} g(x), & \text{if } a < x < b; \\ 0, & \text{if } x = a. \end{cases}$$

Note that F and G are both continuous on $[a,b)$ and differentiable on (a,b). Thus, for each x in (a,b), we are able to apply the Cauchy mean value theorem to F and G on $[a,x]$ to obtain c_x in (a,x) such that

$$\frac{F'(c_x)}{G'(c_x)} = \frac{F(x) - F(a)}{G(x) - G(a)} = \frac{F(x) - 0}{G(x) - 0} = \frac{F(x)}{G(x)} = \frac{f(x)}{g(x)}.$$

Therefore, since Theorem 12.20, the squeeze theorem, tells us that $\lim_{x\to a^+} c_x = a$, and since $F'(t) = f'(t)$ and $G'(t) = g'(t)$ for all t in (a,b), we have

$$\lim_{x\to a^+} \frac{f(x)}{g(x)} = \lim_{x\to a^+} \frac{F'(c_x)}{G'(c_x)} = \lim_{x\to a^+} \frac{f'(c_x)}{g'(c_x)} = \lim_{x\to a^+} \frac{f'(x)}{g'(x)} = L.$$

Now we show that if $\lim_{x\to a^+} g(x) = \infty$ and $\lim_{x\to a^+} \frac{f'(x)}{g'(x)} = L$ for some real number L, then it must follow that $\lim_{x\to a^+} \frac{f(x)}{g(x)} = L$; the arguments are similar if $\lim_{x\to a^+} g(x) = -\infty$ or $\lim_{x\to a^+} \frac{f'(x)}{g'(x)} = \pm\infty$.

Consider any positive number ε. Select a number K for which $L - \varepsilon < K < L$. As $\lim_{x\to a^+} \frac{f'(x)}{g'(x)} = L$, there exists a positive number r_1 such that $r_1 < b - a$ and whenever x is in $(a, a + r_1)$, it follows that $\frac{f'(x)}{g'(x)} > K$. Let t be any number in $(a, a + r_1)$. Since $\lim_{x\to a^+} g(x) = \infty$, there is no loss of generality in assuming that $g(x) > 0$ and $g(x) > g(t)$ for all x in (a,t). Thus, for each x in (a,t), we may apply the Cauchy mean value theorem to f and g on $[x,t]$ to obtain c_x in (x,t) such that

$$K < \frac{f'(c_x)}{g'(c_x)} = \frac{f(t) - f(x)}{g(t) - g(x)}.$$

From $K < \frac{f(t)-f(x)}{g(t)-g(x)}$ it follows that

$$\frac{f(x)}{g(x)} > \frac{f(t)}{g(x)} + K\left(1 - \frac{g(t)}{g(x)}\right).$$

Again, using the assumption that $\lim_{x \to a^+} g(x) = \infty$, and keeping in mind that t is being treated as a constant when taking a limit as $x \to a^+$, we have

$$\lim_{x \to a^+} \frac{f(t)}{g(x)} = 0 = \lim_{x \to a^+} \frac{g(t)}{g(x)},$$

from which it follows that

$$\lim_{x \to a^+} \frac{f(x)}{g(x)} > \lim_{x \to a^+} \left[\frac{f(t)}{g(x)} + K\left(1 - \frac{g(t)}{g(x)}\right)\right] = K.$$

Thus, there exists a positive number r for which $r < r_1$ and for all x in $(a, a+r)$, it follows that

$$\frac{f(x)}{g(x)} > K.$$

Thus, we have shown that for all x in $(a, a + r)$, we have $\frac{f(x)}{g(x)} > L - \varepsilon$.

In a similar manner (see Exercise 16.32), it is possible to show that there exists a positive number s such that for all x in $(a, a + s)$, we have $\frac{f(x)}{g(x)} < L + \varepsilon$.

Taking $\delta = \min\{r, s\}$, it follows that for any x in $(a, a+\delta)$, we have $\frac{f(x)}{g(x)}$ in $(L-\varepsilon, L+\varepsilon)$. Therefore, $\lim_{x \to a^+} \frac{f(x)}{g(x)} = L$. □

Exercise 16.31. Prove that under the differentiability assumptions in our statement of Theorem 16.22, L'Hôpital's rule, if $\lim_{x \to a^+} f(x) = 0$ and $\lim_{x \to a^+} g(x) = 0$ and $\lim_{x \to a^+} \frac{f'(x)}{g'(x)} = \infty$, it follows that $\lim_{x \to a^+} \frac{f(x)}{g(x)} = \infty$.

Exercise 16.32. Toward the end of our proof of the case in L'Hôpital's rule, where the condition $\lim_{x \to a^+} g(x) = \infty$ holds, we claim that in a manner similar to that in which we obtained r it is possible to obtain s satisfying a stated condition. Provide the details of this argument.

Example 16.24. Note that

$$\lim_{x \to 1} \frac{x^2 - 1}{x - 1}$$

possesses the indeterminate form $\frac{0}{0}$, both the numerator and denominator of $\frac{x^2-1}{x-1}$ are everywhere differentiable, and the derivative 1 of the denominator $x - 1$ is nonzero. Applying L'Hôpital's rule, we conclude that

$$\lim_{x \to 1} \frac{x^2 - 1}{x - 1} = \lim_{x \to 1} \frac{2x}{1} = 2.$$

Example 16.25. Observe that

$$\lim_{x\to\infty} \frac{5x^2+8}{2x^2+9}$$

has the indeterminate form $\frac{\infty}{\infty}$. Both the numerator and the denominator of $\frac{5x^2+8}{2x^2+9}$ are everywhere differentiable, with the derivative $4x$ of the denominator nonzero on $(0,\infty)$. Thus, L'Hôpital's rule may be applied, yielding

$$\lim_{x\to\infty} \frac{5x^2+8}{2x^2+9} = \lim_{x\to\infty} \frac{10x}{4x} = \lim_{x\to\infty} \frac{5}{2} = \frac{5}{2}.$$

Sometimes L'Hôpital's rule must be applied more than once to obtain the value of a limit.

Example 16.26. The limit

$$\lim_{x\to 0} \frac{\sin(x)-x}{x^2}$$

is indeterminate of form $\frac{0}{0}$. The numerator $\sin(x)-x$ and denominator x^2 are everywhere differentiable, with the derivative $2x$ of the denominator nonzero when $x \neq 0$. Applying L'Hôpital's rule twice, we obtain

$$\lim_{x\to 0} \frac{\sin(x)-x}{x^2} = \lim_{x\to 0} \frac{\cos(x)-1}{2x} = \lim_{x\to 0} \frac{-\sin(x)}{2} = 0.$$

Here it is important that we recognize that L'Hôpital's rule can be applied a second time because

$$\lim_{x\to 0} \frac{\cos(x)-1}{2x}$$

possesses the indeterminate form $\frac{0}{0}$, the numerator and denominator of $\frac{\cos(x)-1}{2x}$ are everywhere differentiable, and the derivative 2 of the denominator is nonzero.

Some care must be taken to ensure that the hypotheses under which L'Hôpital's rule may be applied are met before the rule is invoked.

Example 16.27. Note that $\lim_{x\to 0} \frac{x^2+1}{x^2+2}$ cannot be evaluated via L'Hôpital's rule because it possesses neither of the indeterminate forms $\frac{0}{0}$ or $\frac{\pm\infty}{\pm\infty}$. If we were to attempt to *incorrectly* employ this rule here, we would *wrongly* conclude that

$$\lim_{x\to 0} \frac{x^2+1}{x^2+2}$$

could be calculated as

$$\lim_{x\to 0} \frac{2x}{2x} = \lim_{x\to 0} 1 = 1,$$

whereas, in actuality, as we are taking the limit of a rational function at a point in the function's domain, the value of the limit can be obtained using direct substitution:

$$\lim_{x\to 0} \frac{x^2+1}{x^2+2} = \frac{0^2+1}{0^2+2} = \frac{1}{2}.$$

Exercise 16.33. Where possible, use L'Hôpital's rule to evaluate the given limit. If L'Hôpital's rule cannot be applied, use another method to evaluate the limit or explain why the limit does not exist.

(a) $\lim_{x\to -2} \frac{x^2+2x}{x^3+8x^2+21x+18}$

(b) $\lim_{x\to -\infty} \frac{x^3}{\sqrt{x^6+10}}$

(c) $\lim_{x\to 1} \frac{x^2-1}{x^2+1}$

Exercise 16.34. Note that $\lim_{x\to\infty} \frac{x-1}{2x+\cos(x)}$ possesses the indeterminate form $\frac{\infty}{\infty}$, so it would appear that we may apply L'Hôpital's rule. Try doing so and then explain why it turns out that L'Hôpital's rule cannot be used to evaluate this limit. Then explain why the value of this limit is $\frac{1}{2}$.

Exercise 16.35. Define the functions f and g on $[0, 1]$ so that

$$f(x) = \begin{cases} x\sin(\frac{1}{x}), & \text{if } 0 < x \le 1; \\ 0, & \text{if } x = 0; \end{cases}$$

and

$$g(x) = x.$$

Note that both f and g are differentiable on the interval $(0, 1)$, with $g'(x) \neq 0$ for all x in $(0, 1)$, and $\lim_{x\to 0^+} f(x) = \lim_{x\to 0^+} g(x) = 0$. Yet, $\lim_{x\to 0^+} \frac{f(x)}{g(x)}$ does not exist. Why does this situation not contradict L'Hôpital's rule?

---

17 Taylor polynomials

Historically, one very important application of the derivative has been its use in approximating function values. Here we demonstrate how the existence of higher-order derivatives for a function leads to polynomial approximations for the function's values. We also show how it is sometimes possible to obtain a bound for the error resulting from such approximations.

Taylor polynomials and the approximation of function values

Given a nonnegative integer n, a function $p : \mathbb{R} \to \mathbb{R}$ for which

$$p(x) = a_n x^n + a_{n-1}x^{n-1} + \cdots + a_2 x^2 + a_1 x + a_0$$

for some real numbers $a_0, a_1, a_2, \ldots, a_n$, where $a_n \neq 0$, is called a **polynomial function of degree** n. We also define the constant zero function on $\mathbb{R}$ to be the **polynomial function of degree** -1. Thus, when we refer to polynomial functions whose degree is no greater than a certain value, we are including the constant zero function.

For hundreds of years, polynomial functions have been used to approximate the values of other functions because evaluating a polynomial is relatively easy, involving only the operations of addition and multiplication. So imagine we wish to approximate the value of a function f at an input x and we know the value $f(x_0)$ at a nearby input x_0.

Example 17.1. Consider the principal square root function f defined by $f(x) = \sqrt{x}$. Note that $f(4) = \sqrt{4} = 2$. We would like to approximate $f(5) = \sqrt{5}$.

If no other information is available, it makes some sense to use $f(x_0)$ as an estimate for $f(x)$ when x is close to x_0, especially if the function f is continuous, as in such a circumstance we would expect inputs close to one another to generate outputs that are close to one another. That is, the constant polynomial function P_0, defined by

$$P_0(x) = f(x_0)$$

could be used to approximate the values of f at points x "close to" x_0.

Example 17.1 (Continued). Note that in this case, as $f(x) = \sqrt{x}$ and $x_0 = 4$, we have

$$P_0(x) = f(4) = 2.$$

Thus, we would estimate that

$$\sqrt{5} \approx P_0(5) = 2.$$

The graphs of f and the local approximating function P_0 are displayed in Figure 17.1.

https://doi.org/10.1515/9783111429564-017

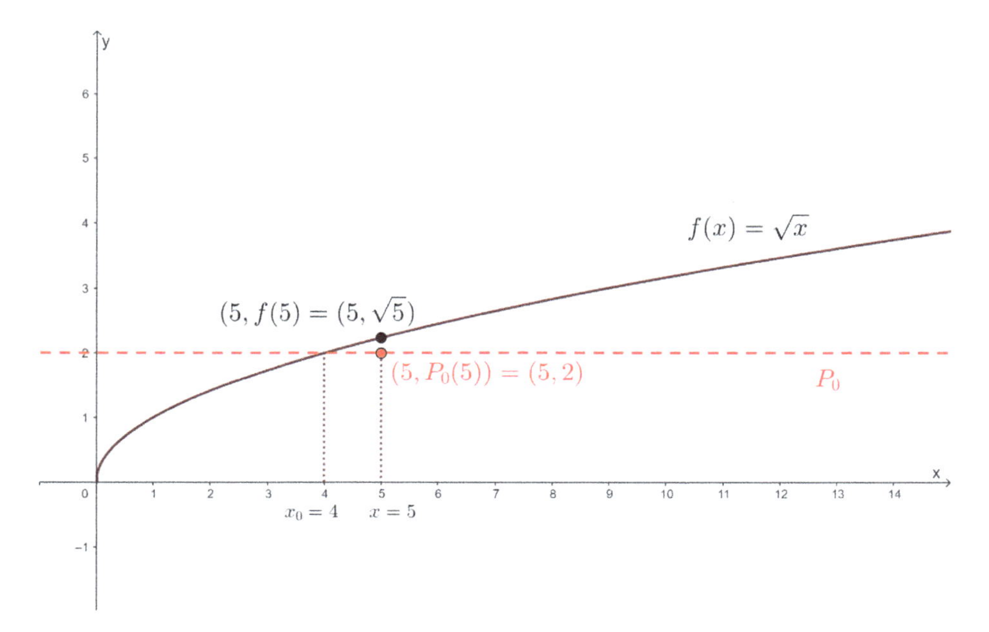

Figure 17.1: Graphs of function f and local approximation P_0 from Example 17.1.

If the function f is differentiable at x_0 and we know the value of $f'(x_0)$, then we expect the linear polynomial function P_1 defined by

$$P_1(x) = f(x_0) + f'(x_0)(x - x_0)$$

to produce better approximations. Note that P_1 is the local linearization of f at x_0 (i. e., the tangent line approximation), defined in Chapter 15. Also observe that $P_1(x_0) = f(x_0)$ and $P_1'(x_0) = f'(x_0)$.

Example 17.1 (Continued). Note that $f'(x) = \frac{1}{2\sqrt{x}}$ so that $f'(x_0) = f'(4) = \frac{1}{4}$. Thus,

$$P_1(x) = f(4) + f'(4)(x - 4) = 2 + \frac{1}{4}(x - 4).$$

We would then estimate that

$$\sqrt{5} \approx P_1(5) = 2 + \frac{1}{4}(5 - 4) = 2.25.$$

Our work here is identical to what we did in Example 15.5, except that the function P_1 was named L and referred to as the local linearization. The graphs of f and the local approximating function P_1 are displayed in Figure 17.2.

It seems conceivable that we could obtain an even better approximation to values of a function f by constructing a degree 2 (i. e., quadratic) polynomial function P_2 that agrees with f at x_0 and whose first two derivatives agree with the first two derivatives,

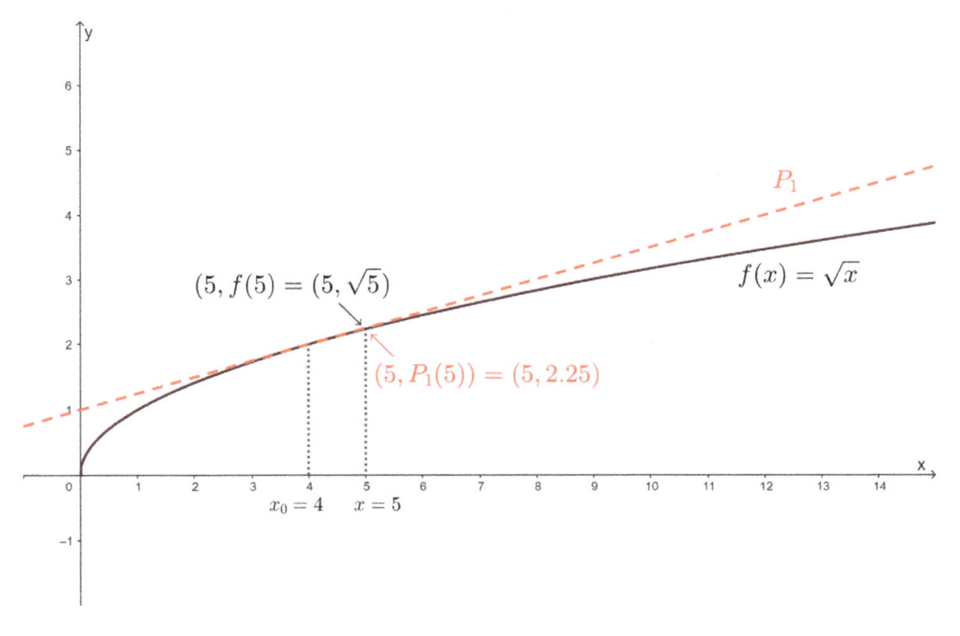

Figure 17.2: Graph of function f and local approximation P_1 from Example 17.1.

respectively, of f at x_0. Intuitively this is because the resulting parabola should better conform to the graph of f, at least "near" x_0.

Exercise 17.1. Suppose f is a function defined on an open interval that includes x_0 and that both $f'(x_0)$ and $f''(x_0)$ exist. Let P_2 be the degree 2 polynomial function defined by

$$P_2(x) = f(x_0) + f'(x_0)(x - x_0) + \frac{f''(x_0)}{2}(x - x_0)^2.$$

Verify that $P_2(x_0) = f(x_0)$, that $P_2'(x_0) = f'(x_0)$, and that $P_2''(x_0) = f''(x_0)$.

Example 17.1 (Continued). We apply the result of Exercise 17.1 to the situation in which we are trying to estimate the value of $\sqrt{5}$. We have already found that $f(4) = 2$ and $f'(4) = \frac{1}{4}$. Noting that $f''(x) = -\frac{1}{4x\sqrt{x}}$, we find that $f''(4) = -\frac{1}{32}$. Hence,

$$P_2(x) = f(4) + f'(4)(x - 4) + \frac{f''(4)}{2}(x - 4)^2 = 2 + \frac{1}{4}(x - 4) - \frac{1}{64}(x - 4)^2,$$

from which we obtain the estimate

$$\sqrt{5} \approx P_2(5) = 2 + \frac{1}{4}(5 - 4) - \frac{1}{64}(5 - 4)^2 = 2.234375.$$

The graphs of f and the local approximating function P_2 appear in Figure 17.3.

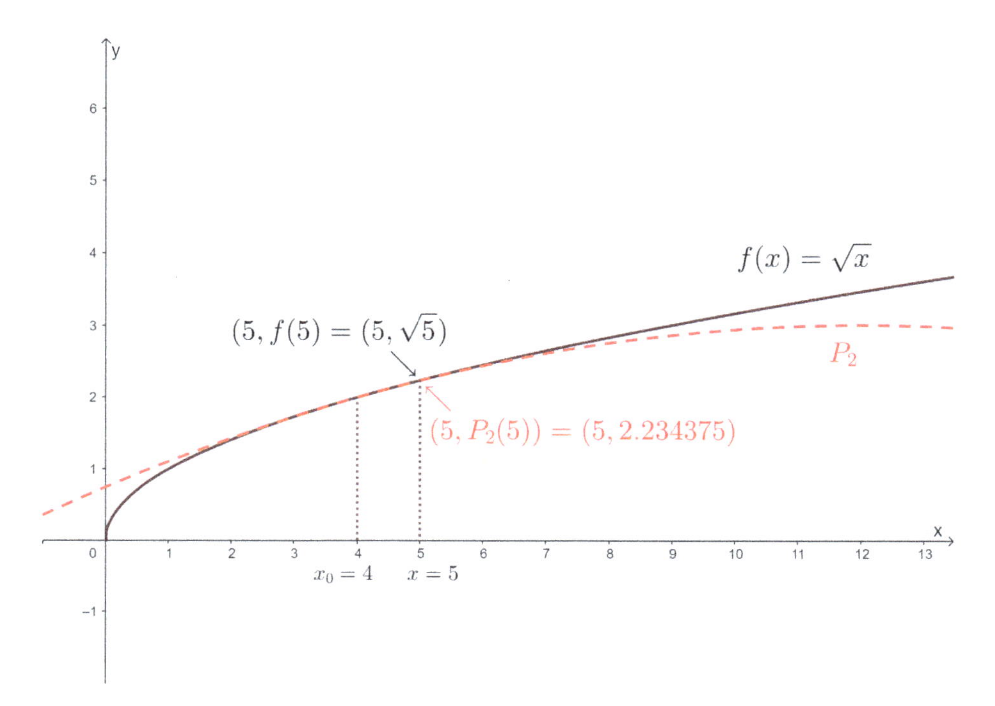

Figure 17.3: Graph of function f and local approximation P_2 from Example 17.1.

In fact, it would appear that we could obtain a better local polynomial approximation to a function f by constructing a polynomial function that agrees with f at x_0 and whose first n derivatives agree with the first n derivatives of f at x_0, assuming, of course, that f has an nth derivative at x_0. Intuitively at least, the greater the value of n, the better we expect the polynomial approximation to be. Note that the function P_n defined so that

$$P_n(x) = f(x_0) + f'(x_0)(x - x_0) + \frac{f''(x_0)}{2!}(x - x_0)^2 + \cdots + \frac{f^{(n)}(x_0)}{n!}(x - x_0)^n$$

is a polynomial function of degree at most n that agrees with f at x_0 and whose first n derivatives agree with the first n derivatives of f at x_0. It is called the **nth Taylor polynomial for (of) f at x_0**.

The English mathematician Brook Taylor introduced these polynomials in 1715, though other mathematicians had effectively used them in specific instances prior to Taylor's more general presentation.

Example 17.1 (Continued). We have already found that the zeroth Taylor polynomial for the principal square root function at 4 is P_0 given by

$$P_0(x) = 2,$$

the first Taylor polynomial for the principal square root function at 4 is P_1 given by

$$P_1(x) = 2 + \frac{1}{4}(x - 4),$$

and the second Taylor polynomial for the principal square root function at 4 is P_2 given by

$$P_2(x) = 2 + \frac{1}{4}(x - 4) - \frac{1}{64}(x - 4)^2.$$

To construct, say, the fifth Taylor polynomial P_5 for the principal square root function at 4, we need the values of f and its first five derivatives, where $f(x) = \sqrt{x}$. We already determined that $f'(x) = \frac{1}{2\sqrt{x}}$ and $f''(x) = -\frac{1}{4x\sqrt{x}}$. A little more calculation reveals that

$$f'''(x) = \frac{3}{8x^2\sqrt{x}},$$
$$f^{(4)}(x) = -\frac{15}{16x^3\sqrt{x}},$$

and

$$f^{(5)}(x) = \frac{105}{32x^4\sqrt{x}},$$

so that

$$f'''(4) = \frac{3}{256},$$
$$f^{(4)}(4) = -\frac{15}{2048},$$

and

$$f^{(5)}(4) = \frac{105}{16,384}.$$

Thus, the fifth Taylor polynomial for the principal square root function at 4 is

$$\begin{aligned} P_5(x) &= f(4) + f'(4)(x - 4) + \frac{f''(4)}{2!}(x - 4)^2 + \frac{f'''(4)}{3!}(x - 4)^3 + \frac{f^{(4)}(4)}{4!}(x - 4)^4 \\ &\quad + \frac{f^{(5)}(4)}{5!}(x - 4)^5 \\ &= 2 + \frac{1}{4}(x - 4) - \frac{1}{64}(x - 4)^2 + \frac{1}{512}(x - 4)^3 - \frac{5}{16,384}(x - 4)^4 \\ &\quad + \frac{7}{131,072}(x - 4)^5. \end{aligned}$$

We then obtain the estimate

$$\begin{aligned}\sqrt{5} &\approx P_5(5)\\ &= 2 + \frac{1}{4}(5-4) - \frac{1}{64}(5-4)^2 + \frac{1}{512}(5-4)^3 - \frac{5}{16,384}(5-4)^4 + \frac{7}{131,072}(5-4)^5\\ &= 2.2360763549804.\end{aligned}$$

An online calculator gives $\sqrt{5} \approx 2.2360679774998$, so P_5 produces an approximation to $\sqrt{5}$ that is accurate to four decimal places. The graphs of f and the local approximating function P_5 are shown in Figure 17.4.

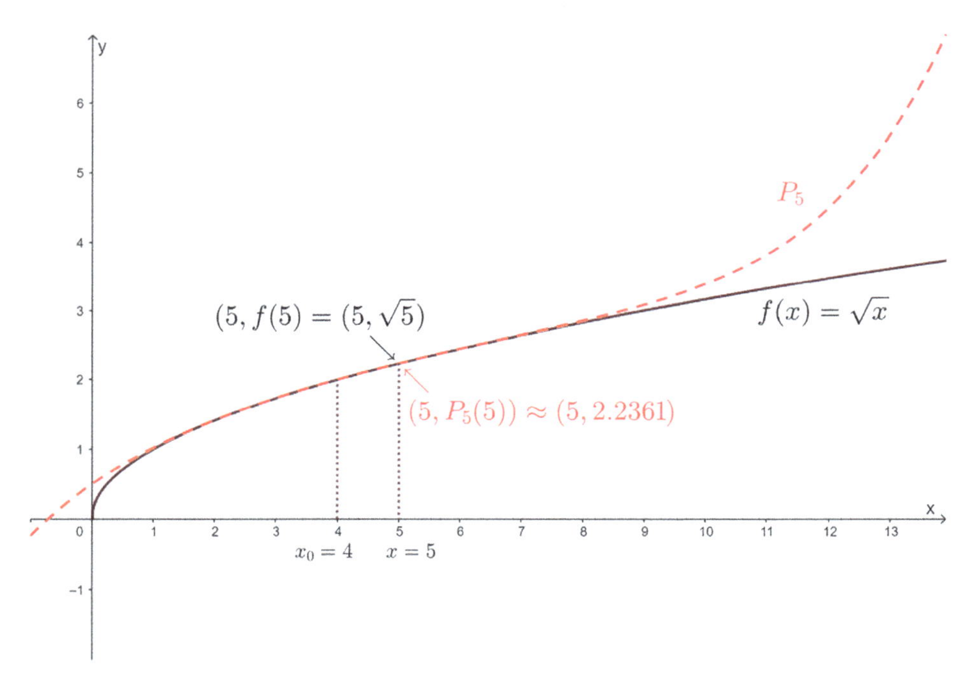

Figure 17.4: Graph of function f and local approximation P_5 from Example 17.1.

Exercise 17.2. Obtain the third and fourth Taylor polynomials at $x_0 = 4$ for the principal square root function. Then use each of them to approximate $\sqrt{5}$. Be sure to make use of the derivative calculations provided in Example 17.1.

Exercise 17.3. Obtain the Taylor polynomials P_0, P_1, P_2, and P_3 at $x_0 = 1$ for the principal square root function. Then use each of them to approximate $\sqrt{1.1}$.

Exercise 17.4. Consider the function f defined so that $f(x) = \frac{1}{x}$.
(a) Find the Taylor polynomials P_0, P_1, P_2, and P_3 for f at $x_0 = 1$.
(b) Approximate $f(1.5)$ using the Taylor polynomial P_3. How close is your approximation to the true value of $f(1.5)$?

It is important to understand that a Taylor polynomial approximation may not be very good at approximating a function value once the input is "far" from the original input about which the polynomial was constructed.

Example 17.1 (Continued). If we use the fifth Taylor polynomial at 4 for the principal square root function to estimate $\sqrt{16}$, we get

$$\begin{aligned}\sqrt{16} &\approx P_5(16)\\ &= 2 + \frac{1}{4}(16-4) - \frac{1}{64}(16-4)^2 + \frac{1}{512}(16-4)^3 - \frac{5}{16,384}(16-4)^4\\ &\quad + \frac{7}{131,072}(16-4)^5\\ &= 13.0859375,\end{aligned}$$

which no one is likely to consider a good estimate, since we know $\sqrt{16} = 4$. The problem is that 16 is not close enough to 4 for the function P_5 to yield a good approximation. Taking another look at the graphs of the principal square root function and P_5 in Figure 17.4 indicates a significant, and ever widening, divergence between them for values of x larger than 10.

Exercise 17.5. Use the Taylor polynomial P_3 you obtained in Exercise 17.4(a) to estimate the numerical value of $\frac{1}{10}$. Would you say the estimate is good? Explain what has happened.

When P_n is the nth Taylor polynomial for f at x_0, we refer to the function R_n defined so that

$$R_n(x) = f(x) - P_n(x)$$

as the **Taylor remainder**. The Taylor remainder represents the error that results when $P_n(x)$ is used to approximate $f(x)$.

The following theorem gives some insight as to why a Taylor polynomial P_n is expected to produce good approximations for function values $f(x)$ when n is relatively large and x is relatively close to the input x_0 at which P_n is constructed.

Theorem 17.2. *Let f be a function whose nth derivative at the input x_0 exists. If P_n is the nth Taylor polynomial for f at x_0, then*

$$\lim_{x \to x_0} \frac{f(x) - P_n(x)}{(x - x_0)^n} = \lim_{x \to x_0} \frac{R_n(x)}{(x - x_0)^n} = 0.$$

We ask you to prove this result in Exercise 17.6. The reason it implies $P_n(x)$ must be close to $f(x)$ for x sufficiently close to x_0 is because the only way for $\frac{f(x)-P_n(x)}{(x-x_0)^n}$ to tend toward zero as x approaches x_0, when it is also clear that the denominator $(x-x_0)^n$ tends toward zero, is for the numerator $f(x) - P_n(x)$ to approach zero.

Exercise 17.6. Use induction to prove Theorem 17.2.

It is often convenient to employ Σ-notation in order to write the nth Taylor polynomial as

$$P_n(x) = \sum_{k=0}^{n} \frac{f^{(k)}(x_0)}{k!}(x - x_0)^k,$$

where we recall that, by definition, $f^{(0)} = f$ and $0! = 1$. Doing so, however, involves a slight abuse of notation, because when $k = 0$ and $x = x_0$, we get

$$\frac{f^{(k)}(x_0)}{k!}(x - x_0)^k = \frac{f^{(0)}(x_0)}{0!}(x_0 - x_0)^0 = \frac{f(x_0)}{1} \cdot 0^0,$$

which we intend to represent $f(x_0)$, but which incorporates the undefined expression 0^0. In order to avoid a more awkward expression such as

$$P_n(x) = f(x_0) + \sum_{k=1}^{n} \frac{f^{(k)}(x_0)}{k!}(x - x_0)^k,$$

we agree to interpret 0^0 as 1 in this setting, though generally speaking 0^0 shall remain undefined.

Taylor's theorem

We now consider the question of how good an approximation is produced by a Taylor polynomial. Taylor's theorem, first proved by Joseph-Louis Lagrange (1736–1813), gives us an actual measure of how closely a Taylor polynomial approximates a function value.

Theorem 17.3 (Taylor's theorem). *Let n be a nonnegative integer, let I be an open interval that includes x_0, and let f be a function for which $f^{(n+1)}$ exists on I. For each x in $I - \{x_0\}$, there exists a number c between x and x_0 such that*

$$f(x) = P_n(x) + R_n(x),$$

where

$$P_n(x) = \sum_{k=0}^{n} \frac{f^{(k)}(x_0)}{k!}(x - x_0)^k$$

is the nth Taylor polynomial for f at x_0 and the resulting Taylor remainder is given by

$$R_n(x) = \frac{f^{(n+1)}(c)}{(n+1)!}(x - x_0)^{n+1}.$$

Proof. First, we consider the special case in which

$$f(x_0) = f'(x_0) = f''(x_0) = \cdots = f^{(n)}(x_0) = 0.$$

Note that to prove Taylor's theorem in this circumstance, we need to show that for each x in $I - \{x_0\}$, there is a number c between x and x_0 such that

$$f(x) = \frac{f^{(n+1)}(c)}{(n+1)!}(x - x_0)^{n+1}.$$

Define the function g on I so that $g(t) = (t - x_0)^{n+1}$, and note that

$$g(x_0) = g'(x_0) = \cdots = g^{(n)}(x_0) = 0,$$

while

$$g^{(n+1)}(t) = (n+1)!.$$

Without loss of generality, assume $x > x_0$. We apply Theorem 16.23, the Cauchy mean value theorem, to the functions f and g on $[x_0, x]$ to obtain c_1 in (x_0, x) for which

$$\frac{f'(c_1)}{g'(c_1)} = \frac{f(x) - f(x_0)}{g(x) - g(x_0)} = \frac{f(x)}{g(x)}.$$

Then we apply the Cauchy mean value theorem to the functions f' and g' on $[x_0, c_1]$ to obtain c_2 in (x_0, c_1) for which

$$\frac{f''(c_2)}{g''(c_2)} = \frac{f'(c_1) - f'(x_0)}{g'(c_1) - g'(x_0)} = \frac{f'(c_1)}{g'(c_1)}.$$

Continuing in this way we eventually obtain c_{n+1} in (x_0, c_n) for which

$$\frac{f^{(n+1)}(c_{n+1})}{g^{(n+1)}(c_{n+1})} = \frac{f^{(n)}(c_n) - f^{(n)}(x_0)}{g^{(n)}(c_n) - g^{(n)}(x_0)} = \frac{f^{(n)}(c_n)}{g^{(n)}(c_n)} = \cdots = \frac{f''(c_2)}{g''(c_2)} = \frac{f'(c_1)}{g'(c_1)} = \frac{f(x)}{g(x)},$$

which implies

$$f(x) = \frac{f^{(n+1)}(c_{n+1})}{g^{(n+1)}(c_{n+1})} g(x) = \frac{f^{(n+1)}(c_{n+1})}{(n+1)!}(x - x_0)^{n+1}.$$

This completes the argument for the special case.

To establish the more general result, note that the Taylor remainder R_n is defined so that

$$R_n(t) = f(t) - P_n(t),$$

and observe that R_n satisfies the conditions

$$R_n(x_0) = R_n'(x_0) = \cdots = R_n^{(n)}(x_0) = 0$$

of the special case we just considered. Thus, for each x in $I - \{x_0\}$, there is a number c between x_0 and x for which

$$R_n(x) = \frac{R_n^{(n+1)}(c)}{(n+1)!}(x - x_0)^{n+1}.$$

But the definition of R_n tells us that

$$R_n^{(n+1)}(c) = f^{(n+1)}(c) - P_n^{(n+1)}(c) = f^{(n+1)}(c) - 0 = f^{(n+1)}(c).$$

Therefore,

$$f(x) = P_n(x) + R_n(x),$$

where

$$R_n(x) = \frac{f^{(n+1)}(c)}{(n+1)!}(x - x_0)^{n+1}.$$ □

In practice, we usually cannot identify the point c, whose existence is provided for in Taylor's theorem, so we try to obtain a bound for the absolute value of $f^{(n+1)}$ on the interval formed by x and x_0.

Example 17.1 (Continued). We have seen that the fifth Taylor polynomial at 4 for the function f, where $f(x) = \sqrt{x}$ yields an approximation

$$P_5(5) = 2.2360763549804$$

for $\sqrt{5}$. Taylor's theorem can be used to determine how far this approximate value could be from the true value of $\sqrt{5}$. According to Taylor's theorem, this error is

$$\frac{f^{(n+1)}(c)}{(n+1)!}(x - x_0)^{n+1} = \frac{f^{(6)}(c)}{6!}(5 - 4)^6,$$

for some number c between 4 and 5, since we are using $x_0 = 4$, $x = 5$, and $n = 5$. It is readily checked that $f^{(6)}(t) = -\frac{945}{64t^5\sqrt{t}}$. As $f^{(6)}(t)$ is always negative, the error

$$\frac{f^{(6)}(c)}{6!}(5-4)^6 = \sqrt{5} - P_5(5)$$

must be negative, and we conclude that our estimate $P_5(5)$ is larger than the true value of $\sqrt{5}$. On the interval $[4, 5]$, the absolute value of $f^{(6)}(t)$ is largest when $t = 4$, with

$$|f^{(6)}(4)| = \frac{945}{64 \cdot 4^5\sqrt{4}} \approx 0.007201 < 0.007202.$$

We may then conclude that $P_5(5)$ overestimates $\sqrt{5}$ by an amount no larger than

$$\frac{0.007202}{6!}(5-4)^6 = 0.00001000\overline{27}.$$

If we assume the online calculator's approximate value of 2.2360679774998 for $\sqrt{5}$ is accurate to the number of decimal places recorded, we may observe that

$$P_5(5) - \sqrt{5} \approx 2.2360763549804 - 2.2360679774998 = 0.0000083774806,$$

which is less than the error bound $0.00001000\overline{27}$ we just obtained.

Exercise 17.7. In Exercise 17.4, you used the Taylor polynomial P_3 at $x_0 = 1$ for the reciprocal function f to approximate $f(1.5)$ and also determined how far the approximation is from the true value of $f(1.5)$. Use Taylor's theorem to obtain a bound for the absolute value of the error and make sure the actual error is consistent with the bound.

In Chapter 25, we connect our work with Taylor polynomials to the notion of a *Taylor series*. Then, in Chapter 29, we explore a more general result that shows how any function continuous on a closed bounded interval can be approximated arbitrarily closely by means of polynomial functions.

18 The Riemann integral

Along with the derivative, the idea of an *integral* is among the most familiar from our study of calculus. There we learned a great deal about how to evaluate integrals as well as how to use them to calculate areas, volumes, lengths of curves, work done by a force, and maybe even probabilities. Our goal in this chapter is to better understand the notion of an integral by creating a precise mathematical definition for the concept and using it to establish some basic properties of integrals.

Geometric motivation for the integral

In calculus, we often think of an integral, for instance,

$$\int_2^7 [4 - (x - 3)^2]\,dx,$$

as representing an area, in this case the area between the graph of the function f defined by

$$f(x) = 4 - (x - 3)^2$$

and the x-axis, along the input interval $[2, 7]$, with the understanding that area above the x-axis is counted positively, while area below the x-axis is counted negatively.

Figure 18.1 reveals that the graph of f lies above the x-axis between $x = 2$ and $x = 5$ (which is easily verified to be an x-intercept) and below the x-axis between $x = 5$ and $x = 7$. Thus, the area under the graph of f lying above the x-axis along the interval $[2, 5]$ is counted positively, but the area above the graph of f lying below the x-axis along the interval $[5, 7]$ is counted negatively.

It is natural to try to approximate the area an integral represents by placing simpler geometric figures over or within the region whose area is being measured, figures whose areas are easily calculated, and then summing the areas of these simpler figures. Since finding areas of rectangles is quite straightforward, we use rectangles as our simpler geometric figures.

For example, if we add the areas of the four rectangles shown in Figure 18.2, counting the areas of the rectangles lying above the x-axis positively and the areas of the rectangles lying below the x-axis negatively, it appears we would obtain a reasonable approximation to the signed area measured by $\int_2^7 [4 - (x - 3)^2]dx$.

Ultimately, the integral itself is defined as a type of limit of these sorts of approximating sums, so we need to have some clear guidelines for forming the sums. With respect to the locations of the rectangles and their widths, the diagram indicates that the rectangles all lie along the interval $[2, 7]$ of interest, with each rectangle's width being the

https://doi.org/10.1515/9783111429564-018

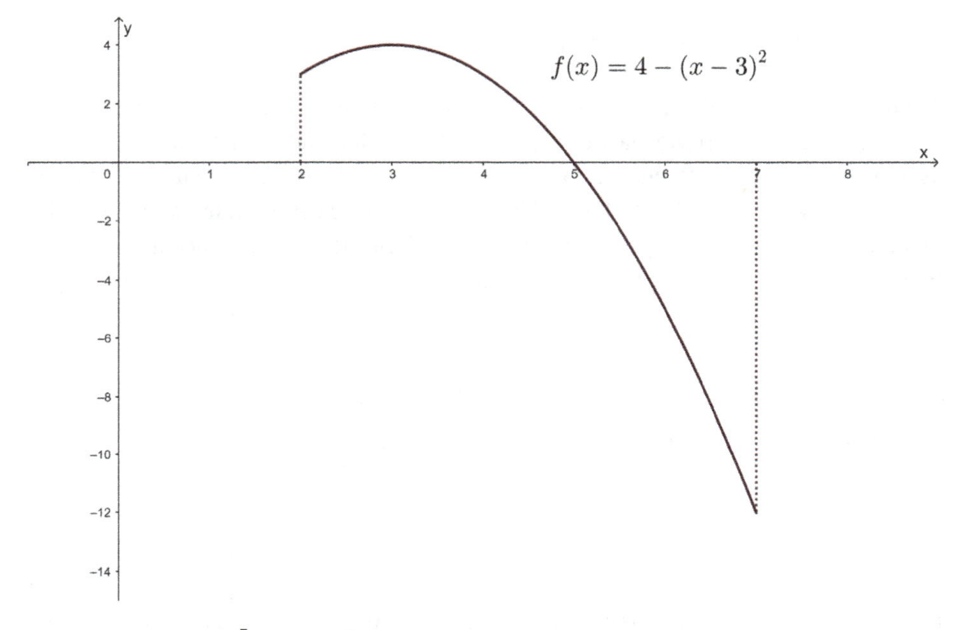

Figure 18.1: Visualizing $\int_2^7 [4 - (x-3)^2]dx$ as "signed" area.

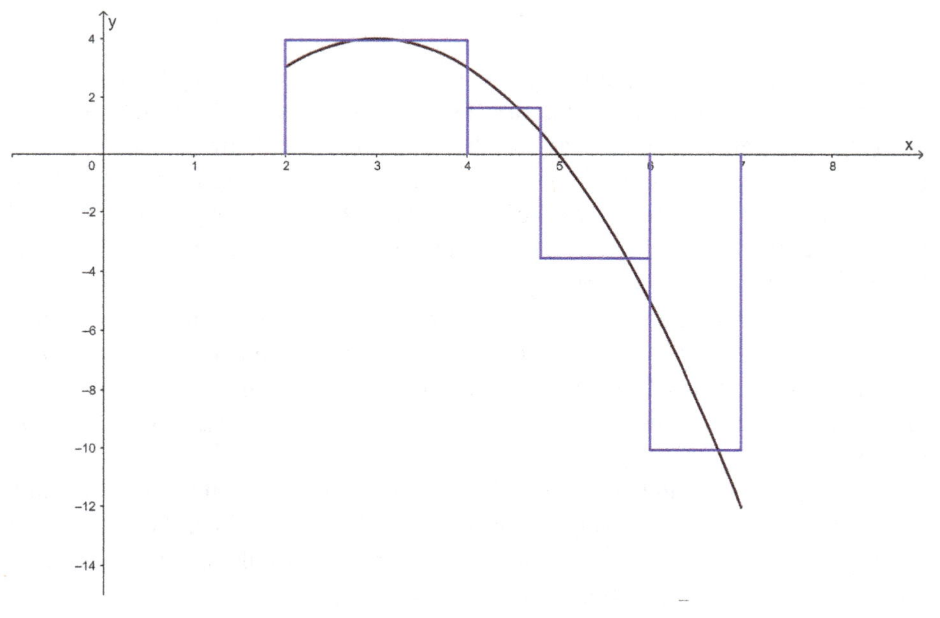

Figure 18.2: A rectangular approximating sum for $\int_2^7 [4 - (x-3)^2]dx$.

length of some subinterval of $[2,7]$, and the sum of the widths of all the rectangles being the length of the interval $[2,7]$. These observations lead us to introduce the idea of a *partition* of an interval.

Partitions and tagged partitions

We assume throughout this chapter that when a closed bounded interval $[a,b]$ is referred to, it is nontrivial, meaning $a \neq b$. A **partition** of a closed bounded interval $[a,b]$ is a finite subset

$$\{p_0, p_1, p_2, \ldots, p_n\}$$

of $[a,b]$ for which

$$a = p_0 < p_1 < p_2 < \cdots < p_n = b.$$

The points of the partition subdivide the interval $[a,b]$ into the subintervals

$$[p_0, p_1], \quad [p_1, p_2], \quad \ldots, \quad [p_{n-1}, p_n].$$

Figure 18.3 displays a generic partition of $[a,b]$ into $n = 5$ subintervals.

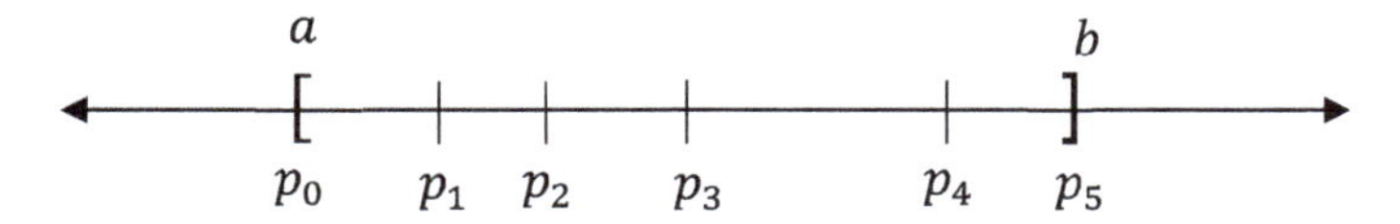

Figure 18.3: A partition of the interval $[a,b]$ with five subintervals.

Example 18.1. Let

$$P = \{2, 4, 4.8, 6, 7\}.$$

Then P is a partition of the interval $[2,7]$ that subdivides $[2,7]$ into the four subintervals

$$[2,4], \quad [4,4.8], \quad [4.8,6], \quad [6,7].$$

These subintervals were used to horizontally locate the four rectangles along the interval $[2,7]$ in Figure 18.2.

Another partition of $[2,7]$ is

$$Q = \{2, 2.5, 3, 3.5, 4, 4.5, 5, 5.5, 6, 6.5, 7\},$$

which separates [2, 7] into the ten subintervals

$$[2, 2.5], \quad [2.5, 3], \quad [3, 3.5], \quad [3.5, 4], \quad [4, 4.5],$$
$$[4.5, 5], \quad [5, 5.5], \quad [5.5, 6], \quad [6, 6.5], \quad [6.5, 7].$$

If $\{p_0, p_1, p_2, \ldots, p_n\}$ is a partition of $[a, b]$ and for each i in $\{1, 2, \ldots, n\}$, the number t_i is in the subinterval $[p_{i-1}, p_i]$, the set

$$\dot{P} = \{([p_{i-1}, p_i], t_i) \mid i \in \{1, 2, \ldots, n\}\}$$

is called a **tagged partition** of $[a, b]$ and the numbers $t_1, t_2, \ldots, t_n$ are referred to as **tags**. We are adopting the use of the "dot" to indicate that a partition has been tagged as it seems to be the most commonly employed of several competing notations.

Figure 18.4 displays a generic tagged partition of $[a, b]$ into $n = 5$ subintervals. Note how the tag t_1 is in the first subinterval $[p_0, p_1]$, the tag t_2 is in the second subinterval $[p_1, p_2]$, and so on. The tag t_3 appears to be the endpoint p_3 of the third subinterval $[p_2, p_3]$, but that is permitted by the definition.

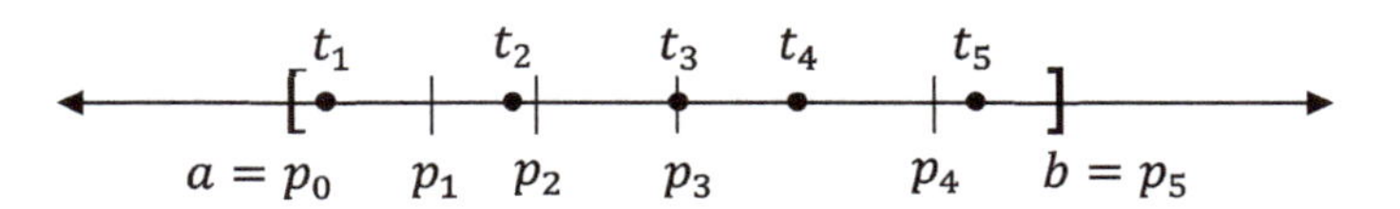

Figure 18.4: A tagged partition of the interval $[a, b]$ with five subintervals.

Example 18.2. Consider again the partition $P = \{2, 4, 4.8, 6, 7\}$ of the interval [2, 7]. Since the number 2.5 is in [2, 4], the number 4 is in [4, 4.8], the number 5 is in [4.8, 6], and the number 6.999 is in [6, 7], it follows that

$$\dot{P}_1 = \{([2, 4], 2.5), ([4, 4.8], 4), ([4.8, 6], 5), ([6, 7], 6.999)\}$$

is a tagged partition of [2, 7], with the number 2.5 serving as the tag for the subinterval [2, 4], the number 4 the tag for [4, 4.8], the number 5 the tag for [4.8, 6], and the number 6.999 the tag for [6, 7].

It is important to realize that a given partition of an interval gives rise to (infinitely) many different tagged partitions. For instance, another tagged partition of [2, 7] associated with the partition $P = \{2, 4, 4.8, 6, 7\}$ is

$$\dot{P}_2 = \{([2, 4], 2.75), ([4, 4.8], 4.55), ([4.8, 6], 5.75), ([6, 7], 6.75)\}.$$

The **norm** of a partition, untagged or tagged, is the length of the longest subinterval determined by the partition. The norm of a partition P is denoted by $\|P\|$.

Example 18.1 (Continued). The norm of the partition

$$P = \{2, 4, 4.8, 6, 7\}$$

of the interval $[2, 7]$ is 2 as this number is the length of the longest subinterval $[2, 4]$ determined by P; hence, $\|P\| = 2$.

Since each subinterval of the partition

$$Q = \{2, 2.5, 3, 3.5, 4, 4.5, 5, 5.5, 6, 6.5, 7\}$$

of $[2, 7]$ has length 0.5, the "longest" subinterval has length 0.5, which makes 0.5 the norm of Q; thus, we may write $\|Q\| = 0.5$.

A partition, untagged or tagged, is called **regular** if all of the subintervals it determines have the same length.

Example 18.1 (Continued). The subintervals determined by the partition

$$Q = \{2, 2.5, 3, 3.5, 4, 4.5, 5, 5.5, 6, 6.5, 7\}$$

of $[2, 7]$ all have the same length 0.5, so Q is a regular partition.

As the subintervals determined by the partition

$$P = \{2, 4, 4.8, 6, 7\}$$

of $[2, 7]$ are of varying lengths, the partition P is not regular.

Exercise 18.1. Consider the interval $[-3, 3]$.

(a) Give the partition of this interval that determines the subintervals $[-3, -1.5]$, $[-1.5, 0]$, $[0, 1.5]$, and $[1.5, 3]$.

(b) Is the partition you gave as your answer to (a) regular?

(c) What is the norm of the partition you gave as your answer to (a)?

Exercise 18.2. Consider the tagged partition

$$\dot{P} = \{([i^2, (i+1)^2], i^2 + 1) \mid i \in \{0, 1, 2\}\}$$

of the interval $[0, 9]$.

(a) From what partition of $[0, 9]$ has $\dot{P}$ been constructed?

(b) Compute $\|\dot{P}\|$.

(c) Is $\dot{P}$ regular?

(d) What is the numerical value of the tag corresponding to the subinterval that is determined by $\dot{P}$ having right endpoint 4?

Riemann sums

The **Riemann sum** of a function $f : [a,b] \to \mathbb{R}$ determined by a tagged partition $\dot{P} = \{([p_{i-1}, p_i], t_i) \mid i \in \{1, 2, \ldots, n\}\}$ of the interval $[a,b]$ is

$$S(f, \dot{P}) = \sum_{i=1}^{n} f(t_i) \cdot (p_i - p_{i-1}).$$

The term *Riemann sum* honors the mathematician Bernhard Riemann (1826–1866), who in 1854 developed the first rigorous definition of the integral of a function on an interval.

We may geometrically interpret a Riemann sum as a sum of "areas" of rectangles, one rectangle for each subinterval determined by the given partition, with the "base" of a rectangle sitting along the subinterval on the x-axis and the "height" equal to the value of the given function at the tag specified for the particular subinterval. We have put the words "areas," "base," and "height" in quotes because the value of the given function at a certain tag may be negative, thus resulting in the rectangle extending downward so that what we are referring to as the rectangle's "base" would actually be its "top"; in this instance, the "height" and "area" of the rectangle would both be assigned negative values.

The initial importance of Riemann sums is their use in approximating the value of an integral. Ultimately, Riemann sums are used to formally define what is meant by an integral.

Example 18.3. Consider again the tagged partition

$$\dot{P}_2 = \{([2,4], 2.75), ([4,4.8], 4.55), ([4.8,6], 5.75), ([6,7], 6.75)\}$$

of the interval $[2,7]$. Let the function $f : [2,7] \to \mathbb{R}$ be defined so that

$$f(x) = 4 - (x-3)^2.$$

Then the Riemann sum of f determined by $\dot{P}_2$ is

$$\begin{aligned} S(f, \dot{P}_2) &= f(2.75) \cdot (4-2) + f(4.55) \cdot (4.8-4) + f(5.75) \cdot (6-4.8) \\ &\quad + f(6.75) \cdot (7-6) \\ &= 3.9375 \cdot 2 + 1.5975 \cdot 0.8 + (-3.5625) \cdot 1.2 + (-10.0625) \cdot 1 \\ &= -5.1845. \end{aligned}$$

This Riemann sum is depicted in Figure 18.2, along with the graph of f, and provides us with an approximation to the value of

$$\int_2^7 [4 - (x-3)^2]\,dx.$$

It would seem that the approximation could be improved by using a partition that creates more and shorter subintervals of the interval $[2, 7]$.

There is a variety of special types of Riemann sums. Given a function $f : [a, b] \to \mathbb{R}$ and a tagged partition $\dot{P}$ of the interval $[a, b]$, the Riemann sum $S(f, \dot{P})$ is called

(1) a **left endpoint (Riemann) sum** if the tag for each subinterval is the subinterval's left endpoint;
(2) a **right endpoint (Riemann) sum** if the tag for each subinterval is the subinterval's right endpoint;
(3) a **midpoint (Riemann) sum** if the tag for each subinterval is the subinterval's midpoint;
(4) a **lower (Riemann) sum** if the tag for each subinterval yields the minimum value of f on the subinterval;
(5) an **upper (Riemann) sum** if the tag for each subinterval yields the maximum value of f on the subinterval.

In some instances, a lower sum or upper sum may be undefined because a function may not achieve a minimum or maximum value on one or more of the subintervals determined by a partition.

Exercise 18.3. Consider the function $f : [-1, 5] \to \mathbb{R}$ defined by

$$f(x) = x^2 - 10.$$

In each case, evaluate the Riemann sum of f corresponding to the specified tagged partition of $[-1, 5]$. Also make a hand-drawn sketch in which you show the graph of f along with the rectangles associated with the Riemann sum.

(a) $\{([-1, 0], -1), ([0, 3], 2), ([3, 5], 4)\}$
(b) the right endpoint sum corresponding to the regular partition of $[-1, 5]$ into three subintervals
(c) the midpoint sum corresponding to the regular partition of $[-1, 5]$ into two subintervals
(d) the lower sum corresponding to the partition $\{-1, 0, 3, 5\}$ of $[-1, 5]$

Just as is done in calculus courses, we have used the geometric concept of area to motivate the notion of integral and we have geometrically interpreted Riemann sums as sums of areas of rectangles. Keep in mind, though, that these visual representations are meant only to guide our thinking. Our definitions, for instance, that of *Riemann sum*, are purely analytic and do not depend on the notion of area, and neither do the proofs of the theorems we investigate that incorporate the notion of integrability. However, we continue to sometimes employ area representations so that you can more easily relate the rigorous approach to the integral presented here to what you learned in calculus.

You might be wondering why we do not just employ regular partitions all the time. Allowing for the use of partitions in which the subintervals do not all have the same length can aid in the calculation of a Riemann sum in certain circumstances.

Example 18.4. Let the function $g : [0,1] \to \mathbb{R}$ be defined so that $g(x) = \sqrt{x}$. Note that the right endpoint sum corresponding to the regular partition of $[0,1]$ into five subintervals is

$$\sqrt{\frac{1}{5}} \cdot \frac{1}{5} + \sqrt{\frac{2}{5}} \cdot \frac{1}{5} + \sqrt{\frac{3}{5}} \cdot \frac{1}{5} + \sqrt{\frac{4}{5}} \cdot \frac{1}{5} + \sqrt{1} \cdot \frac{1}{5},$$

whereas the irregular partition $\{0, \frac{1}{25}, \frac{4}{25}, \frac{9}{25}, \frac{16}{25}, 1\}$ yields a right endpoint sum

$$\begin{aligned} &\sqrt{\frac{1}{25}} \cdot \frac{1}{25} + \sqrt{\frac{4}{25}} \cdot \frac{3}{25} + \sqrt{\frac{9}{25}} \cdot \frac{5}{25} + \sqrt{\frac{16}{25}} \cdot \frac{7}{25} + \sqrt{1} \cdot \frac{9}{25} \\ &= \frac{1}{5} \cdot \frac{1}{25} + \frac{2}{5} \cdot \frac{3}{25} + \frac{3}{5} \cdot \frac{5}{25} + \frac{4}{5} \cdot \frac{7}{25} + 1 \cdot \frac{9}{25} \\ &= \frac{19}{25} \end{aligned}$$

that is easier to compute.

Exercise 18.4. Let n be a natural number and, for each i in $\{0, 1, 2, \ldots, n\}$, let $p_i = \frac{i^2}{n^2}$. Let P_n be the irregular partition $\{p_0, p_1, p_2, \ldots, p_n\}$ of $[0,1]$ and let the function g be as in Example 18.4. Verify that the right endpoint Riemann sum of g determined by P_n can be simplified to $\frac{1}{n^3} \sum_{i=1}^{n} (2i^2 - i)$.

A Precise definition of the Riemann integral

We know that Riemann sums can be used to approximate the value of an integral. Intuitively, we can obtain better approximations by using partitions for which the resulting subintervals are smaller in length.

Example 18.3 (Continued). Again, suppose that the function f, where $f(x) = 4 - (x-3)^2$ is being considered on the interval $[2,7]$. The right endpoint sum corresponding to the regular partition of $[2,7]$ into five subintervals is displayed in Figure 18.5. It evaluates to -10 and provides an approximation for $\int_2^7 [4 - (x-3)^2]dx$.

We can obtain a more accurate approximation by using a regular partition consisting of more subintervals. For instance, the right endpoint Riemann sum corresponding to the regular partition of $[2,7]$ into 50 subintervals (Figure 18.6) evaluates to -2.425.

Thus, a Riemann sum created using a partition that results in all of the subintervals determined by the partition being small in length should yield a good approximation to the value of an integral. In order to control the lengths of the subintervals determined by a partition, we can simply require that the partition's norm, which is the length of the longest subinterval, be suitably small.

These observations suggest that the integral itself might be conceived of as the limiting value of a sequence of Riemann sums whose norms become arbitrarily small.

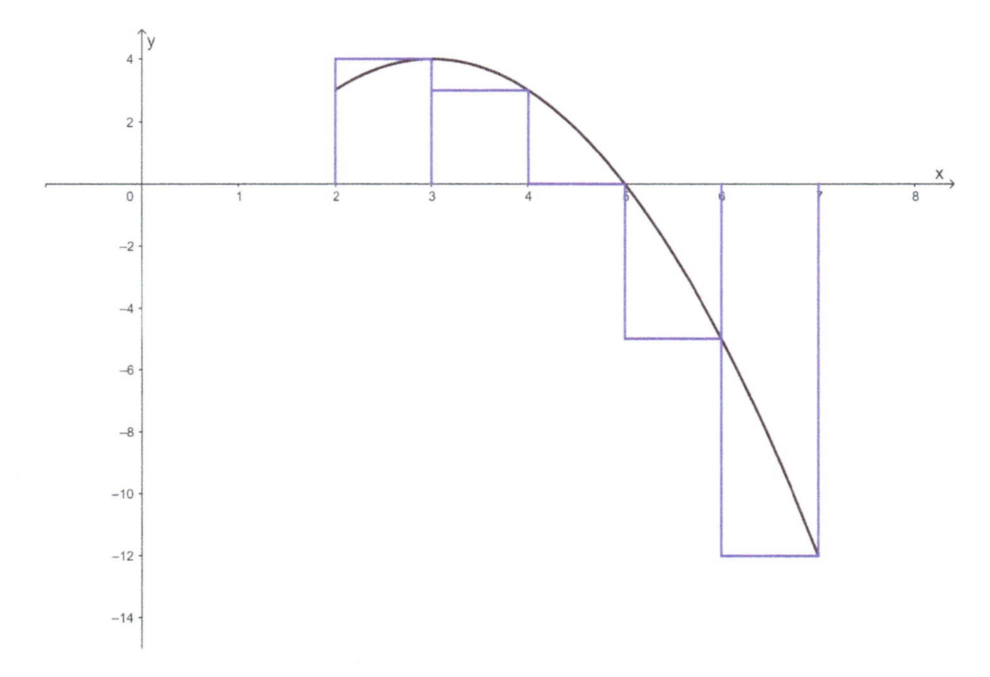

Figure 18.5: A right endpoint sum corresponding to a regular partition having five subintervals.

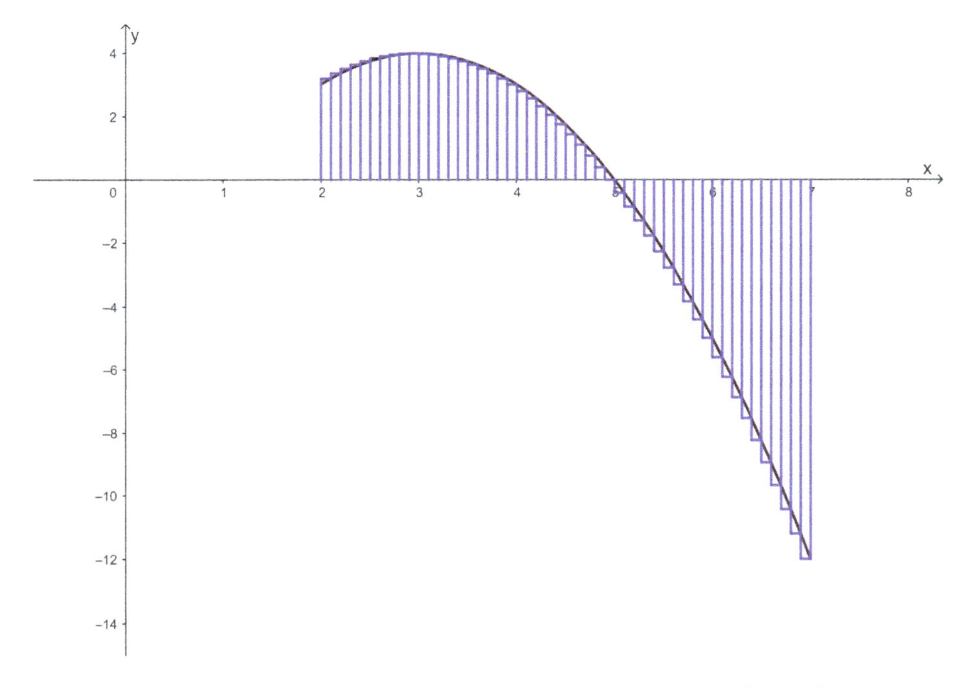

Figure 18.6: A right endpoint sum corresponding to a regular partition having 50 subintervals.

Example 18.3 (Continued). In approximating the value of

$$\int_2^7 [4 - (x-3)^2]\,dx,$$

we considered right endpoint Riemann sums corresponding to regular partitions of the interval $[2,7]$ into five and then fifty subintervals, obtaining the approximations

$$R_5 = -10$$

and

$$R_{50} = -2.425$$

to the integral, where R_n denotes the right endpoint Riemann sum corresponding to the regular partition of $[2,7]$ into n subintervals. Since

$$\|R_n\| = \frac{7-2}{n} \to 0,$$

it seems intuitively plausible that $\int_2^7 [4-(x-3)^2]dx$ would be the limit of the sequence (R_n).

There are, though, a couple of concerns with this line of reasoning. One is the question of whether the specified sequence (R_n) of Riemann sums does indeed have a limiting value. A second is that, even if the sequence has a limit, there are still many other sequences of Riemann sums that could be considered in place of the sequence (R_n). What if left endpoint Riemann sums or midpoint sums were used instead of right endpoint sums? What if the tags are chosen in some other way? What if the partitions are not regular? Does every such sequence of Riemann sums have a limiting value and are all of these limits the same?

Given a function f defined on an interval $[a,b]$, we see there are infinitely many sequences of Riemann sums with norms tending toward zero, so in order for the integral of f on $[a,b]$ to be defined, we anticipate that all such sequences must tend toward a common limiting value. The first of the two equivalent conditions given in the following theorem is simply a statement of this requirement, and is probably the most conceptually appealing distillation of the definition of the integral. The second condition provides an alternative, but very similar, characterization of the definition that does not directly involve sequences.

As we point out after proving the theorem, either condition may be regarded as the definition of *Riemann integrability* (both conditions originate with Riemann). The choice of which condition to apply often depends on the particular situation. In the next chapter, we consider several other conditions that are also equivalent to these two particular conditions, so there are actually quite a number of different ways one can go about establishing integrability.

Theorem 18.5. *Given a function $f : [a,b] \to \mathbb{R}$ and a real number L, the following are equivalent:*
(i) *for any sequence $(\dot{P}_n)$ of tagged partitions of $[a,b]$ for which the sequence $(\|\dot{P}_n\|)$ converges to 0, it follows that the sequence $(S(f,\dot{P}_n))$ converges to L;*
(ii) *for every positive number ε, there exists a positive number δ such that whenever $\dot{P}$ is a tagged partition of $[a,b]$ for which $\|\dot{P}\| < \delta$, it follows that $|S(f,\dot{P}) - L| < \varepsilon$.*

Proof. (i) $\Rightarrow$ (ii) Assume (ii) does not hold. Then there is a positive number ε such that, for every natural number n, there exists a tagged partition $\dot{P}_n$ of $[a,b]$ for which $\|\dot{P}_n\| < \frac{1}{n}$ and $|S(f,\dot{P}_n) - L| \geq \varepsilon$. Thus, $\|\dot{P}_n\| \to 0$ and $S(f,\dot{P}_n) \nrightarrow L$.

(ii) $\Rightarrow$ (i) Assume (ii) holds and consider any sequence $(\dot{P}_n)$ of tagged partitions of $[a,b]$ for which $\|\dot{P}_n\| \to 0$. Suppose $\varepsilon > 0$. Then from our hypothesis (ii), there is a positive number δ for which whenever $\dot{P}$ is a tagged partition of $[a,b]$ for which $\|\dot{P}\| < \delta$, it follows that $|S(f,\dot{P}) - L| < \varepsilon$. As $\|\dot{P}_n\| \to 0$, there exists an index N such that whenever $n \geq N$, we have $\|\dot{P}_n\| < \delta$, which then implies that $|S(f,\dot{P}_n) - L| < \varepsilon$. Therefore, $S(f,\dot{P}_n) \to L$. □

Condition (ii) says that the distance between Riemann sums and the number L can be made arbitrarily small by choosing tagged partitions of appropriately small norm. As was the case for the definitions of *function limit* and *continuity of a function*, the value of δ usually depends on the value of ε, with different values of ε giving rise to different values of δ.

Definition 18.6 (Definitions of integrability and the integral). A function $f : [a,b] \to \mathbb{R}$ is **(Riemann) integrable** on the interval $[a,b]$ if there is a real number L, called the **(Riemann) integral** of f on $[a,b]$ and denoted by $\int_a^b f$, which together with f satisfies either of the equivalent conditions (i) or (ii) stated in Theorem 18.5.

In the expression

$$\int_a^b f,$$

the symbol $\int$ is an **integral sign**, the interval $[a,b]$ is the **interval of integration**, the number a is the **lower endpoint of integration**, the number b is the **upper endpoint of integration**, and the function f is the **integrand**. We sometimes write $\int_a^b f(x)dx$ in place of $\int_a^b f$, especially for functions f defined by formulas $f(x)$.

The definition of the Riemann integral is purely analytic and does not depend on the notion of area. As the only integral we consider in this book is the Riemann integral, we often abbreviate *Riemann integral* to just *integral*.

Example 18.7. Later (see Exercise 28.14), you will demonstrate that

$$\int_0^{\pi} \sin = 2.$$

In this case, the sine function is the integrand and the interval of integration is $[0, \pi]$, with 0 being the lower endpoint of integration and π the upper endpoint of integration. Because this integral evaluates to a real number (namely 2), we can say that the sine function is Riemann integrable on the interval $[0, \pi]$.

Using the definition of the integral to determine whether a function is integrable on an interval requires that we are able to identify a number that serves as the integral's value (you may recall the same sort of dilemma is faced when working with the definitions of limit of a function and limit of a sequence). Sometimes a geometric calculation or an examination of Riemann sums can produce an educated guess as to an integral's value, but, of course, one cannot expect such a strategy to be effective all the time.

Moreover, the definition of integral is often challenging to employ even with relatively simple functions. We provide an illustration of the use of the definition, but note that it is actually the theorems regarding integrability that we eventually derive from the definition that serve as the primary means for determining whether a function is integrable.

Example 18.8. Simple geometry based on the diagram in Figure 18.7 suggests that the function f defined so that $f(x) = x$ is integrable on $[a, b]$ with $\int_a^b f = \frac{1}{2}(b^2 - a^2)$.

The diagram in Figure 18.8 depicts the midpoint sum for f corresponding to a regular partition of $[a, b]$ into six subintervals.

For each rectangle, the small triangular region where it extends above the graph of f is congruent to the small triangular region where the top of the rectangle lies below the graph of f. Thus, it appears this midpoint sum is equal to the exact area measured by the integral.

We now show that the midpoint sum for f corresponding to *any* partition, regular or irregular, of $[a, b]$ is equal to $\frac{1}{2}(b^2 - a^2)$. Given any partition $\{p_0, p_1, p_2, \ldots, p_n\}$ of $[a, b]$, the midpoint sum is

$$\begin{aligned}
\sum_{i=1}^{n} f\left(\frac{p_{i-1} + p_i}{2}\right) \cdot (p_i - p_{i-1}) &= \sum_{i=1}^{n} \frac{p_{i-1} + p_i}{2} \cdot (p_i - p_{i-1}) \\
&= \frac{1}{2} \sum_{i=1}^{n} (p_i^2 - p_{i-1}^2) \\
&= \frac{1}{2}[(p_1^2 - p_0^2) + (p_2^2 - p_1^2) + (p_3^2 - p_2^2) + \cdots + (p_n^2 - p_{n-1}^2)] \\
&= \frac{1}{2}(b^2 - a^2),
\end{aligned}$$

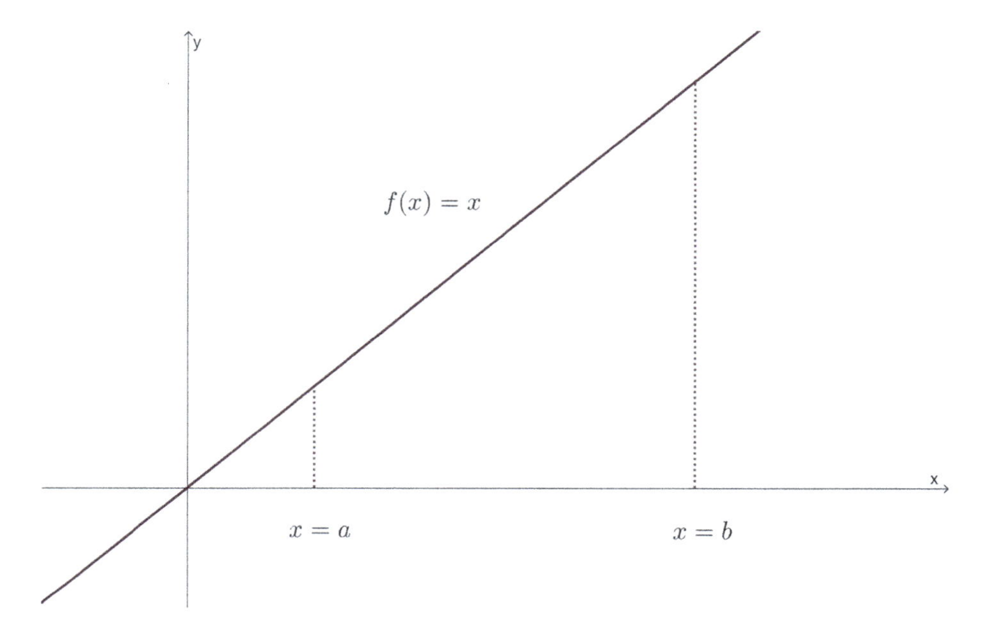

Figure 18.7: The region whose area is measured by the integral from Example 18.8.

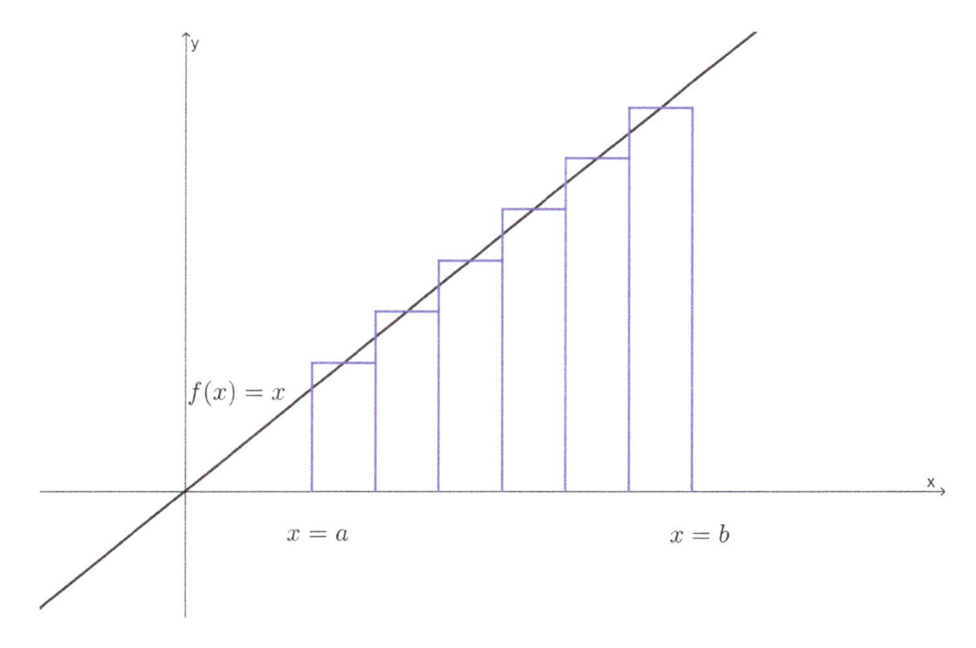

Figure 18.8: A midpoint sum for the integral from Example 18.8.

since $p_0 = a$ and $p_n = b$, and the terms involving the other partition points subtract out. This observation about the value of *every* midpoint sum for f can now be used in applying the definition of Riemann integrability to verify that $\int_a^b f = \frac{1}{2}(b^2 - a^2)$.

Consider any positive number ε, let $\delta = \frac{\varepsilon}{b-a}$, which is positive, and consider any tagged partition $\dot{P} = \{([p_{i-1}, p_i], t_i) \mid i \in \{1, 2, \ldots, n\}\}$ of $[a, b]$ for which $\|\dot{P}\| < \delta$. Substituting for $\frac{1}{2}(b^2 - a^2)$ the midpoint sum based on the partition underlying $\dot{P}$, we see that

$$\begin{aligned}
\left|S(f, \dot{P}) - \frac{1}{2}(b^2 - a^2)\right| &= \left|\sum_{i=1}^{n} f(t_i) \cdot (p_i - p_{i-1}) - \sum_{i=1}^{n} f\left(\frac{p_i + p_{i-1}}{2}\right) \cdot (p_i - p_{i-1})\right| \\
&= \left|\sum_{i=1}^{n}\left(t_i - \frac{p_i + p_{i-1}}{2}\right) \cdot (p_i - p_{i-1})\right| \\
&\le \sum_{i=1}^{n}\left|t_i - \frac{p_i + p_{i-1}}{2}\right| \cdot |p_i - p_{i-1}| \\
&< \sum_{i=1}^{n} \delta \cdot (p_i - p_{i-1}) \\
&= \delta(b - a) \\
&= \frac{\varepsilon}{b - a} \cdot (b - a) \\
&= \varepsilon,
\end{aligned}$$

where $|t_i - \frac{p_i + p_{i-1}}{2}| < \delta$ because t_i and $\frac{p_i + p_{i-1}}{2}$ are both in the same subinterval $[p_{i-1}, p_i]$ determined by $\dot{P}$, and $\|\dot{P}\| < \delta$. Thus, by definition, f is Riemann integrable on $[a, b]$ and $\int_a^b f = \frac{1}{2}(b^2 - a^2)$.

Geometry also suggests that constant functions are integrable, with the integral's value being the product of the constant and the length of the interval of integration.

Theorem 18.9 (Constant functions are integrable). *Let c be a real number and define the function $f : [a, b] \to \mathbb{R}$ so that $f(x) = c$. Then f is Riemann integrable and $\int_a^b f = c(b - a)$.*

Exercise 18.5. Use the definition of the integral to prove Theorem 18.9.

As the wording of the definition of the Riemann integral would seem to imply, there is at most one number L that can serve as the integral of a particular function f on a specified interval $[a, b]$.

Theorem 18.10 (Uniqueness of the value of an integral). *For a given function f and interval $[a, b]$, if $\int_a^b f$ exists, then its value is unique.*

Exercise 18.6. Prove Theorem 18.10.

The next example illustrates that not all functions are integrable.

Example 18.11. We show the **Dirichlet function** $f : [0,1] \to \mathbb{R}$ defined by

$$f(x) = \begin{cases} 1, & \text{if } x \text{ rational;} \\ 0, & \text{if } x \text{ irrational;} \end{cases}$$

is not integrable on $[0,1]$ by exhibiting two sequences of Riemann sums that have different limiting values, even though in each case the norms of the tagged partitions on which the Riemann sums are based tend toward zero. This function was introduced by Peter Gustav Lejeune Dirichlet in 1829.

For each natural number n, let $\dot{P}_n$ and $\dot{Q}_n$ be tagged regular partitions of $[0,1]$ into n subintervals, but chosen so that all tags t_i are irrational for $\dot{P}_n$ and all tags u_i are rational for $\dot{Q}_n$ (this is, of course, possible to do because of the density of both the rational numbers and the irrational numbers in $\mathbb{R}$). Then for each n, we have $\|\dot{P}_n\| = \|\dot{Q}_n\| = \frac{1}{n}$, so that $\|\dot{P}_n\| \to 0$ and $\|\dot{Q}_n\| \to 0$. But as

$$S(f, \dot{P}_n) = \sum_{i=1}^{n} f(t_i) \cdot \frac{1}{n} = \sum_{i=1}^{n} 0 \cdot \frac{1}{n} = 0,$$

and

$$S(f, \dot{Q}_n) = \sum_{i=1}^{n} f(u_i) \cdot \frac{1}{n} = \sum_{i=1}^{n} 1 \cdot \frac{1}{n} = 1,$$

it follows that $S(f, \dot{P}_n) \to 0$, whereas $S(f, \dot{Q}_n) \to 1$.

The existence of the sequences $(S(f, \dot{P}_n))$ and $(S(f, \dot{Q}_n))$ of Riemann sums that converge to different limits, but for which the norms of the tagged partitions $\dot{P}_n$ and $\dot{Q}_n$ both tend toward zero as n grows ever larger, tells us that f is not integrable on $[0,1]$.

Exercise 18.7. Consider the function g from Exercise 13.17. There, you showed this function is continuous at every irrational number in $[0,1]$, but discontinuous at every rational number in $[0,1]$. Prove that g is integrable on $[0,1]$ and $\int_a^b g = 0$. (The function g is known as **Thomae's function**, named after Carl Johannes Thomae (1840–1921), who created it.)

It may be expected that a function that is integrable on an interval must be bounded on that interval, as otherwise it would seem feasible to create a sequence of Riemann sums that becomes unbounded even as the norms of the partitions on which the Riemann sums are based are becoming arbitrarily small.

Theorem 18.12 (Integrable functions are bounded). *If the function f is Riemann integrable on the interval $[a,b]$, then f is bounded on $[a,b]$.*

Proof. We show that if f is unbounded on $[a,b]$, then f is not integrable on $[a,b]$. Assume f takes arbitrarily large positive values on $[a,b]$; you consider the case where f takes arbitrarily small negative values in Exercise 18.8.

For each natural number n, let P_n be the regular partition of $[a,b]$ into n subintervals. By hypothesis, for each n, there is a subinterval I_n of $[a,b]$ determined by P_n on which f takes arbitrarily large values. We tag P_n to form $\dot{P}_n$ by using the right endpoints of the subintervals as the tags, except that the tag t_n for I_n is chosen so that

$$f(t_n) > \frac{n - R_n + f(r_n)\|P_n\|}{\|P_n\|},$$

where R_n is the right endpoint sum corresponding to P_n and r_n is the right endpoint of I_n.

Observe that $\|\dot{P}_n\| = \frac{b-a}{n} \to 0$, but for each n, we have

$$S(f,\dot{P}_n) = R_n - f(r_n)\|P_n\| + f(t_n)\|P_n\| > R_n - f(r_n)\|P_n\| + n - R_n + f(r_n)\|P_n\| = n,$$

so that the sequence $(S(f,\dot{P}_n))$ diverges to ∞. Thus, by (i) of Theorem 18.5, f is not integrable on $[a,b]$. □

Exercise 18.8. Complete the proof of Theorem 18.12 by considering the case in which f takes arbitrarily small negative values on $[a,b]$.

Elementary properties of the integral

We now provide analytic arguments supporting several properties of the integral familiar from calculus courses. These properties may all be anticipated by means of the area interpretation of the integral, and for each one we ask you to illustrate the underlying geometry. Also note that in some cases we are able to use integrability of a function on an interval to prove either integrability of another function on that interval or integrability of the same function on a different interval.

The first property we examine guarantees the integrability of any sum of integrable functions, with the integral of the sum being the sum of the integrals of the summands.

Theorem 18.13 (Sum rule for integrals). *If the functions f and g are Riemann integrable on the interval $[a,b]$, then the function $f+g$ is also Riemann integrable on $[a,b]$, and $\int_a^b (f+g) = \int_a^b f + \int_a^b g$.*

Proof. Assume f and g are integrable on $[a,b]$. Observe that for any tagged partition $\dot{P} = \{([p_{i-1},p_i],t_i) \mid i \in \{1,2,\ldots,n\}\}$ of $[a,b]$, we have

$$\begin{aligned}
S(f+g,\dot{P}) &= \sum_{i=1}^{n}(f+g)(t_i)\cdot(p_i-p_{i-1}) \\
&= \sum_{i=1}^{n}(f(t_i)+g(t_i))\cdot(p_i-p_{i-1}) \\
&= \sum_{i=1}^{n}f(t_i)\cdot(p_i-p_{i-1})+\sum_{i=1}^{n}g(t_i)\cdot(p_i-p_{i-1}) \\
&= S(f,\dot{P})+S(g,\dot{P}).
\end{aligned}$$

Suppose $\varepsilon > 0$. Then there exist positive numbers δ_1 and δ_2 such that for any tagged partition $\dot{P}$ of $[a,b]$, if $\|\dot{P}\| < \delta_1$, it follows that $|S(f,\dot{P}) - \int_a^b f| < \frac{\varepsilon}{2}$, and if $\|\dot{P}\| < \delta_2$, it follows that $|S(g,\dot{P}) - \int_a^b g| < \frac{\varepsilon}{2}$. Now consider any tagged partition $\dot{P}$ of $[a,b]$ for which $\|\dot{P}\| < \min\{\delta_1,\delta_2\}$. Applying our earlier observation that $S(f+g,\dot{P}) = S(f,\dot{P})+S(g,\dot{P})$, we deduce that

$$\begin{aligned}
\left|S(f+g,\dot{P})-\left(\int_a^b f+\int_a^b g\right)\right| &= \left|S(f,\dot{P})+S(g,\dot{P})-\int_a^b f-\int_a^b g\right| \\
&\le \left|S(f,\dot{P})-\int_a^b f\right|+\left|S(g,\dot{P})-\int_a^b g\right| \\
&< \frac{\varepsilon}{2}+\frac{\varepsilon}{2} \\
&= \varepsilon.
\end{aligned}$$

Hence, by definition, $f+g$ is Riemann integrable on $[a,b]$ and $\int_a^b(f+g) = \int_a^b f + \int_a^b g$. □

Exercise 18.9. According to the sum rule for integrals,

$$\int_0^2 (5+x)dx = \int_0^2 5dx + \int_0^2 xdx.$$

Draw diagrams in which each of the three integrals in this equation is interpreted as measuring the area of a region. Then use the diagrams to give a geometric explanation for why the equation is true.

The next property guarantees the integrability of any constant multiple of an integrable function and provides for the capability of "pulling out" constant factors from the integrand.

Theorem 18.14 (Constant multiple rule for integrals). *If c is a real number and the function f is Riemann integrable on the interval $[a,b]$, then the function cf is also Riemann integrable on $[a,b]$, and $\int_a^b cf = c\int_a^b f$.*

Exercise 18.10. According to the constant multiple rule for integrals,

$$\int_0^2 4x\,dx = 4\int_0^2 x\,dx.$$

Draw diagrams in which each of the two integrals in this equation is interpreted as measuring the area of a region. Then use the diagrams to give a geometric explanation for why the equation is true.

Exercise 18.11. Prove Theorem 18.14.

Exercise 18.12. Use Theorems 18.13 and 18.14 to deduce that if the functions f and g are Riemann integrable on $[a,b]$, then the function $f-g$ is Riemann integrable on $[a,b]$ and $\int_a^b (f-g) = \int_a^b f - \int_a^b g$.

If $c_1, c_2, \ldots, c_n$ are constants and $f_1, f_2, \ldots, f_n$ are functions, the function

$$c_1 f_1 + c_2 f_2 + \cdots + c_n f_n$$

is called a **linear combination** of $f_1, f_2, \ldots, f_n$. In other words, a **linear combination** of functions is any sum of constant multiples of the functions. Together the sum and constant multiple rules imply that any linear combination of Riemann integrable functions on $[a,b]$ is also Riemann integrable on $[a,b]$. For this reason, these two rules are known collectively as the **linearity properties** of the integral.

Exercise 18.13. Suppose the function f is defined on $[a,b]$ and has the property that $f(x) = 0$ except for finitely many points in $[a,b]$. Show that f is Riemann integrable on $[a,b]$ and $\int_a^b f = 0$.

Exercise 18.14. Suppose the functions f and g both have domain $[a,b]$, that g is Riemann integrable on $[a,b]$, and that $f(x) = g(x)$ except for finitely many points in $[a,b]$. Use the result from Exercise 18.13 to show that f is Riemann integrable on $[a,b]$ and $\int_a^b f = \int_a^b g$.

Geometrically, an integrable function whose values, input by input throughout an interval, are at least as large as the corresponding values of another integrable function, should have an integral that is at least as large as the integral of the other function.

Theorem 18.15 (Monotonicity of the integral). *If the functions f and g are Riemann integrable on the interval $[a,b]$, and $f(x) \le g(x)$ for all x in $[a,b]$, then $\int_a^b f \le \int_a^b g$.*

Proof. Suppose f and g are integrable on $[a,b]$, and $f(x) \le g(x)$ for all x in $[a,b]$. Then $(g-f)(x) \ge 0$ for all x in $[a,b]$ and, because of the linearity of the integral, $g-f$ is Riemann integrable on $[a,b]$ and $\int_a^b (g-f) = \int_a^b g - \int_a^b f$. Also, as $(g-f)(x) \geq 0$ for all x in $[a,b]$, it follows, using the result of Exercise 18.15, that $\int_a^b (g-f) \geq 0$. Hence, $\int_a^b f \leq \int_a^b g$. □

Exercise 18.15. Show that if the function f is Riemann integrable on $[a,b]$ and $f(x) \geq 0$ for all x in $[a,b]$, then $\int_a^b f \geq 0$.

Exercise 18.16. Assume the function f is Riemann integrable on $[a,b]$ and let B be a positive number for which $|f(x)| \leq B$ for all x in $[a,b]$.
(a) Prove that if $\dot{P}$ is any tagged partition of $[a,b]$, then $|S(f,\dot{P})| \leq B(b-a)$. (This result shows that the set of all Riemann sums of f determined by all tagged partitions of $[a,b]$ is bounded.)
(b) Prove that $|\int_a^b f| \leq B(b-a)$.

Exercise 18.17. In the next chapter we prove that continuity of a function on a closed bounded interval implies integrability of the function on this interval. Suppose the functions f and g are continuous on $[a,b]$ and $g(x) > 0$ for all x in $[a,b]$. Prove that there is a number c in $[a,b]$ for which $\int_a^b fg = f(c)\int_a^b g$.

We now establish the property that permits us to express a given integral as the sum of integrals over subintervals of the interval of integration.

Theorem 18.16 (Additivity of the integral). *Given a function $f : [a,b] \to \mathbb{R}$ and a number c between a and b, the following are equivalent:*
(i) *the function f is Riemann integrable on the interval $[a,b]$;*
(ii) *the function f is Riemann integrable on both of the intervals $[a,c]$ and $[c,b]$.*

Furthermore, if either of these equivalent conditions holds, it follows that

$$\int_a^b f = \int_a^c f + \int_c^b f.$$

Before giving the proof of this theorem, we want to comment on the strategy we employ to prove integrability of f on $[a,c]$ in the first part of the proof.

In attempting to establish the integrability of a function on an interval, we emphasize that it is not enough to show that a *single* sequence of Riemann sums based on tagged partitions whose norms are tending toward zero has a limit. For instance, Example 18.11 above shows that the sequence of upper sums of the Dirichlet function corresponding to regular partitions of $[0,1]$ into ever more subintervals converges to 1, while the sequence of lower sums corresponding to the same regular partitions converges to 0. Integrability of a function on an interval requires that there is a common limiting value for *every* sequence of Riemann sums based on tagged partitions whose norms are becoming arbitrarily small.

However, to identify a candidate for the value of the integral of a function that we suspect is integrable on an interval, we can examine a particular sequence of Riemann sums for which the norms of the underlying partitions tend toward zero. In employing this strategy, we do not necessarily know that this particular sequence of Riemann sums converges, but if we are able to show the sequence is bounded, we may then apply Theorem 11.5, the Bolzano–Weierstrass theorem, to conclude it has a convergent subsequence. Integrability follows if we can then show that every sequence of Riemann sums based on tagged partitions whose norms tend toward zero converges to the limiting value of the convergent subsequence of the particular sequence of Riemann sums we started with.

Proof of Theorem 18.16. Assume f is integrable on $[a, b]$; we show f must be integrable on $[a, c]$ (the argument that f must be integrable on $[c, b]$ is similar).

Choose a sequence $(\dot{Q}_n)$ of tagged partitions of $[a, c]$ for which $\|\dot{Q}_n\| \to 0$. As f is integrable on $[a, b]$, it follows that f is bounded on $[a, b]$, hence, bounded on the subinterval $[a, c]$. Therefore, the sequence $(S(f, \dot{Q}_n))$ of Riemann sums is necessarily bounded (*do you see why?*). As $(S(f, \dot{Q}_n))$ is bounded, the Bolzano–Weierstrass theorem tells us this sequence of Riemann sums has a convergent subsequence $(S(f, \dot{Q}_{k_n}))$ with limit L.

Let $(\dot{P}_n)$ be any sequence of tagged partitions of $[a, c]$ for which $\|\dot{P}_n\| \to 0$. To show f is integrable on $[a, c]$ with integral L, we must show $S(f, \dot{P}_n) \to L$. So consider any positive number ε.

For each natural number n, let $\dot{P}'_n$ be any tagged partition of $[a, b]$ whose restriction to $[a, c]$ is $\dot{P}_n$ and for which $\|\dot{P}'_n\| = \|\dot{P}_n\|$. Then, for each natural number n, let $\dot{Q}'_n$ be the tagged partition of $[a, b]$ whose restriction to $[a, c]$ is $\dot{Q}_{k_N}$ and whose restriction to $[c, b]$ is the same as the restriction of $\dot{P}'_n$ to $[c, b]$. Because $\dot{P}'_n$ and $\dot{Q}'_n$ determine the same subintervals of $[c, b]$, and for each of these subintervals, they employ the same tag, we may conclude that

$$S(f, \dot{P}_n) - S(f, \dot{Q}_{k_n}) = S(f, \dot{P}'_n) - S(f, \dot{Q}'_n).$$

Since f is integrable on $[a, b]$, and both $\|\dot{P}'_n\| \to 0$ and $\|\dot{Q}'_n\| \to 0$, there are indices N_1 and N_2 such that, whenever $n \geq N_1$, it follows that

$$\left| S(f, \dot{P}'_n) - \int_a^b f \right| < \frac{\varepsilon}{3}$$

and, whenever $n \geq N_2$, it follows that

$$\left| S(f, \dot{Q}'_n) - \int_a^b f \right| < \frac{\varepsilon}{3}.$$

Also, since $S(f, \dot{Q}_{k_n}) \to L$, there is an index N_3 such that whenever $n \geq N_3$, it follows that

$$|S(f,\dot{Q}_{k_n}) - L| < \frac{\varepsilon}{3}.$$

Let $N = \max\{N_1, N_2, N_3\}$ and suppose $n \geq N$. Then

$$\begin{aligned}
|S(f,\dot{P}_n) - L| &= |S(f,\dot{P}_n) - S(f,\dot{Q}_{k_n}) + S(f,\dot{Q}_{k_n}) - L| \\
&= |S(f,\dot{P}'_n) - S(f,\dot{Q}'_n) + S(f,\dot{Q}_{k_n}) - L| \\
&\leq \left|S(f,\dot{P}'_n) - \int_a^b f + \int_a^b f - S(f,\dot{Q}'_n)\right| + |S(f,\dot{Q}_{k_n}) - L| \\
&\leq \left|S(f,\dot{P}'_n) - \int_a^b f\right| + \left|\int_a^b f - S(f,\dot{Q}'_n)\right| + |S(f,\dot{Q}_{k_n}) - L| \\
&< \frac{\varepsilon}{3} + \frac{\varepsilon}{3} + \frac{\varepsilon}{3} = \varepsilon.
\end{aligned}$$

Therefore, $S(f,\dot{P}_n) \to L$, and we have shown that f is integrable on $[a,c]$.

Now assume f is integrable on both $[a,c]$ and $[c,b]$; we show f must be integrable on $[a,b]$ and that $\int_a^b f = \int_a^c f + \int_c^b f$. To do so, consider any sequence $(\dot{P}_n)$ of tagged partitions of $[a,b]$ for which $\|\dot{P}_n\| \to 0$; our argument is complete when we have shown $S(f,\dot{P}_n) \to \int_a^c f + \int_c^b f$. So consider any positive number ε.

For each natural number n, construct a tagged partition $\dot{Q}_n$ of $[a,c]$ and a tagged partition $\dot{R}_n$ of $[c,b]$ as follows. If c is a partition point of $\dot{P}_n$, we let $\dot{Q}_n$ be the restriction of $\dot{P}_n$ to $[a,c]$ and let $\dot{R}_n$ be the restriction of $\dot{P}_n$ to $[c,b]$. If c is not a partition point of

$$\dot{P}_n = \{([p_{i-1}, p_i], t_i) \mid i \in \{1, 2, \ldots, m\}\},$$

then $c \in (p_{k-1}, p_k)$ for some k, and we let

$$\dot{Q}_n = \{([p_{i-1}, p_i], t_i) \mid i < k\} \cup \{([p_{k-1}, c], c)\}$$

and

$$\dot{R}_n = \{([p_{i-1}, p_i], t_i) \mid i > k\} \cup \{([c, p_k], c)\}.$$

Observe that for every n, we have $\|\dot{Q}_n\| \leq \|\dot{P}_n\|$ and $\|\dot{R}_n\| \leq \|\dot{P}_n\|$, so it follows that $\|\dot{Q}_n\| \to 0$ and $\|\dot{R}_n\| \to 0$.

Since f is integrable on $[a,c]$ and $\|\dot{Q}_n\| \to 0$, there is an index N_1 such that whenever $n \geq N_1$, it follows that

$$\left|S(f,\dot{Q}_n) - \int_a^c f\right| < \frac{\varepsilon}{3}.$$

Similarly, since f is integrable on $[c, b]$ and $\|\dot{R}_n\| \to 0$, there is an index N_2 such that whenever $n \geq N_2$, it follows that

$$\left| S(f, \dot{R}_n) - \int_c^b f \right| < \frac{\varepsilon}{3}.$$

Also, since integrability implies boundedness, we know f must be bounded on both $[a, c]$ and $[c, b]$, so there is a positive number B such that $|f(x)| \leq B$ for all x in $[a, b]$. Then, as $\|\dot{P}_n\| \to 0$, there is an index N_3 such that whenever $n \geq N_3$, it follows that

$$\|\dot{P}_n\| < \frac{\varepsilon}{6B}.$$

Observe that if c is a partition point of $\dot{P}_n$, we have

$$S(f, \dot{P}_n) - S(f, \dot{Q}_n) - S(f, \dot{R}_n) = 0 < \frac{\varepsilon}{3},$$

and if c is not a partition point of $\dot{P}_n$, we have

$$\begin{aligned} S(f, \dot{P}_n) - S(f, \dot{Q}_n) - S(f, \dot{R}_n) &= f(t_k)(p_k - p_{k-1}) - f(c)(c - p_{k-1}) - f(c)(p_k - c) \\ &= (f(t_k) - f(c))(p_k - p_{k-1}) \\ &\leq 2B\|\dot{P}_n\| \\ &< 2B \cdot \frac{\varepsilon}{6B} \\ &= \frac{\varepsilon}{3}. \end{aligned}$$

Let $N = \max\{N_1, N_2, N_3\}$ and suppose $n \geq N$. Then

$$\begin{aligned} &\left| S(f, \dot{P}_n) - \left(\int_a^c f + \int_c^b f \right) \right| \\ &\quad \leq |S(f, \dot{P}_n) - S(f, \dot{Q}_n) - S(f, \dot{R}_n)| + \left| S(f, \dot{Q}_n) - \int_a^c f \right| + \left| S(f, \dot{R}_n) - \int_c^b f \right| \\ &\quad < \frac{\varepsilon}{3} + \frac{\varepsilon}{3} + \frac{\varepsilon}{3} \\ &\quad = \varepsilon. \end{aligned}$$

Therefore, $S(f, \dot{P}_n) \to \int_a^c f + \int_c^b f$. □

Example 18.17. We can use the additivity of the integral to evaluate

$$\int_{-3}^{2} f,$$

where $f(x) = |x|$. We have

$$\int_{-3}^{2} f = \int_{-3}^{0} f + \int_{0}^{2} f = \int_{-3}^{0} -g + \int_{0}^{2} g,$$

where $g(x) = x$, according to the definition of absolute value. Thus,

$$\int_{-3}^{2} f = \int_{-3}^{0} -g + \int_{0}^{2} g = -\int_{-3}^{0} g + \int_{0}^{2} g = -\frac{1}{2}[0^2 - (-3)^2] + \frac{1}{2}[2^2 - 0^2] = 6.5,$$

where we have made use of the result from Example 18.8.

The definition of the integral can be extended as follows. Given a function f that is Riemann integrable on the interval $[a, b]$ and a number c in $[a, b]$, we define

$$\int_{c}^{c} f = 0$$

and

$$\int_{b}^{a} f = -\int_{a}^{b} f.$$

With these definitions we can show that the additivity property of the integral, that

$$\int_{a}^{b} f = \int_{a}^{c} f + \int_{c}^{b} f,$$

does not actually require that the number c lie between a and b.

Theorem 18.18 (Generalized additivity). *If the function f is Riemann integrable on the interval $[a, b]$ and u, v, and w are all in $[a, b]$, then $\int_u^w f = \int_u^v f + \int_v^w f$.*

Proof. Theorem 18.16 gives us the desired result provided that $u < v < w$. Each of the other possible cases can be considered on its own.

For instance, if any two of u, v, and w are equal, the fact that an integral for which the endpoints of integration are equal has been defined to be 0 gives us the desired conclusion.

Also, if $u < w < v$, then the additivity property tells us that

$$\int_u^v f = \int_u^w f + \int_w^v f.$$

However, since $w < v$, by definition

$$\int_v^w f = -\int_w^v f,$$

so that $\int_u^v f = \int_u^w f + \int_w^v f$ may be rewritten as

$$\int_u^v f = \int_u^w f - \int_v^w f,$$

which then implies that

$$\int_u^w f = \int_u^v f + \int_v^w f,$$

which is what we wanted.

The other possibilities can be handled similarly. □

Exercise 18.18. Suppose $\int_2^5 3f = 4$, $\int_3^5 f = -2$, and $\int_3^5 g = 10$. Evaluate each of the following.

(a) $\int_2^5 f$

(b) $\int_5^3 f$

(c) $\int_3^5 (f + g)$

(d) $\int_3^5 (f - g)$

(e) $\int_3^3 f$

(f) $\int_3^5 (g + 2)$

(g) $\int_2^3 f$

(h) $\int_3^0 g + \int_0^5 g$

Exercise 18.19. Let the function f be continuous and integrable on the interval $[0, 1]$ with the property that $\int_0^c f = \int_c^1 f$ for every c in $[0, 1]$. Prove that $f(x) = 0$ for all x in $[0, 1]$.

19 Criteria for integrability

Our experience with the integral in calculus courses mostly focused on methods for calculating them, along with situations in which they represented quantities of more practical interest, for example, the mass of an object or a business's total revenue over some specified period of time.

However, as we saw in Example 18.11, not all functions are integrable, so we now expend some effort to demonstrate that certain types of functions, specifically, continuous functions, monotone functions, and step functions, are integrable. Creating an inventory of functions known to be integrable then means they are available for use in applications and that it is reasonable to attempt to evaluate integrals involving them. We also survey several equivalent characterizations of the notion of integrability, culminating with Henri Lebesgue's succinct criterion for identifying whether a function is integrable.

Integrability of continuous functions

The continuous functions form the most important class of integrable functions. The proof that a continuous function defined on a closed bounded interval is integrable on that interval relies on the fact that continuity on a closed bounded interval implies uniform continuity on that interval (Theorem 14.20).

Theorem 19.1 (Continuous functions are integrable). *If the function $f : [a,b] \to \mathbb{R}$ is continuous, then f is Riemann integrable on $[a,b]$.*

Proof. Assume the function $f : [a,b] \to \mathbb{R}$ is continuous. Choose a sequence $(\dot{Q}_n)$ of tagged partitions of $[a,b]$ for which $\|\dot{Q}_n\| \to 0$. As $(S(f,\dot{Q}_n))$ is bounded, Theorem 11.5, the Bolzano–Weierstrass theorem, tells us this sequence has a convergent subsequence $(S(f,\dot{Q}_{k_n}))$ with limit L.

Let $(\dot{P}_n)$ be any sequence of tagged partitions of $[a,b]$ for which $\|\dot{P}_n\| \to 0$. To show f is integrable on $[a,b]$ with integral L, we show $S(f,\dot{P}_n) \to L$. Consider any positive number ε. As $S(f,\dot{Q}_{k_n}) \to L$, there is an index N_1 such that when $n \geq N_1$, it follows that

$$\left|S(f,\dot{Q}_{k_n}) - L\right| < \frac{\varepsilon}{3}.$$

Also, the continuity of f on $[a,b]$ implies, via Theorem 14.20, that f is uniformly continuous on $[a,b]$, so there is a positive number δ such that when x and y are in $[a,b]$ and $|x-y| < \delta$, we may conclude that

$$\left|f(x) - f(y)\right| < \frac{\varepsilon}{3(b-a)}.$$

Additionally, since $\|\dot{P}_n\| \to 0$ and $\|\dot{Q}_{k_n}\| \to 0$, there are indices N_2 and N_3 such that, when $n \geq N_2$, it follows that $\|\dot{P}_n\| < \delta$, and when $n \geq N_3$, it follows that $\|\dot{Q}_{k_n}\| < \delta$.

https://doi.org/10.1515/9783111429564-019

Let $N = \max\{N_1, N_2, N_3\}$ and for each natural number n for which $n \geq N$, let P_n and Q_{k_n} be the untagged partitions underlying, respectively, the tagged partitions $\dot{P}_n$, and $\dot{Q}_{k_n}$. Then let $R_n = P_n \cup Q_{k_n}$, which is also a partition of $[a, b]$, and choose any tags to create the tagged partition $\dot{R}_n$ having subintervals determined from R_n. Suppose

$$\dot{P}_n = \{([p_{i-1}, p_i], t_i) \mid i \in \{1, 2, \ldots, m\}\}$$

and

$$\dot{R}_n = \{([r_{i-1}, r_i], s_i) \mid i \in \{1, 2, \ldots, j\}\}.$$

As $P_n \subseteq R_n$, an interval $[p_{i-1}, p_i]$ determined by P_n is a union of intervals determined by R_n, say

$$[p_{i-1}, p_i] = [r_{u-1}, r_u] \cup [r_u, r_{u+1}] \cup \cdots \cup [r_{v-1}, r_v].$$

Observe that the contribution D_i to $S(f, \dot{P}_n) - S(f, \dot{R}_n)$ determined by these intervals is

$$\begin{aligned} D_i &= f(t_i)(p_i - p_{i-1}) \\ &\quad - (f(s_u)(r_u - r_{u-1}) + f(s_{u+1})(r_{u+1} - r_u) + \cdots + f(s_v)(r - r_{v-1})) \\ &= (f(t_i) - f(s_u))(r_u - r_{u-1}) + (f(t_i) - f(s_{u+1}))(r_{u+1} - r_u) \\ &\quad + \cdots + (f(t_i) - f(s_v))(r_v - r_{v-1}). \end{aligned}$$

Since $n \geq N_2$, we have $\|\dot{R}_n\| \leq \|\dot{P}_n\| < \delta$, so each of the differences $r_u - r_{u-1}, r_{u+1} - r_u, \ldots, r_v - r_{v-1}$ is less than δ, and our earlier deduction based on the uniform continuity of f on $[a, b]$ then tells us that each of $|f(t_i) - f(s_u)|, |f(t_i) - f(s_{u+1})|, \ldots, |f(t_i) - f(s_v)|$ is less than $\frac{\varepsilon}{3(b-a)}$. Hence,

$$\begin{aligned} D_i &< \frac{\varepsilon}{3(b-a)} \cdot (r_u - r_{u-1}) + \frac{\varepsilon}{3(b-a)} \cdot (r_{u+1} - r_u) + \cdots + \frac{\varepsilon}{3(b-a)} \cdot (r_v - r_{v-1}) \\ &= \frac{\varepsilon}{3(b-a)} \cdot (p_i - p_{i-1}). \end{aligned}$$

We may then conclude that

$$\begin{aligned} S(f, \dot{P}_n) - S(f, \dot{R}_n) &= \sum_{i=1}^{m} D_i \\ &< \sum_{i=1}^{m} \frac{\varepsilon}{3(b-a)} \cdot (p_i - p_{i-1}) \\ &= \frac{\varepsilon}{3(b-a)} \sum_{i=1}^{m} (p_i - p_{i-1}) \\ &= \frac{\varepsilon}{3(b-a)} \cdot (b-a) \\ &= \frac{\varepsilon}{3}. \end{aligned}$$

In exactly the same manner, since $\|\dot{R}_n\| \leq \|\dot{Q}_{k_n}\| < \delta$, we are able to determine that

$$S(f, \dot{P}_n) - S(f, \dot{Q}_{k_n}) < \frac{\varepsilon}{3}.$$

Therefore, for any n for which $n \geq N$, we have

$$\begin{aligned}|S(f, \dot{P}_n) - L| &\leq |S(f, \dot{P}_n) - S(f, \dot{R}_n)| + |S(f, \dot{R}_n) - S(f, \dot{Q}_{k_n})| + |S(f, \dot{Q}_{k_n}) - L| \\ &< \frac{\varepsilon}{3} + \frac{\varepsilon}{3} + \frac{\varepsilon}{3} \\ &= \varepsilon,\end{aligned}$$

so we have shown that f is integrable on $[a, b]$. □

Exercise 19.1. Suppose the function $f : [a, b] \to \mathbb{R}$ is continuous, that $f(x) \geq 0$ for all x in $[a, b]$, and that $\int_a^b f = 0$.

(a) Prove that $f(x) = 0$ for all x in $[a, b]$. *Hint*: Try contradiction.

(b) Give an example to show that if the function $f : [a, b] \to \mathbb{R}$ is continuous on $[a, b)$, but is discontinuous at b, and $f(x) \geq 0$ for all x in $[a, b]$, it is still possible for $\int_a^b f = 0$.

Exercise 19.2. Suppose the functions f and g are continuous on $[a, b]$, and that $\int_a^b f = \int_a^b g$. Show that there exists a number c in $[a, b]$ such that $f(c) = g(c)$. *Hint*: Try contradiction and apply the intermediate value theorem to $g - f$ on $[a, b]$. Also, make use of the extreme value theorem.

Exercise 19.3. Show that if the function f is continuous on $[a, b]$, then there exists a number c in $[a, b]$ such that $f(c) = \frac{1}{b-a} \int_a^b f$. This result is known as the *mean value theorem for integrals*. *Hint*: Consider using both the extreme value theorem and the intermediate value theorem.

Once we know that a function f is integrable on $[a, b]$, *every* sequence of Riemann sums for which the norms of the underlying tagged partitions are becoming arbitrarily small must converge to $\int_a^b f$, so we can identify the value of $\int_a^b f$ by determining the limit of *any single* such sequence.

Example 19.2. The function $f : [1, 5] \to \mathbb{R}$ defined by $f(x) = x^3$ is a polynomial function, hence, continuous and therefore integrable on $[1, 5]$. Thus, we may evaluate $\int_1^5 f$ by taking the limit, as $n \to \infty$, of, say, the sequence of right endpoint Riemann sums corresponding to the regular partition of the interval $[1, 5]$ into n subintervals. Calculating gives us

$$\begin{aligned}\int_1^5 f &= \lim_{n\to\infty}\sum_{i=1}^{n} f\left(1+\frac{4i}{n}\right)\cdot\frac{4}{n}\\ &= \lim_{n\to\infty}\sum_{i=1}^{n}\left(1+\frac{4i}{n}\right)^3\cdot\frac{4}{n}\\ &= \lim_{n\to\infty}\sum_{i=1}^{n}\left(1+\frac{12i}{n}+\frac{48i^2}{n^2}+\frac{64i^3}{n^3}\right)\cdot\frac{4}{n}\\ &= \lim_{n\to\infty}\left(4+\frac{48}{n^2}\sum_{i=1}^{n} i+\frac{192}{n^3}\sum_{i=1}^{n} i^2+\frac{256}{n^4}\sum_{i=1}^{n} i^3\right)\\ &= \lim_{n\to\infty}\left(4+\frac{48}{n^2}\cdot\frac{n(n+1)}{2}+\frac{192}{n^3}\cdot\frac{n(n+1)(2n+1)}{6}+\frac{256}{n^4}\cdot\frac{n^2(n+1)^2}{4}\right)\\ &= 4+24+64+64\\ &= 156,\end{aligned}$$

where we have made use of several summation properties and formulas presented in Theorems 3.20 and 3.21.

Exercise 19.4. Define $f : [1,3] \to \mathbb{R}$ so that $f(x) = x^2 + 1$. Verify that $\int_1^3 f = \frac{32}{3}$ by calculating the limit as $n \to \infty$ of the sequence of right endpoint Riemann sums determined by regular partitions of $[1,3]$ into n subintervals.

Exercise 19.5. Define $g : [0,1] \to \mathbb{R}$ so that $g(x) = \sqrt{x}$. In Exercise 18.4, you showed that the right endpoint Riemann sum determined by the irregular partition P_n described there is given by $\frac{1}{n^3}\sum_{i=1}^{n}(2i^2 - i)$. Verify that $\int_0^1 g = \frac{2}{3}$ by calculating the limit as $n \to \infty$ of these right endpoint Riemann sums.

Integrability of monotone functions

Another important class of integrable functions is formed by the monotone functions, those functions that are nondecreasing or nonincreasing. While many monotone functions are continuous, continuity is not necessary for a function to be monotone, so establishing that monotone functions are integrable properly enlarges the class of functions that we can integrate.

Example 19.3. The function $f : [0,8] \to \mathbb{R}$ defined so that

$$f(x) = \begin{cases} x, & \text{if } 0 \le x < 5; \\ x^2, & \text{if } 5 \le x \le 8; \end{cases}$$

is monotone, being increasing on $[0,8]$, but is discontinuous at 5 (Figure 19.1).

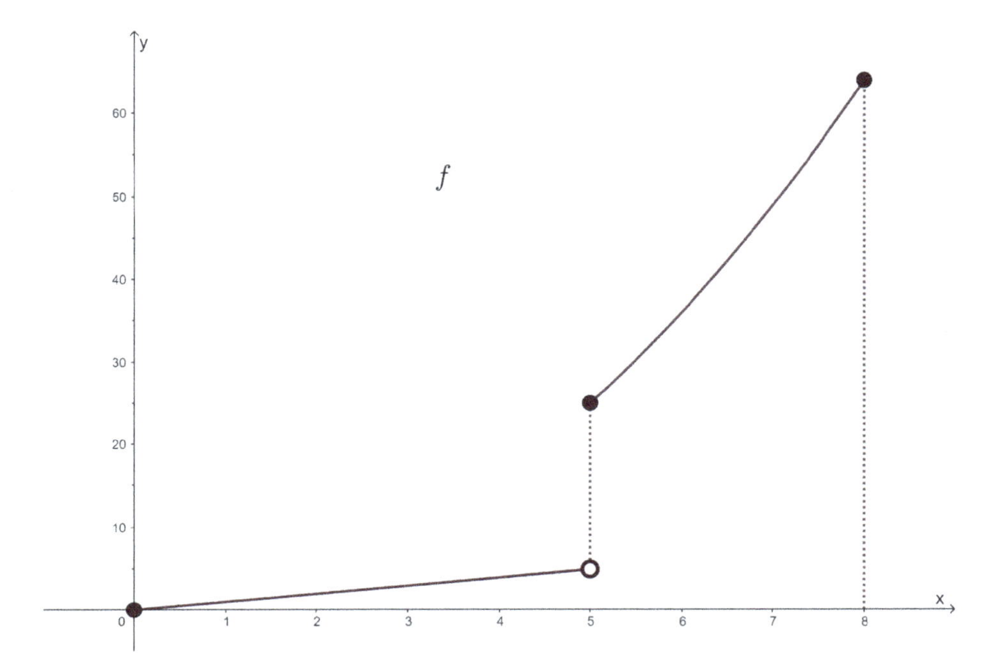

Figure 19.1: Graph of the function from Example 19.3.

The area interpretation of the integral certainly suggests that f is integrable on $[0,8]$, since the region whose area is measured by the integral is bounded. In fact, assuming integrability, the additivity of the integral would permit us to obtain $\int_0^8 f$ by adding together $\int_0^5 f$ and $\int_5^8 f$.

Our proof that monotone functions are integrable applies several ideas that are also more broadly valuable.

Given a function f that is bounded on $[a,b]$ and a partition $P = \{p_0, p_1, \ldots, p_n\}$ of $[a,b]$, we define the **lower Darboux sum** $L(f,P)$ and the **upper Darboux sum** $U(f,P)$ so that

$$L(f,P) = \sum_{i=1}^{n} m_i(p_i - p_{i-1})$$

and

$$U(f,P) = \sum_{i=1}^{n} M_i(p_i - p_{i-1}),$$

where $m_i = \inf\{f(x) \mid p_{i-1} \le x \le p_i\}$ and $M_i = \sup\{f(x) \mid p_{i-1} \le x \le p_i\}$.

Lower and upper Darboux sums need not be Riemann sums, because a bounded function defined on $[a,b]$ may not achieve its infimum or supremum on one or more of the subintervals determined by a partition of $[a, b]$ (see Exercise 14.24(c) for an example of a function that has an infimum on $[-1, 3]$ that is not achieved as a value of the function).

If a function achieves an infimum on each subinterval determined by a partition, the lower Darboux sum is a Riemann sum, what we previously referred to as a lower Riemann sum. Similarly, if a function achieves a supremum on each subinterval determined by a partition, the upper Darboux sum is a Riemann sum, the upper Riemann sum. For instance, as the extreme value theorem tells us that a continuous function achieves both a maximum and a minimum on any closed, bounded interval, the lower and upper Darboux sums of a continuous function on $[a, b]$ are Riemann sums no matter what partition of $[a, b]$ has been specified.

It is also the case that the lower and upper Darboux sums of a monotone function defined on $[a, b]$ are Riemann sums, regardless of what partition of $[a, b]$ is taken. For instance, if the function f is nondecreasing on $[a, b]$, then f is nondecreasing on each subinterval determined by a partition of $[a, b]$, hence, achieves a minimum value at the subinterval's left endpoint and a maximum value at the subinterval's right endpoint.

Exercise 19.6. Suppose the function f is nonincreasing on $[a, b]$ and that P is a partition of $[a, b]$. Show that the lower and upper Darboux sums of f determined by P are actually Riemann sums.

The lower and upper Darboux sums corresponding to a given partition provide bounds on the Riemann sums determined from tagged partitions formed from the given partition.

Theorem 19.4. *Suppose the function $f : [a, b] \to \mathbb{R}$ is bounded and P is a partition of the interval $[a, b]$. Then for any tagged partition $\dot{P}$ of $[a, b]$ having the same subintervals as those determined by P, we have $L(f, P) \le S(f, \dot{P}) \le U(f, P)$.*

Exercise 19.7. Prove Theorem 19.4.

A partition Q of an interval $[a, b]$ is called a **refinement** of a partition P of $[a, b]$ if all of the partition points of P are included among all the partition points of Q. A refinement of a partition is also said to **refine** the partition.

Example 19.5. Each of

$$P = \{3, 4, 4.5, 5, 7\},$$

$$Q = \{3, \pi, 4, 4.5, 5, 6.25, 6.75, 7\},$$

and

$$R = \{3, 5, 5.5, 6, 7\}$$

is a partition of $[3,7]$. The partition Q is a refinement of the partition P since Q includes all the points of P (so Q refines P). But neither P nor R is a refinement of the other since neither of them includes all the points of the other.

Exercise 19.8. Show that if P and Q are both partitions of $[a,b]$, then $P \cup Q$ is a refinement of each of P and Q.

Exercise 19.9. Suppose the function $f : [a,b] \to \mathbb{R}$ is bounded, both P and Q are partitions of $[a,b]$, and Q refines P. Show that $L(f,P) \le L(f,Q) \le U(f,Q) \le U(f,P)$.

We are now ready to prove that monotone functions are integrable.

Theorem 19.6 (Monotone functions are integrable). *If the function $f : [a,b] \to \mathbb{R}$ is monotone, then f is Riemann integrable on $[a,b]$.*

Proof. We prove that if f is nondecreasing on $[a,b]$, then f is integrable on $[a,b]$; the proof when f is nonincreasing is similar. First note that if $f(b) = f(a)$, the monotonicity of f would then imply that f is constant on $[a,b]$, hence, integrable on $[a,b]$ by Theorem 18.9. So we assume $f(a) < f(b)$.

Let (Q_n) be any sequence of partitions of $[a,b]$ such that, for any n, the partition Q_{n+1} refines the partition Q_n, and $\|Q_n\| = \frac{1}{n+1}$ so that $\|Q_n\| \to 0$. Then by Exercise 19.9, the sequence $(L(f,Q_n))$ is nondecreasing and bounded above, hence, by Theorem 10.18, the monotone convergence theorem, converges to some real number L. We use (ii) of Theorem 18.5 to show that f is integrable on $[a,b]$ with integral L.

Consider any positive number ε. Use the convergence of $(L(f,Q_n))$ to L along with the Archimedean property to obtain a natural number N for which

$$|L(f,Q_N) - L| < \frac{\varepsilon}{3}$$

and

$$\frac{1}{N} < \frac{\varepsilon}{3(f(b) - f(a))}.$$

Define $\delta = \frac{\varepsilon}{3(f(b)-f(a))}$, which is a positive number. Let $\dot{P}$ be any tagged partition of $[a,b]$ for which $\|\dot{P}\| < \delta$, and let P be the (untagged) partition underlying $\dot{P}$. From Theorem 19.4 we may conclude that

$$L(f,P) \le S(f,\dot{P}) \le U(f,P),$$

and, since $P \cup Q_N$ is a refinement of P, from Exercise 19.9, we may conclude that

$$L(f,P) \le L(f,P \cup Q_N) \le U(f,P \cup Q_N) \le U(f,P).$$

Also, as a result of Exercise 19.11 below, we have

$$U(f,P) - L(f,P) \le (f(b) - f(a))\|P\|.$$

It then follows that

$$\begin{aligned}
|S(f,\dot{P}) - L(f,P \cup Q_N)| &\le (f(b) - f(a))\|P\| \\
&< (f(b) - f(a))\delta \\
&= (f(b) - f(a)) \cdot \frac{\varepsilon}{3(f(b) - f(a))} \\
&= \frac{\varepsilon}{3}.
\end{aligned}$$

Similarly, since $P \cup Q_N$ is a refinement of Q_N, we have

$$L(f,Q_N) \le L(f,P \cup Q_N) \le U(f,P \cup Q_N) \le U(f,Q_N)$$

and

$$U(f,Q_N) - L(f,Q_N) \le (f(b) - f(a))\|Q_N\|,$$

so that

$$\begin{aligned}
L(f,P \cup Q_N) - L(f,Q_N) &\le (f(b) - f(a))\|Q_N\| \\
&= (f(b) - f(a)) \cdot \frac{1}{N+1} \\
&< (f(b) - f(a)) \cdot \frac{1}{N} \\
&< (f(b) - f(a)) \cdot \frac{\varepsilon}{3(f(b) - f(a))} \\
&= \frac{\varepsilon}{3}.
\end{aligned}$$

Taking everything we have established into account, we now have

$$\begin{aligned}
|S(f,\dot{P}) - L| &\le |S(f,\dot{P}) - L(f,P \cup Q_N)| + |L(f,P \cup Q_N) - L(f,Q_N)| + |L(f,Q_N) - L| \\
&< \frac{\varepsilon}{3} + \frac{\varepsilon}{3} + \frac{\varepsilon}{3} \\
&= \varepsilon,
\end{aligned}$$

which completes the proof. □

Exercise 19.10. Theorem 19.6 tells us that the function f from Example 19.3 is integrable on $[0, 8]$. Find the value of $\int_0^8 f$.

Exercise 19.11. Prove that if the function f is nondecreasing on $[a,b]$ and P is any partition of $[a,b]$, then $U(f,P) - L(f,P) \leq (f(b) - f(a))\|P\|$.

Integrability of step functions

A function $f : [a,b] \to \mathbb{R}$ is called a **step function** if it takes only finitely many values and each of these values is taken on a finite number of subintervals of $[a,b]$.

Example 19.7. The function $f : [-12, 12] \to \mathbb{R}$ defined so that

$$f(x) = \begin{cases} \pi, & \text{if } -12 \leq x \leq -8; \\ e^2, & \text{if } -8 < x < 0; \\ 5, & \text{if } x = 0 \text{ or } 7 < x \leq 12; \\ -1, & \text{if } 0 < x \leq 7; \end{cases}$$

is a step function because it takes only four values and each of these values is taken on a finite number of subintervals of $[-12, 12]$. Specifically, the value π is taken on the interval $[-12, -8]$, the value e^2 is taken on the interval $(-8, 0)$, the value 5 is taken on the

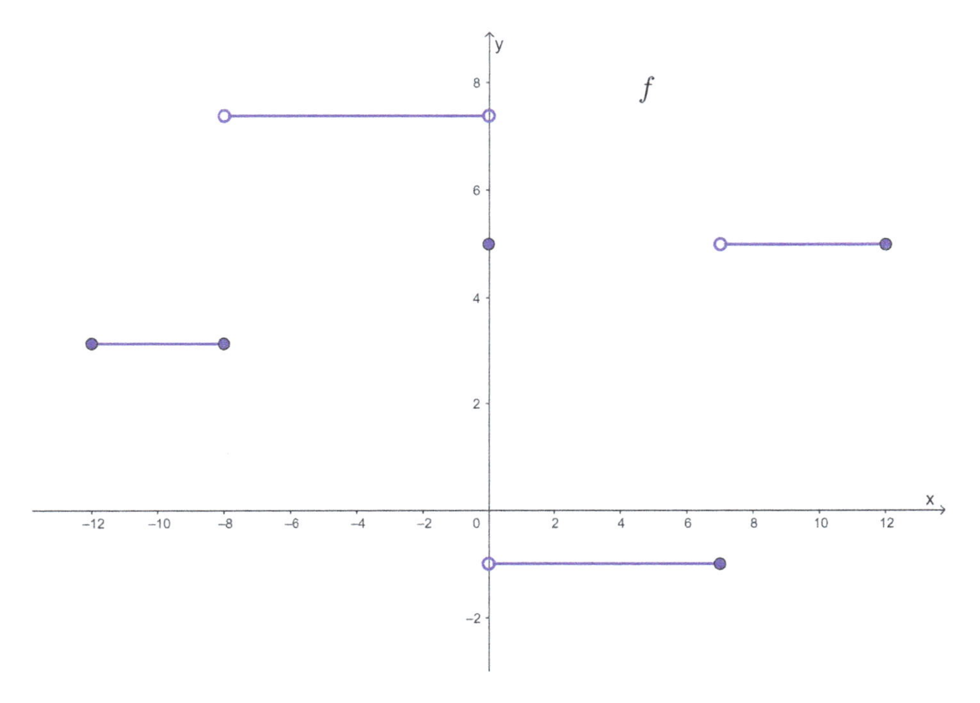

Figure 19.2: Graph of the function from Example 19.7.

two intervals $\{0\}$ and $(7, 12]$, and the value -1 is taken on the interval $(0, 7]$. The graph of f is displayed in Figure 19.2.

As illustrated in Example 19.7, unless it is constant, a step function is not continuous and need not be monotone. The fact that step functions are integrable (proof assigned as Exercise 19.12) therefore further enlarges the class of integrable functions.

Theorem 19.8 (Step functions are integrable). *If f is a step function defined on the interval $[a, b]$, then f is Riemann integrable on $[a, b]$.*

Exercise 19.12. Let S be an arbitrary subset of $[a, b]$ and define the function $\chi_S : [a, b] \to \mathbb{R}$ so that

$$\chi_S(x) = \begin{cases} 1, & \text{if } x \in S; \\ 0, & \text{if } x \notin S. \end{cases}$$

(The symbol χ is the lowercase Greek letter *chi*.)

(a) Suppose u is in $[a, b]$. Show that $\chi_{\{u\}}$ is Riemann integrable on $[a, b]$ and $\int_a^b \chi_{\{u\}} = 0$. *Hint*: Apply the result from Exercise 18.13.

(b) Suppose u and v are in $[a, b]$, and $u < v$. Show that $\chi_{[u,v]}$ is Riemann integrable on $[a, b]$ and $\int_a^b \chi_{[u,v]} = v - u$.

(c) Given that u and v are in $[a, b]$, and $u < v$, write each of $\chi_{[u,v)}$, $\chi_{(u,v]}$, and $\chi_{(u,v)}$ as linear combinations of the types of functions whose integrability has been verified in (a) and (b). Then explain why this implies that each of $\chi_{[u,v)}$, $\chi_{(u,v]}$, and $\chi_{(u,v)}$ is Riemann integrable on $[a, b]$, with

$$\int_a^b \chi_{[u,v)} = \int_a^b \chi_{(u,v]} = \int_a^b \chi_{(u,v)} = v - u.$$

(d) Let $f : [a, b] \to \mathbb{R}$ be an arbitrary step function, which means there is a natural number n, along with subintervals $I_1, I_2, \ldots, I_n$ of $[a, b]$ and real numbers $c_1, c_2, \ldots, c_n$, such that all of the following are true:
- $[a, b] = I_1 \cup I_2 \cup \cdots \cup I_n$;
- for each i in $\{2, 3, \ldots, n\}$, all the elements of I_{i-1} are less than all the elements of I_i;
- for each i in $\{1, 2, \ldots, n\}$, we have $f(x) = c_i$ for all x in I_i.

Write f as a linear combination of functions of the types dealt with in (a), (b), and (c) above. Then explain why this implies that f is Riemann integrable on $[a, b]$, with

$$\int_a^b f = \sum_{i=1}^n c_i(v_i - u_i),$$

where u_i and v_i are, respectively, the left and right endpoints of the interval I_i.

Exercise 19.13. Consider the sign function sgn defined in Example 14.10.

(a) How do we know sgn is integrable on $[-3, \pi]$?

(b) Evaluate $\int_{-3}^{\pi} \text{sgn}$.

Exercise 19.14. Let f be a function defined on $[a,b]$ and consider any tagged partition $\dot{P}$ of $[a,b]$. Show that $S(f,\dot{P}) = \int_a^b g$ for some step function g defined on $[a,b]$.

The Cauchy criterion for integrability

There are a number of conditions equivalent to Riemann integrability that offer new perspectives and insight on the concept. These conditions also present us with alternative means for establishing the integrability of a particular function or class of functions. We explore several of these conditions.

The so-called *Cauchy criterion* tells us that integrability is equivalent to the ability to make all Riemann sums arbitrarily close to one another once the norms of the partitions on which they are based are suitably small.

Theorem 19.9 (The Cauchy criterion for integrability). *Given a function $f : [a,b] \to \mathbb{R}$, the following are equivalent:*

(i) *f is Riemann integrable on $[a,b]$;*

(ii) *for every positive number ε, there exists a positive number δ such that whenever $\dot{P}$ and $\dot{Q}$ are tagged partitions of $[a,b]$ for which $\|\dot{P}\| < \delta$ and $\|\dot{Q}\| < \delta$, it follows that $|S(f,\dot{P}) - S(f,\dot{Q})| < \varepsilon$.*

Proof. (i) $\Rightarrow$ (ii) Assigned as Exercise 19.15.

(ii) $\Rightarrow$ (i) Assume (ii) holds. Then we may recursively construct a sequence (δ_n) of positive numbers such that $\delta_1 \geq \delta_2 \geq \delta_3 \geq \dots$ and whenever $\dot{P}$ and $\dot{Q}$ are tagged partitions of $[a,b]$ for which $\|\dot{P}\| < \delta_n$ and $\|\dot{Q}\| < \delta_n$, it follows that $|S(f,\dot{P}) - S(f,\dot{Q})| < \frac{1}{n}$.

Now for each natural number n, let $\dot{P}_n$ be a tagged partition of $[a,b]$ for which $\|\dot{P}_n\| < \delta_n$. We show that $(S(f,\dot{P}_n))$ is a Cauchy sequence. Given a positive number ε, use the Archimedean property to obtain a natural number N such that $\frac{1}{N} < \varepsilon$. Then if $m \geq N$ and $n \geq N$, we have $\|\dot{P}_m\| < \delta_m \leq \delta_N$ and $\|\dot{P}_n\| < \delta_n \leq \delta_N$, and so we may conclude that $|S(f,\dot{P}_n) - S(f,\dot{P}_m)| < \frac{1}{N} < \varepsilon$. Thus, $(S(f,\dot{P}_n))$ is a Cauchy sequence and, therefore, using Theorem 11.16, it follows that $S(f,\dot{P}_n) \to L$ for some real number L. To complete our proof, we need only show that $L = \int_a^b f$.

Consider any positive number ε. Since $S(f,\dot{P}_n) \to L$, there is an index N_1 such that whenever $n \geq N_1$, it follows that $|S(f,\dot{P}_n) - L| < \frac{\varepsilon}{2}$. Now choose a natural number N_2 so that $N_2 \geq N_1$ and $\frac{1}{N_2} < \frac{\varepsilon}{2}$. Let $\dot{Q}$ be any tagged partition of $[a,b]$ for which $\|\dot{Q}\| < \delta_{N_2}$. Then

$$\begin{aligned} |S(f,\dot{Q}) - L| &= |S(f,\dot{Q}) - S(f,\dot{P}_{N_2}) + S(f,\dot{P}_{N_2}) - L| \\ &\leq |S(f,\dot{Q}) - S(f,\dot{P}_{N_2})| + |S(f,\dot{P}_{N_2}) - L| \\ &< \frac{1}{N_2} + \frac{\varepsilon}{2} \end{aligned}$$

$$< \frac{\varepsilon}{2} + \frac{\varepsilon}{2}$$
$$= \varepsilon.$$

Hence, by definition, L is the Riemann integral of f on $[a, b]$. □

Exercise 19.15. Prove (i) ⇒ (ii) of Theorem 19.9.

Below we use the Cauchy criterion to establish Riemann's criterion for integrability. First, though, by means of logical negation and the Archimedean property, we reformulate the Cauchy criterion to make it more directly applicable to showing a function is not integrable on an interval.

Theorem 19.10. *Given a function $f : [a, b] \to \mathbb{R}$, the following are equivalent:*

(i) *f is not Riemann integrable on $[a, b]$;*

(ii) *there exists a positive number ε such that for every natural number n, there exist tagged partitions $\dot{P}_n$ and $\dot{Q}_n$ of $[a, b]$ for which $\|\dot{P}_n\| < \frac{1}{n}$ and $\|\dot{Q}_n\| < \frac{1}{n}$, but $|S(f, \dot{P}_n) - S(f, \dot{Q}_n)| \geq \varepsilon$.*

Exercise 19.16. Use Theorem 19.10 to show the Dirichlet function from Example 18.11 is not integrable on $[0, 1]$.

Exercise 19.17. Define $f : [0, 1] \to \mathbb{R}$ so that

$$f(x) = \begin{cases} x + 1, & \text{if } x \text{ rational;} \\ -(x + 1), & \text{if } x \text{ irrational.} \end{cases}$$

Use Theorem 19.10 to show that f is not integrable on $[0, 1]$.

Riemann's criterion for integrability

In 1875, Darboux developed an alternate definition of the integral that turns out to be equivalent to Riemann's definition. Darboux's definition is captured in what has come to be known as *Riemann's criterion*, which states that integrability is equivalent to the ability to find partitions of the interval of integration for which the upper and lower Darboux sums become arbitrarily close to each other. This criterion is intuitively credible because, according to Theorem 19.4, all Riemann sums corresponding to tagged partitions built off the same untagged partition lie between the lower and upper Darboux sums determined by the untagged partition.

Since an integrable function must be bounded, there is no harm in assuming boundedness of the function whenever we are laying out a criterion for integrability.

Theorem 19.11 (Riemann's criterion for integrability). *Given a bounded function* $f : [a, b] \to \mathbb{R}$*, the following are equivalent:*
(i) *f is Riemann integrable on $[a, b]$;*
(ii) *for every positive number ε, there exists a partition P_ε such that $U(f, P_\varepsilon) - L(f, P_\varepsilon) < \varepsilon$.*

Proof. (i) $\Rightarrow$ (ii) Suppose f is integrable on $[a, b]$ and consider any positive number ε. Then, by Cauchy's criterion for integrability, there is a positive number δ such that whenever $\dot{P}$ and $\dot{Q}$ are tagged partitions of $[a, b]$ for which $\|\dot{P}\| < \delta$ and $\|\dot{Q}\| < \delta$, it follows that

$$|S(f, \dot{P}) - S(f, \dot{Q})| < \frac{\varepsilon}{2}.$$

Let $P_\varepsilon = \{p_0, p_1, \dots, p_n\}$ be any partition for which $\|P_\varepsilon\| < \delta$. Then it follows that for any tagged partitions $\dot{P}$ and $\dot{Q}$ having the same subintervals as those determined by P_ε, we have

$$|S(f, \dot{P}) - S(f, \dot{Q})| < \frac{\varepsilon}{2}.$$

For each i in $\{1, 2, \dots, n\}$, let

$$m_i = \inf\{f(x) \mid p_{i-1} \le x \le p_i\}$$

and

$$M_i = \sup\{f(x) \mid p_{i-1} \le x \le p_i\}.$$

Since $m_i + \frac{\varepsilon}{4(b-a)}$ is not a lower bound for f on $[p_{i-1}, p_i]$, there exists a number s_i in $[p_{i-1}, p_i]$ such that

$$f(s_i) < m_i + \frac{\varepsilon}{4(b-a)},$$

and since $M_i - \frac{\varepsilon}{4(b-a)}$ is not an upper bound for f on $[p_{i-1}, p_i]$, there exists a number t_i in $[p_{i-1}, p_i]$ such that

$$f(t_i) > M_i - \frac{\varepsilon}{4(b-a)}.$$

The tagged partitions

$$\dot{P}_1 = \{([p_{i-1}, p_i], s_i) \mid i \in \{1, 2, \dots, n\}\}$$

and

$$\dot{P}_2 = \{([p_{i-1}, p_i], t_i) \mid i \in \{1, 2, \dots, n\}\}$$

of $[a,b]$ each have the same subintervals as those determined by P_ε, so it follows that

$$|S(f,\dot{P}_2) - S(f,\dot{P}_1)| < \frac{\varepsilon}{2}.$$

We may now observe that

$$\begin{aligned}
U(f,P) - L(f,P) &\le U(f,P) - S(f,\dot{P}_2) + |S(f,\dot{P}_2) - S(f,\dot{P}_1)| + S(f,\dot{P}_1) - L(f,P) \\
&= \sum_{i=1}^{n}(M_i - f(t_i))(p_i - p_{i-1}) + |S(f,\dot{P}_2) - S(f,\dot{P}_1)| \\
&\quad + \sum_{i=1}^{n}(f(s_i) - m_i)(p_i - p_{i-1}) \\
&< \frac{\varepsilon}{4(b-a)} \cdot (b-a) + \frac{\varepsilon}{2} + \frac{\varepsilon}{4(b-a)} \cdot (b-a) \\
&= \varepsilon.
\end{aligned}$$

(ii) $\Rightarrow$ (i) Assume (ii) holds and consider any positive number ε. Then there exists a partition $P_\varepsilon = \{p_0, p_1, \dots, p_n\}$ such that

$$U(f,P_\varepsilon) - L(f,P_\varepsilon) < \frac{\varepsilon}{3}.$$

For each i in $\{1,2,\dots,n\}$, let

$$m_i = \inf\{f(x) \mid p_{i-1} \le x \le p_i\}$$

and

$$M_i = \sup\{f(x) \mid p_{i-1} \le x \le p_i\}.$$

Define functions l and u on $[a,b]$ so that $l(x) = m_i$ and $u(x) = M_i$ when x is in (p_{i-1}, p_i) and also so that $l(x) = u(x) = f(x)$ when x is in P_ε. Observe that

$$l(x) \le f(x) \le u(x)$$

for all x. As l and u are step functions, they are integrable, and it follows using Exercise 19.12(d) that

$$\int_a^b l = L(f,P_\varepsilon)$$

and

$$\int_a^b u = U(f,P_\varepsilon).$$

The integrability of l and u also means that there are positive numbers δ_l and δ_u such that for any tagged partition $\dot{Q}$ of $[a,b]$, when $\|\dot{Q}\| < \delta_l$, it follows that

$$\left|S(l,\dot{Q}) - L(f,P_\varepsilon)\right| < \frac{\varepsilon}{3}$$

and when $\|\dot{Q}\| < \delta_u$, it follows that

$$\left|S(u,\dot{Q}) - U(f,P_\varepsilon)\right| < \frac{\varepsilon}{3}.$$

Now let $\delta = \min\{\delta_l, \delta_u\}$ and consider any tagged partitions $\dot{Q}_1$ and $\dot{Q}_2$ of $[a,b]$ for which $\|\dot{Q}_1\| < \delta$ and $\|\dot{Q}_2\| < \delta$. First focusing on $\dot{Q}_1$, we use our previous deduction to conclude that

$$\left|S(l,\dot{Q}_1) - L(f,P_\varepsilon)\right| < \frac{\varepsilon}{3}$$

and

$$\left|S(u,\dot{Q}_1) - U(f,P_\varepsilon)\right| < \frac{\varepsilon}{3},$$

which then implies that

$$L(f,P_\varepsilon) - \frac{\varepsilon}{3} < S(l,\dot{Q}_1)$$

and

$$S(u,\dot{Q}_1) < U(f,P_\varepsilon) + \frac{\varepsilon}{3}.$$

Also, as $l(x) \le f(x) \le u(x)$ for all x, we have

$$S(l,\dot{Q}_1) \le S(f,\dot{Q}_1) \le S(u,\dot{Q}_1)$$

so that

$$L(f,P_\varepsilon) - \frac{\varepsilon}{3} < S(f,\dot{Q}_1) < U(f,P_\varepsilon) + \frac{\varepsilon}{3}.$$

Because $\dot{Q}_2$ satisfies the same hypotheses as $\dot{Q}_1$, we may similarly conclude that

$$L(f,P_\varepsilon) - \frac{\varepsilon}{3} < S(f,\dot{Q}_2) < U(f,P_\varepsilon) + \frac{\varepsilon}{3}.$$

Hence,

$$\left|S(f,\dot{Q}_1) - S(f,\dot{Q}_2)\right| < U(f,P_\varepsilon) - L(f,P_\varepsilon) + \frac{2\varepsilon}{3} < \frac{\varepsilon}{3} + \frac{2\varepsilon}{3} = \varepsilon,$$

and so by the Cauchy criterion, f is integrable on $[a,b]$. □

The **oscillation** of a bounded function f defined on a nontrivial closed bounded interval $[a,b]$, denoted $\operatorname{osc}(f,[a,b])$, is defined so that

$$\operatorname{osc}(f,[a,b]) = \sup\{f(x) \mid a \le x \le b\} - \inf\{f(x) \mid a \le x \le b\}$$

and gives a measure of the total amount by which the values of f vary from each other on $[a,b]$.

Exercise 19.18. Define the function $f : [-1,3] \to \mathbb{R}$ so that

$$f(x) = \begin{cases} x^2, & \text{if } x \ne 0; \\ 1, & \text{if } x = 0. \end{cases}$$

Find the oscillation of f on $[-1,3]$.

Exercise 19.19. Suppose S is a nonempty bounded subset of $\mathbb{R}$. Show that

$$\sup\big\{|x-y| \mid x,y \in S\big\} = \sup(S) - \inf(S).$$

Thus, $\operatorname{osc}(f,[a,b]) = \sup\{|f(x) - f(y)| \mid x,y \in [a,b]\}$.

The following theorem relates the oscillations of a bounded function on the subintervals determined by a partition of $[a,b]$ to the difference between the upper and lower Darboux sums of the function determined by the partition.

Theorem 19.12. *Let f be a bounded function defined on the interval $[a,b]$ and let $P = \{p_0, p_1, \ldots, p_n\}$ be any partition of $[a,b]$. Then*

$$\sum_{i=1}^{n} \operatorname{osc}(f,[p_{i-1},p_i])(p_i - p_{i-1}) = U(f,P) - L(f,P).$$

Exercise 19.20. Prove Theorem 19.12.

Riemann's criterion tells us that integrability of a function on an interval requires the existence of a sequence of partitions of the interval for which the differences between the upper and lower Darboux sums generated from these partitions tends toward zero. This observation suggests that an integrable function should always allow for a partition of the interval of integration which restricts relatively large oscillations in the function values to subintervals whose total length is relatively small (so that, geometrically, extremes in height are compensated for by limitations in length).

Theorem 19.13. *Given a bounded function $f : [a,b] \to \mathbb{R}$, the following are equivalent:*

(i) *f is Riemann integrable on $[a, b]$;*
(ii) *for any positive numbers M and T, there exists a partition P of $[a, b]$ such that when the lengths of the subintervals determined by P on which the oscillation of f is larger than M are added, the resulting sum is less than T.*

Exercise 19.21. Use Theorems 19.11 and 19.12 to prove Theorem 19.13. *Hint*: For (i) $\Rightarrow$ (ii), use Theorem 19.11 to obtain a partition P of $[a, b]$ for which $U(f, P) - L(f, P) < MT$.

Lebesgue's criterion for integrability

The final integrability criterion we shall examine is *Lebesgue's criterion*, which provides the definitive answer to the question of which functions are Riemann integrable. Before stating and proving this result, we need to introduce what it means for a set to have *measure zero*.

A subset S of $\mathbb{R}$ is said to have **measure zero** if for every positive number ε, there is a countable collection of bounded open intervals (a_i, b_i) for which

$$S \subseteq \bigcup_i (a_i, b_i)$$

and

$$\sum_i (b_i - a_i) < \varepsilon.$$

Thus, a set having measure zero is small in the sense that we are always able to find a sequence of open intervals whose union contains the set and for which the sum of the lengths of the intervals is as small a positive number as desired. Because *countable* means *finite* or *countably infinite*, both the union and summation here may, depending on the particular situation, be taken over a finite or countably infinite set of indices.

Example 19.14. Every finite subset of $\mathbb{R}$ has measure zero. The empty set has measure zero as, for any choice of a positive number ε, it is a subset of the single open interval $(-\frac{\varepsilon}{3}, \frac{\varepsilon}{3})$ having length $\frac{2\varepsilon}{3}$. Now let $S = \{s_1, s_2, \ldots, s_n\}$ be a finite subset of $\mathbb{R}$ having n elements. Consider any positive number ε and note that

$$S \subseteq \bigcup_{i=1}^{n} \left(s_i - \frac{\varepsilon}{3n}, s_i + \frac{\varepsilon}{3n} \right)$$

and

$$\sum_{i=1}^{n} \left(s_i + \frac{\varepsilon}{3n} - \left(s_i - \frac{\varepsilon}{3n} \right) \right) = \frac{2\varepsilon}{3} < \varepsilon.$$

So, by definition, S has measure zero.

Every countably infinite subset $S = \{s_1, s_2, s_3 \dots\}$ of $\mathbb{R}$ also has measure zero because, given a positive number ε, we may observe that

$$S \subseteq \bigcup_{i=1}^{\infty}\left(s_i - \frac{\varepsilon}{2^{i+2}}, s_i + \frac{\varepsilon}{2^{i+2}}\right)$$

and

$$\sum_{i=1}^{\infty}\left(s_i + \frac{\varepsilon}{2^{i+2}} - \left(s_i - \frac{\varepsilon}{2^{i+2}}\right)\right) = \sum_{i=1}^{\infty}\frac{\varepsilon}{2^{i+1}} = \frac{\varepsilon}{2}\sum_{i=1}^{\infty}\frac{1}{2^i} = \frac{\varepsilon}{2}\cdot 1 < \varepsilon,$$

where here we have used the results from Theorem 10.25 and Example 10.22.

Even some uncountable subsets of $\mathbb{R}$ have measure zero (see Exercise 19.23).

Exercise 19.22. Show that
(a) any subset of a set having measure zero also has measure zero;
(b) any countable union of sets having measure zero also has measure zero.

In his doctoral dissertation of 1902, Henri Lebesgue (1875–1941) put forth the following succinct characterization of Riemann integrability, which effectively says that a function is integrable as long as it does not have too many discontinuities [3]. Lebesgue also developed what is now referred to as the *Lebesgue integral*, a generalization of the Riemann integral that interacts more favorably with other sorts of limiting processes.

Theorem 19.15 (Lebesgue's criterion for integrability). *Given a bounded function $f : [a, b] \to \mathbb{R}$, the following are equivalent:*
(i) *f is Riemann integrable on $[a, b]$;*
(ii) *the set of points in $[a, b]$ where f is discontinuous has measure zero.*

Proof. (i) $\Rightarrow$ (ii) Suppose f is integrable on $[a, b]$. Consider any positive number ε. For each natural number n, use Theorem 19.13 to obtain a partition P_n of $[a, b]$ for which when the lengths of the subintervals determined by P_n on which the oscillation of f is greater than $\frac{1}{n}$ are added, the resulting sum is less than $\frac{\varepsilon}{2^{n+1}}$. For each n in $\mathbb{N}$, let the subintervals determined by P_n on which the oscillation of f is greater than $\frac{1}{n}$ be $[y_1(n), z_1(n)], [y_2(n), z_2(n)], \dots, [y_{k_n}(n), z_{k_n}(n)]$. Then, for each n in $\mathbb{N}$ and each i in $\{1, 2, \dots, k_n\}$, let

$$G_i(n) = \left(y_i(n) - \frac{\varepsilon}{k_n \cdot 2^{n+2}}, z_i(n) + \frac{\varepsilon}{k_n \cdot 2^{n+2}}\right).$$

Suppose f is discontinuous at p. Then there exists a natural number n and an interval I determined by P_n such that

$$\operatorname{osc}(f, I) > \frac{1}{n}.$$

Hence, the set of all points at which f is discontinuous is a subset of

$$\bigcup_{n=1}^{\infty}\left[\bigcup_{i=1}^{k_n} G_i(n)\right],$$

which is a countable union of open bounded intervals. Moreover,

$$\begin{aligned}
\sum_{n=1}^{\infty}&\left[\sum_{i=1}^{k_n}\left(z_i(n) + \frac{\varepsilon}{k_n \cdot 2^{n+2}} - \left(y_i(n) - \frac{\varepsilon}{k_n \cdot 2^{n+2}}\right)\right)\right] \\
&= \sum_{n=1}^{\infty}\left[\sum_{i=1}^{k_n}(z_i(n) - y_i(n)) + \sum_{i=1}^{k_n} \frac{\varepsilon}{k_n \cdot 2^{n+1}}\right] \\
&< \sum_{n=1}^{\infty}\left(\frac{\varepsilon}{2^{n+1}} + \frac{\varepsilon}{2^{n+1}}\right) \\
&= \sum_{n=1}^{\infty} \frac{\varepsilon}{2^n} \\
&= \varepsilon \sum_{n=1}^{\infty} \frac{1}{2^n} \\
&= \varepsilon \cdot 1 \\
&= \varepsilon,
\end{aligned}$$

so that the sum of the lengths of the countably many open intervals $G_i(n)$ is less than ε (note the use of the results from Theorem 10.25 and Example 10.22). Therefore, the set of points where f is discontinuous has measure zero.

(ii) $\Rightarrow$ (i) Assume that the set D of points at which f is discontinuous has measure zero and consider any positive number ε. Then there is a collection $\{G_k \mid k \in J\}$ of open bounded intervals G_k indexed by a finite or countably infinite indexing set J for which

$$D \subseteq \bigcup_{k \in J} G_k$$

and

$$\sum_{k \in J} l_k < \frac{\varepsilon}{4B},$$

where l_k is the length of G_k and B is a positive number that bounds f on $[a, b]$.

Consider any number q in $[a, b]$. If $q \notin D$, the continuity of f at q guarantees that there is a positive number δ_q such that

$$|f(x) - f(q)| < \frac{\varepsilon}{4(b-a)},$$

when $x \in F_q$, where

$$F_q = [q - \delta_q, q + \delta_q] \cap [a, b].$$

It then follows that

$$\operatorname{osc}(f, F_q) \leq \frac{\varepsilon}{2(b-a)}.$$

On the other hand, if $q \in D$, then $q \in G_k$ for some k, and as G_k is open, there is a positive number δ_q such that $F_q \subseteq G_k$, where

$$F_q = [q - \delta_q, q + \delta_q] \cap [a, b],$$

and, necessarily,

$$\operatorname{osc}(f, F_q) \leq 2B.$$

Let $\dot{P} = \{([p_{i-1}, p_i], t_i) \mid i \in \{1, 2, \ldots, n\}\}$ be any tagged partition of $[a, b]$ for which $[p_{i-1}, p_i] \subseteq F_{t_i}$ for each $i \in \{1, 2, \ldots, n\}$ and let P be the partition of $[a, b]$ having the same subintervals as $\dot{P}$. Note that as $[p_{i-1}, p_i] \subseteq F_{t_i}$, it follows that

$$\operatorname{osc}(f, [p_{i-1}, p_i]) \leq \operatorname{osc}(f, F_{t_i}).$$

Let I_D be the subset of $I = \{1, 2, \ldots, n\}$ consisting of those indices i for which f is discontinuous at the tag t_i of $[p_{i-1}, p_i]$. From our earlier conclusions, it follows that

$$\operatorname{osc}(f, [p_{i-1}, p_i]) \leq 2B$$

when $i \in I_D$, and

$$\operatorname{osc}(f, [p_{i-1}, p_i]) \leq \frac{\varepsilon}{2(b-a)}$$

when $i \in I - I_D$. Also, since for each i in I_D, the subinterval $[p_{i-1}, p_i]$ is a subset of F_{t_i}, which in turn is a subset of some G_k, it follows that

$$\sum_{i \in I_D} (p_i - p_{i-1}) \leq \sum_{k \in J} l_k.$$

As the partition P has been generated from the positive number ε, if we are able to show $U(f, P) - L(f, P) < \varepsilon$, it then follows via Riemann's criterion, Theorem 19.11, that f is integrable on $[a, b]$. Observe that

$$
\begin{aligned}
U(f,P) - L(f,P)& \\
&= \sum_{i=1}^{n} \operatorname{osc}(f,[p_{i-1},p_i])(p_i - p_{i-1}) \\
&= \sum_{i\in I_D} \operatorname{osc}(f,[p_{i-1},p_i])(p_i - p_{i-1}) + \sum_{i\in I-I_D} \operatorname{osc}(f,[p_{i-1},p_i])(p_i - p_{i-1}) \\
&\le \sum_{i\in I_D} \operatorname{osc}(f,F_{t_i})(p_i - p_{i-1}) + \sum_{i\in I-I_D} \operatorname{osc}(f,F_{t_i})(p_i - p_{i-1}) \\
&\le 2B \sum_{i\in I_D} (p_i - p_{i-1}) + \sum_{i\in I-I_D} \operatorname{osc}(f,F_{t_i})(p_i - p_{i-1}) \\
&\le 2B \sum_{k\in J} l_k + \sum_{i\in I-I_D} \operatorname{osc}(f,F_{t_i})(p_i - p_{i-1}) \\
&< 2B\cdot \frac{\varepsilon}{4B} + \frac{\varepsilon}{2(b-a)}\cdot (b-a) \\
&= \varepsilon.
\end{aligned}
$$

□

We already proved that continuous functions and step functions are Riemann integrable, but Lebesgue's theorem makes these results seem trivial, as the set of discontinuities of a continuous function is empty and the set of discontinuities of a step function is finite, and all finite subsets of $\mathbb{R}$ have measure zero.

Exercise 19.23. Recall the Cantor set from Example 8.4.

(a) Show that the Cantor set is uncountable. *Hint*: The elements of the Cantor set are precisely the numbers in $[0,1]$ having a ternary representation $(t_1, t_2, t_3, \dots)$ for which each t_n is in $\{0,2\}$ (see Exercise 8.5). Use contradiction to argue that there are uncountably many of these.

(b) Show that the Cantor set has measure zero. *Hint*: Note that in the construction presented in Example 8.4, the Cantor set is a subset of each F_n, and F_n is the union of 2^{n-1} closed bounded intervals each having length $\frac{1}{3^{n-1}}$.

We can use Lebesgue's theorem to obtain further integrability results. For instance, when an integrable function is acted upon by a continuous function, the resulting composite function is integrable.

Theorem 19.16 (The composition theorem). *If the function f is Riemann integrable on an interval $[a,b]$ and the function g is continuous on an interval $[c,d]$ that contains the range of f, then $g\circ f$ is Riemann integrable on $[a,b]$.*

Exercise 19.24. Use Theorem 19.15 to prove Theorem 19.16.

Exercise 19.25. Give an example to show that in Theorem 19.16 the hypothesis that g is continuous is necessary.

Exercise 19.26. Suppose the function f is integrable on $[a, b]$. Explain why $|f|$ is integrable on $[a, b]$ and show that $|\int_a^b f| \leq \int_a^b |f|$.

In Theorem 18.13, we determined that the sum of functions integrable on the same interval is integrable on that interval. Lebesgue's theorem can help us show that the product of integrable functions is integrable, though we must take some care as the integral of a product of functions is not usually equal to the product of the integrals of the functions.

Theorem 19.17. *If the functions f and g are Riemann integrable on the interval $[a, b]$, then so is their product fg.*

Exercise 19.27. Prove Theorem 19.17. *Hint*: Since the squaring function is continuous, the composition theorem tells us that the square of an integrable function is integrable. Observe that $(f+g)^2 = f^2 + 2fg + g^2$.

20 The fundamental theorem of calculus

Both Newton and Leibniz viewed differentiation and integration as inverses of each other, primarily because the functions they integrated had antiderivatives, which then permitted the application of the fundamental theorem of calculus in the evaluation of the integrals. In this chapter, we prove this theorem, formally define the natural logarithmic and natural exponential functions, verify two classic methods of integration, and show how unboundedness of either the interval of integration or the integrand can sometimes be accommodated.

Antiderivatives

An **antiderivative** of a function f defined on an interval $[a, b]$ is a function $F : [a, b] \to \mathbb{R}$ for which $F'(x) = f(x)$ for all x in $[a, b]$. In other words, an antiderivative of a given function is a function that differentiates into the given function.

Example 20.1. The sine function is an antiderivative of the cosine function on the interval $[0, 2\pi]$, as for every x in $[0, 2\pi]$, it is true that

$$\sin'(x) = \cos(x).$$

Actually, the sine function is an antiderivative of the cosine function on any closed bounded interval of real numbers.

The function F defined so that $F(x) = \sin(x) + 4$ is also an antiderivative of the cosine function on $[0, 2\pi]$, since

$$F'(x) = \cos(x) + 0 = \cos(x)$$

for all x in $[0, 2\pi]$.

This example reveals that a given function can have more than one antiderivative on an interval. In fact, when an antiderivative exists, any function that differs from this specific antiderivative by a constant is also an antiderivative (and there are no other antiderivatives).

Theorem 20.2. *Let the function F be an antiderivative of the function f on the interval $[a, b]$. Given a function G that is defined on $[a, b]$, the following are equivalent:*
(i) *G is an antiderivative of f on $[a, b]$;*
(ii) *$G(x) = F(x) + C$ for some constant C.*

This result has already been established as part of Theorem 16.15, the constant difference theorem.

https://doi.org/10.1515/9783111429564-020

Exercise 20.1. Define the functions f and F so that $f(x) = 1$ and $F(x) = |x|$. Explain why
(a) F is an antiderivative of f on $[1, 8]$;
(b) F is an antiderivative of $-f$ on $[-8, -1]$;
(c) F is not an antiderivative of f on $[-1, 1]$.

Exercise 20.2. In Chapter 15, we established the power rule for derivatives $\frac{d}{dx}[x^r] = rx^{r-1}$ in the case where the exponent r is rational. Use this fact to find an antiderivative of the function defined by the equation $y = x^r$ on $[0, b]$ if b is a positive real number and r is a rational number different from -1.

The fundamental theorem of calculus: evaluating integrals

The fundamental theorem of calculus provides us with a widely applicable means for evaluating many, though not all, integrals without having to resort to the (usually difficult) process of finding a limit of Riemann sums.

Theorem 20.3 (The fundamental theorem of calculus, part one). *Suppose the function f is Riemann integrable on the interval $[a, b]$, the function F is continuous on $[a, b]$, and that $F'(x) = f(x)$ for all but finitely many points x in $[a, b]$. Then*

$$\int_a^b f = F(b) - F(a).$$

Proof. The finitely many points of $[a, b]$ where $F'(x) \neq f(x)$ can be accommodated by using these points to create finitely many subintervals of $[a, b]$ on which the following proof can be applied, and then employing the additivity of the integral (see Exercise 20.3 below). Thus, we shall assume that $F'(x) = f(x)$ for all x in (a, b). Consider any positive number ε. Since f is integrable on $[a, b]$, there exists a positive number δ so that when $\dot{P}$ is a tagged partition of $[a, b]$ for which $\|\dot{P}\| < \delta$, it follows that

$$\left| S(f, \dot{P}) - \int_a^b f \right| < \varepsilon.$$

Let $P = \{p_0, p_1, \ldots, p_n\}$ be a partition of $[a, b]$ having norm less than δ. For each i in $\{1, 2, \ldots, n\}$, as F is continuous on $[p_{i-1}, p_i]$ and differentiable on (p_{i-1}, p_i), we may apply the mean value theorem to obtain tags c_i in (p_{i-1}, p_i) for which

$$F'(c_i) = \frac{F(p_i) - F(p_{i-1})}{p_i - p_{i-1}}.$$

Hence, for the tagged partition

$$\dot{P} = \{([p_{i-1}, p_i], c_i) \mid i \in \{1, 2, \dots, n\}\}$$

just created from P, since we have assumed that $F'(c_i) = f(c_i)$, we must have

$$\begin{aligned} F(b) - F(a) &= \sum_{i=1}^{n} [F(p_i) - F(p_{i-1})] \\ &= \sum_{i=1}^{n} F'(c_i) \cdot (p_i - p_{i-1}) \\ &= \sum_{i=1}^{n} f(c_i) \cdot (p_i - p_{i-1}) \\ &= S(f, \dot{P}). \end{aligned}$$

Therefore, from our earlier conclusion based on the integrability of f on $[a, b]$, we may now deduce that

$$\left| F(b) - F(a) - \int_a^b f \right| < \varepsilon.$$

As the positive number ε is anonymous, this inequality holds for any positive ε and the desired conclusion follows. □

As in calculus courses, we write

$$F(x)\Big|_a^b$$

to mean $F(b) - F(a)$.

If F is an antiderivative of f on $[a, b]$, the fundamental theorem can be applied to evaluate

$$\int_a^b f$$

provided f is integrable on $[a, b]$, as in such a situation the continuity of F follows immediately from its differentiability.

Corollary 20.4. *If the function f is Riemann integrable on the interval $[a, b]$ and the function F is an antiderivative of f on $[a, b]$, then*

$$\int_a^b f = F(b) - F(a).$$

Example 20.5. Note that the function $F : [1,4] \to \mathbb{R}$, where $F(x) = \frac{1}{3}x^3$ is an antiderivative on $[1,4]$ of the continuous, hence, integrable, function $f : [1,4] \to \mathbb{R}$ defined so that $f(x) = x^2$. Thus, according to the fundamental theorem

$$\int_1^4 x^2 dx = \int_1^4 f = F(4) - F(1) = \frac{1}{3} \cdot 4^3 - \frac{1}{3} \cdot 1^3 = 21.$$

Exercise 20.3. Suppose the function f is integrable on $[a,b]$, the function F is continuous on $[a,b]$, and that $F'(x) = f(x)$ for all x in $[a,b]$ except for a number p that is in (a,b), where $F'(p) \neq f(p)$. Indicate how the additivity of the integral, along with what was explicitly established in our proof of Theorem 20.3, can be used to show that $\int_a^b f = F(b) - F(a)$ in this situation.

It is not always possible to use the fundamental theorem when trying to evaluate an integral.

Example 20.6. The fundamental theorem cannot be used to evaluate

$$\int_0^{0.5} \sqrt{1 - x^4} dx$$

as it has been proven, though the proof is beyond the scope of this book, that there is no *elementary function* F for which $F'(x) = \sqrt{1 - x^4}$ for all but finitely many points of the interval $[0, 0.5]$ of integration. We revisit this example below in connection with a brief discussion of *elementary functions*.

The fundamental theorem does not require that F be an antiderivative of f in order to be applied, as it allows for a finite number of points in the interval of integration at which either F' is undefined or for which $F' \neq f$. Exercise 20.4 illustrates this scenario.

Exercise 20.4. Define the functions f and F on $\mathbb{R}$ so that

$$f(x) = \begin{cases} x^3, & \text{if } x < -2 \text{ or } x > 2; \\ -x^3, & \text{if } -2 \leq x \leq 2; \end{cases}$$

and

$$F(x) = \frac{1}{4}\left|x^4 - 16\right|.$$

(a) For what values of x is $F'(x) = f(x)$?
(b) Explain why F is not an antiderivative of f on $[-4, 2]$.
(c) Verify that the fundamental theorem can be used to evaluate $\int_{-4}^{2} f$ and then perform the integration.

It may seem quite strange, but there are derivatives that are not Riemann integrable; in other words, F being an antiderivative of f on $[a, b]$ does not guarantee that the function f is Riemann integrable on $[a, b]$. We present a famous example of this phenomenon.

Example 20.7. Define the functions f and F on $[0, 1]$ so that for all x in $(0, 1]$,

$$f(x) = 2x \sin\left(\frac{1}{x^2}\right) - \frac{2\cos(\frac{1}{x^2})}{x}$$

and

$$F(x) = x^2 \sin\left(\frac{1}{x^2}\right),$$

while

$$f(0) = F(0) = 0.$$

It follows that $F'(x) = f(x)$ for all x in $[0, 1]$ (*exercise*), thus making F an antiderivative of f on $[0, 1]$. But since f is unbounded on $[0, 1]$ (*exercise*), it follows that f is not Riemann integrable on $[0, 1]$. In other words, in this situation $\int_0^1 f$ does not exist, so it would be incorrect to attempt to compute it as $F(1) - F(0)$, even though the latter calculation can certainly be performed and F is an antiderivative of f on $[0, 1]$.

Exercise 20.5. Verify the two statements labeled as exercises in Example 20.7.

Integration by parts

The fundamental theorem, together with the product rule for computing derivatives, yields the familiar formula for computing an integral "by parts."

Theorem 20.8 (Integration by parts). *If the functions u and v are differentiable on $[a, b]$, with each of u' and v' integrable on $[a, b]$, then*

$$\int_a^b uv' = uv\Big|_a^b - \int_a^b vu'.$$

Proof. Assume u and v are differentiable on $[a, b]$, with u' and v' integrable on $[a, b]$. As the sum and product of integrable functions are integrable, it follows that the function uv', the function vu', and the function $uv' + vu' = (uv)'$ is each integrable on $[a, b]$. Therefore,

$$\int_a^b uv' + \int_a^b vu' = \int_a^b (uv' + vu') = \int_a^b (uv)' = uv\Big|_a^b,$$

where the last equality results from the fundamental theorem of calculus. Thus,

$$\int_a^b uv' = uv\Big|_a^b - \int_a^b vu'. \qquad \square$$

Example 20.9. We use integration by parts to evaluate

$$\int_0^\pi x\cos(x)dx.$$

The functions u and v defined so that $u(x) = x$ and $v(x) = \sin(x)$ are differentiable on $[0,\pi]$, and their derivatives $u'(x) = 1$ and $v'(x) = \cos(x)$ are continuous, hence integrable, on $[0,\pi]$, so the integration by parts formula may be applied. Calculating we obtain

$$\begin{aligned}\int_0^\pi x\cos(x)dx &= \int_0^\pi uv' \\ &= uv\Big|_0^\pi - \int_0^\pi vu' \\ &= x\sin(x)\Big|_0^\pi - \int_0^\pi \sin(x)dx \\ &= \pi\cdot\sin(\pi) - 0\cdot\sin(0) + \cos(\pi) - \cos(0) \\ &= \pi\cdot 0 - 0\cdot 0 + (-1) - 1 \\ &= -2.\end{aligned}$$

Exercise 20.6. Use integration by parts to evaluate $\int_{-\pi}^{\pi} x^2\sin(x)dx$.

Exercise 20.7. Let n be a nonnegative integer, let I be an open interval that includes the number x_0, and let f be a function for which each of $f', f'', \ldots, f^{(n+1)}$ exists on I and is Riemann integrable on every closed bounded subinterval of I. Use induction and integration by parts to prove that

$$R_n(x) = \frac{1}{n!}\int_{x_0}^{x} f^{(n+1)}(t)(x-t)^n dt,$$

where

$$R_n(x) = f(x) - P_n(x)$$

is the remainder resulting from the use of the nth Taylor polynomial

$$P_n(x) = \sum_{k=0}^{n} \frac{f^{(k)}(x_0)}{k!}(x - x_0)^k$$

for f at x_0 to approximate $f(x)$.

Indefinite integrals

Given a function f that is Riemann integrable on the interval $[a, b]$, the **indefinite integral of f** is the function $F : [a, b] \to \mathbb{R}$ defined so that

$$F(x) = \int_a^x f.$$

The number a is sometimes referred to as the **base** of this indefinite integral.

Example 20.10. The function f, where $f(x) = x^2$ is integrable on the interval $[3, 100]$, so the indefinite integral $F : [3, 100] \to \mathbb{R}$ of f with base 3 is defined so that

$$F(x) = \int_3^x f.$$

We can use the fundamental theorem to calculate the integral on the right side of this last equation, obtaining

$$F(x) = \int_3^x f = \int_3^x t^2 dt = \frac{1}{3}x^3 - 9.$$

Thus, for instance, $F(6) = \frac{1}{3} \cdot 6^3 - 9 = 63$.

Exercise 20.8. Let G be the indefinite integral of the cosine function on the interval $[\frac{\pi}{2}, 3\pi]$.
(a) Evaluate $G(\frac{\pi}{2})$.
(b) Evaluate $G(\pi)$.
(c) Find a simplified formula for $G(x)$ in general.

Exercise 20.9. Suppose f is Riemann integrable on $[a, b]$ and let F be the indefinite integral of f on $[a, b]$. Express each of the following in terms of F.
(a) $\int_c^x f$, where $a < c < b$

(b) $\int_x^b f$

(c) $\int_{3x}^{4x} 2f$, assuming all of x, $3x$, and $4x$ are in $[a, b]$

Exercise 20.10. Prove that if f is Riemann integrable on $[a, b]$, then the indefinite integral F of f is continuous on $[a, b]$.

The fundamental theorem of calculus: constructing antiderivatives

The indefinite integral of a continuous function always turns out to be an antiderivative of that function.

Theorem 20.11 (The fundamental theorem of calculus, part two). *If the function $f : [a, b] \to \mathbb{R}$ is continuous, then the indefinite integral F of f is differentiable on $[a, b]$ and for every x in $[a, b]$, we have*

$$F'(x) = \frac{d}{dx}\left[\int_a^x f\right] = f(x).$$

Proof. Assume $f : [a, b] \to \mathbb{R}$ is continuous, hence, integrable, on $[a, b]$ and suppose p is in $[a, b]$. Consider any positive number ε. Since f is continuous at p, there exists a positive number δ such that whenever $x \in [a, b]$ and $|x - p| < \delta$, it follows that $|f(x) - f(p)| < \varepsilon$. Now for any x in $[a, b]$ satisfying $p < x < p + \delta$, the additivity of the integral tells us that

$$\int_p^x f = \int_a^x f - \int_a^p f = F(x) - F(p).$$

Furthermore, for such x,

$$(f(p) - \varepsilon) \cdot (x - p) < F(x) - F(p) = \int_p^x f < (f(p) + \varepsilon) \cdot (x - p),$$

as the continuity of f at p tells us that $f(x) \in (f(p) - \varepsilon, f(p) + \varepsilon)$ and the length of the interval of integration for $\int_p^x f$ is $x - p$. But we can rewrite the above inequality as

$$-\varepsilon < \frac{F(x) - F(p)}{x - p} - f(p) < \varepsilon,$$

which tells us that

$$\left|\frac{F(x) - F(p)}{x - p} - f(p)\right| < \varepsilon.$$

Thus,

$$\lim_{x\to p^+} \frac{F(x) - F(p)}{x - p} = f(p).$$

In a similar manner, it can be shown that

$$\lim_{x\to p^-} \frac{F(x) - F(p)}{x - p} = f(p).$$

Thus, we may conclude that

$$F'(p) = f(p).$$

(Note that if $p = a$ or $p = b$, only a single one-sided limit needs to be considered.) □

Example 20.10 (Continued). Let F be the indefinite integral of $f(x) = x^2$ on $[3, 100]$. Observe that

$$F'(x) = \frac{d}{dx}\left[\int_3^x f\right] = \frac{d}{dx}\left[\int_3^x t^2 dt\right] = \frac{d}{dx}\left[\frac{1}{3}x^3 - 9\right] = x^2 = f(x),$$

which is in agreement with part two of the fundamental theorem.

Example 20.12. If $f(x) = \int_4^{x^2} \sin(5t)dt$, then the fundamental theorem, together with the chain rule, tells us that

$$f'(x) = \frac{d}{dx}\left[\int_4^{x^2} \sin(5t)dt\right] = \sin(5x^2) \cdot \frac{d}{dx}[x^2] = 2x\sin(5x^2).$$

The **elementary functions** of mathematics include the polynomial, rational, trigonometric, inverse trigonometric, exponential, and logarithmic functions, along with arithmetic combinations of these functions and compositions of these functions. It turns out that an elementary function may have antiderivatives, but that none of these antiderivatives are able to be expressed as an elementary function. The second part of the fundamental theorem of calculus provides a means of constructing antiderivatives for all continuous functions.

Example 20.6 (Continued). The function $f : [0, 0.5] \to \mathbb{R}$, where $f(x) = \sqrt{1 - x^4}$, is continuous, being the composition of a polynomial function and the principal square root function, each of which is continuous. Even though it can be shown, using mathematical concepts beyond the scope of this book, that f has no antiderivative on $[0, 0.5]$ that can be expressed in the form of an elementary function, part two of the fundamental theorem tells us that the function $F : [0, 0.5] \to \mathbb{R}$, where

$$F(x) = \int_0^x f = \int_0^x \sqrt{1-t^4}dt,$$

is an antiderivative of f on $[0, 0.5]$. It is the inability to remove the operation of integration in the definition of F that makes F a non-elementary function.

When a function has an antiderivative that is an elementary function, the antiderivative is often referred to as an **elementary antiderivative**. The constant difference theorem tells us that when a function has an elementary antiderivative, all of the function's antiderivatives must be elementary (as they all differ from each other by a constant).

Exercise 20.11. Find a formula for the derivative of the given function.
(a) $F : [1,3] \to \mathbb{R}$, where $F(x) = \int_1^x \sqrt{t}dt$
(b) $G : [-1,1] \to \mathbb{R}$, where $G(x) = \int_{-1}^{\sin(x)} 4tdt$
(c) $H : [0,1] \to \mathbb{R}$, where $H(x) = \int_{x^2}^{x^3} \cos(t)dt$

Exercise 20.12. Redo Exercise 18.19, this time using Theorem 20.11, part two of the fundamental theorem of calculus.

Selecting a different base generally yields a different antiderivative of the integrand, but the conclusion of part two of the fundamental theorem still holds.

Corollary 20.13. *If the function $f : [a,b] \to \mathbb{R}$ is continuous and $c \in [a,b]$, then*

$$\frac{d}{dx}\left[\int_c^x f\right] = f(x).$$

Proof. Applying the additivity of the integral and the second part of the fundamental theorem of calculus, we have

$$\frac{d}{dx}\left[\int_c^x f\right] = \frac{d}{dx}\left[\int_c^a f + \int_a^x f\right] = 0 + f(x) = f(x). \qquad \square$$

The natural logarithmic function

As the reciprocal function

$$y = \frac{1}{x}$$

is continuous on $(0, \infty)$, the second part of the fundamental theorem of calculus tells us it must have an antiderivative, and so it is that the **natural logarithmic function** $\ln : (0, \infty) \to \mathbb{R}$ is defined so that

$$\ln(x) = \int_1^x \frac{1}{t} dt.$$

Thus, from a purely analytic perspective, natural logarithms are generated by means of the Riemann integral.

Theorem 20.14. *The natural logarithmic function* $\ln$ *is differentiable on its domain* $(0, \infty)$ *with* $\ln'(x) = \frac{1}{x}$.

Proof. Note that $\ln'(x) = \frac{d}{dx}[\int_1^x \frac{1}{t} dt] = \frac{1}{x}$. □

Exercise 20.13. Show that $\ln(1) = 0$.

Exercise 20.14. Let the function $f : [2, 5] \to \mathbb{R}$ be defined so that $f(x) = 2x + \frac{2}{x}$. Which of the following are antiderivatives of f on $[2, 5]$?
(a) $F(x) = x^2$
(b) $F(x) = x^2 + \ln(x^2)$
(c) $F(x) = x^2 + \ln(x^2 + 1)$
(d) $F(x) = x^2 + \ln(x^2) + 1$
(e) $F(x) = x^2 + \ln(3x^2)$

To prove differentiable functions are equal on an interval, it is enough to show they have the same derivative throughout the interval and take the same value at a single input (see Theorem 16.15). We employ this strategy in establishing (1) of the following theorem.

Theorem 20.15 (Some fundamental properties of the natural logarithmic function).
(1) $\ln(xy) = \ln(x) + \ln(y)$ *for any positive numbers* x *and* y.
(2) $\ln(x^r) = r\ln(x)$ *for any positive number* x *and any real number* r.
(3) *The natural logarithmic function* $\ln$ *is increasing on its domain* $(0, \infty)$.
(4) *The range of the natural logarithmic function* $\ln$ *is* $\mathbb{R}$.

Proof. (1) Fix a positive number y and define the functions f and g on $(0, \infty)$ by

$$f(x) = \ln(xy)$$

and

$$g(x) = \ln(x) + \ln(y).$$

Note that for every x in $\mathbb{R}$, we have

$$f'(x) = g'(x) = \frac{1}{x},$$

so f and g must differ by a constant. However, since

$$f(1) = g(1) = \ln(y),$$

we may actually conclude that $f(x) = g(x)$ for all positive numbers x.

(2) You are asked to establish this result for the case in which r is rational in Exercise 20.15. Later, in Exercise 20.20, after the natural exponential function has been defined, you prove the general result.

(3) Since $\ln'(x) = \frac{1}{x} > 0$ for all positive numbers x, we may conclude, via Theorem 16.9, the monotonicity theorem, that ln is increasing on $(0, \infty)$.

(4) Consider any real number y. If $y = 0$, then we already know $\ln(1) = 0 = y$.

Now suppose $y > 0$. Since ln is increasing, $\ln(1) = 0$, and ln is continuous (being differentiable), the intermediate value theorem tells us that $\ln(x) = y$ has a solution if we can find a number t such that $\ln(t) > y$. Note that for each natural number n, we have

$$\ln(2^n) = n\ln(2)$$

using the result (2). Also, since ln is increasing, it follows that $\ln(2) > \ln(1) = 0$. So, as $\frac{y}{\ln(2)}$ is a positive number, the Archimedean property tells us there is a natural number N such that

$$N > \frac{y}{\ln(2)},$$

from which it then follows that

$$\ln(2^N) > y.$$

Finally, if $y < 0$, then $-y > 0$, so that from our work above, we know there exists a positive number x such that $\ln(x) = -y$. Hence, $\frac{1}{x}$ is a positive number and

$$\ln\left(\frac{1}{x}\right) = \ln(x^{-1}) = (-1)\cdot\ln(x) = (-1)\cdot(-y) = y. \qquad \square$$

From the arguments presented in support of Theorem 20.15(4), we immediately obtain the following results.

Corollary 20.16. *The natural logarithmic function satisfies both limiting properties* $\lim_{x\to\infty}\ln(x) = \infty$ *and* $\lim_{x\to 0^+}\ln(x) = -\infty$.

Exercise 20.15. Prove Theorem 20.15(2) in the special case in which r is rational.

Euler's number and the natural exponential function

As the natural logarithmic function is increasing on its domain, it has an inverse that is also an increasing function. According to Theorem 15.15, this inverse must also be differentiable because the natural logarithmic function is differentiable and its derivative is never zero.

Theorem 20.17. *The natural logarithmic function* ln *has an inverse* $\ln^{-1} : \mathbb{R} \to (0, \infty)$ *that is differentiable on* $\mathbb{R}$.

As all real numbers lie within the range of ln, there must be an input to the natural logarithmic function that yields 1 as the output. The unique solution to the equation $\ln(x) = 1$ is called **Euler's number** and is denoted by e.

Exercise 20.16. Justify each of the following.
(a) e>1.
(b) For every rational number q, we must have $\ln^{-1}(q) = e^q$.

We intend to show that $\ln^{-1}(x) = e^x$ for every real number x, not just rational numbers, but in our argument we need to be able to interchange the action of a function with the action of the supremum operator. Unfortunately, this sort of interchange is not always permissible.

Example 20.18. Consider the function $f : (-\infty, 0) \to \mathbb{R}$ defined so that $f(x) = -\frac{1}{x}$. Observe that

$$\sup((-\infty, 0)) = 0,$$

but

$$f(0) = f(\sup((-\infty, 0)))$$

is undefined. Moreover,

$$f[(-\infty, 0)] = (0, \infty),$$

a set that has no supremum. Thus, it would be incorrect to say that $f(\sup((-\infty, 0)))$ is equal to $\sup(f[(-\infty, 0)])$ because neither of them actually exists.

Even when both $f(\sup(B))$ and $\sup(f[B])$ exist, they may not be equal. Take, for instance, the function $f : [-2, 2] \to \mathbb{R}$ defined by

$$f(x) = \begin{cases} x^2, & \text{if } x \neq 0; \\ 5, & \text{if } x = 0. \end{cases}$$

Note that $\sup([-2,2]) = 2$ so that

$$f(\sup([-2,2])) = f(2) = 4,$$

but $f[[-2,2]] = (0,4] \cup \{5\}$ so that

$$\sup f([-2,2]) = 5.$$

Happily, though, there are circumstances under which we may interchange function evaluation and supremum.

Theorem 20.19. *Let A be a nonempty closed subset of $\mathbb{R}$ that is bounded above. If the function $f : A \to \mathbb{R}$ is nondecreasing and continuous on A, then for each nonempty subset B of A,*

$$f(\sup(B)) = \sup(f[B]).$$

Proof. As A is a nonempty subset of $\mathbb{R}$ that is bounded above, we know $\sup(A)$ exists. Moreover, as A is closed, we know $\sup(A) \in A$ meaning $\sup(A) = \max(A)$.

Now suppose the function $f : A \to \mathbb{R}$ is nondecreasing and continuous on A. As f is nondecreasing, for every a in A we have $f(a) \leq f(\max(A))$. Thus, $f[A]$ has a maximum element $\max(f[A]) = f(\max(A))$.

Let B be any nonempty subset of A. As A is bounded above, so is B, and therefore $p = \sup(B)$ exists. Since A is closed, it follows that $p \in A$, as otherwise some neighborhood of p would include no points of A, hence, no points of B, making it impossible for p to be the supremum of B. Also, as B is nonempty, $f[B]$ is also nonempty, and as $f[B] \subseteq f[A]$ and $f[A]$ is bounded above, it follows that $f[B]$ is bounded above. Thus, $f[B]$ has a supremum. Having verified that both $\sup(B)$ and $\sup(f[B])$ exist, we are now ready to show that

$$f(\sup(B)) = \sup(f[B]).$$

First, consider any b in B. Then $b \leq \sup(B)$ so that, as f is nondecreasing, $f(b) \leq f(\sup(B))$. Therefore, $f(\sup(B))$ is an upper bound for $f[B]$, and it follows that $\sup(f[B]) \leq f(\sup(B))$.

Now, consider any nondecreasing sequence (b_n) in B for which $b_n \to \sup(B)$. As f is continuous, it follows that $f(b_n) \to f(\sup(B))$. Also, though, as f is nondecreasing, we know that $f(b_n) \to s$, where $s = \sup(\{f(b_n) | n \in \mathbb{N}\})$. As the limit of a sequence is unique, we may conclude that $f(\sup(B)) = s \leq \sup(f[B])$. □

Exercise 20.17. Suppose $b > 1$ and $r < s$. Prove that $b^r \leq b^s$. *Hint*: Apply the result established in Exercise 4.34 that when $b > 1$ and $r > 0$, we must have $b^r \geq 1$.

Exercise 20.18. Define the function f on $\mathbb{R}$ so that $f(x) = e^x$.
(a) Use Exercise 20.17 to explain why f is nondecreasing.
(b) Prove that f is continuous.

Theorem 20.20. *For every real number x, we have $\ln^{-1}(x) = e^x$.*

Proof. We already know the result holds for rational numbers x by Exercise 20.16(b). Let x be any irrational number and note that $x = \sup\{q \mid q \in \mathbb{Q} \text{ and } q < x\}$. Then, as $\ln^{-1}$ is increasing and continuous (being differentiable) on the closed set $(-\infty, x]$, which is bounded above, we may apply Theorem 20.19, to conclude that

$$\ln^{-1}(x) = \ln^{-1}(\sup\{q \mid q \in \mathbb{Q} \text{ and } q < x\}) = \sup(\{\ln^{-1}(q) \mid q \in \mathbb{Q} \text{ and } q < x\}).$$

But since $\ln^{-1}(q) = e^q$ holds for rational numbers q, we have

$$\ln^{-1}(x) = \sup(\{e^q \mid q \in \mathbb{Q} \text{ and } q < x\}).$$

Now let f be the function defined on $\mathbb{R}$ so that $f(x) = e^x$. It follows from Exercise 20.18 that f is nondecreasing and continuous on $(-\infty, x]$, a set which is closed and bounded above. Thus, Theorem 20.19 permits us to conclude that

$$\sup(\{e^q \mid q \in \mathbb{Q} \text{ and } q < x\}) = e^{\sup(\{q \mid q \in \mathbb{Q} \text{ and } q < x\})}.$$

Therefore, we now have

$$\ln^{-1}(x) = e^{\sup(\{q \mid q \in \mathbb{Q} \text{ and } q < x\})} = e^x. \qquad \square$$

The **natural exponential function** exp is defined as the inverse of the natural logarithmic function. Thus,

$$\exp(x) = \ln^{-1}(x) = e^x,$$

for every real number x. Since the domain of the natural logarithmic function is the set $(0, \infty)$ of positive real numbers, this set is also the range of the natural exponential function. Moreover, as $\lim_{x\to\infty} \ln(x) = \infty$ and $\lim_{x\to 0^+} \ln(x) = -\infty$, we immediately obtain the results stated in the following theorem.

Theorem 20.21. *The natural exponential function* exp *satisfies both limiting properties*

$$\lim_{x\to\infty} \exp(x) = \lim_{x\to\infty} e^x = \infty$$

and

$$\lim_{x\to-\infty} \exp(x) = \lim_{x\to-\infty} e^x = 0.$$

The natural exponential function's most characteristic property is that its derivative is itself.

Theorem 20.22. *The natural exponential function* exp *is differentiable on* $\mathbb{R}$ *with* $\exp' = \exp$. *That is, if* $y = e^x$, *then* $y' = e^x$.

Exercise 20.19. Prove Theorem 20.22 by applying Theorem 15.15 concerning the derivatives of inverse functions.

Exercise 20.20. Complete the proof of Theorem 20.15(2) by providing an argument that applies to any real number exponent r.

In Exercise 10.32, you demonstrated that the sequence whose nth term is $(1 + \frac{1}{n})^n$ has a limiting value that lies between 2 and 3. The following theorem reveals this limit to be e, and as a consequence, we deduce that $2 < e < 3$.

Theorem 20.23. *Euler's number e can be calculated as* $\lim_{x\to\infty}(1 + \frac{1}{x})^x$. *It follows that* $2 < e < 3$.

Exercise 20.21. Verify that

$$\lim_{x\to\infty}\left(1 + \frac{1}{x}\right)^x = e$$

without using L'Hôpital's rule. *Hint*: Note that $(1+\frac{1}{x})^x = e^{\ln(1+\frac{1}{x})^x}$. Consider $\lim_{x\to\infty} \ln(1+\frac{1}{x})^x$ and try rewriting this limit so that it defines the derivative of a specific function at a specific input.

Taylor's theorem offers a means for calculating a decimal approximation for e that achieves any prescribed degree of accuracy.

Theorem 20.24. *For any natural number n, the value of Euler's number e can be approximated as*

$$\sum_{k=0}^{n} \frac{1}{k!},$$

and the resulting approximation is less than $\frac{3}{(n+1)!}$ *from the true value of e.*

Proof. Since $\exp'(x) = e^x$, it follows that all higher-order derivatives of exp exist and $\exp^{(k)}(x) = e^x$ for every natural number k. Hence, for every k, we have

$$\exp^{(k)}(0) = e^0 = 1.$$

Consider any natural number n. The nth Taylor polynomial for exp at 0 is

$$P_n(x) = \sum_{k=0}^{n} \frac{\exp^{(k)}(0)}{k!} x^k = \sum_{k=0}^{n} \frac{x^k}{k!},$$

so by Theorem 17.3, Taylor's theorem,

$$\exp(1) = e^1 = e = \sum_{k=0}^{n} \frac{1^k}{k!} + \frac{e^c}{(n+1)!} \cdot 1^{n+1} = \sum_{k=0}^{n} \frac{1}{k!} + \frac{e^c}{(n+1)!}$$

for some number c between 0 and 1. The precise value of c depends on the value of n, but in any case

$$1 = e^0 < e^c < e^1 < 3,$$

so that the error term $\frac{e^c}{(n+1)!}$ has the property that

$$\frac{e^c}{(n+1)!} < \frac{3}{(n+1)!}$$

and tends toward zero as $n \to \infty$. Therefore,

$$e \approx \sum_{k=0}^{n} \frac{1}{k!}$$

and this approximation is less than $\frac{3}{(n+1)!}$ units from the true value of e. □

Example 20.25. According to Theorem 20.24, if we want to calculate e to within 0.00001 of its true value, we need to find n so that $\frac{3}{(n+1)!} < 0.00001$, that is, $(n+1)! > 300{,}000$. Observing that $8! = 40{,}320$ and $9! = 362{,}880$, we conclude that

$$\sum_{k=0}^{8} \frac{1}{k!} = 1 + 1 + \frac{1}{2} + \frac{1}{6} + \frac{1}{24} + \frac{1}{120} + \frac{1}{720} + \frac{1}{5040} + \frac{1}{40{,}320} = 2.71827877$$

yields an approximation of e correct to four decimal places (the digit in the hundred-thousandths place for e is actually 8).

In Chapter 27, we show that e is irrational.

Exercise 20.22. Imagine you are teaching a calculus class and you claim that the derivative of the natural exponential function is itself. You want to provide your students with some justification for this claim. You let $f(x) = e^x$ and note that

$$f'(x) = \lim_{h \to 0} \frac{e^{x+h} - e^x}{h} \approx \frac{e^{x+0.01} - e^x}{0.01}.$$

If you then define the function g so that $g(x) = \frac{e^{x+0.01} - e^x}{0.01}$, describe what you might then ask your students to do in order to make your point. What should your students observe? What might you tell a student to do if he or she wants to be "more sure" of your claim?

Exercise 20.23. Show that $b^r = e^{r\ln(b)}$ for any positive number b and any real number r. (This result provides us with another way of understanding what it means to raise a positive number to an irrational power. For instance, we may regard 2^π as $e^{\pi\ln(2)}$.)

Exercise 20.24. Use the result of Exercise 20.23 to prove each of the following.

(a) If b is a positive number and the function $f : \mathbb{R} \to \mathbb{R}$ is defined so that $f(x) = b^x$, then $f'(x) = b^x \cdot \ln(b)$.

(b) For any fixed real number r, if the function $f : (0, \infty) \to \mathbb{R}$ is defined so that $f(x) = x^r$, then $f'(x) = rx^{r-1}$. (This establishes that the power rule for computing derivatives can be applied even when the exponent is an irrational number.)

(c) For any positive number p, it follows that $\lim_{x\to\infty} x^p = \infty$.

Exercise 20.25. Define the function $f : \mathbb{R}^+ \to \mathbb{R}$ so that $f(x) = x^x$. Explain why f is differentiable on $\mathbb{R}^+$ and find a formula for $f'(x)$.

Exercise 20.26. Use integration by parts to evaluate $\int_1^e \ln(x)dx$.

Exercise 20.27. Find a formula for the derivative of the given function.

(a) $F : [1, e] \to \mathbb{R}$, where $F(x) = \int_1^x \ln(t)dt$

(b) $G : [-5, 5] \to \mathbb{R}$, where $G(x) = \int_{4x}^3 e^t dt$

L'Hôpital's rule revisited

Recall that the evaluation of a limit that exhibits one of the indeterminate forms $\frac{0}{0}$ or $\frac{\pm\infty}{\pm\infty}$ is often facilitated by means of L'Hôpital's rule, Theorem 16.22. The next exercise illustrates how L'Hôpital's rule can be used in the presence of exponential and logarithmic functions.

Exercise 20.28. Where possible, use L'Hôpital's rule to evaluate the given limit. If L'Hôpital's rule cannot be applied, use another method to evaluate the limit or explain why the limit does not exist.

(a) $\lim_{x\to\infty} \frac{x^4}{e^{2x}}$

(b) $\lim_{x\to-\infty} \frac{x^4}{e^{2x}}$

(c) $\lim_{x\to\infty} \frac{\ln(x)}{\sqrt{x}}$

There are other indeterminate forms besides $\frac{0}{0}$ and $\frac{\pm\infty}{\pm\infty}$. Among them are the following:

$$\infty - \infty, \quad 0 \cdot \infty, \quad \infty^0, \quad 1^\infty, \quad 0^0.$$

While we do not verify that these forms are indeterminate, we want you to be aware that L'Hôpital's rule is often useful in evaluating limits involving these forms.

Example 20.26. Observe that $\lim_{x\to 0^+} x\ln(x)$ possesses the indeterminate form $0\cdot(-\infty)$. Now

$$\lim_{x\to 0^+} x\ln(x) = \lim_{x\to 0^+} \frac{\ln(x)}{\frac{1}{x}},$$

where the latter limit is seen to possess the indeterminate form $\frac{-\infty}{\infty}$ to which L'Hôpital's rule applies (you should check that all required hypotheses are satisfied). You are asked to complete the evaluation of this limit in Exercise 20.29.

Exercise 20.29. Complete the evaluation of $\lim_{x\to 0^+} x\ln(x)$ begun in Example 20.26.

Example 20.27. The limit

$$\lim_{x\to\infty}\left(1+\frac{1}{x}\right)^x$$

possesses the indeterminate form 1^∞. Assume for the moment that this limit exists and let $L = \lim_{x\to\infty}(1+\frac{1}{x})^x$. It follows that

$$\ln(L) = \ln\left(\lim_{x\to\infty}\left(1+\frac{1}{x}\right)^x\right) = \lim_{x\to\infty}\ln\left(1+\frac{1}{x}\right)^x,$$

the second equality resulting from the fact that the natural logarithmic function is continuous. Using the property $\ln(b^r) = r\ln(b)$ of logarithms, we then have

$$\ln(L) = \lim_{x\to\infty} x\ln\left(1+\frac{1}{x}\right) = \lim_{x\to\infty}\frac{\ln(1+\frac{1}{x})}{\frac{1}{x}}.$$

Note that this last limit has the form $\frac{0}{0}$, so we can try to evaluate it by L'Hôpital's rule (make sure you agree that all necessary hypotheses are satisfied). Calculating gives us

$$\ln(L) = \lim_{x\to\infty}\frac{\ln(1+\frac{1}{x})}{\frac{1}{x}} = \lim_{x\to\infty}\frac{\frac{1}{1+\frac{1}{x}}\cdot(-\frac{1}{x^2})}{-\frac{1}{x^2}} = \lim_{x\to\infty}\frac{1}{1+\frac{1}{x}} = 1.$$

Therefore, as $\ln(L) = 1$, it follows that $\lim_{x\to\infty}(1+\frac{1}{x})^x = L = e^1 = e$. In fact, the limiting value of $(1+\frac{1}{x})^x$ as x grows without bound is sometimes taken as the definition of Euler's number e. (Compare with Exercise 10.32.)

Exercise 20.30. Evaluate the following limits.
(a) $\lim_{x\to\infty}(x - \ln(x))$ *Hint*: First evaluate $\lim_{x\to\infty} e^{x-\ln(x)}$
(b) $\lim_{x\to 0^+} x^x$
(c) $\lim_{x\to\infty}(1 + \frac{1}{x})^{x^2}$

Integration via change of variables

We can use part two of the fundamental theorem of calculus to verify another widely applicable integration technique known as *change of variables* or *substitution.*

Theorem 20.28 (Change of variables theorem). *Suppose the function u has a continuous derivative u' on the interval $[a,b]$, the function f is continuous on the interval $[c,d]$, and $u(x)$ is in $[c,d]$ for all x in $[a,b]$. Then*

$$\int_a^b (f \circ u) \cdot u' = \int_{u(a)}^{u(b)} f.$$

Proof. Define functions g and h on $[a,b]$ so that

$$g(x) = \int_a^x (f \circ u) \cdot u'$$

and

$$h(x) = \int_{u(a)}^{u(x)} f.$$

Then, using Theorem 20.11, part two of the fundamental theorem of calculus, we have

$$g'(x) = ((f \circ u) \cdot u')(x) = (f \circ u)(x) \cdot u'(x) = f(u(x)) \cdot u'(x) = h'(x).$$

But, as we also have

$$g(a) = \int_a^a (f \circ u) \cdot u' = 0 = \int_{u(a)}^{u(a)} f = h(a),$$

we may conclude, via Theorem 16.15, the constant difference theorem, that $g = h$ on $[a,b]$, from which the desired result follows immediately. □

Example 20.29. We can use a change of variables to calculate

$$\int_1^4 \frac{\sin(\sqrt{x})}{\sqrt{x}} dx.$$

Define the functions f and u so that $f(x) = \sin(x)$ and $u(x) = \sqrt{x}$. Then $u'(x) = \frac{1}{2\sqrt{x}}$ and we see that u' is continuous on $[1,4]$. Also, when x is in $[1,4]$, it follows that $u(x)$ is in $[1,2]$, which is an interval on which f is continuous. Thus, the change of variables theorem for integrals tells us that

$$\int_1^4 \sin(\sqrt{x}) \cdot \frac{1}{2\sqrt{x}} dx = \int_1^4 (f \circ u) \cdot u' = \int_{u(1)}^{u(4)} f = \int_1^2 \sin(x) dx = \cos(1) - \cos(2).$$

Therefore,

$$\int_1^4 \frac{\sin(\sqrt{x})}{\sqrt{x}} dx = 2 \int_1^4 \frac{\sin(\sqrt{x})}{2\sqrt{x}} dx = 2\cos(1) - 2\cos(2).$$

Exercise 20.31. Use a change of variables to evaluate $\int_e^{e^2} \frac{1}{x \ln(x)} dx$.

Exercise 20.32. Suppose f is integrable on $[a,b]$, let t be any real number, and define the function g on $[a+t, b+t]$ so that $g(x) = f(x-t)$. Prove that g is integrable on $[a,b]$ and $\int_{a+t}^{b+t} g = \int_a^b f$. (Thus, the integral is invariant with respect to horizontal translations.)

The following theorem, not usually covered in calculus courses, offers another option for evaluating some integrals via change of variables.

Theorem 20.30 (Second change of variables theorem). *Suppose the function u is differentiable with $u'(x) \neq 0$ for all x in the interval $[a,b]$, the function v is the inverse of u on $I = u[[a,b]]$, and the function f is continuous on I. Then*

$$\int_a^b (f \circ u) = \int_{u(a)}^{u(b)} fv'.$$

Example 20.31. Neither of the "obvious" possible substitutions, $u = \sqrt{x}$ or $u = 1 + \sqrt{x}$, is helpful in attempting to evaluate

$$\int_1^4 \frac{1}{1+\sqrt{x}} dx$$

via Theorem 20.28, so we try a change of variables based on Theorem 20.30.

Taking u and f to be the functions defined by $u(x) = \sqrt{x}$ and $f(x) = \frac{1}{1+x}$, we note that $u'(x) = \frac{1}{2\sqrt{x}}$ is defined and nonzero on the interval $[1,4]$ of integration, the function v defined on $[1,2]$ by $v(x) = x^2$ is the inverse of u on $[1,4]$, and the function f, being a rational function whose only point of discontinuity is -1, is continuous on $[1,2]$.

Thus, all of the hypotheses necessary for applying Theorem 20.30 are met, and since $v'(x) = 2x$, we have

$$\begin{aligned}\int_1^4 \frac{1}{1+\sqrt{x}} dx &= \int_1^4 (f \circ u) \\ &= \int_{u(1)}^{u(4)} fv' \\ &= \int_{\sqrt{1}}^{\sqrt{4}} \frac{1}{1+x} \cdot 2x dx \\ &= 2\int_1^2 \left(1 - \frac{1}{1+x}\right) dx \\ &= 2[x - \ln(1+x)]\Big|_{x=1}^2 \\ &= 2 + 2\ln(2) - 2\ln(3).\end{aligned}$$

Exercise 20.33. Use Theorem 20.30 to evaluate $\int_{-8}^{-1} \frac{\sqrt[3]{x}}{1-\sqrt[3]{x}} dx$.

Exercise 20.34. Prove Theorem 20.30.

Improper integrals

The notion of an integral can be extended to include some situations in which the interval of integration is unbounded or includes an infinite discontinuity (or perhaps both).

Unbounded intervals of integration arise in a variety of areas. In the mathematical study of probability, the standard normal random variable Z has probability density function f_Z that is defined for all real numbers so that

$$f_Z(x) = \frac{e^{-0.5x^2}}{\sqrt{2\pi}}$$

and whose graph is the famous bell curve displayed in Figure 20.1.

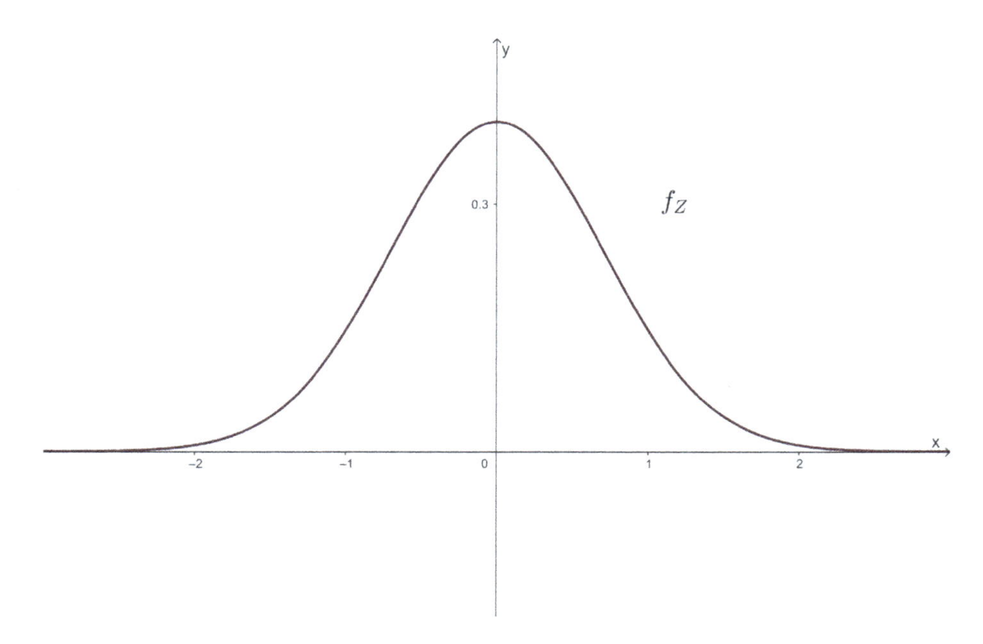

Figure 20.1: Graph of the standard normal probability density function f_Z.

Because $\lim_{x\to\pm\infty} f_Z(x) = 0$, the graph of f_Z approaches the x-axis asymptotically as x increases or decreases without bound. Also, because probabilities associated with Z are identified with areas under the graph of f_Z, it is required that the total area under the graph of f_Z be equal to 1, a requirement that is usually formulated as

$$\int_{-\infty}^{\infty} \frac{e^{-0.5x^2}}{\sqrt{2\pi}} dx = 1,$$

a statement involving an "integral" for which the interval of integration is the entire real line $(-\infty, \infty)$. To make sense of such an integral we employ the following definitions.

If a is a fixed number and the function f is Riemann integrable on the interval $[a, b]$ for every number b such that $b > a$, the **improper integral** $\int_a^{\infty} f$ is defined so that

$$\int_a^{\infty} f = \lim_{b\to\infty} \int_a^b f,$$

provided this limit exists. If $\lim_{b\to\infty}\int_a^b f = \infty$, we write $\int_a^\infty f = \infty$, and if $\lim_{b\to\infty}\int_a^b f = -\infty$, we write $\int_a^\infty f = -\infty$.

If b is a fixed number and the function f is Riemann integrable on the interval $[a,b]$ for every number a such that $a < b$, the **improper integral** $\int_{-\infty}^b f$ is defined so that

$$\int_{-\infty}^{b} f = \lim_{a\to-\infty}\int_a^b f,$$

provided this limit exists. If $\lim_{a\to-\infty}\int_a^b f = \infty$, we write $\int_{-\infty}^b f = \infty$, and if $\lim_{a\to-\infty}\int_a^b f = -\infty$, we write $\int_{-\infty}^b f = -\infty$.

If the function f is Riemann integrable on every interval $[a,b]$, we define the **improper integral** $\int_{-\infty}^\infty f$ so that

$$\int_{-\infty}^{\infty} f = \int_{-\infty}^{0} f + \int_{0}^{\infty} f,$$

provided both improper integrals on the right side of this equation exist. If both $\int_{-\infty}^0 f = \infty$ and $\int_0^\infty f = \infty$, we write $\int_{-\infty}^\infty f = \infty$, and if both $\int_{-\infty}^0 f = -\infty$ and $\int_0^\infty f = -\infty$, we write $\int_{-\infty}^\infty f = -\infty$.

Generally, we say that an improper integral **converges** if its value exists as a real number; otherwise, we say it **diverges**.

Example 20.32. The improper integral $\int_1^\infty \frac{1}{x^2}dx$ converges because

$$\int_1^\infty \frac{1}{x^2}dx = \lim_{b\to\infty}\int_1^b \frac{1}{x^2}dx = \lim_{b\to\infty} -\frac{1}{x}\bigg|_{x=1}^{b} = \lim_{b\to\infty}\left(1-\frac{1}{b}\right) = 1.$$

On the other hand, the improper integral $\int_1^\infty \frac{1}{x}dx$ diverges because

$$\int_1^\infty \frac{1}{x}dx = \lim_{b\to\infty}\int_1^b \frac{1}{x}dx = \lim_{b\to\infty} \ln(x)\big|_{x=1}^{b} = \lim_{b\to\infty}\ln(b) = \infty.$$

From a geometric perspective, the unbounded region lying below the graph of $y = \frac{1}{x^2}$, above the x-axis, and to the right of the line $x = 1$ has finite area equal to 1, while the unbounded region lying below the graph of $y = \frac{1}{x}$, above the x-axis, and to the right of the line $x = 1$ has infinite area (Figure 20.2).

The x-axis is a horizontal asymptote for the graphs of both $y = \frac{1}{x^2}$ and $y = \frac{1}{x}$. Apparently, the rate at which $\frac{1}{x^2}$ approaches 0 as x grows without bound is rapid enough

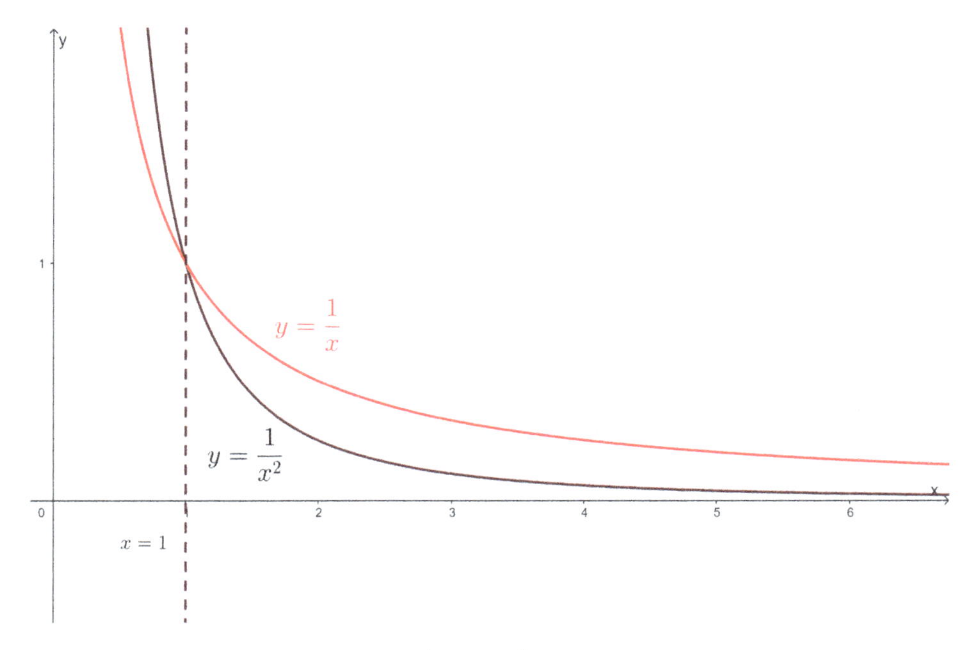

Figure 20.2: The graphs of the functions $y = \frac{1}{x^2}$ and $y = \frac{1}{x}$ along the interval $[1, \infty)$.

that $\int_1^\infty \frac{1}{x^2} dx$ converges, whereas the rate at which $\frac{1}{x}$ approaches 0 as x grows without bound is slow enough that $\int_1^\infty \frac{1}{x} dx$ diverges.

Exercise 20.35. Let a and p be positive numbers.
(a) Show that $\int_a^\infty \frac{1}{x^p} dx$ converges if $p > 1$, and find the value to which the improper integral converges.
(b) Show that $\int_a^\infty \frac{1}{x^p} dx$ diverges if $p \leq 1$.

We should never attempt to evaluate an improper integral of the form $\int_{-\infty}^\infty f$ by calculating $\lim_{t\to\infty} \int_{-t}^t f$ as this limit may exist even though the improper integral diverges.

Exercise 20.36. Show that $\lim_{t\to\infty} \int_{-t}^t x dx = 0$, but $\int_{-\infty}^\infty x dx$ diverges.

Exercise 20.37. Suppose $\int_{-\infty}^\infty f$ converges. Show that for any number c we have

$$\int_{-\infty}^{\infty} f = \int_{-\infty}^{c} f + \int_{c}^{\infty} f.$$

Exercise 20.38. The average speed of the molecules in an ideal gas is defined to be

$$\overline{x} = \sqrt{\frac{2}{\pi}}\left(\sqrt{\frac{M}{CT}}\right)^{3}\int_{0}^{\infty} x^{3}e^{-Mx^{2}/(2CT)}\,dx,$$

where M, T, and C are positive constants, the molecular weight, temperature, and gas constant, respectively, and x is the molecular speed.

(a) Use integration by parts with $u = x^2$ and $v' = xe^{-Mx^2/(2CT)}$ to show that

$$\int_{0}^{b} x^{3}e^{-Mx^{2}/(2CT)}\,dx = \frac{2C^{2}T^{2}}{M^{2}} - \frac{CTMb^{2} + 2C^{2}T^{2}}{M^{2}e^{Mb^{2}/(2CT)}}.$$

(b) Use L'Hôpital's rule to show that $\lim_{b\to\infty} \frac{CTMb^2+2C^2T^2}{M^2e^{Mb^2/(2CT)}} = 0$.

(c) Use the conclusions in (a) and (b) to show that $\int_0^\infty x^3e^{-Mx^2/(2CT)}\,dx = \frac{2C^2T^2}{M^2}$.

(d) Use the given definition of $\overline{x}$ along with the conclusion in (c) to show that

$$\overline{x} = \sqrt{\frac{8CT}{\pi M}}.$$

Convergent improper integrals for which the interval of integration is unbounded result in unbounded regions having finite area. A region in the plane can also be unbounded if it is demarcated in part by the graph of a function and a vertical asymptote of the function's graph. This observation leads to the following definitions for “integrals” where the interval of integration includes an infinite discontinuity of the integrand.

If the function f has an infinite discontinuity at a, but is Riemann integrable on the interval $[t, b]$ for every number t such that $a < t < b$, the **improper integral** $\int_a^b f$ is defined so that

$$\int_{a}^{b} f = \lim_{t\to a^{+}} \int_{t}^{b} f,$$

provided this limit exists. If $\lim_{t\to a^+} \int_t^b f = \infty$, we write $\int_a^b f = \infty$, and if $\lim_{t\to a^+} \int_t^b f = -\infty$, we write $\int_a^b f = -\infty$.

If the function f has an infinite discontinuity at b, but is Riemann integrable on the interval $[a, t]$ for every number t such that $a < t < b$, the **improper integral** $\int_a^b f$ is defined so that

$$\int_{a}^{b} f = \lim_{t\to b^{-}} \int_{a}^{t} f,$$

provided this limit exists. If $\lim_{t\to b^-}\int_a^t f=\infty$, we write $\int_a^b f=\infty$, and if $\lim_{t\to b^-}\int_a^t f=-\infty$, we write $\int_a^b f=-\infty$.

If the function f has an infinite discontinuity at c, but is Riemann integrable on $[a,t]$ for every number t such that $a<t<c$ and also Riemann integrable on $[t,b]$ for every number t such that $c<t<b$, we define the **improper integral** $\int_a^b f$ so that

$$\int_a^b f=\int_a^c f+\int_c^b f,$$

provided both improper integrals on the right side of this equation exist. If both $\int_a^c f=\infty$ and $\int_c^b f=\infty$, we write $\int_a^b f=\infty$, and if both $\int_a^c f=-\infty$ and $\int_c^b f=-\infty$, we write $\int_a^b f=-\infty$.

Example 20.33. The improper integral $\int_0^4 \frac{1}{\sqrt{x}}dx$ converges because

$$\int_0^4 \frac{1}{\sqrt{x}}dx=\lim_{t\to 0^+}\int_t^4 \frac{1}{\sqrt{x}}dx=\lim_{t\to 0^+} 2\sqrt{x}\Big|_{x=t}^4=\lim_{t\to 0^+}(4-2\sqrt{t})=4.$$

On the other hand, the improper integral $\int_1^2 \frac{1}{(2x-4)^2}dx$ diverges because

$$\int_1^2 \frac{1}{(2x-4)^2}dx=\lim_{t\to 2^-}\int_1^t \frac{1}{(2x-4)^2}dx=\lim_{t\to 2^-}\frac{-1}{2(2x-4)}\Big|_{x=1}^t=\lim_{t\to 2^-}\left(\frac{-1}{2(2t-4)}-\frac{1}{4}\right)=\infty.$$

Exercise 20.39. Determine whether $\int_0^9 \frac{1}{\sqrt[3]{x-1}}dx$ converges or diverges. If it converges, find its value.

Exercise 20.40. Let x be an arbitrary positive number and consider

$$\int_0^\infty t^{x-1}e^{-t}\,dt.$$

This integral is improper because the interval of integration is unbounded, but note that when $0<x<1$, the integral is improper in both senses we have discussed, as 0 is an infinite discontinuity of the integrand. Verify that $\int_0^\infty t^{x-1}e^{-t}\,dt$ converges. *Hint*: First show $\int_0^1 t^{x-1}e^{-t}\,dt$ exists; the case where $x\ge 1$ is immediate, but for the case where $0<x<1$, make use of the fact that $0<e^{-t}\le 1$ for t in $[0,1]$. Then show $\int_1^\infty t^{x-1}e^{-t}\,dt$ converges by using L'Hôpital's rule to find a number c for which $c\ge 1$ and $t^{x-1}\le e^{t/2}$ whenever $t\ge c$.

Exercise 20.41. The **gamma function** Γ, which is applied in the study of probability, is defined so that

$$\Gamma(x) = \int_0^\infty t^{x-1} e^{-t}\, dt$$

for every positive number x (the improper integral here converges by Exercise 20.40).

(a) Show that $\Gamma(1) = 1$.

(b) Show that $\Gamma(x + 1) = x\Gamma(x)$ for every positive number x.

(c) Explain why $\Gamma(n + 1) = n!$ for every nonnegative integer n.

21 Approximating the value of an integral

Even when an integral is known to exist, it may be difficult or impossible to determine its exact value. However, it is nearly always possible to approximate the value of an integral by either a Riemann sum or some other type of approximating sum. Moreover, we can often give a bound on the error involved in such an approximation, which can then allow for an integral to be approximated to a specified degree of accuracy.

Limitations on the ability to calculate an integral

Since a given integrand need not have an elementary antiderivative, the fundamental theorem of calculus, Theorem 20.3, is not always applicable when trying to evaluate an integral.

Example 21.1. We already mentioned in Example 20.6 that the function f, where

$$f(x) = \sqrt{1 - x^4}$$

has no elementary antiderivative, so the fundamental theorem of calculus is of no use in attempting to calculate

$$\int_0^{0.5} \sqrt{1 - x^4}\, dx.$$

Moreover, it may happen in an applied problem that we have only some isolated values of a function, acquired experimentally, rather than a formula for the function, yet we would still like to evaluate (approximately, at least) an integral involving this function.

Example 21.2. A county sheriff records the speed of a passing car using a radar gun, obtaining the speeds given in Table 21.1, measured at quarter-second intervals with time $t = 0$ corresponding to the initial reading.

The total distance the car travels during this two-second time interval is given by

$$\int_0^2 f,$$

where $f(t)$ is the car's speed in feet per second at time t seconds. Since we only know the car's speed at several isolated moments in time, we do not have enough information to evaluate this integral exactly. But we could approximate the integral's value via a Riemann sum that uses the available speed values.

https://doi.org/10.1515/9783111429564-021

Table 21.1: Data for Example 21.2.

t (seconds)	Speed (feet per second)
0	99.88
0.25	99.59
0.50	99.15
0.75	98.41
1.00	97.24
1.25	95.63
1.50	93.87
1.75	91.96
2.00	89.61

Denoting the left and right Riemann sums of f corresponding to the regular partition of $[0, 2]$ into eight subintervals by L_8 and R_8, respectively, we obtain

$$\begin{aligned} L_8 &= f(0) \cdot 0.25 + f(0.25) \cdot 0.25 + f(0.50) \cdot 0.25 + f(0.75) \cdot 0.25 \\ &\quad + f(1.00) \cdot 0.25 + f(1.25) \cdot 0.25 + f(1.50) \cdot 0.25 + f(1.75) \cdot 0.25 \\ &= 99.88 \cdot 0.25 + 99.59 \cdot 0.25 + 99.15 \cdot 0.25 + 98.41 \cdot 0.25 \\ &\quad + 97.24 \cdot 0.25 + 95.63 \cdot 0.25 + 93.87 \cdot 0.25 + 91.96 \cdot 0.25 \\ &= 193.9325 \end{aligned}$$

and

$$\begin{aligned} R_8 &= f(0.25) \cdot 0.25 + f(0.50) \cdot 0.25 + f(0.75) \cdot 0.25 + f(1.00) \cdot 0.25 \\ &\quad + f(1.25) \cdot 0.25 + f(1.50) \cdot 0.25 + f(1.75) \cdot 0.25 + f(2.00) \cdot 0.25 \\ &= 99.59 \cdot 0.25 + 99.15 \cdot 0.25 + 98.41 \cdot 0.25 + 97.24 \cdot 0.25 \\ &\quad + 95.63 \cdot 0.25 + 93.87 \cdot 0.25 + 91.96 \cdot 0.25 + 89.61 \cdot 0.25 \\ &= 191.3650. \end{aligned}$$

If we assume the car is slowing down throughout the two-second time interval, we would then be able to conclude that

$$191.365 < \int_0^2 f < 193.9325,$$

meaning the actual distance the car travels during the two seconds the radar gun was producing speed measurements would be between 191.3650 feet and 193.9325 feet.

As illustrated in Example 21.2, the definition of the integral as a limit of Riemann sums permits us to approximate an integral's value using a Riemann sum. We use the notations L_n, R_n, and M_n to represent, respectively, the left endpoint, right endpoint, and midpoint Riemann sums corresponding to a partition of the interval of integration into

n subintervals all having the same length. As n grows larger, the Riemann sum approximations more closely approach the true value of the integral.

Example 21.3. In Example 18.3, we approximated

$$\int_2^7 [4 - (x-3)^2]\,dx$$

using a right endpoint Riemann sum corresponding to five subintervals of equal length, obtaining $R_5 = -10$.

We were able to improve our approximation by using fifty subintervals of equal length, which yielded $R_{50} = -2.425$.

Of course, this integral can be readily evaluated using the fundamental theorem of calculus, so there is really no need to calculate Riemann sum approximations in this particular situation. We did so earlier to motivate the idea that an integral is a limit of Riemann sums, and are reminding you of this fact now to promote the idea that any integral can be approximated by Riemann sums.

Exercise 21.1. Calculate the exact value of $\int_2^7 [4 - (x-3)^2]dx$. How does the value of the integral compare with the approximation $R_{50} = -2.425$?

Estimating the error resulting from approximating the value of an integral

If the number A is an approximation for a quantity Q, we define the resulting (**absolute**) **error** to be the distance $|A - Q|$ between the approximation and the quantity. In practice we usually cannot precisely determine the error resulting from an approximation A for a quantity Q because we do not know the value of Q. But any time one approximates a value, such as that of an integral, it is still desirable to understand as much as possible about the potential error introduced by the approximation. In many situations, we are able to find an upper bound for the error, which then represents a worst-case scenario.

When a Riemann sum is used to approximate an integral, the difference between the upper and lower Darboux sums corresponding to the partition on which the Riemann sum is based serves as an upper bound for the resulting error.

Theorem 21.4. *If the function f is Riemann integrable on the interval $[a, b]$ and $\dot{P}$ is a tagged partition of $[a, b]$, then*

$$\left| S(f, \dot{P}) - \int_a^b f \right| \leq U(f, P) - L(f, P),$$

where P is the untagged partition determining the same subintervals as $\dot{P}$.

Exercise 21.2. Prove Theorem 21.4. *Hint*: Make use of Theorem 19.4.

Exercise 21.3. Referring to Example 21.2, use Theorem 21.4 to obtain an upper bound for the error that results from using L_8 to approximate $\int_0^2 f$.

Midpoint approximating sums

The use of midpoint Riemann sums to approximate the integral of a function for which the second derivative exists and is bounded on the interval of integration yields the following error bound formula.

Theorem 21.5. *Let f be a function that is Riemann integrable on the interval $[a, b]$ and let M_n be the midpoint Riemann sum for $\int_a^b f$ corresponding to the regular partition of $[a, b]$ yielding n subintervals.*

If f'' exists and its absolute value is bounded by the positive number B on $[a, b]$, then

$$\left| M_n - \int_a^b f \right| \le \frac{B(b-a)^3}{24n^2}.$$

Proof. For each i in $\{1, 2, \ldots, n\}$, let m_i be the midpoint of the ith subinterval determined by the regular partition P of $[a, b]$ yielding n subintervals, each subinterval having length $h = \frac{b-a}{n}$, and define the function F_i on $[0, \frac{h}{2}]$ so that

$$F_i(x) = f(m_i) \cdot 2x - \int_{m_i - x}^{m_i + x} f.$$

Note that

$$F_i\left(\frac{h}{2}\right) = f(m_i) \cdot h - \int_{m_i - \frac{h}{2}}^{m_i + \frac{h}{2}} f$$

represents the difference between the term of the midpoint Riemann sum M_n corresponding to the ith subinterval of P and the integral of f on this ith subinterval. Consequently,

$$\left| M_n - \int_a^b f \right| = \left| \sum_{i=1}^{n} \left(f(m_i) \cdot h - \int_{m_i - \frac{h}{2}}^{m_i + \frac{h}{2}} f \right) \right| = \left| \sum_{i=1}^{n} F_i\left(\frac{h}{2}\right) \right|.$$

Our goal is to bound each quantity $F_i(\frac{h}{2})$, in other words, each term in the above summation, thereby ultimately bounding the potential error when M_n is used to approximate $\int_a^b f$.

Observe that for each i, we have

$$F_i(0) = 0$$

and, via the fundamental theorem of calculus,

$$\begin{aligned}
F_i'(x) &= \frac{d}{dx}\left[f(m_i) \cdot 2x - \int_{m_i - x}^{m_i} f - \int_{m_i}^{m_i + x} f \right] \\
&= 2f(m_i) + \frac{d}{dx}\left[\int_{m_i}^{m_i - x} f - \int_{m_i}^{m_i + x} f \right] \\
&= 2f(m_i) + f(m_i - x) \cdot (-1) - f(m_i + x) \\
&= 2f(m_i) - f(m_i - x) - f(m_i + x).
\end{aligned}$$

Thus, for each i, it follows that

$$F_i'(0) = 0$$

and

$$F_i''(x) = f'(m_i - x) - f'(m_i + x).$$

Then, as f' is differentiable on $[a, b]$, for each i in $\{1, 2, \ldots, n\}$ and each x in $(0, \frac{h}{2}]$, the mean value theorem tells us there exists a number $c_i(x)$ for which

$$m_i - x < c_i(x) < m_i + x$$

and

$$f''(c_i(x)) = \frac{f'(m_i + x) - f'(m_i - x)}{2x}.$$

By hypothesis $|f''(x)| \leq B$ for all x in $[a, b]$, so we may then conclude that for each x in $[0, \frac{h}{2}]$ we have

$$|F_i''(x)| = |f'(m_i - x) - f'(m_i + x)| \leq 2Bx.$$

Applying the fundamental theorem of calculus once again, along with the result of Exercise 19.26 and the fact that $F_i'(0) = 0$, we then determine that

$$|F_i'(x)| = |F_i'(x) - F_i'(0)| = \left|\int_0^x F_i''\right| \leq \int_0^x |F_i''(t)|dt \leq \int_0^x 2Btdt = Bx^2,$$

which, also using the fact that $F_i(0) = 0$, implies that

$$\left|F_i\left(\frac{h}{2}\right)\right| = \left|F_i\left(\frac{h}{2}\right) - F_i(0)\right| = \left|\int_0^{\frac{h}{2}} F_i'\right| \leq \int_0^{\frac{h}{2}} |F_i'(x)|dx \leq \int_0^{\frac{h}{2}} Bx^2dx = \frac{Bh^3}{24}.$$

Hence,

$$\left|M_n - \int_a^b f\right| = \left|\sum_{i=1}^n F_i\left(\frac{h}{2}\right)\right| \leq \sum_{i=1}^n \left|F_i\left(\frac{h}{2}\right)\right| \leq n \cdot \frac{Bh^3}{24} = \frac{B(b-a)^3}{24n^2}. \qquad \square$$

Example 21.1 (Continued). Calculating the midpoint Riemann sum approximation M_4 for $\int_0^{0.5} \sqrt{1-x^4}\,dx$ gives us

$$\begin{aligned}
M_4 &= \sqrt{1-\left(\frac{1}{16}\right)^4} \cdot \frac{1}{8} + \sqrt{1-\left(\frac{3}{16}\right)^4} \cdot \frac{1}{8} + \sqrt{1-\left(\frac{5}{16}\right)^4} \cdot \frac{1}{8} + \sqrt{1-\left(\frac{7}{16}\right)^4} \cdot \frac{1}{8} \\
&\approx 1.000 \cdot \frac{1}{8} + 1.000 \cdot \frac{1}{8} + 0.995 \cdot \frac{1}{8} + 0.982 \cdot \frac{1}{8} \\
&\approx 0.497,
\end{aligned}$$

where we have approximated the required square roots to three decimal places (we can use the method given in Exercise 10.29 to approximate the square root of a positive number to any desired level of accuracy).

The integrand for this integral is the function f defined by $f(x) = \sqrt{1-x^4}$, and it is straightforward to verify that

$$\begin{aligned}
f'(x) &= -\frac{2x^3}{\sqrt{1-x^4}}, \\
f''(x) &= \frac{2x^6 - 6x^2}{(\sqrt{1-x^4})^3},
\end{aligned}$$

and

$$f'''(x) = -\frac{12x^5 + 12x}{(\sqrt{1-x^4})^5}.$$

As f''' is negative on the interval $[0, 0.5]$ of integration, it follows that f'' is decreasing on $[0, 0.5]$. So, as $f''(0) = 0$ and $f''(0.5) \approx -1.618$, we may conclude that $|f''(x)| < 1.62$ for all x in $[0, 0.5]$. Thus, by Theorem 21.5, we have

$$\left| M_4 - \int_0^{0.5} \sqrt{1 - x^4}\, dx \right| \le \frac{1.62(0.5 - 0)^3}{24 \cdot 4^2} \approx 0.0000527 < 0.00006,$$

meaning our approximation $M_4 \approx 0.497$ is less than 0.00006 units from the true value of $\int_0^{0.5} \sqrt{1 - x^4}\, dx$. In particular, then, the approximation is accurate to the three decimal places we have recorded, and this accuracy is achieved with a partition of the interval of integration into only four subintervals.

Exercise 21.4. Calculate the midpoint Riemann sum approximation M_{30} for $\int_0^2 e^{x^2}\, dx$, using 2.718 as an approximation for e. Then find an upper bound for how far M_{30} could be from the integral's true value. (The integrand here is another function that has no elementary antiderivative.)

We can use the error bound formula given in Theorem 21.5 to determine the least value of n, the number of subintervals produced from a regular partition of the interval of integration, that guarantees the midpoint approximation M_n for an integral lies within a specified error tolerance.

Exercise 21.5. Find a value of n for which the approximation M_n for $\int_0^2 e^{x^2}\, dx$ is no more than 0.1 units from the true value of the integral.

Trapezoidal approximating sums

Geometrically, a Riemann sum approximation to an integral $\int_a^b f$ is a sum of signed areas of rectangles, one rectangle for each subinterval. The Riemann sum can then be interpreted as the integral of the piecewise constant step function whose graph essentially consists of the "tops" of the rectangles (if the rectangle lies below the x-axis, the top is really the "bottom"). The diagram in Figure 21.1 illustrates this scenario.

In other words, a Riemann sum approximation to $\int_a^b f$ uses a step function to approximate the integrand f, and the approximation itself is the integral of this step function.

There is an extensive literature devoted to the approximation of integrals and many of the approximation methods have their roots in geometry. For instance, rather than build rectangles along each subinterval, as is done with Riemann sums, we could instead build trapezoids. The "top" of the trapezoid built along the ith subinterval $[p_{i-1}, p_i]$ connects the points $(p_{i-1}, f(p_{i-1}))$ and $(p_i, f(p_i))$ by a straight line segment (Figure 21.2).

In this approach, the integrand f is being approximated by a piecewise linear function (Figure 21.3) whose graph consists of the "tops" of the trapezoids (if the trapezoid lies partially or fully below the x-axis, at least part of the top is really the "bottom").

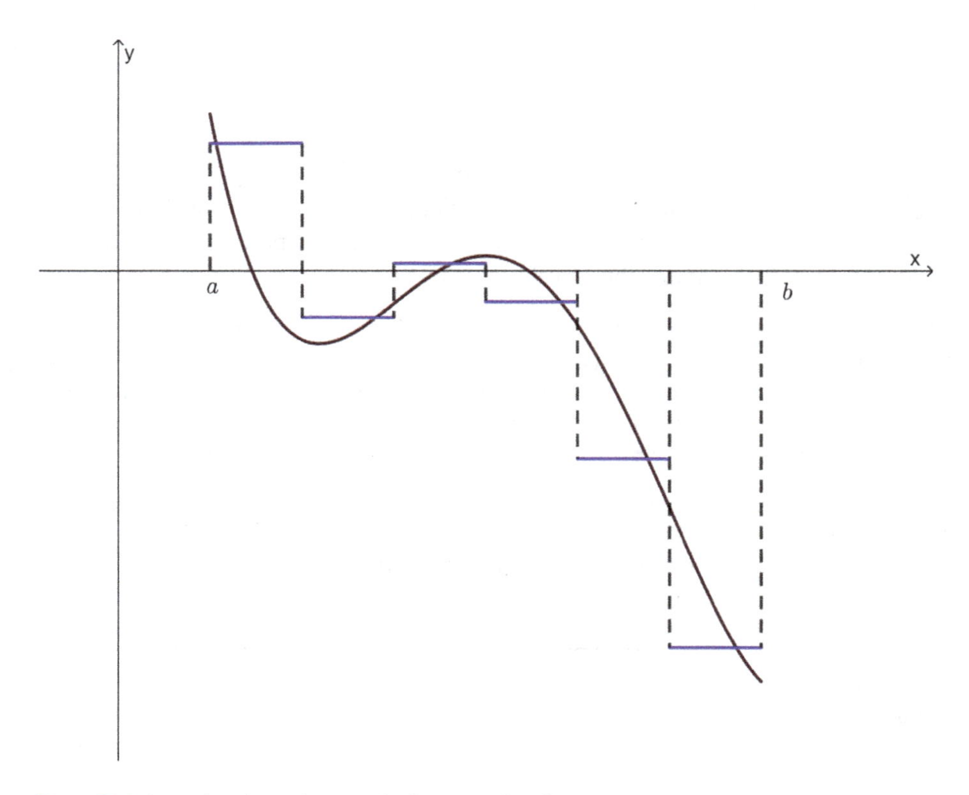

Figure 21.1: Approximating an integrand using a step function.

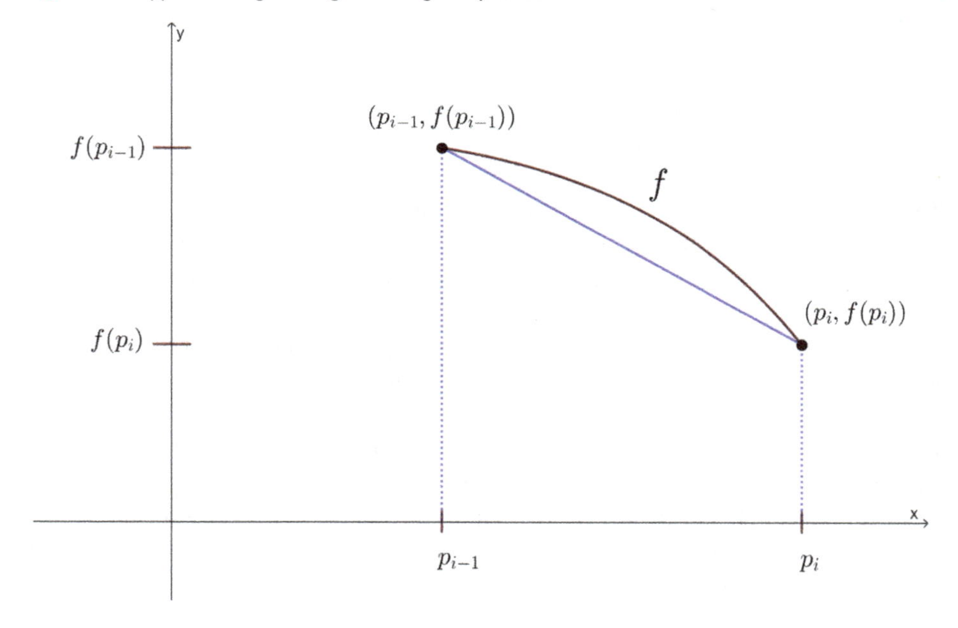

Figure 21.2: Trapezoid built along $[p_{i-1}, p_i]$ and reaching vertically to the graph of function f.

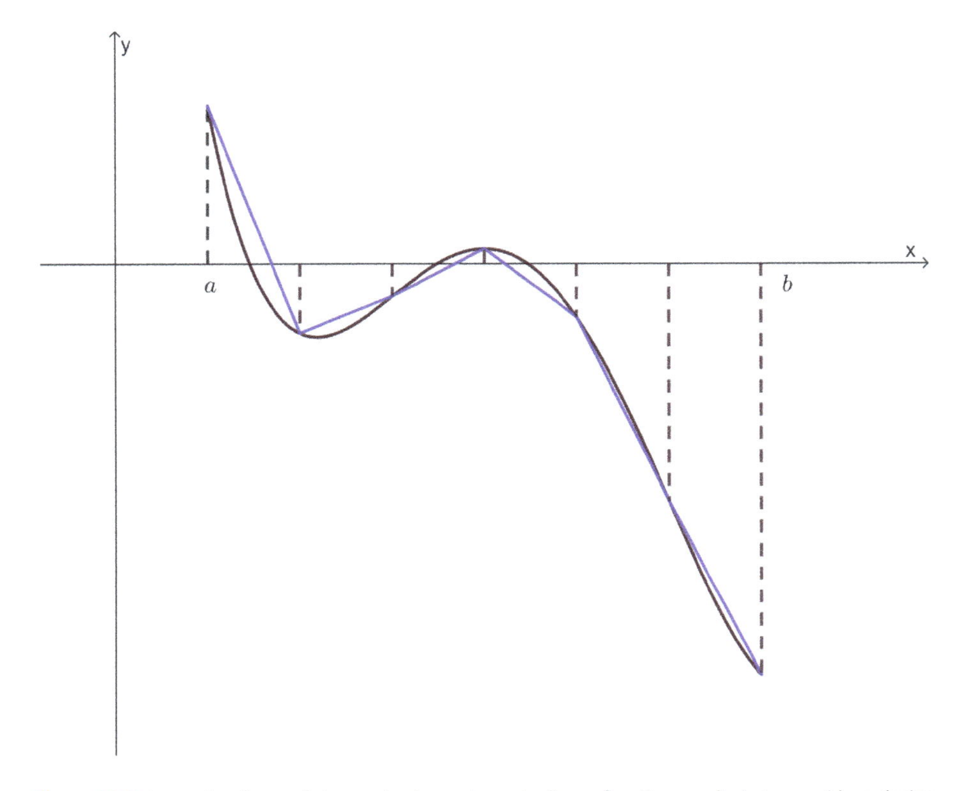

Figure 21.3: Approximating an integrand using a piecewise linear function results in trapezoids replacing rectangles in the resulting approximating sum.

The so-called **trapezoidal approximating sum** T_n for $\int_a^b f$ corresponding to the partition of $[a, b]$ yielding n subintervals of equal length $h = \frac{b-a}{n}$ is obtained by adding the areas of trapezoids built along the subintervals, with the trapezoid built along the ith subinterval

$$[a + (i - 1)h, a + ih]$$

having parallel vertical sides of heights $f(a+(i-1)h)$ and $f(a+ih)$ that are at a distance h from each other (Figure 21.4). Thus, since the area of a trapezoid is half the product of the sum of the lengths of the parallel sides and the distance between them (see Exercise 21.6), we see that

$$T_n = \sum_{i=1}^{n} \frac{1}{2}(f(a + (i - 1)h) + f(a + ih)) \cdot h.$$

The trapezoid built along a particular subinterval may lie partly or entirely below the x-axis, in which case, one or both of its vertical sides' heights are negative. In the

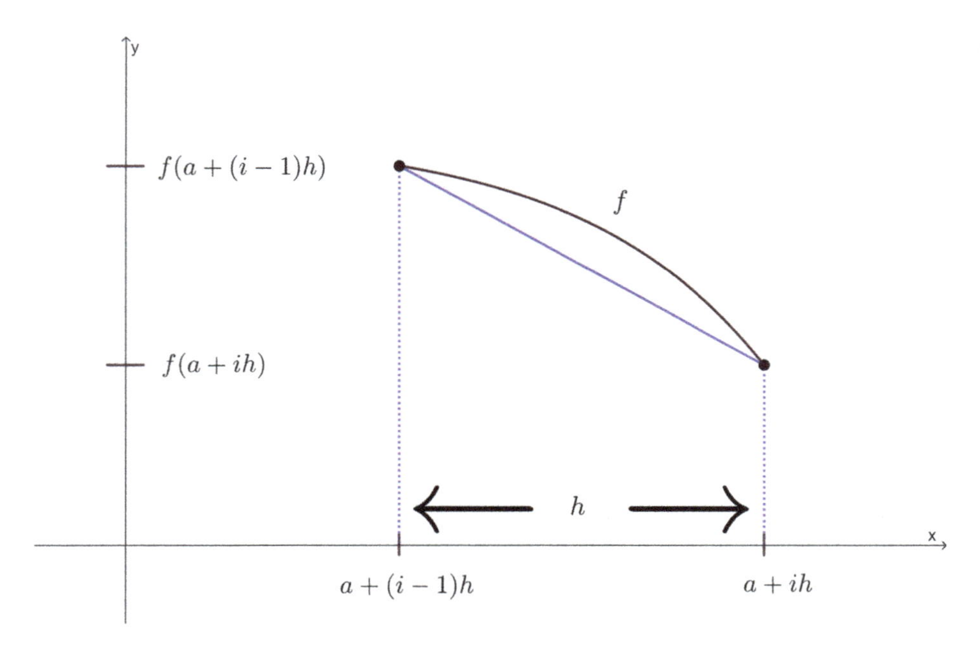

Figure 21.4: Another view of the trapezoid built along the ith subinterval in the trapezoidal approximating sum for $\int_a^b f$.

case where the trapezoid is partially above and partially below the x-axis, the "trapezoid" actually consists of two triangles, one lying above the x-axis and one lying below the x-axis. Remember that, geometrically, the integral measures signed area, which is why sometimes, in this context, heights can be negative and what we are referring to as trapezoids do not always look like traditional trapezoids.

Exercise 21.6. Use the area formulas for rectangles and triangles to verify the formula

$$A = \frac{1}{2}(u+v)d$$

for the area A of a trapezoid having parallel sides of lengths u and v whose distance apart is d.

For consistency of language, we sometimes refer to a left endpoint Riemann sum as a **left endpoint approximating sum**, a right endpoint Riemann sum as a **right endpoint approximating sum**, and a midpoint Riemann sum as a **midpoint approximating sum**.

It turns out that the trapezoidal approximating sum T_n for $\int_a^b f$ is the average of the left endpoint and right endpoint approximating sums L_n and R_n, a result we can certainly anticipate by thinking geometrically.

Theorem 21.6. *If L_n, R_n, and T_n are, respectively, the left endpoint, right endpoint, and trapezoidal approximating sums for $\int_a^b f$ corresponding to the regular partition of $[a,b]$ yielding n subintervals, then $T_n = \frac{L_n+R_n}{2}$. It then follows that*

$$T_n = \frac{h}{2}\left[f(a) + 2\sum_{i=1}^{n-1} f(a+ih) + f(b)\right],$$

where $h = \frac{b-a}{n}$.

Exercise 21.7. Prove Theorem 21.6.

Exercise 21.8. Calculate the trapezoidal approximating sum T_{30} for $\int_0^2 e^{x^2}\,dx$, using 2.718 as an approximation for e.

In a manner similar to that used to obtain an error bound formula for the midpoint approximating sum M_n, we can determine an error bound formula for the trapezoidal approximating sum T_n.

Theorem 21.7. *Let f be a function that is Riemann integrable on the interval $[a,b]$ and let T_n be the trapezoidal approximating sum for $\int_a^b f$ corresponding to the regular partition of $[a,b]$ yielding n subintervals.*

If f'' exists and its absolute value is bounded by the positive number B on $[a,b]$, then

$$\left|T_n - \int_a^b f\right| \le \frac{B(b-a)^3}{12n^2}.$$

Proof. Let $P = \{p_0, p_1, p_2, \ldots, p_n\}$ be the regular partition of $[a,b]$ yielding n subintervals and note that for each i in $\{0,1,2,\ldots,n\}$, we have $p_i = a + ih$, where $h = \frac{b-a}{n}$. For each i in $\{1,2,\ldots,n\}$, define the function F_i on $[0,h]$ so that

$$F_i(x) = \frac{1}{2}(f(p_{i-1}) + f(p_{i-1}+x)) \cdot x - \int_{p_{i-1}}^{p_{i-1}+x} f.$$

Note that

$$F_i(h) = \frac{1}{2}(f(p_{i-1}) + f(p_i)) \cdot h - \int_{p_{i-1}}^{p_i} f$$

represents the difference between the term of the trapezoidal approximating sum T_n corresponding to the ith subinterval of P and the integral of f on this ith subinterval. Consequently,

$$\left|T_n - \int_a^b f\right| = \left|\sum_{i=1}^{n}\left(\frac{1}{2}(f(p_{i-1}) + f(p_i)) \cdot h - \int_{p_{i-1}}^{p_i} f\right)\right| = \left|\sum_{i=1}^{n} F_i(h)\right|.$$

You are asked to complete the proof in Exercise 21.9. □

Notice that the error bound for T_n is exactly twice the error bound for M_n, with the same hypothesis that the second derivative of the integrand be bounded on the interval of integration. Hence, for the same regular partition, the midpoint sum tends to produce more accurate approximations for an integral than the trapezoidal sum.

Exercise 21.9. Complete the proof of Theorem 21.7 by calculating each of

$$F_i(0), \quad F_i'(x), \quad F_i'(0), \quad \text{and} \quad F_i''(x),$$

and then applying the fundamental theorem of calculus to show that

$$\left|F_i'(x)\right| \le \frac{1}{4}Bx^2$$

so that

$$\left|F_i(h)\right| \le \frac{Bh^3}{12}.$$

Simpson's rule

Riemann sum approximations to an integral use piecewise constant functions (degree zero polynomials) to approximate the integrand, while trapezoidal approximations use piecewise linear functions (degree one polynomials). It would seem reasonable to next consider employing piecewise quadratic functions (degree two polynomials) to approximate the integrand. Because the graph of a quadratic function is curved rather than straight, we would expect a parabolic arc to more closely conform to the graph of the integrand, thus resulting in a better approximation to the integral.

Given three points on the graph of a function, we can show there is a unique polynomial function of degree two or less whose graph also contains the three points.

Theorem 21.8. *Let f be a function defined at three distinct inputs x_1, x_2, and x_3, with $f(x_1) = y_1$, $f(x_2) = y_2$, and $f(x_3) = y_3$. Then there is a unique polynomial function p of degree no greater than two such that $p(x_1) = y_1$, $p(x_2) = y_2$, and $p(x_3) = y_3$.*

Exercise 21.10. Prove Theorem 21.8. Note that if the points (x_1,y_1), (x_2,y_2), and (x_3,y_3) are collinear, since their x-coordinates are assumed to be distinct, there is a unique linear function whose graph is the line containing all three points. So it is really a matter of showing that if these three points are noncollinear, then there is a unique quadratic function whose graph contains all of them.

Simpson's rule, first developed by Thomas Simpson (1710–1761), is a method for approximating $\int_a^b f$ that, for a regular partition of $[a,b]$ yielding an even number n of subintervals, groups the subintervals into adjacent pairs (the first and second, the third and fourth, and so on) and then uses Theorem 21.8 to replace the integrand f over each pair of adjacent subintervals with the polynomial approximating function of degree no larger than two whose graph passes through the points on the graph of f corresponding to the three endpoints of the adjacent subintervals. For example, in Figure 21.5, the graph of f is shown in black, while the graph in blue consists of parts of three parabolic arcs, the first passing through the points A, B, and C, the second through C, D, and E, and the last through E, F, and G.

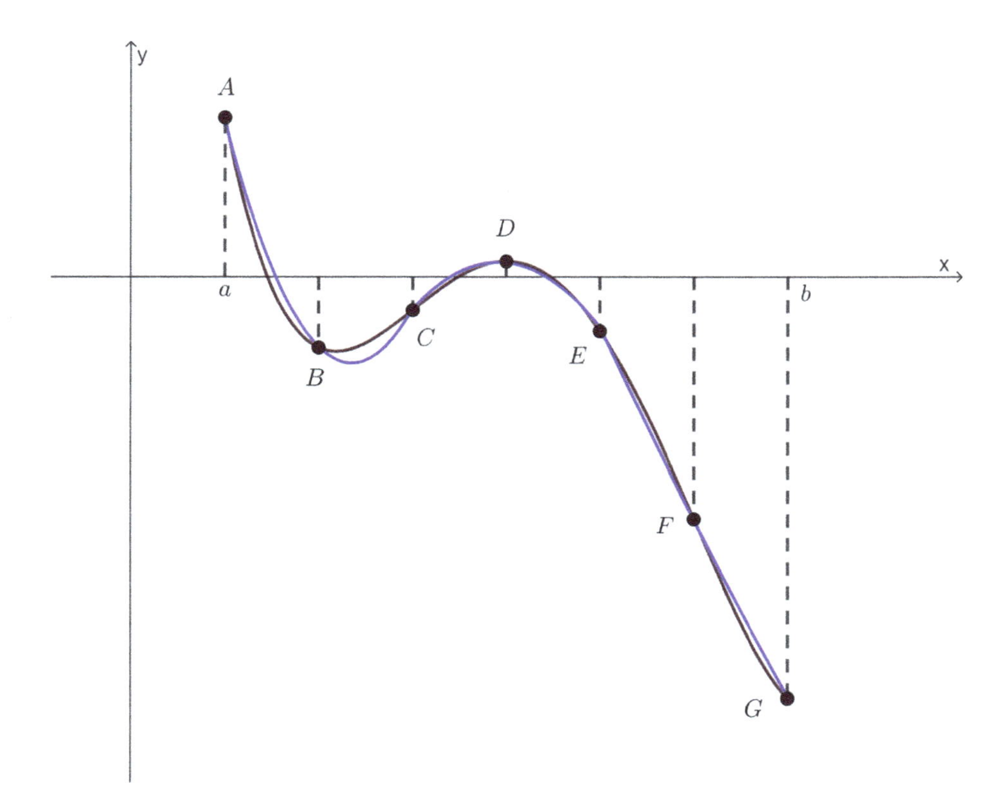

Figure 21.5: Approximating an integrand using a piecewise quadratic function as in Simpson's rule.

The signed area between the graph of the resulting piecewise function and the x-axis along the interval $[a,b]$ is the **Simpson's rule approximating sum** S_n for $\int_a^b f$.

Example 21.1 (Continued). As previously noted, the fundamental theorem of calculus cannot be used to evaluate

$$\int_0^{0.5} \sqrt{1-x^4}\, dx$$

because the integrand does not have an elementary antiderivative. Here, we obtain the Simpson's rule approximation S_4 for this integral, which as $n = 4$ requires determining formulas for two parabolic arcs, one for the first pair of subintervals, $[0, 0.125]$ and $[0.125, 0.25]$, and another for the second pair, $[0.25, 0.375]$ and $[0.375, 0.5]$.

Taking $f(x) = \sqrt{1-x^4}$, observe that

$$\begin{aligned} f(0) &= 1, \\ f(0.125) &\approx 0.9999, \\ f(0.25) &\approx 0.998, \\ f(0.375) &\approx 0.990, \end{aligned}$$

and

$$f(0.5) \approx 0.968,$$

where we have supplied approximations accurate to the nearest thousandth of a unit for values of f that are irrational (Exercise 10.29 could be used to approximate the needed square roots).

Thus, for the first pair of adjacent subintervals, we are looking for a function $q_1 : [0, 0.25] \to \mathbb{R}$, where $q_1(x) = ax^2 + bx + c$ for some constants a, b, and c, and for which $q_1(0) = 1$, $q_1(0.125) = 0.9999$, and $q_1(0.25) = 0.998$. Solving for a, b, and c produces $q_1(x) = -0.064x^2 + 0.008x + 1$.

Similarly, for the second pair of adjacent subintervals, we are looking for a function $q_2 : [0.25, 0.5] \to \mathbb{R}$, where $q_2(x) = Ax^2 + Bx + C$ for some constants A, B, and C, and for which $q_2(0.25) = 0.998$, $q_2(0.375) = 0.990$, and $q_2(0.5) = 0.968$. Solving for A, B, and C produces $q_2(x) = -0.448x^2 + 0.216x + 0.972$.

Figure 21.6 displays the graph of the integrand f along the interval $[0, 0.5]$ of integration, the graph of the quadratic approximating function q_1 along the interval $[0, 0.25]$, and the graph of the quadratic approximating function q_2 along the interval $[0.25, 0.5]$.

The Simpson's rule approximation S_4 for $\int_0^{0.5} \sqrt{1-x^4}\, dx$ is the sum of the areas between the x-axis and the graphs of q_1 and q_2, which is easily calculated using the fundamental theorem of calculus:

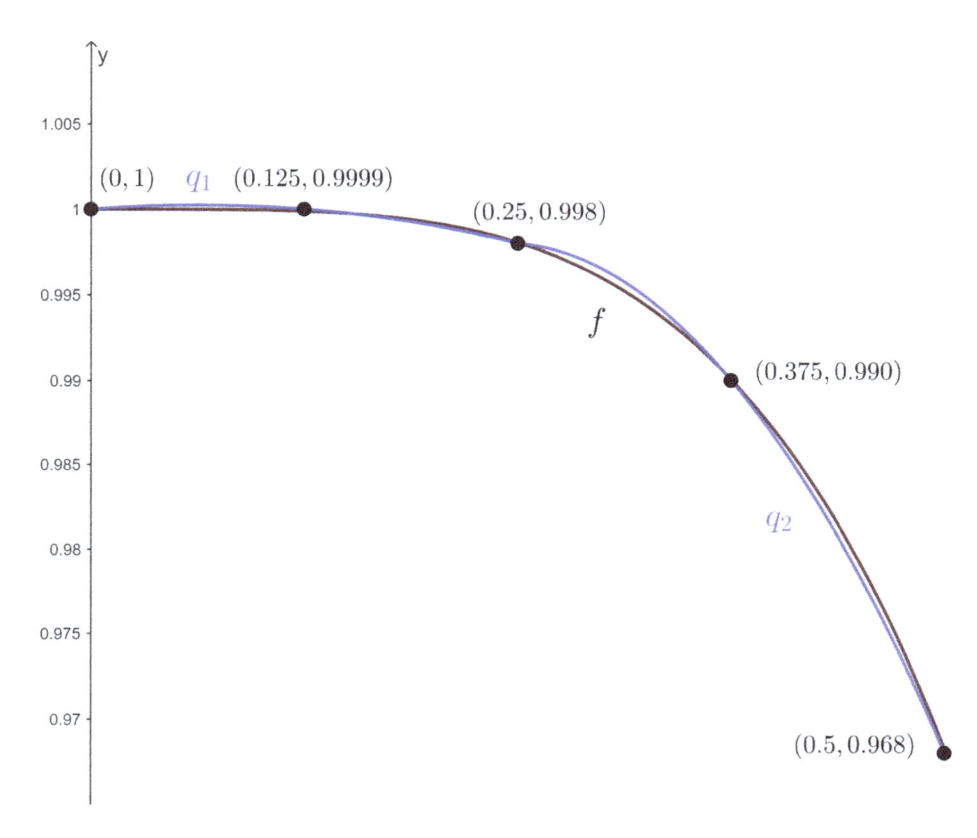

Figure 21.6: Graphs of the functions f, q_1, and q_2 from Example 21.1.

$$\begin{aligned}
S_4 &= \int_0^{0.25} q_1 + \int_{0.25}^{0.5} q_2 \\
&= \int_0^{0.25} (-0.064x^2 + 0.008x + 1)dx + \int_{0.25}^{0.5} (-0.448x^2 + 0.216x + 0.972)dx \\
&= \left[-0.064\frac{x^3}{3} + 0.008\frac{x^2}{2} + x\right]_{x=0}^{0.25} + \left[-0.448\frac{x^3}{3} + 0.216\frac{x^2}{2} + 0.972x\right]_{x=0.25}^{0.5} \\
&\approx 0.497.
\end{aligned}$$

Exercise 21.11. Verify the calculation of the coefficients for the functions q_1 and q_2 of Example 21.1.

We can derive a formula that expresses the Simpson's rule approximation for an integral in terms of the integrand's values at the endpoints of the subintervals into which the interval of integration has been partitioned. The formula means that when applying Simpson's rule, we need not calculate the actual quadratic approximating functions along pairs of adjacent subintervals as we did previously in Example 21.1.

Theorem 21.9. *The Simpson's Rule approximation for*

$$\int_a^b f$$

based on the regular partition of $[a, b]$ *yielding an even number* n *of subintervals of equal length* $h = \frac{b-a}{n}$ *is*

$$S_n = \frac{h}{3}\left[f(a) + 4\sum_{i=1}^{\frac{n}{2}} f(a + (2i-1)h) + 2\sum_{i=1}^{\frac{n}{2}-1} f(a + 2ih) + f(b)\right].$$

Proof. First, consider the special case where the two adjacent subintervals, for which an approximation

$$q(x) = ux^2 + vx + w$$

is to be determined are $[-h, 0]$ and $[0, h]$. The resulting contribution to the Simpson's rule approximation along $[-h, h]$ is

$$\int_{-h}^{h} q = \int_{-h}^{h} (ux^2 + vx + w)dx = \frac{h}{3}(2uh^2 + 6w).$$

Then, as we must have

$$f(-h) = u(-h)^2 + v(-h) + w = uh^2 - vh + w,$$
$$f(0) = u(0)^2 + v(0) + w = w,$$

and

$$f(h) = uh^2 + vh + w,$$

it follows that

$$f(-h) + 4f(0) + f(h) = 2uh^2 + 6w.$$

Hence, the integral from above may be expressed as

$$\int_{-h}^{h} q = \frac{h}{3}[f(-h) + 4f(0) + f(h)].$$

Now, for each i in $\{1, 2, \ldots, \frac{n}{2}\}$, let q_i be the Simpson's rule approximation of f on the ith pair of subintervals $[a + (2i - 2)h, a + (2i - 1)h]$ and $[a + (2i - 1)h, a + 2ih]$, so that

$$q_i(a + (2i - 2)h) = f(a + (2i - 2)h),$$
$$q_i(a + (2i - 1)h) = f(a + (2i - 1)h),$$

and

$$q_i(a + 2ih) = f(a + 2ih).$$

The function q_i is a polynomial function of degree no greater than two, hence, so is the function q_i^* defined on $[-h, h]$ so that

$$q_i^*(x) = q_i(x + a + (2i - 1)h),$$

and we may also observe that

$$q_i^*(-h) = q_i(a + (2i - 2)h) = f(a + (2i - 2)h),$$
$$q_i^*(0) = q_i(a + (2i - 1)h) = f(a + (2i - 1)h),$$

and

$$q_i^*(h) = q_i(a + 2ih) = f(a + 2ih).$$

Geometrically, the graph of q_i^* is obtained by translating the graph of q_i horizontally. Exercise 20.32 analytically encapsulates the intuitive geometric idea that the value of an integral does not change if the graph of the integrand is horizontally translated. Hence, for each i in $\{1, 2, \ldots, \frac{n}{2}\}$, using the conclusion we reached in the special case initially considered,

$$\int_{a+2(i-1)h}^{a+2ih} q_i = \int_{-h}^{h} q_i^* = \frac{h}{3}[f(a + (2i - 2)h) + 4f(a + (2i - 1)h) + f(a + 2ih)].$$

The Simpson's rule approximation for $\int_a^b f$ is therefore

$$\begin{aligned} S_n &= \sum_{i=1}^{\frac{n}{2}} \int_{a+2(i-1)h}^{a+2ih} q_i \\ &= \sum_{i=1}^{\frac{n}{2}} \frac{h}{3}[f(a + (2i - 2)h) + 4f(a + (2i - 1)h) + f(a + 2ih)] \\ &= \frac{h}{3}\left[4\sum_{i=1}^{\frac{n}{2}} f(a + (2i - 1)h) + \sum_{i=1}^{\frac{n}{2}} f(a + (2i - 2)h) + \sum_{i=1}^{\frac{n}{2}} f(a + 2ih)\right] \end{aligned}$$

$$= \frac{h}{3}\left[4\sum_{i=1}^{\frac{n}{2}} f(a+(2i-1)h) + f(a) + \sum_{i=2}^{\frac{n}{2}} f(a+(2i-2)h) \right.$$
$$\left. + \sum_{i=1}^{\frac{n}{2}-1} f(a+2ih) + f(b)\right]$$
$$= \frac{h}{3}\left[4\sum_{i=1}^{\frac{n}{2}} f(a+(2i-1)h) + f(a) + \sum_{i=1}^{\frac{n}{2}-1} f(a+2ih) \right.$$
$$\left. + \sum_{i=1}^{\frac{n}{2}-1} f(a+2ih) + f(b)\right]$$
$$= \frac{h}{3}\left[f(a) + 4\sum_{i=1}^{\frac{n}{2}} f(a+(2i-1)h) + 2\sum_{i=1}^{\frac{n}{2}-1} f(a+2ih) + f(b)\right]. \qquad \square$$

Example 21.1 (Continued). Using the formula from Theorem 21.9 to calculate the Simpson's rule approximation S_4 for $\int_0^{0.5} \sqrt{1-x^4}\,dx$, we have

$$f(x) = \sqrt{1-x^4}, \quad n = 4, \quad \frac{n}{2} = 2, \quad a = 0, \quad b = \frac{1}{2}, \quad \text{and} \quad h = \frac{1}{8},$$

so that

$$S_4 = \frac{1/8}{3}\left[f(0) + 4\sum_{i=1}^{2} f\left((2i-1)\cdot\frac{1}{8}\right) + 2\sum_{i=1}^{1} f\left(2i\cdot\frac{1}{8}\right) + f\left(\frac{1}{2}\right)\right]$$
$$= \frac{1}{24}\left[f(0) + 4\cdot\left(f\left(\frac{1}{8}\right) + f\left(\frac{3}{8}\right)\right) + 2\cdot f\left(\frac{1}{4}\right) + f\left(\frac{1}{2}\right)\right]$$
$$\approx \frac{1}{24}[1 + 4\cdot(1.000 + 0.990) + 2\cdot 0.998 + 0.968]$$
$$\approx 0.497,$$

which matches our earlier "direct" calculation of S_4.

Exercise 21.12. Use Theorem 21.9 to calculate the Simpson's rule approximating sum S_{30} for $\int_0^2 e^{x^2}\,dx$, using 2.718 as an approximation for e.

Next, we obtain an upper bound for the error resulting from the use of Simpson's rule. The proof is similar to those of Theorems 21.5 and 21.7.

Theorem 21.10. *Let f be a function that is Riemann integrable on the interval $[a,b]$ and let S_n be the Simpson's rule approximating sum for $\int_a^b f$ corresponding to the regular partition of $[a,b]$ yielding an even number n of subintervals.*

If $f^{(4)}$ exists and its absolute value is bounded by the positive number B on $[a,b]$, then

$$\left| S_n - \int_a^b f \right| \leq \frac{B(b-a)^5}{180n^4}.$$

Proof. Assume $f^{(4)}$ exists and its absolute value is bounded by the positive number B on $[a,b]$. Let $P = \{p_0, p_1, p_2, \ldots, p_n\}$ be the regular partition of $[a,b]$ yielding an even number n of subintervals and let $h = \frac{b-a}{n}$. For each i in $\{1, 2, \ldots, \frac{n}{2}\}$, define the function F_i on $[0,h]$ so that

$$F_i(x) = \frac{x}{3}[f(p_{2i-1} - x) + 4f(p_{2i-1}) + f(p_{2i-1} + x)] - \int_{p_{2i-1}-x}^{p_{2i-1}+x} f.$$

Based on what we discovered within our proof of Theorem 21.9, we note that

$$F_i(h) = \frac{h}{3}[f(p_{2i-2}) + 4f(p_{2i-1}) + f(p_{2i})] - \int_{p_{2i-2}}^{p_{2i}} f$$

represents the difference between the term of the Simpson's rule approximating sum S_n corresponding to the ith pair of subintervals $[p_{2i-2}, p_{2i-1}]$ and $[p_{2i-1}, p_{2i}]$ determined by P and the integral of f on these subintervals. Consequently,

$$\left| S_n - \int_a^b f \right| = \left| \sum_{i=1}^{\frac{n}{2}} F_i(h) \right|.$$

You are asked to complete the proof in Exercise 21.13. □

Exercise 21.13. Complete the proof of Theorem 21.10. You will need to find formulas for each of $F_i'(x)$, $F_i''(x)$, and $F_i'''(x)$, and verify that for every i in $\{1, 2, \ldots, \frac{n}{2}\}$ it is true that

$$F_i(0) = F_i'(0) = F_i''(0) = F_i'''(0) = 0.$$

Then argue that for every i in $\{1, 2, \ldots, \frac{n}{2}\}$ and every x in $(0,h]$, the mean value theorem may be applied to the function f''' on the interval $[p_{2i-1} - x, p_{2i-1} + x]$. Continue in a manner similar to what was done in the proof of Theorem 21.5.

Example 21.1 (Continued). Earlier, we gave the third derivative of the function f, where $f(x) = \sqrt{1 - x^4}$. It can then be shown that the fourth derivative exists on $[0, 0.5]$ and

$$f^{(4)}(x) = -\frac{60x^8 + 168x^4 + 12}{(\sqrt{1-x^4})^7}.$$

In Exercise 21.14, you show $f^{(4)}$ is decreasing on $[0, 0.5]$. So, as

$$f^{(4)}(0) = -12$$

and

$$f^{(4)}(0.5) \approx -28.496 < -28.5,$$

it follows that, for all x in $[0, 0.5]$, we have $|f^{(4)}(x)| < 28.5$. So, according to Theorem 21.10,

$$\left| S_4 - \int_0^{0.5} \sqrt{1 - x^4}\, dx \right| \leq \frac{28.5(0.5 - 0)^5}{180 \cdot 4^4} \approx 0.0000193 < 0.00002,$$

meaning the Simpson's rule approximation $S_4 = 0.497$ is within 0.00002 units of the true value of $\int_0^{0.5} \sqrt{1 - x^4}\, dx$.

Exercise 21.14. Referring to Example 21.1, verify that $f^{(4)}$ is decreasing on $[0, 0.5]$.

Of the approximate integration methods we have examined, Simpson's rule tends to produce the least error (for the same number of subintervals of the interval of integration).

Example 21.1 (Continued). While we obtained the same approximate value 0.497 for $\int_0^{0.5} \sqrt{1 - x^4}\, dx$ in calculating both M_4 and S_4, the upper bound 0.00002 for the error associated with the Simpson's Rule approximation S_4 is less than the upper bound 0.00006 for the error associated with the midpoint approximation M_4.

Exercise 21.15. Consider once again $\int_0^2 e^{x^2}\, dx$.

(a) In Exercise 21.12, you calculated the Simpson's Rule approximating sum S_{30} for this integral. Find an upper bound for how far S_{30} could be from the integral's true value.

(b) Find a value of n for which the approximation S_n is less than 0.0001 units from the integral's true value.

Exercise 21.16. Referring to Example 21.2, use Simpson's Rule to estimate the total distance the car travels over this two-second time interval. Is there evidence that the car's average speed over this time interval exceeded the speed limit of 65 mph?

Exercise 21.17. Suppose the function f is integrable on the interval $[a, b]$ and we want to approximate $\int_a^b f$. Prove that for any natural number n, we have $S_{2n} = \frac{1}{3}T_n + \frac{2}{3}M_n$. (Thus, the Simpson's Rule approximation for $2n$ equal length subintervals is a weighted average of the trapezoidal and midpoint approximating sums for n equal length subintervals.)

---

22 Infinite series

In Chapter 10, we introduced the notion of an *infinite series* as a special kind of sequence obtained from another sequence, the nth term of the series being the sum of the first n terms of the sequence on which it is built. In Chapter 19, we used infinite series to define what it means for a subset of $\mathbb{R}$ to have measure zero and to prove Lebesgue's characterization of Riemann integrability. Infinite series were the primary object of study in the early history of analysis, up through Cauchy's work in the early nineteenth century, and remain an indispensable tool in many areas of mathematics, including analysis and probability. In this chapter, we initiate a more comprehensive study of infinite series.

Infinite series and convergence

The fundamental goal in the study of infinite series is to give meaning, where possible, to summations of the form

$$a_1 + a_2 + a_3 + \cdots,$$

which we usually write using Σ-notation as

$$\sum_{n=1}^{\infty} a_n,$$

in which there is an unending sequence of numbers being added. By way of concrete motivation, consider the nonterminating decimal expression

$$0.\overline{1} = 0.1111\ldots$$

We studied this kind of representation for real numbers near the end of Chapter 6, but even without the more formal treatment presented there, it certainly seems reasonable to at least intuitively interpret this decimal representation as the "infinite addition"

$$0.1 + 0.01 + 0.001 + 0.0001 + \cdots.$$

The following definition describes what is really happening mathematically when we attempt to add all the infinitely many numbers in a sequence.

Definition 22.1. Given a real sequence (a_n), the **(infinite) series**

$$a_1 + a_2 + a_3 + \cdots = \sum_{n=1}^{\infty} a_n$$

with summands $a_1, a_2, a_3, \ldots$ is defined to be the sequence

https://doi.org/10.1515/9783111429564-022

$$(a_1, a_1 + a_2, a_1 + a_2 + a_3, \ldots),$$

whose nth term is

$$a_1 + a_2 + \cdots + a_n = \sum_{k=1}^{n} a_k,$$

which is also referred to as the **nth partial sum** of the series. Thus, by definition, the series $\sum_{n=1}^{\infty} a_n$ is the sequence $(\sum_{k=1}^{n} a_k)$ consisting of its partial sums.

If the sequence of partial sums of the series $\sum_{n=1}^{\infty} a_n$ converges to a real number S, we say the series **converges**, we refer to S as the **sum** of the series and say the series **converges to** S, and we may write $\sum_{n=1}^{\infty} a_n = S$.

If the sequence of partial sums diverges, we say the series itself **diverges**. In particular, if the sequence of partial sums diverges to ∞, we say the series **diverges to** ∞ and we may write $\sum_{n=1}^{\infty} a_n = \infty$, while if the sequence of partial sums diverges to $-\infty$, we say the series **diverges to** $-\infty$ and we may write $\sum_{n=1}^{\infty} a_n = -\infty$.

Example 22.2. Consider again the series

$$0.1 + 0.01 + 0.001 + 0.0001 + \cdots = \sum_{n=1}^{\infty} (0.1)^n$$

we used as motivation above. The summands of this series are the numbers

$$0.1, \quad 0.01, \quad 0.001, \quad \ldots$$

that are being added, but the series itself is the sequence

$$(0.1,\ 0.1 + 0.01,\ 0.1 + 0.01 + 0.001,\ \ldots)$$

of partial sums determined by the sequence of summands. In particular,

$$0.1$$

is the first partial sum of the series,

$$0.1 + 0.01 = 0.11$$

is the second partial sum of the series,

$$0.1 + 0.01 + 0.001 = 0.111$$

is the third partial sum of the series, and so on. In other words, the series $\sum_{n=1}^{\infty}(0.1)^n$ is actually the sequence that begins

$$(0.1,\ 0.11,\ 0.111,\ \dots).$$

Note that the nth partial sum of the series is

$$\sum_{k=1}^{n}(0.1)^k = 0.1 + (0.1)^2 + (0.1)^3 + \cdots + (0.1)^n.$$

We can simplify this sum using the formula

$$\sum_{i=0}^{n} r^i = 1 + r + r^2 + r^3 + \cdots + r^n = \frac{1 - r^{n+1}}{1 - r}$$

for summing consecutive powers given in Theorem 3.21(6). Doing so, we determine that the nth partial sum of the series $\sum_{n=1}^{\infty}(0.1)^n$ is

$$\sum_{k=1}^{n}(0.1)^k = \frac{1 - (0.1)^{n+1}}{1 - 0.1} - 1.$$

Then, as

$$\lim_{n\to\infty}(0.1)^{n+1} = 0,$$

it follows that the sequence of partial sums of the series $\sum_{n=1}^{\infty}(0.1)^n$ converges to

$$\lim_{n\to\infty}\left(\frac{1 - (0.1)^{n+1}}{1 - 0.1} - 1\right) = \frac{1 - 0}{0.9} - 1 = \frac{1}{9}.$$

By definition, an infinite series is its sequence of partial sums, so we may conclude that the sum of the series $\sum_{n=1}^{\infty}(0.1)^n$ is $\frac{1}{9}$, which is why we may write

$$\sum_{n=1}^{\infty}(0.1)^n = 0.1 + 0.01 + 0.001 + 0.0001 + \cdots = \frac{1}{9}.$$

Also, using the nonterminating decimal expression

$$0.\overline{1} = 0.1111\dots$$

as shorthand for the series

$$0.1 + 0.01 + 0.001 + 0.0001 + \cdots,$$

we see why it is true that

$$0.\overline{1} = 0.1111\dots = \frac{1}{9}.$$

As the wording of Definition 22.1 suggests, infinite series can diverge in various ways.

Example 22.3. The series

$$\sum_{n=1}^{\infty} 2^n$$

is the sequence

$$(2,\ 2+4,\ 2+4+8,\ 2+4+8+16,\ \dots) = (2,\ 6,\ 14,\ 30,\ \dots)$$

whose terms are the partial sums of the series. Since this sequence diverges to ∞, we say that the series $\sum_{n=1}^{\infty} 2^n$ diverges to ∞ and may write $\sum_{n=1}^{\infty} 2^n = \infty$.

Example 22.4. For the series

$$\sum_{n=1}^{\infty} (-1)^n,$$

the first partial sum is

$$(-1)^1 = -1,$$

the second partial sum is

$$(-1)^1 + (-1)^2 = 0,$$

the third partial sum is

$$(-1)^1 + (-1)^2 + (-1)^3 = -1,$$

the fourth partial sum is

$$(-1)^1 + (-1)^2 + (-1)^3 + (-1)^4 = 0,$$

and we see that the sequence of partial sums, which by definition is the given series, is

$$(-1, 0, -1, 0, \dots).$$

This sequence of partial sums diverges, as the odd-indexed terms form a (constant) subsequence that converges to -1, while the even-indexed terms form a (constant) subsequence that converges to 0. Thus, the series $\sum_{n=1}^{\infty} (-1)^n$ diverges.

Determining whether an infinite series converges or diverges can sometimes be aided by rewriting the summands.

Example 22.5. The series

$$\sum_{n=1}^{\infty} \frac{1}{(n+1)(n+2)}$$

is the sequence

$$\left(\frac{1}{2\cdot 3}, \frac{1}{2\cdot 3} + \frac{1}{3\cdot 4}, \frac{1}{2\cdot 3} + \frac{1}{3\cdot 4} + \frac{1}{4\cdot 5}, \dots\right),$$

where

$$\frac{1}{2\cdot 3} = \frac{1}{6}$$

is the first partial sum of the series,

$$\frac{1}{2\cdot 3} + \frac{1}{3\cdot 4} = \frac{1}{4}$$

is the second partial sum of the series,

$$\frac{1}{2\cdot 3} + \frac{1}{3\cdot 4} + \frac{1}{4\cdot 5} = \frac{3}{10}$$

is the third partial sum of the series, and so on. In other words, the given series is the sequence that begins

$$\left(\frac{1}{6}, \frac{1}{4}, \frac{3}{10}, \dots\right).$$

Observing that

$$\frac{1}{(n+1)(n+2)} = \frac{1}{n+1} - \frac{1}{n+2}$$

we see that this series may also be viewed as

$$\sum_{n=1}^{\infty} \left(\frac{1}{n+1} - \frac{1}{n+2}\right),$$

so the sequence of partial sums defining the series can also be expressed as

$$\left(\frac{1}{2} - \frac{1}{3}, \frac{1}{2} - \frac{1}{3} + \frac{1}{3} - \frac{1}{4}, \frac{1}{2} - \frac{1}{3} + \frac{1}{3} - \frac{1}{4} + \frac{1}{4} - \frac{1}{5}, \dots\right),$$

with the nth partial sum being

$$\left(\frac{1}{2} - \frac{1}{3}\right) + \left(\frac{1}{3} - \frac{1}{4}\right) + \left(\frac{1}{4} - \frac{1}{5}\right) + \dots + \left(\frac{1}{n+1} - \frac{1}{n+2}\right) = \frac{1}{2} - \frac{1}{n+2}.$$

The cancelation of terms in this sum leading to its simplification is an effect known as **telescoping**. At this point, we recognize the original series $\sum_{n=1}^{\infty} \frac{1}{(n+1)(n+2)}$ is actually the sequence $(\frac{1}{2} - \frac{1}{n+2})$, a sequence that converges to $\frac{1}{2}$. Thus, the series converges to a sum of $\frac{1}{2}$, and we may write

$$\sum_{n=1}^{\infty} \frac{1}{(n+1)(n+2)} = \sum_{n=1}^{\infty} \left(\frac{1}{n+1} - \frac{1}{n+2}\right) = \frac{1}{2}.$$

Exercise 22.1. Find the first four partial sums of the given series.
(a) $\sum_{n=1}^{\infty} \frac{5^n}{3^{n+1}}$ *Hint*: Make use of the formula from Theorem 3.21(6).
(b) $\sum_{n=1}^{\infty} (n^2 + (-1)^n)$
(c) $\sum_{n=1}^{\infty} (\frac{4}{2^n} - \frac{4}{2^{n+1}})$

Exercise 22.2. Determine which of the series from Exercise 22.1 converge and which diverge. For each series that converges, find its sum.

Exercise 22.3. Prove that the series $\sum_{n=1}^{\infty} (a_{n+1} - a_n)$ converges if and only if the sequence (a_n) converges.

Exercise 22.4. Determine whether the series $\sum_{n=1}^{\infty} \ln(\frac{n+1}{n})$ converges or diverges.

The following two theorems concerning arithmetic properties of convergent infinite series were established in Chapter 10 (see Theorems 10.24 and 10.25, along with Exercise 10.34). The first states that the limit of a sum or difference of convergent series is the sum or difference, respectively, of the series' limits, the second that the limit of a constant multiple of a convergent series is the product of that constant and the limit of the series.

Theorem 22.6 (Sum and difference rules for convergent series). *Suppose the series $\sum_{n=1}^{\infty} a_n$ and $\sum_{n=1}^{\infty} b_n$ both converge, with $\sum_{n=1}^{\infty} a_n = S$ and $\sum_{n=1}^{\infty} b_n = T$. Then each of the series $\sum_{n=1}^{\infty} (a_n + b_n)$ and $\sum_{n=1}^{\infty} (a_n - b_n)$ also converges, with*

$$\sum_{n=1}^{\infty} (a_n + b_n) = \sum_{n=1}^{\infty} a_n + \sum_{n=1}^{\infty} b_n = S + T$$

and

$$\sum_{n=1}^{\infty} (a_n - b_n) = \sum_{n=1}^{\infty} a_n - \sum_{n=1}^{\infty} b_n = S - T.$$

Theorem 22.7 (Constant multiple rule for convergent series). *Suppose the series* $\sum_{n=1}^{\infty} a_n$ *converges, with* $\sum_{n=1}^{\infty} a_n = S$. *Then, for any real number constant* c*, the series* $\sum_{n=1}^{\infty} ca_n$ *converges, with*

$$\sum_{n=1}^{\infty} ca_n = c \sum_{n=1}^{\infty} a_n = cS.$$

The constant multiple rule can be viewed as an infinite extension of the distributive property, as when the series are written out in expanded form, the rule tells us that

$$c(a_1 + a_2 + a_3 + \cdots) = ca_1 + ca_2 + ca_3 + \cdots.$$

Exercise 22.5. Suppose (a_n) and (b_n) are sequences for which $\sum_{n=1}^{\infty}(a_n + 3b_n) = 7$ and $\sum_{n=1}^{\infty}(a_n - b_n) = -5$. Find the sums of the series $\sum_{n=1}^{\infty} a_n$ and $\sum_{n=1}^{\infty} b_n$.

Exercise 22.6. Show that if the series $\sum_{n=1}^{\infty} a_n$ diverges and c is a nonzero real number, then the series $\sum_{n=1}^{\infty} ca_n$ also diverges.

For convenience, there are times when we may choose to begin the indexing for an infinite summation with an integer different from 1.

Example 22.8. To express

$$\frac{1}{8} + \frac{1}{16} + \frac{1}{32} + \frac{1}{64} + \cdots$$

using Σ-notation, we may write

$$\sum_{n=3}^{\infty} \frac{1}{2^n},$$

although the same series can also be expressed as

$$\sum_{n=1}^{\infty} \frac{1}{2^{n+2}}.$$

Changing the lower (starting) index of summation for an infinite series does not affect whether or not the series converges, although such a change may alter the sum if the original series converges.

Theorem 22.9 (Lower index of summation irrelevant to convergence). *Given an infinite series* $\sum_{n=N}^{\infty} a_n$ *whose starting index is the integer* N*, along with any integer* M *larger than* N*, the series* $\sum_{n=N}^{\infty} a_n$ *and the series* $\sum_{n=M}^{\infty} a_n$ *both converge or both diverge. Moreover, if* $\sum_{n=N}^{\infty} a_n$ *has sum* S*, then* $\sum_{n=M}^{\infty} a_n$ *has sum* $S - \sum_{k=N}^{M-1} a_k$.

Exercise 22.7. Prove Theorem 22.9.

Example 22.5 (Continued). We have already determined that the series

$$\sum_{n=1}^{\infty} \frac{1}{(n+1)(n+2)}$$

converges to $\frac{1}{2}$. According to Theorem 22.9, it follows that each of the series

$$\sum_{n=5}^{\infty} \frac{1}{(n+1)(n+2)}$$

and

$$\sum_{n=0}^{\infty} \frac{1}{(n+1)(n+2)}$$

must also converge, though neither has sum $\frac{1}{2}$. Determining their sums is left to you as Exercise 22.8.

Exercise 22.8. Use the fact that $\sum_{n=1}^{\infty} \frac{1}{(n+1)(n+2)} = \frac{1}{2}$ to find the sum of the given series.
(a) $\sum_{n=5}^{\infty} \frac{1}{(n+1)(n+2)}$
(b) $\sum_{n=0}^{\infty} \frac{1}{(n+1)(n+2)}$

Exercise 22.9. Referring to Exercise 22.5, suppose it is also known that $\sum_{n=2}^{\infty} a_n = 5$. Find the value of a_1.

Recall that a sequence (a_n) is a **Cauchy sequence** if for every positive number ε, there is an index N such that when both $m \geq N$ and $n \geq N$, it follows that $| a_m - a_n | < \varepsilon$ (i. e., the distance between terms of the sequence can be made smaller than any positive number by moving far enough out in the sequence). Theorem 11.16, the Cauchy criterion for convergence of a sequence tells us that the Cauchy sequences are precisely the convergent sequences. The following theorem, whose proof is left as an exercise, reformulates the Cauchy criterion in the context of an infinite series.

Theorem 22.10 (The Cauchy criterion for convergence of a series). *A series $\sum_{n=1}^{\infty} a_n$ converges if and only if for every positive number ε, there is an index N such that for any natural numbers m and n for which $n \geq m \geq N$, it follows that $|\sum_{k=m}^{n} a_k| < \varepsilon$.*

In other words, convergence of a series requires that by moving out to a large enough index, the absolute value of the sum of finitely many consecutive summands of the series can always be made arbitrarily small. The value of the index N usually varies with the value of the positive number ε, with smaller values of ε typically giving rise to larger values of N.

Example 22.11. We showed in Example 11.17 that the **harmonic series**

$$\sum_{n=1}^{\infty} \frac{1}{n}$$

diverges because its sequence of partial sums is not a Cauchy sequence. Here we provide a variation on that argument that directly applies Theorem 22.10.

To use this theorem to demonstrate divergence, we must find a particular positive number ε for which, no matter how large an index is chosen, we can always find finitely many consecutive summands beyond this index whose sum exceeds the originally specified ε. Note that for $\varepsilon = \frac{1}{2}$ and any natural number N, it follows that

$$\frac{1}{N+1} + \frac{1}{N+2} + \cdots + \frac{1}{2N} > N \cdot \frac{1}{2N} = \frac{1}{2} = \varepsilon,$$

and so we have shown that in this instance the Cauchy criterion for convergence of a series cannot be fulfilled, hence, the series $\sum_{n=1}^{\infty} \frac{1}{n}$ diverges.

Exercise 22.10. Prove Theorem 22.10.

Exercise 22.11. Use the Cauchy criterion to show the series $\sum_{n=2}^{\infty} (\frac{1}{n-1} - \frac{1}{n})$ converges.

The only way an infinite series can converge is if the sequence of summands, the numbers being added in the series, converges to 0. However, while this condition is necessary for convergence, it is not sufficient.

Theorem 22.12 (The divergence test). *If the series $\sum_{n=1}^{\infty} a_n$ converges, then $a_n \to 0$. Equivalently, if $a_n \nrightarrow 0$, then the series $\sum_{n=1}^{\infty} a_n$ diverges.*

Proof. Assume the series $\sum_{n=1}^{\infty} a_n$ converges and consider any positive number ε. Then by the Cauchy Criterion, there is an index N such that whenever $n \geq N$, it follows that

$$|a_n - 0| = |a_n| = \left| \sum_{k=n}^{n} a_k \right| < \varepsilon,$$

from which we may conclude that $a_n \to 0$. □

Example 22.3 (Continued). As $2^n \nrightarrow 0$, the divergence test provides another way of verifying that the series $\sum_{n=1}^{\infty} 2^n$ diverges.

Example 22.4 (Continued). The sequence whose nth term is $(-1)^n$ does not converge as it endlessly alternates between the numerical values -1 and 1; in particular, then, it does not converge to 0. Hence, as we discovered earlier, the series $\sum_{n=1}^{\infty}(-1)^n$ diverges.

Just because $a_n \to 0$ does not mean the series $\sum_{n=1}^{\infty} a_n$ converges. When $a_n \nrightarrow 0$, we know for certain that $\sum_{n=1}^{\infty} a_n$ diverges, but when $a_n \to 0$, it is possible for $\sum_{n=1}^{\infty} a_n$ to converge or diverge.

Example 22.11 (Continued). Even though $\frac{1}{n} \to 0$, we have shown that the series $\sum_{n=1}^{\infty} \frac{1}{n}$ diverges.

Exercise 22.12. According to the divergence test, which of the following series must diverge?
(a) $\sum_{n=1}^{\infty} \frac{3n^2}{n^2+2}$
(b) $\sum_{n=1}^{\infty}(1.01)^n$
(c) $\sum_{n=1}^{\infty} \frac{1}{\sqrt{n}}$
(d) $\sum_{n=1}^{\infty} \frac{(-1)^n}{n^2}$

Geometric series

It is often a nontrivial matter to determine whether a given infinite series converges. Moreover, even when we are able to establish that a series converges, we may not have a way to determine its sum (there are many series that are known to be convergent for which mathematicians have not yet been able to find the sum of the series).

The so-called *geometric series* form the only significant class of series where it is known which of them converge and how to find their sums when they do converge. Furthermore, we shall see that geometric series can sometimes be used to determine whether another series converges, and they play an important role in connection with *power series*, a topic we take up in Chapter 25.

Given real numbers a and r, the infinite series

$$a + ar + ar^2 + ar^3 + \cdots = \sum_{n=0}^{\infty} ar^n$$

is called a **geometric series with ratio** r. Thus, the defining feature of a geometric series is the existence of a constant r that when multiplied by any summand of the series yields the very next summand. When $r = 0$, we allow ourselves to interpret $r^0 = 0^0$, usually taken to be undefined, to have the value 1, so that the series becomes

$$a \cdot 1 + a \cdot 0 + a \cdot 0^2 + a \cdot 0^3 + \cdots = a.$$

Example 22.2 (Continued). Note that

$$\sum_{n=1}^{\infty}(0.1)^n = 0.1 + 0.01 + 0.001 + 0.0001 + \cdots$$
$$= 0.1 + 0.1 \cdot (0.1)^1 + 0.1 \cdot (0.1)^2 + 0.1 \cdot (0.1)^3 + \cdots$$

is a geometric series with ratio $r = 0.1$ and initial summand $a = 0.1$.

A geometric series converges precisely when its initial summand is 0 or its ratio has absolute value less than 1.

Theorem 22.13 (Convergence of geometric series). *A geometric series $\sum_{n=0}^{\infty} ar^n$ for which $a \neq 0$ converges with sum $\frac{a}{1-r}$ if $|r| < 1$ and diverges if $|r| \geq 1$.*

Proof. Consider a geometric series $\sum_{n=0}^{\infty} ar^n$ for which $a \neq 0$.

First, suppose $|r| < 1$. For each nonnegative integer n, let

$$s_n = \sum_{k=0}^{n} ar^k = a + ar + ar^2 + \cdots + ar^n.$$

Then

$$rs_n = ar + ar^2 + \cdots + ar^n + ar^{n+1} = \sum_{k=1}^{n+1} ar^k.$$

It follows that

$$s_n - rs_n = (a + ar + ar^2 + \cdots + ar^n) - (ar + ar^2 + \cdots + ar^n + ar^{n+1})$$
$$= a - ar^{n+1}$$

so that

$$s_n = \frac{a(1 - r^{n+1})}{1 - r}.$$

Thus,

$$\sum_{n=0}^{\infty} ar^n = \lim_{n\to\infty} s_n = \lim_{n\to\infty} \frac{a(1 - r^{n+1})}{1 - r} = \frac{a}{1 - r},$$

since, with $|r| < 1$, we have $r^{n+1} \to 0$.

Now suppose $|r| \geq 1$. Then as we have also assumed $a \neq 0$, it follows that $ar^n \nrightarrow 0$, so by the divergence test, $\sum_{n=0}^{\infty} ar^n$ diverges. □

Example 22.2 (Continued). We may use the formula for a convergent geometric series to conclude, just as we did earlier, that

$$\sum_{n=1}^{\infty}(0.1)^n = \sum_{n=0}^{\infty} 0.1(0.1)^n = \frac{0.1}{1-0.1} = \frac{1}{9}.$$

Exercise 22.13. What can you conclude about the series in Exercise 22.12(b)?

Exercise 22.14. Determine whether or not the given geometric series converges. If it does, find its sum.
(a) $\sum_{n=0}^{\infty} 7(\frac{4}{9})^n$
(b) $\sum_{n=0}^{\infty}(\frac{6}{5})^n$
(c) $\sum_{n=0}^{\infty} \frac{3^n}{4^{n+1}}$
(d) $\sum_{n=2}^{\infty} 2(-\frac{1}{3})^n$

More about the decimal representation of real numbers

We now continue the discussion of decimal representations of real numbers. In Chapter 6, we explained how the nested intervals property, Theorem 6.7, guarantees that every nonzero real number has a unique **nonterminating decimal representation** of the form

$$\pm N.d_1 d_2 d_3 \ldots,$$

where N is a nonnegative integer, each d_n is in the set $D = \{0,1,2,3,4,5,6,7,8,9\}$ of decimal digits, and infinitely many of the digits d_n are nonzero. Between the examples and exercises, we found that

$$\frac{3}{4} = 0.74\overline{9},$$
$$-\frac{15}{4} = -3.74\overline{9},$$
$$1 = 0.\overline{9},$$

and

$$\frac{1}{27} = 0.\overline{037}.$$

Here we want to use our understanding of infinite series and their convergence to show that $N.d_1 d_2 d_3 \ldots$ is really shorthand for

$$N + d_1 \cdot \frac{1}{10} + d_2 \cdot \frac{1}{10^2} + d_3 \cdot \frac{1}{10^3} + \cdots,$$

then go on to indicate how rational numbers and irrational numbers are distinguished from each other by the type of nonterminating decimal representations they possess.

Theorem 22.14 (Nonterminating decimal representations of real numbers). *Let p be any positive real number and let*

$$N.d_1d_2d_3\ldots$$

be the unique nonterminating decimal representation of p, meaning N is a (nonnegative) integer and each d_n is a decimal digit, with infinitely many of the digits d_n being nonzero. Then

$$p = N + d_1 \cdot \frac{1}{10} + d_2 \cdot \frac{1}{10^2} + d_3 \cdot \frac{1}{10^3} + \cdots = N + \sum_{n=1}^{\infty} d_n \cdot \frac{1}{10^n}$$

and

$$-p = -\left(N + d_1 \cdot \frac{1}{10} + d_2 \cdot \frac{1}{10^2} + d_3 \cdot \frac{1}{10^3} + \cdots\right) = -\left(N + \sum_{n=1}^{\infty} d_n \cdot \frac{1}{10^n}\right).$$

This theorem tells us that the nonterminating decimal representations for real numbers we created using the nested intervals property behave as mathematically expected according to the base ten system of numeration.

Example 22.15. Thus, the mathematical meaning of the nonterminating decimal representation of $\frac{3}{4}$ is

$$0.74\overline{9} = 7 \cdot \frac{1}{10} + 4 \cdot \frac{1}{10^2} + \sum_{n=3}^{\infty} 9 \cdot \frac{1}{10^n}.$$

Proof of Theorem 22.14. The result for negative real numbers follows immediately from that for positive real numbers. So consider any positive real number p and let $N.d_1d_2d_3\ldots$ be its nonterminating decimal representation. In the construction presented at the end of Chapter 6 (see Example 6.10 for a concrete illustration) that indicates how to obtain the nonterminating decimal representation $0.d_1d_2d_3\ldots$ for $p - N$, the nested intervals property is used to determine that

$$\{p - N\} = \bigcap_{n=1}^{\infty} [a_n, b_n],$$

where in particular $a_n = \sum_{k=1}^{n} \frac{d_k}{10^k}$. Since the intervals $[a_n, b_n]$ are nested, it follows from Exercise 11.4 that $a_n \to p - N$. Then as a_n is the nth partial sum of the series $\sum_{n=1}^{\infty} \frac{d_n}{10^n}$, we may conclude that this series converges to $p - N$. Hence, $p = N + \sum_{n=1}^{\infty} \frac{d_n}{10^n}$. □

Exercise 22.15. We have seen that the nonterminating decimal representation of $\frac{1}{27}$ is $0.\overline{037}$. Write $0.\overline{037}$ as an infinite series. Then verify that the sum of this infinite series is $\frac{1}{27}$.

Informally, a nonterminating decimal representation is *periodic* if it eventually repeats, endlessly, the same finite block of digits. For example, $0.74\overline{9}$ is periodic because it eventually just continually repeats the digit 9, and $0.\overline{037}$ is periodic as it endlessly repeats the block 037.

More precisely, the nonterminating decimal representation $\pm N.d_1d_2d_3\ldots$ of a nonzero real number is called **periodic** if there exists a nonnegative integer m and a positive integer j such that whenever $n > m$ we have $d_n = d_{n+j}$. Given that $\pm N.d_1d_2d_3\ldots$ is periodic, if the values of m and j in the defining condition are taken as small as possible, then $d_{m+1}d_{m+2}\ldots d_{m+j}$ is called the **period** and, if $m > 0$, then $d_1d_2\ldots d_m$ is called the **pre-period**. Thus, when $\pm N.d_1d_2d_3\ldots$ is periodic with pre-period $d_1d_2\ldots d_m$ and period $d_{m+1}d_{m+2}\ldots d_{m+j}$, we have

$$\pm N.d_1d_2d_3\ldots = \pm N.d_1d_2\ldots d_m\overline{d_{m+1}d_{m+2}\ldots d_{m+j}}.$$

Nonterminating decimal representations that are not periodic are called **nonperiodic**.

Example 22.16. The nonterminating decimal representation $0.74\overline{9}$ of $\frac{3}{4}$ is periodic with period 9 and pre-period 74. The nonterminating decimal representation $0.\overline{037}$ of $\frac{1}{27}$ is periodic with period 037 and no pre-period.

Example 22.17. The nonterminating decimal representation

$$0.0101000100000001\ldots = \sum_{n=1}^{\infty} \frac{1}{10^{(2^n)}}$$

is nonperiodic.

Exercise 22.16. Identify the pre-period (if any) and period for each of the following periodic decimal representations.

(a) $16.0012\overline{83}$

(b) $0.\overline{9}$

(c) $250.4\overline{953577}$

We can show that the nonzero rational numbers have periodic nonterminating decimal representations, while the irrational numbers have nonperiodic nonterminating decimal representations.

Theorem 22.18. *The nonterminating decimal representation of a nonzero real number is periodic if and only if the number is rational.*

Proof. As a negative number p is rational if and only if the positive number $-p$ is rational, it is sufficient to establish the result for every positive number p.

($\Rightarrow$) Suppose the nonterminating decimal representation $N.d_1d_2d_3\ldots$ of the positive real number p is periodic. Then so is the nonterminating decimal representation $0.d_1d_2d_3\ldots$ of $p - N$ and it suffices to show that $p - N$ is rational.

Assume the pre-period of $0.d_1d_2d_3\ldots$, if there is one, is $d_1d_2\ldots d_m$ and the period is $d_{m+1}d_{m+2}\ldots d_{m+j}$. If we let

$$a = \frac{d_{m+1}}{10^{m+1}} + \frac{d_{m+2}}{10^{m+2}} + \cdots + \frac{d_{m+j}}{10^{m+j}},$$

we have

$$p - N = \sum_{n=1}^{m} \frac{d_n}{10^n} + \sum_{n=0}^{\infty} a\left(\frac{1}{10^j}\right)^n,$$

where

$$\sum_{n=0}^{\infty} a\left(\frac{1}{10^j}\right)^n$$

is a convergent geometric series, since here the positive ratio $\frac{1}{10^j}$ is less than 1. Further, since a is a rational number, the sum S of $\sum_{n=0}^{\infty} a(\frac{1}{10^j})^n$ is also rational, and it follows that

$$p - N = \sum_{n=1}^{m} \frac{d_n}{10^n} + S$$

is rational.

($\Leftarrow$) Suppose p is a positive rational number for which $p = \frac{a}{b}$ for some natural numbers a and b. According to the division algorithm, Theorem 3.7, for each natural number k, there are unique integers Q_k and R_k for which $10^k a = bQ_k + R_k$ and $R_k \in \{0, 1, 2, \ldots, b-1\}$. Our argument proceeds based on whether or not any of the remainders R_k are zero.

First, we consider the possibility that $R_k = 0$ for some k. Then $10^k a = bQ_k$ so that $p = \frac{a}{b} = \frac{Q_k}{10^k}$. As p is positive, the integer Q_k is positive, and it follows that

$$p = \frac{Q_k}{10^k} = N + \frac{d_1}{10} + \frac{d_2}{10^2} + \cdots + \frac{d_k}{10^k},$$

for some nonnegative integer N and decimal digits $d_1, d_2, \ldots, d_k$. Thus, p has the terminating decimal representation $N.d_1d_2\ldots d_k$. If $d_1 = d_2 = \cdots = d_k = 0$, then

$$p = N = (N-1) + 1 = (N-1) + \sum_{n=1}^{\infty} \frac{9}{10^n},$$

where

$$\sum_{n=1}^{\infty} \frac{9}{10^k} = \frac{\frac{9}{10}}{1 - \frac{1}{10}} = 1$$

using the formula for a convergent geometric series. If we let $M = N-1$, then we see that the nonterminating decimal representation of p is $M.\overline{9}$, which is periodic. Otherwise, let $m = \max\{i \mid d_i \neq 0\}$, in which case we have

$$p = N + \frac{d_1}{10} + \frac{d_2}{10^2} + \cdots + \frac{d_m}{10^m} = N + \frac{d_1}{10} + \frac{d_2}{10^2} + \cdots + \frac{d_m - 1}{10^m} + \sum_{n=1}^{\infty} \frac{9}{10^{m+n}},$$

where

$$\sum_{n=1}^{\infty} \frac{9}{10^{m+n}} = \frac{\frac{9}{10^{m+1}}}{1 - \frac{1}{10}} = \frac{1}{10^m},$$

again using the formula for a convergent geometric series. Note that as $d_m \neq 0$, it follows that $d = d_m - 1$ is a decimal digit. Thus, the nonterminating decimal representation of p is $N.d_1 d_2 \ldots d_{m-1} d\overline{9}$, which is periodic.

Having dealt with the case where some remainder is zero, we turn to the possibility that $R_k \neq 0$ for all k. Let $N.d_1 d_2 d_3 \ldots$ be the nonterminating decimal representation of p, so that

$$p = N + \sum_{n=1}^{\infty} \frac{d_n}{10^n}.$$

Thus, for each natural number k, we have

$$10^k p = \left(N \cdot 10^k + \sum_{n=1}^{k} d_n \cdot 10^{k-n} \right) + \sum_{n=1}^{\infty} \frac{d_{k+n}}{10^n}.$$

Using our earlier conclusion, based on the division algorithm, that $10^k a = bQ_k + R_k$, it follows that

$$10^k p = Q_k + \frac{R_k}{b}.$$

Consequently,

$$Q_k = N \cdot 10^k + \sum_{n=1}^{k} d_n \cdot 10^{k-n}$$

and

$$\frac{R_k}{b} = \sum_{n=1}^{\infty} \frac{d_{k+n}}{10^n}.$$

Hence, the nonterminating decimal representation for $10^k p$ is

$$Q_k.d_{k+1}d_{k+2}d_{k+3}\dots$$

and the nonterminating decimal representation for $\frac{R_k}{b}$ is

$$0.d_{k+1}d_{k+2}d_{k+3}\dots.$$

Since there are only finitely many values that the nonzero remainders R_k can assume (specifically, the integers 1 up through $b-1$), they cannot be distinct for all natural numbers k. Let i and j be the least natural numbers for which $R_i = R_j$ and $i < j$. Then $\frac{R_i}{b} = \frac{R_j}{b}$ so that

$$0.d_{i+1}d_{i+2}d_{i+3}\dots = 0.d_{j+1}d_{j+2}d_{j+3}\dots,$$

and, hence, by the uniqueness of nonterminating decimal representations,

$$d_{i+k} = d_{j+k}$$

for all natural numbers k. Therefore, when $n \geq i + 1$, so that $n - i \geq 1$, it follows that

$$d_{n+(j-i)} = d_{j+(n-i)} = d_{i+(n-i)} = d_n,$$

so, by definition, the nonterminating decimal representation $N.d_1d_2d_3\dots$ for p is periodic. □

Example 22.16 (Continued). Note that the nonterminating decimal representations $0.74\overline{9}$ and $0.\overline{037}$, both of which are periodic, denote rational numbers, respectively, $\frac{3}{4}$ and $\frac{1}{27}$.

Example 22.17 (Continued). While there is a pattern to the digits in the nonterminating decimal representation

$$0.0101000100000001\dots = \sum_{n=1}^{\infty} \frac{1}{10^{(2^n)}},$$

the pattern is not one in which a specific finite block of digits is eventually endlessly repeated, so the representation is non-periodic and denotes an irrational number.

Exercise 22.17. Tell which decimal representations represent rational numbers and which represent irrational numbers.

(a) 16.001283

(b) $16.001\overline{283}$

(c) $0.83083008300083\dots = \sum_{n=1}^{\infty} \frac{83}{10^{0.5n^2+1.5n}}$

Exercise 22.18. Express the periodic decimal representation $0.68\overline{521}$ in the form of a fraction.

Exercise 22.19. Find the nonterminating decimal representation of the rational number $\frac{57}{66}$.

Exercise 22.20. (See also Exercise 6.16.) Recall that the **binary digits** are 0 and 1. It can be shown that every number p in the interval $(0,1]$ has a **nonterminating binary representation** $0.b_1b_2b_3\ldots$, where each b_n is a binary digit, infinitely many of these digits are nonzero, and $p = \sum_{n=1}^{\infty} b_n \cdot \frac{1}{2^n}$. When a number p in $(0,1]$ can be represented as $\sum_{k=1}^{n} b_k \cdot \frac{1}{2^k}$ for some binary digits $b_1, b_2, \ldots, b_n$, we refer to $0.b_1b_2\ldots b_n$ as the **terminating binary representation** of p; not all numbers in $(0,1]$ have terminating binary representations.
(a) Find both terminating and nonterminating binary representations for $\frac{1}{2}$.
(b) Find the rational number (in fraction form) having nonterminating binary representation $0.\overline{01}$.

Exercise 22.21. (See also Exercise 6.17.) The ternary digits are 0, 1, and 2. It can be shown that every number p in the interval $(0,1]$ has a **nonterminating ternary representation** $0.t_1t_2t_3\ldots$, where each t_n is a ternary digit, infinitely many of these digits are nonzero, and $p = \sum_{n=1}^{\infty} t_n \cdot \frac{1}{3^n}$. When a number p in $(0,1]$ can be represented as $\sum_{k=1}^{n} t_k \cdot \frac{1}{3^k}$ for some ternary digits $t_1, t_2, \ldots, t_n$, we refer to $0.t_1t_2\ldots t_n$ as the **terminating ternary representation** of p; not all numbers in $(0,1]$ have terminating ternary representations.
(a) Find both terminating and nonterminating ternary representations for $\frac{1}{9}$.
(b) Find the rational number (in fraction form) having nonterminating ternary representation $0.\overline{01}$.
(c) The Cantor set was defined in Example 8.4. Describe the ternary representations of numbers in the Cantor set. (Compare with Exercise 8.5(b).)

Nonnegative series

A series $\sum_{n=1}^{\infty} a_n$ all of whose summands a_n are nonnegative is called a **nonnegative series**. There are a number of results that apply to nonnegative series, but not always to series more generally. The most fundamental of these results is that a series having no negative summands converges precisely when its sequence of partial sums is bounded, a consequence of the monotone convergence theorem.

Theorem 22.19 (Convergence of nonnegative series). *A nonnegative series converges if and only if its sequence of partial sums is bounded. When a nonnegative series converges, the sum of the series is the least upper bound of the set of partial sums of the series.*

Example 22.5 (Continued). Earlier we showed that the series

$$\sum_{n=1}^{\infty} \frac{1}{(n+1)(n+2)}$$

converges to $\frac{1}{2}$. The summands of this series are the numbers $\frac{1}{(n+1)(n+2)}$ where n is a natural number. As all the summands are nonnegative, it follows that $\frac{1}{2}$ must be the least upper bound of the set

$$\left\{ \sum_{k=1}^{n} \frac{1}{(k+1)(k+2)} \;\middle|\; n \in \mathbb{N} \right\}$$

of partial sums of the series. This conclusion is not really surprising, because we had already determined that

$$\sum_{k=1}^{n} \frac{1}{(k+1)(k+2)} = \frac{1}{2} - \frac{1}{n+2},$$

and it is clear that

$$\sup\left\{ \frac{1}{2} - \frac{1}{n+2} \;\middle|\; n \in \mathbb{N} \right\} = \frac{1}{2}.$$

Exercise 22.22. Use Theorem 10.18, the monotone convergence theorem, to prove Theorem 22.19.

Exercise 22.23. Use induction to prove that $\sum_{k=1}^{2^n} \frac{1}{k} > \frac{n+1}{2}$ for every natural number n. Then explain how this shows the partial sums of the harmonic series $\sum_{n=1}^{\infty} \frac{1}{n}$ are unbounded, which provides another argument for why the harmonic series diverges.

Exercise 22.24. Show that the conclusions of Theorem 22.19 hold even when a series $\sum_{n=1}^{\infty} a_n$ is just **eventually nonnegative**, meaning there is an index N for which a_n is nonnegative whenever $n \geq N$.

Exercise 22.25. Show that if $\sum_{n=1}^{\infty} a_n$ is a convergent nonnegative series and (a_{k_n}) is a subsequence of (a_n), then the series $\sum_{n=1}^{\infty} a_{k_n}$ converges.

Exercise 22.26. Suppose (a_n) is a bounded nonincreasing sequence and (b_n) is a bounded nondecreasing sequence. For each natural number n, let $c_n = a_n + b_n$. Prove that the series $\sum_{n=1}^{\infty} |c_{n+1} - c_n|$ converges.

It is sometimes possible to resolve the question of whether or not a nonnegative series converges by comparing its summands with those of another nonnegative series whose convergence status is known. If the corresponding summands are no larger than those of some convergent nonnegative series, the series must converge, and if the corresponding summands are no smaller than those of some divergent nonnegative series, the series must diverge.

Theorem 22.20 (The comparison test). *Let (a_n) and (b_n) be sequences, and suppose there is a natural number N for which $0 \leq a_n \leq b_n$ whenever $n \geq N$.*

(1) *If the series* $\sum_{n=1}^{\infty} b_n$ *converges, then the series* $\sum_{n=1}^{\infty} a_n$ *converges.*
(2) *If the series* $\sum_{n=1}^{\infty} a_n$ *diverges, then the series* $\sum_{n=1}^{\infty} b_n$ *diverges.*

Example 22.21. The nonnegative geometric series $\sum_{n=1}^{\infty} \frac{1}{4^n}$ has ratio $\frac{1}{4}$, so it converges. Thus, as for each natural number n, we have $\frac{1}{1+4^n} < \frac{1}{4^n}$, it follows by comparison that the nonnegative series $\sum_{n=1}^{\infty} \frac{1}{1+4^n}$ also converges.

Exercise 22.27. Show how Theorem 22.20, the comparison test, follows from Theorem 22.19.

Exercise 22.28. Use the result of Exercise 22.11 along with Theorem 22.20 to show that the series $\sum_{n=1}^{\infty} \frac{1}{n^2}$ converges.

Exercise 22.29. Let (a_n) be a sequence of positive numbers and suppose the series $\sum_{n=1}^{\infty} a_n$ converges. For each natural number n, define m_n to be $\frac{1}{n} \sum_{k=1}^{n} a_k$, the arithmetic mean of the first n terms of (a_n). Show that the series $\sum_{n=1}^{\infty} m_n$ diverges.

When each summand of one convergent series is no larger than the corresponding summand of another convergent series, the sums of the series must also maintain this ordering.

Theorem 22.22. *Suppose that* $a_n \leq b_n$ *for every natural number n. If both of the series* $\sum_{n=1}^{\infty} a_n$ *and* $\sum_{n=1}^{\infty} b_n$ *converge, with* $\sum_{n=1}^{\infty} a_n = S$ *and* $\sum_{n=1}^{\infty} b_n = T$*, then* $S \leq T$.

Example 22.21 (Continued). Using the formula for the sum of a convergent geometric series, we may conclude that $\sum_{n=1}^{\infty} \frac{1}{4^n} = \frac{1}{3}$. As $\frac{1}{1+4^n} < \frac{1}{4^n}$ for every natural number n, it follows that the sum of the series $\sum_{n=1}^{\infty} \frac{1}{1+4^n}$, which we already demonstrated must converge, is no larger than $\frac{1}{3}$.

Exercise 22.30. Prove Theorem 22.22.

Exercise 22.31. Consider the series $\sum_{n=1}^{\infty} e^{-n}$.
(a) Find the sum of this series.
(b) Explain how we then know that the series $\sum_{n=1}^{\infty} e^{-n^2}$ converges. What do we know about the sum of this series?

An integral, being a limiting value of Riemann sums, represents a form of "continuous" summation. Our next result relates the two forms of summation represented by integrals and infinite series.

Theorem 22.23 (The integral test). *Suppose the function $f : [1, \infty) \to \mathbb{R}$ is nonincreasing and takes only positive values. Then the infinite series $\sum_{n=1}^{\infty} f(n)$ converges if and only if the improper integral $\int_1^{\infty} f$ converges.*

Proof. Since f is monotone, it is integrable on $[1, n+1]$ for each natural number n. Also, since f is nonincreasing the monotonicity of the integral implies that, for every n, we have

$$\sum_{k=1}^{n} f(k) \geq \int_{1}^{n+1} f \geq \sum_{k=2}^{n+1} f(k).$$

This inequality implies the sequence of partial sums of $\sum_{n=1}^{\infty} f(n)$ is bounded if and only if the sequence $(\int_1^n f)$ is bounded. The desired conclusion follows from the fact that $\sum_{n=1}^{\infty} f(n)$, being a nonnegative series, converges if and only if it is bounded. □

Exercise 22.32. Suppose the function $f : [1, \infty) \to \mathbb{R}$ is nonincreasing and takes only positive values, so the hypotheses of the integral test are satisfied. Assume that both $\sum_{n=1}^{\infty} f(n)$ and $\int_1^{\infty} f$ converge.
(a) Show by example that it is possible for $\sum_{n=1}^{\infty} f(n)$ and $\int_1^{\infty} f$ to have different values.
(b) Prove that for each natural number n we must have

$$\int_{n+1}^{\infty} f \leq \sum_{k=1}^{\infty} f(k) - \sum_{k=1}^{n} f(k) \leq \int_{n}^{\infty} f.$$

Thus, provided we can calculate each of the improper integrals here, the difference between their values yields an upper bound for how far the nth partial sum of the series $\sum_{n=1}^{\infty} f(n)$ is from the sum of the series.

For each positive real number p, the series $\sum_{n=1}^{\infty} \frac{1}{n^p}$ is called the ***p*-series**. As we have seen, the 1-series $\sum_{n=1}^{\infty} \frac{1}{n}$ is also called the **harmonic series**. Collectively, the p-series form another significant class of infinite series, and we can use the integral test to determine which p-series converge.

Theorem 22.24 (Convergence of p-series). *The p-series $\sum_{n=1}^{\infty} \frac{1}{n^p}$ converges when $p > 1$ and diverges when $0 < p \leq 1$.*

Exercise 22.33. Use the results from Exercise 20.35 along with the integral test to prove Theorem 22.24.

Example 22.11 (Continued). Theorem 22.24 provides another means for concluding that the harmonic series $\sum_{n=1}^{\infty} \frac{1}{n}$ diverges.

Example 22.25. According to Theorem 22.24, the series $\sum_{n=1}^{\infty} \frac{1}{n^2}$ converges while the series $\sum_{n=1}^{\infty} \frac{1}{\sqrt{n}} = \sum_{n=1}^{\infty} \frac{1}{n^{1/2}}$ diverges.

There is no general formula for determining the sum of a convergent p-series. In fact, the sums have been found for relatively few specific values of p where $p > 1$. For example, Leonhard Euler (1707–1783) showed that $\sum_{n=1}^{\infty} \frac{1}{n^2} = \frac{\pi^2}{6}$.

Another form of comparison that can sometimes be used to determine whether a series converges is provided by the following theorem.

Theorem 22.26 (The limit comparison test). *Suppose (a_n) and (b_n) are sequences of positive numbers for which $\lim_{n\to\infty} \frac{a_n}{b_n} = L$ for some positive number L. Then the two series $\sum_{n=1}^{\infty} a_n$ and $\sum_{n=1}^{\infty} b_n$ both converge or both diverge.*

Proof. Assuming (a_n) and (b_n) are sequences of positive numbers for which $\lim_{n\to\infty} \frac{a_n}{b_n} = L$, where L is a positive number, then there exists an index N such that

$$\frac{1}{2}L < \frac{a_n}{b_n} < \frac{3}{2}L$$

whenever $n \geq N$. Thus, for such n, we have

$$0 < \frac{1}{2}Lb_n < a_n < \frac{3}{2}Lb_n.$$

Applying Theorem 22.20, the comparison test, we see that

(1) if $\sum_{n=1}^{\infty} a_n$ converges, it follows that $\sum_{n=1}^{\infty} \frac{1}{2}Lb_n$ converges, hence, that $\sum_{n=1}^{\infty} b_n$ converges; and,

(2) if $\sum_{n=1}^{\infty} a_n$ diverges, it follows that $\sum_{n=1}^{\infty} \frac{3}{2}Lb_n$ diverges, hence, that $\sum_{n=1}^{\infty} b_n$ diverges. □

It is often possible to apply the limit comparison test with an appropriate p-series in order to determine the convergence or divergence of a nonnegative series whose summands involve algebraic quotients. Because higher powers of n grow faster as $n \to \infty$, one strategy that can help us identify an appropriate p-series is to form the fraction $\frac{n^u}{n^v}$, where n^u and n^v are the highest powers of n appearing in, respectively, the numerator and denominator of the nth summand of the series whose convergence status is to be determined, then reduce this fraction to the form $\frac{1}{n^p}$.

Example 22.27. We can use the limit comparison test to argue why the series $\sum_{n=2}^{\infty} \frac{n+5}{n^2-1}$ diverges. Identifying the highest powers of n that appear in the numerator and denominator of $\frac{n+5}{n^2-1}$ we form the fraction $\frac{n}{n^2}$, which reduces to $\frac{1}{n}$. So we compare the given series to the series $\sum_{n=2}^{\infty} \frac{1}{n}$ via the limit comparison test.

For $n \geq 2$, let $a_n = \frac{n+5}{n^2-1}$ and let $b_n = \frac{1}{n}$. Note that $\sum_{n=2}^{\infty} b_n$ diverges as otherwise the harmonic series $\sum_{n=1}^{\infty} \frac{1}{n}$ would converge. Now

$$\lim_{n\to\infty} \frac{a_n}{b_n} = \lim_{n\to\infty} \frac{\frac{n+5}{n^2-1}}{\frac{1}{n}} = \lim_{n\to\infty} \frac{n^2+5n}{n^2-1} = \lim_{n\to\infty} \frac{1+\frac{5}{n}}{1-\frac{1}{n^2}} = 1,$$

a positive number. Thus, by the limit comparison test, we may conclude that $\sum_{n=2}^{\infty} \frac{n+5}{n^2-1}$ diverges.

Exercise 22.34. Use the limit comparison test to determine whether the given series converges.
(a) $\sum_{n=1}^{\infty} \frac{1}{(n+5)^2}$
(b) $\sum_{n=1}^{\infty} \frac{1}{\sqrt{n^3}+1}$
(c) $\sum_{n=1}^{\infty} \frac{3}{4+7n}$

Exercise 22.35. Show by example that if (a_n) and (b_n) are sequences of positive numbers for which the given condition holds, then it is possible for one of the series $\sum_{n=1}^{\infty} a_n$ and $\sum_{n=1}^{\infty} b_n$ to converge and the other to diverge.
(a) $\lim_{n\to\infty} \frac{a_n}{b_n} = 0$
(b) $\lim_{n\to\infty} \frac{a_n}{b_n} = \infty$

Exercise 22.36. Show that if (a_n) and (b_n) are sequences of positive numbers for which $\lim_{n\to\infty} \frac{a_n}{b_n} = 0$ and the series $\sum_{n=1}^{\infty} b_n$ converges, then the series $\sum_{n=1}^{\infty} a_n$ also converges.

We present one more convergence test that applies to nonnegative series whose summands are nonincreasing.

Theorem 22.28 (The Cauchy condensation test). *Suppose (a_n) is a nonincreasing sequence of nonnegative numbers. Then the series $\sum_{n=1}^{\infty} a_n$ converges if and only if the series $\sum_{n=0}^{\infty} 2^n a_{2^n}$ converges.*

Proof. First suppose $n < 2^m$. As (a_n) is a nonincreasing and nonnegative, it follows that

$$\begin{aligned} &a_1 + 2a_2 + 4a_4 + 8a_8 + \cdots + 2^m a_{2^m} \\ &\quad \geq a_1 + (a_2 + a_3) + (a_4 + a_5 + a_6 + a_7) + (a_8 + \cdots + a_{15}) + \cdots \\ &\qquad \cdots + (a_{2^m} + \cdots + a_{2^{m+1}-1}) \\ &\quad \geq a_1 + a_2 + a_3 + a_4 + \cdots + a_n. \end{aligned}$$

Thus, if the partial sums of $\sum_{n=1}^{\infty} a_n$ are unbounded, so are the partial sums of $\sum_{n=0}^{\infty} 2^n a_{2^n}$.

Now suppose $n > 2^m$. Again, as (a_n) is a nonincreasing and nonnegative, it follows that

$$\begin{aligned} &a_1 + a_2 + a_3 + a_4 + \cdots + a_n \\ &\quad = a_1 + a_2 + (a_3 + a_4) + (a_5 + a_6 + a_7 + a_8) + (a_9 + \cdots + a_{16}) + \cdots \\ &\qquad \cdots + (a_{2^{m-1}+1} + \cdots + a_{2^m}) + (a_{2^m+1} + \cdots + a_n) \\ &\quad \geq a_1 + 1a_2 + 2a_4 + 4a_8 + \cdots + 2^{m-1} a_{2^m} \end{aligned}$$

$$= \frac{1}{2}(2a_1 + 2a_2 + 4a_4 + 8a_8 + \cdots + 2^m a_{2^m})$$
$$\geq \frac{1}{2}(a_1 + 2a_2 + 4a_4 + 8a_8 + \cdots + 2^m a_{2^m}).$$

Thus, if the partial sums of $\sum_{n=1}^{\infty} a_n$ are bounded, so are the partial sums of $\sum_{n=0}^{\infty} 2^n a_{2^n}$.

As both $\sum_{n=1}^{\infty} a_n$ and $\sum_{n=0}^{\infty} 2^n a_{2^n}$ are nonnegative series, by Theorem 22.19, they converge if and only if their partial sums are bounded. Therefore, we may now conclude that $\sum_{n=1}^{\infty} a_n$ converges if and only if $\sum_{n=0}^{\infty} 2^n a_{2^n}$ converges. □

Exercise 22.37. Use the Cauchy condensation test to verify each of the following.

(a) $\sum_{n=2}^{\infty} \frac{1}{n \ln(n)}$ diverges.

(b) $\sum_{n=2}^{\infty} \frac{1}{n(\ln(n))^p}$ converges when $p > 1$.

---

23 Absolute convergence and conditional convergence

Whether a series that includes both positive and negative summands converges can be difficult to pin down, so in this chapter we discuss several convergence tests we can attempt to apply to such series. We also indicate how the series formed from a given series by replacing the summands with their absolute values can sometimes help to resolve the question of whether the original series converges.

Indirect application of the convergence tests for nonnegative series

In the previous chapter, we presented several tests that can be used to determine whether an infinite series whose summands are nonnegative converges. Because the constant multiple rule tells us that if the series $\sum_{n=1}^{\infty} a_n$ converges, so does the series $\sum_{n=1}^{\infty}(-1)a_n$, these tests can be indirectly applied to determine whether a series having only negative summands converges.

Example 23.1. Applying Theorem 22.24 with $p = 3.5$, we see that

$$\sum_{n=1}^{\infty} \frac{1}{n^3\sqrt{n}}$$

is a convergent p-series, so we may then conclude that the series

$$\sum_{n=1}^{\infty} \frac{-1}{n^3\sqrt{n}} = \sum_{n=1}^{\infty}(-1)\frac{1}{n^3\sqrt{n}},$$

all of whose summands are negative, also converges.

Furthermore, keeping in mind that whether a series converges does not depend on any finite number of the initial summands, we realize the convergence tests for nonnegative series can also be useful in analyzing series having only a finite number of summands of a particular sign, negative or positive (i. e., series whose summands are eventually nonnegative or eventually nonpositive).

Example 23.2. Only the first two summands of the series

$$\sum_{n=1}^{\infty} \frac{2n-5}{n^2}$$

are negative. Applying the limit comparison test, Theorem 22.26, with the two positive series $\sum_{n=3}^{\infty} \frac{2n-5}{n^2}$ and $\sum_{n=3}^{\infty} \frac{1}{n}$, we calculate

https://doi.org/10.1515/9783111429564-023

$$\lim_{n\to\infty} \frac{\frac{2n-5}{n^2}}{\frac{1}{n}} = \lim_{n\to\infty}\left(2 - \frac{5}{n}\right) = 2,$$

which is a positive number, and so as the series $\sum_{n=3}^{\infty} \frac{1}{n}$ diverges, so does the series $\sum_{n=3}^{\infty} \frac{2n-5}{n^2}$. It then follows from Theorem 22.9 that $\sum_{n=1}^{\infty} \frac{2n-5}{n^2}$ diverges, too.

Exercise 23.1. Determine whether or not the given series converges, providing evidence to support your conclusions.

(a) $\sum_{n=1}^{\infty} \frac{-4}{3^n}$

(b) $\sum_{n=1}^{\infty} \frac{n^2-100}{(n\sqrt{n})^3}$

(c) $\sum_{n=1}^{\infty} \frac{1}{n-|3-n|}$

Absolutely convergent series and conditionally convergent series

We have fewer results that can help us analyze series having not only infinitely many positive summands but also infinitely many negative summands. A geometric series $\sum_{n=0}^{\infty} ar^n$ that has a negative ratio r is of this type, since the summands alternate sign, and Theorem 22.13 can be used to determine whether such a series converges. But, more generally, even the indirect application of most of the convergence tests we have explored thus far requires the summands to be eventually positive or eventually negative.

Beyond geometric series having negative ratios, there are other series with infinitely many positive and infinitely many negative summands, some that converge, others that diverge.

Example 23.3. We can show the **alternating harmonic series**

$$1 - \frac{1}{2} + \frac{1}{3} - \frac{1}{4} + \frac{1}{5} - \frac{1}{6} + \cdots = \sum_{n=1}^{\infty} \frac{(-1)^{n+1}}{n},$$

whose summands alternate sign, converges. For each n, let

$$s_n = \sum_{k=1}^{n} \frac{(-1)^{k+1}}{k},$$

which is the nth partial sum of the series.

Note that, for any n, we have

$$s_{2(n+1)-1} = s_{2n-1} - \frac{1}{2n} + \frac{1}{2n+1} < s_{2n-1},$$

which implies the subsequence (s_{2n-1}) of (s_n) is decreasing. Also, $s_1 = 1$ and, when $n \geq 2$, we have

$$s_{2n-1} = \left(1 - \frac{1}{2}\right) + \sum_{k=2}^{n-1}\left(\frac{1}{2k-1} - \frac{1}{2k}\right) + \frac{1}{2n-1} \geq 1 - \frac{1}{2} = \frac{1}{2},$$

so that (s_{2n-1}) is bounded below by 1. Thus, by Theorem 10.18, the monotone convergence theorem, we may conclude that $s_{2n-1} \to S$ for some number S.

Similarly, for any n, we have

$$s_{2(n+1)} = s_{2n} + \frac{1}{2n+1} - \frac{1}{2n+2} > s_{2n},$$

which implies the subsequence (s_{2n}) of (s_n) is increasing. Also, $s_2 = 1 - \frac{1}{2} = \frac{1}{2}$ and, when $n \geq 2$, we have

$$s_{2n} = 1 + \sum_{k=2}^{n}\left(-\frac{1}{2k-2} + \frac{1}{2k-1}\right) - \frac{1}{2n} < 1,$$

so that (s_{2n}) is bounded above by 1. Thus, once again applying the monotone convergence theorem, we conclude that $s_{2n} \to T$ for some number T.

Consequently,

$$0 = \lim_{n\to\infty} -\frac{1}{2n} = \lim_{n\to\infty}(s_{2n} - s_{2n-1}) = \lim_{n\to\infty} s_{2n} - \lim_{n\to\infty} s_{2n-1} = T - S,$$

so that $S = T$. Then, as both of the subsequences (s_{2n-1}) and (s_{2n}) of the sequence (s_n) converge to the same limit, Exercise 11.5 permits us to conclude that the sequence itself converges to this same limit. But since (s_n) is the sequence of partial sums of the alternating harmonic series $\sum_{n=1}^{\infty}\frac{(-1)^{n+1}}{n}$, we may deduce that the series converges.

Example 23.4. The series

$$1 + \frac{1}{2} - \frac{1}{3} + \frac{1}{4} + \frac{1}{5} - \frac{1}{6} + \cdots = \sum_{n=1}^{\infty}\frac{(-1)^{f(n)}}{n},$$

where

$$f(n) = \begin{cases} 1, & \text{if } n \text{ is divisible by 3;} \\ 0, & \text{otherwise;} \end{cases}$$

has infinitely many positive summands and infinitely many negative summands. Let s_n be the nth partial sum of this series. We show that the subsequence (s_{3n}) of the sequence of partial sums diverges, which then implies the sequence of partial sums, that is, the series itself, diverges.

For each n, define

$$t_n = s_{3n} = \sum_{k=1}^{n}\left(\frac{1}{3k-2} + \frac{1}{3k-1} - \frac{1}{3k}\right) = \sum_{k=1}^{n}\frac{9k^2-2}{3k(3k-1)(3k-2)},$$

and note that t_n is the nth partial sum of the series

$$\sum_{n=1}^{\infty} \frac{9n^2 - 2}{3n(3n-1)(3n-2)}.$$

This series has positive summands and diverges according to the limit comparison test because

$$\lim_{n\to\infty} \frac{\frac{9n^2-2}{3n(3n-1)(3n-2)}}{\frac{1}{n}} = \lim_{n\to\infty} \frac{9n^2-2}{3(3n-1)(3n-2)} = \lim_{n\to\infty} \frac{9-\frac{2}{n^2}}{3(3-\frac{1}{n})(3-\frac{2}{n})} = \frac{1}{3},$$

which is a positive number, and we know the series $\sum_{n=1}^{\infty} \frac{1}{n}$ diverges. Hence, (s_{3n}) diverges, and it follows that the original series itself diverges.

It would not be easy to instinctively predict the conclusions we reached for the series given in Examples 23.3 and 23.4. Because the series

$$1 - \frac{1}{2} + \frac{1}{3} - \frac{1}{4} + \frac{1}{5} - \frac{1}{6} + \cdots$$

converges, the subtractions must somehow offset the additions, but because the series

$$1 + \frac{1}{2} - \frac{1}{3} + \frac{1}{4} + \frac{1}{5} - \frac{1}{6} + \cdots$$

diverges, not enough is being subtracted to create a similar offset. A mere inspection of the summands of the two series, though, would not help us to immediately distinguish why the former converges and the latter does not.

There are, however, series that have infinitely many positive and infinitely many negative summands, whose convergence can be anticipated.

Example 23.5. Since the 2-series

$$\sum_{n=1}^{\infty} \frac{1}{n^2} = 1 + \frac{1}{4} + \frac{1}{9} + \frac{1}{16} + \frac{1}{25} + \cdots$$

converges (see Theorem 22.24), it would seem that replacing the even-indexed summands of this nonnegative series with their opposites to produce the series

$$\sum_{n=1}^{\infty} \frac{(-1)^{n+1}}{n^2} = 1 - \frac{1}{4} + \frac{1}{9} - \frac{1}{16} + \frac{1}{25} - \cdots$$

would result in another convergent series. Essentially, the convergence of the original nonnegative series requires its partial sums to eventually become arbitrarily close to one another (the Cauchy criterion, Theorem 22.10), so the replacement of some summands by their opposites would only make the partial sums of the new series lie even closer together.

This example suggests that when convergence of a series with infinitely many positive and infinitely many negative summands is in question, it may be advantageous to examine the related nonnegative series obtained by taking the absolute value of the summands. One reason for doing so is revealed by the following theorem.

Theorem 23.6. *If the infinite series $\sum_{n=1}^{\infty}|a_n|$ converges, so does $\sum_{n=1}^{\infty}a_n$.*

Proof. Assume $\sum_{n=1}^{\infty}|a_n|$ converges and consider any positive number ε. The Cauchy criterion, Theorem 22.10, implies there is an index N such that for any natural numbers m and n for which $n \geq m \geq N$, it follows that

$$\sum_{k=m}^{n}|a_k| < \varepsilon.$$

But as

$$\left|\sum_{k=m}^{n}a_k\right| \leq \sum_{k=m}^{n}|a_k|,$$

the Cauchy criterion now tells us that $\sum_{n=1}^{\infty}a_n$ converges. □

Note that Theorem 23.6 provides the formal rationale for the explanation provided in Example 23.5 as to why the series $\sum_{n=1}^{\infty}\frac{(-1)^{n+1}}{n^2}$ converges. Example 23.3 demonstrates that a series $\sum_{n=1}^{\infty}a_n$ can converge without the related series $\sum_{n=1}^{\infty}|a_n|$ converging.

When an infinite series $\sum_{n=1}^{\infty}a_n$ converges and the series $\sum_{n=1}^{\infty}|a_n|$ also converges, we say the series $\sum_{n=1}^{\infty}a_n$ **converges absolutely** or is **absolutely convergent**. When $\sum_{n=1}^{\infty}a_n$ converges but $\sum_{n=1}^{\infty}|a_n|$ diverges, we say the series $\sum_{n=1}^{\infty}a_n$ **converges conditionally** or is **conditionally convergent**. Theorem 23.6 tells us that *absolute convergence always implies convergence.*

The terminology "conditionally convergent" is a bit unfortunate, as it can suggest that convergence is "conditional" on some other factor, which is not at all what is meant. There should be no doubt that a conditionally convergent series converges; the terminology simply means that while the series converges, the series created by replacing the summands with their absolute values does not.

A convergent series either converges absolutely or converges conditionally, but not both. A divergent series does not converge, hence, does not converge absolutely and does not converge conditionally.

Example 23.5 (Continued). The series $\sum_{n=1}^{\infty}\frac{(-1)^{n+1}}{n^2}$ converges absolutely because the series

$$\sum_{n=1}^{\infty}\left|\frac{(-1)^{n+1}}{n^2}\right| = \sum_{n=1}^{\infty}\frac{1}{n^2}$$

converges, being a p-series for which $p > 1$. Here we are deducing the convergence of $\sum_{n=1}^{\infty}\frac{(-1)^{n+1}}{n^2}$ from the convergence of $\sum_{n=1}^{\infty}\frac{1}{n^2}$ via Theorem 23.6.

Example 23.3 (Continued). Since the alternating harmonic series

$$\sum_{n=1}^{\infty} \frac{(-1)^{n+1}}{n} = 1 - \frac{1}{2} + \frac{1}{3} - \frac{1}{4} + \frac{1}{5} - \frac{1}{6} + \cdots$$

converges, but the harmonic series

$$\sum_{n=1}^{\infty} \left| \frac{(-1)^{n+1}}{n} \right| = \sum_{n=1}^{\infty} \frac{1}{n} = 1 + \frac{1}{2} + \frac{1}{3} + \frac{1}{4} + \frac{1}{5} + \frac{1}{6} + \cdots$$

diverges, the series $\sum_{n=1}^{\infty} \frac{(-1)^{n+1}}{n}$ converges conditionally.

Example 23.4 (Continued). We found that the series

$$1 + \frac{1}{2} - \frac{1}{3} + \frac{1}{4} + \frac{1}{5} - \frac{1}{6} + \cdots$$

diverges, so it converges neither absolutely nor conditionally.

Exercise 23.2. Prove that there are no conditionally convergent geometric series.

Exercise 23.3. If all the summands of an infinite series are negative, can the series be conditionally convergent? Either give an example or show that it is not possible.

Exercise 23.4. Show that if $\sum_{n=1}^{\infty} a_n$ converges absolutely, then $|\sum_{n=1}^{\infty} a_n| \le \sum_{n=1}^{\infty} |a_n|$.

Exercise 23.5. Show that if $\sum_{n=1}^{\infty} a_n$ converges absolutely and (b_n) is a bounded sequence, then $\sum_{n=1}^{\infty} a_n b_n$ converges absolutely. Then give an example to show that if $\sum_{n=1}^{\infty} a_n$ converges conditionally and (b_n) is a bounded sequence, it is possible that $\sum_{n=1}^{\infty} a_n b_n$ diverges.

Dirichlet's test and alternating series

A fairly general convergence test that can often be applied to series having infinitely many positive and infinitely many negative summands is Dirichlet's test, which we present below. To establish this test we use the formula, credited to the short-lived Norwegian mathematician Niels Henrik Abel (1802–1829), stated in the following lemma.

Lemma 23.7 (Abel's partial summation formula). *Let (a_n) and (b_n) be sequences, let $s_0 = 0$, and for each natural number n, let $s_n = \sum_{k=1}^{n} a_k$. If m and n are natural numbers for which $m \le n$, then*

$$\sum_{k=m}^{n} a_k b_k = \sum_{k=m}^{n-1} s_k(b_k - b_{k+1}) + s_n b_n - s_{m-1} b_m.$$

Proof. Observe that under the stated hypotheses,

$$\begin{aligned}
\sum_{k=m}^{n} a_k b_k &= \sum_{k=m}^{n} (s_k - s_{k-1}) b_k \\
&= \sum_{k=m}^{n} s_k b_k - \sum_{k=m}^{n} s_{k-1} b_k \\
&= \sum_{k=m}^{n} s_k b_k - \sum_{k=m-1}^{n-1} s_k b_{k+1} \\
&= \sum_{k=m}^{n-1} s_k b_k + s_n b_n - \sum_{k=m}^{n-1} s_k b_{k+1} - s_{m-1} b_m \\
&= \sum_{k=m}^{n-1} s_k (b_k - b_{k+1}) + s_n b_n - s_{m-1} b_m.
\end{aligned}$$

□

Theorem 23.8 (Dirichlet's test). *If the partial sums of the series $\sum_{n=1}^{\infty} a_n$ form a bounded sequence, and the sequence (b_n) is nonincreasing and converges to 0, then $\sum_{n=1}^{\infty} a_n b_n$ converges.*

Proof. Consider any positive number ε. As the partial sums $s_n = \sum_{k=1}^{n} a_k$ of $\sum_{n=1}^{\infty} a_n$ are bounded, there is a positive number B such that $|s_n| \leq B$ for all n. Note that the hypothesis that (b_n) is nonincreasing and converges to 0 also implies that $b_n \geq 0$ for all n. Thus, there is an index N such that $b_n < \frac{\varepsilon}{2B}$ when $n \geq N$.

Assume that $n \geq m \geq N$. Applying Abel's partial summation formula gives us

$$\begin{aligned}
\left| \sum_{k=m}^{n} a_k b_k \right| &= \left| \sum_{k=m}^{n-1} s_k (b_k - b_{k+1}) + s_n b_n - s_{m-1} b_m \right| \\
&\leq \sum_{k=m}^{n-1} |s_k| (b_k - b_{k+1}) + |s_n| b_n + |s_{m-1}| b_m \\
&\leq \sum_{k=m}^{n-1} B (b_k - b_{k+1}) + B b_n + B b_m \\
&= B(b_m - b_n) + B b_n + B b_m \\
&= 2B b_m \\
&< 2B \cdot \frac{\varepsilon}{2B} \\
&= \varepsilon.
\end{aligned}$$

Therefore, by Theorem 22.10, the Cauchy criterion, the series $\sum_{n=1}^{\infty} a_n b_n$ converges. □

Given a sequence (a_n) of positive numbers, each of the series

$$\sum_{n=1}^{\infty}(-1)^n a_n$$

and

$$\sum_{n=1}^{\infty}(-1)^{n+1} a_n$$

is referred to as an **alternating series**. Alternating series are the most commonly encountered series having infinitely many positive and infinitely many negative summands.

Example 23.9. As all the numbers being added in the series

$$\sum_{n=1}^{\infty}\frac{1}{n} = 1 + \frac{1}{2} + \frac{1}{3} + \frac{1}{4} + \cdots$$

are positive, each of the series

$$\sum_{n=1}^{\infty}\frac{(-1)^n}{n} = -1 + \frac{1}{2} - \frac{1}{3} + \frac{1}{4} - \cdots$$

and

$$\sum_{n=1}^{\infty}\frac{(-1)^{n+1}}{n} = 1 - \frac{1}{2} + \frac{1}{3} - \frac{1}{4} + \cdots$$

is an alternating series. The latter is the alternating harmonic series from Example 23.3.

A straightforward application of Dirichlet's test yields the following extremely useful test for convergence of alternating series.

Theorem 23.10 (The alternating series test). *If (a_n) is a nonincreasing sequence of positive numbers that converges to 0, then both of the alternating series $\sum_{n=1}^{\infty}(-1)^n a_n$ and $\sum_{n=1}^{\infty}(-1)^{n+1} a_n$ converge.*

Proof. The partial sums of the series $\sum_{n=1}^{\infty}(-1)^n$ form the bounded sequence $(-1, 0, -1, 0, \ldots)$, while the partial sums of the series $\sum_{n=1}^{\infty}(-1)^{n+1}$ form the bounded sequence $(1, 0, 1, 0, \ldots)$. Thus, all of the hypotheses of Dirichlet's test are satisfied, so we may conclude that both of the series $\sum_{n=1}^{\infty}(-1)^n a_n$ and $\sum_{n=1}^{\infty}(-1)^{n+1} a_n$ converge. □

Example 23.9 (Continued). As $(\frac{1}{n})$ is a decreasing sequence of positive numbers that converges to 0, the alternating harmonic series $\sum_{n=1}^{\infty}\frac{(-1)^{n+1}}{n}$ and its companion $\sum_{n=1}^{\infty}\frac{(-1)^n}{n}$ both converge according to the alternating series test.

Exercise 23.6. Determine whether the given series converges absolutely, converges conditionally, or diverges.
(a) $\sum_{n=1}^{\infty} \frac{(-4)^n}{5^{n+1}}$
(b) $\sum_{n=1}^{\infty} \frac{(-1)^{n+1}}{2n+1}$
(c) $\sum_{n=1}^{\infty} (-1)^n \frac{\ln(n)}{n}$
(d) $\sum_{n=2}^{\infty} \frac{-1}{\ln(n)}$

Exercise 23.7. Generalize the argument employed in Example 23.3 to show that the alternating harmonic series converges to provide a proof of the alternating series test, Theorem 23.10, that does not use Dirichlet's test.

Exercise 23.8. Let (a_n) be a nonincreasing sequence of positive numbers that converges to 0 and let $S = \sum_{n=1}^{\infty} (-1)^{n+1} a_n$.
(a) Use the proof you obtained in Exercise 23.7 to verify that for each n, we have $|\sum_{k=1}^{n} (-1)^{k+1} a_k - S| \le a_{n+1}$. (This inequality provides a bound on how well the nth partial sum of an alternating series of the type to which the alternating series test applies approximates the sum of the series.)
(b) Use the inequality from (a) to approximate the sum of the alternating harmonic series $\sum_{n=1}^{\infty} \frac{(-1)^{n+1}}{n}$ to within $\frac{1}{10}$ of its true value.

Exercise 23.9. For each natural number n, let $a_n = \sum_{k=1}^{n} \frac{1}{k} - \ln(n)$.
(a) Show that $a_n > 0$ for every n.
(b) Show that the sequence (a_n) is nonincreasing.
(c) Explain how we know, from (a) and (b), that (a_n) converges. (The number to which the sequence (a_n) converges is called **Euler's constant** (not to be confused with Euler's number e) and is usually denoted by the lowercase Greek letter γ (gamma). The approximate value of γ is 0.5772. It is not known whether γ is rational or irrational.)
(d) Let s_n be the nth partial sum of the alternating harmonic series $\sum_{n=1}^{\infty} \frac{(-1)^{n+1}}{n}$. Use induction to verify that $s_{2n} = a_{2n} - a_n + \ln(2)$ for every natural number n.
(e) Explain why it now follows that $\sum_{n=1}^{\infty} \frac{(-1)^{n+1}}{n} = \ln(2)$.

Exercise 23.10. For each natural number n, let $a_n = \frac{1}{n}$ if n is odd and let $a_n = \frac{1}{n^2}$ if n is even. Show that the alternating series $\sum_{n=1}^{\infty} (-1)^{n+1} a_n$ diverges even though $a_n \to 0$. Then explain why this situation does not contradict the alternating series test.

Exercise 23.11. Prove that if $\sum_{n=1}^{\infty} a_n$ converges absolutely, then $\sum_{n=1}^{\infty} a_n^2$ converges. Then give an example to show that if $\sum_{n=1}^{\infty} a_n$ converges conditionally, it is possible that $\sum_{n=1}^{\infty} a_n^2$ diverges.

Exercise 23.12. Use Dirichlet's test to establish the convergence of the series

$$1+\frac{1}{2}-\frac{1}{3}-\frac{1}{4}+\frac{1}{5}+\frac{1}{6}-\frac{1}{7}-\frac{1}{8}+\cdots=\sum_{n=1}^{\infty}\frac{(-1)^{f(n)}}{n},$$

where

$$f(n)=\begin{cases}1, & \text{if } n^2+n \text{ is divisible by 4;}\\ 0, & \text{otherwise.}\end{cases}$$

Exercise 23.13. Use Dirichlet's test to prove that if $\sum_{n=1}^{\infty} a_n$ is a convergent series and (b_n) is a convergent monotone sequence, then the series $\sum_{n=1}^{\infty} a_n b_n$ converges. (This result is known as *Abel's test*.)

The ratio test

Among the most commonly employed tests for absolute convergence of a series are the ratio test and the root test. In Chapter 25 we also use these tests in our study of power series. Our formulation of these tests uses the notions of limit superior and limit inferior, so we recall that the **limit superior** of a sequence (a_n) is

$$\limsup a_n = \lim s_n,$$

where

$$s_n = \sup\{a_k \mid k \geq n\},$$

while the **limit inferior** of (a_n) is

$$\liminf a_n = \lim i_n,$$

where

$$i_n = \inf\{a_k \mid k \geq n\}.$$

When we first introduced these ideas in Chapter 11, we proved that the limit superior of a sequence represents the largest subsequential limit of the sequence and the limit inferior represents the least subsequential limit of the sequence.

Theorem 23.11 (The ratio test). *Suppose (a_n) is a sequence of nonzero numbers.*

(1) *If $\lim_{n\to\infty} |\frac{a_{n+1}}{a_n}| < 1$, then the series $\sum_{n=1}^{\infty} a_n$ converges absolutely. In fact, if $\limsup |\frac{a_{n+1}}{a_n}| < 1$, then $\sum_{n=1}^{\infty} a_n$ converges absolutely.*

(2) *If $\lim_{n\to\infty} |\frac{a_{n+1}}{a_n}| > 1$ or $\lim_{n\to\infty} |\frac{a_{n+1}}{a_n}| = \infty$, then the series $\sum_{n=1}^{\infty} a_n$ diverges. In fact, if $\liminf |\frac{a_{n+1}}{a_n}| > 1$ or $\liminf |\frac{a_{n+1}}{a_n}| = \infty$, then $\sum_{n=1}^{\infty} a_n$ diverges.*

Proof. To establish (1), first recall that when the limit of a sequence exists, that limit is also the sequence's limit superior, so it suffices to show that if $\limsup|\frac{a_{n+1}}{a_n}| < 1$, then $\sum_{n=1}^{\infty} a_n$ converges absolutely. So let $L = \limsup|\frac{a_{n+1}}{a_n}|$ and assume $L < 1$. Choose a real number r so that $L < r < 1$. Then there exists an index N such that when $n \geq N$, it follows that

$$\left|\frac{a_{n+1}}{a_n}\right| < r$$

(see Exercise 23.14 below), that is,

$$|a_{n+1}| < |a_n|r.$$

An induction argument then permits us to conclude that

$$|a_{N+n}| < |a_N|r^n$$

for every natural number n (Exercise 23.14). Since $\sum_{n=1}^{\infty} |a_N|r^n$ is a geometric series with ratio r satisfying $0 < r < 1$, it converges. Thus, using Theorem 22.20, the comparison test, we determine that the series $\sum_{n=1}^{\infty} |a_{N+n}|$ also converges, which then implies the series $\sum_{n=1}^{\infty} |a_n|$ converges, so that the series $\sum_{n=1}^{\infty} a_n$ converges absolutely.

Recall that when the limit of a sequence exists, that limit is also the sequence's limit inferior, and that when a sequence diverges to ∞, so does its limit inferior. Thus, to establish (2), it suffices to show that if $\liminf|\frac{a_{n+1}}{a_n}| > 1$ or $\liminf|\frac{a_{n+1}}{a_n}| = \infty$, then $\sum_{n=1}^{\infty} a_n$ diverges. We handle the case in which $\liminf|\frac{a_{n+1}}{a_n}| > 1$, assigning the case where $\liminf|\frac{a_{n+1}}{a_n}| = \infty$ to the reader (Exercise 23.14).

Let $L = \liminf|\frac{a_{n+1}}{a_n}|$, assume $L > 1$, and choose r so that $1 < r < L$. Then there exists an index N such that when $n \geq N$, it follows that

$$\left|\frac{a_{n+1}}{a_n}\right| > r,$$

that is,

$$|a_{n+1}| > |a_n|r.$$

Since $r > 1$ and the summands of $\sum_{n=1}^{\infty} a_n$ are nonzero, it follows via induction that

$$|a_{N+n}| > |a_N|r^n > |a_N| > 0$$

for every natural number n. Hence, for every natural number n, we may conclude that a_{N+n} is either larger than the positive number $|a_N|$ or smaller than the negative number $-|a_N|$, which implies the sequence (a_{N+n}) does not converge to 0. Therefore, by the divergence test, $\sum_{n=1}^{\infty} a_{N+n}$ diverges, which then implies that $\sum_{n=1}^{\infty} a_n$ also diverges. □

Exercise 23.14. This exercise refers to the proof of the ratio test, Theorem 23.11.

(a) Within the argument for (1), it is claimed that there is an index N such that when $n \geq N$, it follows that $|\frac{a_{n+1}}{a_n}| < r$. Provide details to support this claim.

(b) Within the argument for (1), it is stated that $|a_{N+n}| < |a_N| r^n$ for every natural number n. Use induction to establish this claim.

(c) Provide the argument for (2) when $\liminf |\frac{a_{n+1}}{a_n}| = \infty$.

The ratio test is particularly well-suited to investigating convergence and divergence of series in which the summands involve factorials.

Example 23.12. We can use the ratio test to show that the series $\sum_{n=1}^{\infty} \frac{(-2)^n}{n!}$ converges absolutely. Each summand $a_n = \frac{(-2)^n}{n!}$ is nonzero, and we see that

$$\lim_{n\to\infty} \left| \frac{a_{n+1}}{a_n} \right| = \lim_{n\to\infty} \frac{2^{n+1}}{(n+1)!} \cdot \frac{n!}{2^n} = \lim_{n\to\infty} \frac{2}{n+1} = 0.$$

As this limit is less than 1, we may conclude that $\sum_{n=1}^{\infty} \frac{(-2)^n}{n!}$ converges absolutely.

Exercise 23.15. Use the ratio test to determine whether the given series converges absolutely or diverges.

(a) $\sum_{n=1}^{\infty} \frac{n}{(-2)^n}$

(b) $\sum_{n=1}^{\infty} \frac{(-5)^n}{n^5}$

The ratio test is often employed with series for which all the summands are positive, in which case we can dispense with taking absolute value. One potential advantage of the ratio test over the comparison and limit comparison tests is that there is no need to find another series to which to compare the series of interest.

Example 23.13. All of the summands of the series $\sum_{n=1}^{\infty} \frac{n^n}{(n-1)!}$ are positive. Taking $a_n = \frac{n^n}{(n-1)!}$, we have

$$\lim_{n\to\infty} \frac{a_{n+1}}{a_n} = \lim_{n\to\infty} \frac{(n+1)^{n+1}}{n!} \cdot \frac{(n-1)!}{n^n} = \lim_{n\to\infty} \left(1 + \frac{1}{n}\right)^{n+1} = e,$$

where we have made use of Theorem 20.23. As this limit is greater than 1, we may conclude via the ratio test that $\sum_{n=1}^{\infty} \frac{n^n}{(n-1)!}$ diverges.

Exercise 23.16. Use the ratio test to determine whether the given series converges or diverges.

(a) $\sum_{n=1}^{\infty} \frac{n!}{n^n}$

(b) $\sum_{n=1}^{\infty} \frac{(n!)^3}{(3n)!}$

(c) $\sum_{n=1}^{\infty} \frac{n^n}{e^{n^2}}$

When $\lim_{n\to\infty} |\frac{a_{n+1}}{a_n}| = 1$ it is possible for the series $\sum_{n=1}^{\infty} a_n$ to either converge or diverge. For instance, the 2-series

$$\sum_{n=1}^{\infty} \frac{1}{n^2}$$

converges, with

$$\lim_{n\to\infty} \frac{\frac{1}{(n+1)^2}}{\frac{1}{n^2}} = \lim_{n\to\infty} \frac{n^2}{(n+1)^2} = 1,$$

whereas the 1-series

$$\sum_{n=1}^{\infty} \frac{1}{n}$$

diverges, also with

$$\lim_{n\to\infty} \frac{\frac{1}{n+1}}{\frac{1}{n}} = \lim_{n\to\infty} \frac{n}{n+1} = 1.$$

Thus, the ratio test tells us nothing about whether $\sum_{n=1}^{\infty} a_n$ converges if we discover that $\lim_{n\to\infty} |\frac{a_{n+1}}{a_n}| = 1$.

Exercise 23.17. Show that the ratio test is inconclusive regarding the convergence of any p-series $\sum_{n=1}^{\infty} \frac{1}{n^p}$.

As the examples we have presented thus far illustrate, in attempting to apply the ratio test to a series $\sum_{n=1}^{\infty} a_n$, it is advisable to first ascertain whether $\lim_{n\to\infty} |\frac{a_{n+1}}{a_n}|$ exists or is infinite. One can then go on to investigate $\limsup |\frac{a_{n+1}}{a_n}|$ and $\liminf |\frac{a_{n+1}}{a_n}|$ if necessary.

Example 23.14. Consider the series $\sum_{n=1}^{\infty} a_n$, where

$$a_n = \frac{1}{(5 + (-1)^n)^n}.$$

Since $a_n = \frac{1}{4^n}$ when n is odd and $a_n = \frac{1}{6^n}$ when n is even, it follows that

$$\frac{a_{n+1}}{a_n} = \begin{cases} \frac{1}{6} \cdot (\frac{2}{3})^n, & \text{if } n \text{ is odd;} \\ \frac{1}{4} \cdot (\frac{3}{2})^n, & \text{if } n \text{ is even.} \end{cases}$$

At this point, we realize that the sequence $(\frac{a_{n+1}}{a_n})$ does not converge. However, the subsequence

$$\left(\frac{1}{6} \cdot \left(\frac{2}{3}\right)^n\right)$$

consisting of the odd-indexed terms converges to 0, while the subsequence

$$\left(\frac{1}{4} \cdot \left(\frac{3}{2}\right)^n\right)$$

consisting of the even-indexed terms diverges to ∞. As these two subsequences together exhaust all of the terms of the sequence $(\frac{a_{n+1}}{a_n})$, we may conclude, via Theorems 11.13 and 11.14, that

$$\liminf \frac{a_{n+1}}{a_n} = 0$$

and

$$\limsup \frac{a_{n+1}}{a_n} = \infty.$$

Since this limit superior is not less than 1 and this limit inferior is not greater than 1, the ratio test is inconclusive, and we need to find another way to determine whether the series $\sum_{n=1}^{\infty} \frac{1}{(5+(-1)^n)^n}$ converges (we shall do so very soon).

Exercise 23.18. Define the sequence (a_n) recursively so that

$$a_1 = 1$$

and

$$a_{n+1} = \begin{cases} 0.5a_n, & \text{if } n \text{ is odd;} \\ 0.25a_n, & \text{if } n \text{ is even.} \end{cases}$$

Use the ratio test to determine whether the series $\sum_{n=1}^{\infty} a_n$ converges.

The root test

When the limit superior of the sequence $(\sqrt[n]{|a_n|})$ can be identified, it is often feasible to employ the following result, known as the root test, to determine whether the series $\sum_{n=1}^{\infty} a_n$ converges.

Theorem 23.15 (The root test). *Suppose (a_n) is a real sequence.*

(1) *If $\lim_{n\to\infty} \sqrt[n]{|a_n|} < 1$, then the series $\sum_{n=1}^{\infty} a_n$ converges absolutely. In fact, if $\limsup \sqrt[n]{|a_n|} < 1$, then $\sum_{n=1}^{\infty} a_n$ converges absolutely.*

(2) *If $\lim_{n\to\infty} \sqrt[n]{|a_n|} > 1$ or $\lim_{n\to\infty} \sqrt[n]{|a_n|} = \infty$, then the series $\sum_{n=1}^{\infty} a_n$ diverges. In fact, if $\limsup \sqrt[n]{|a_n|} > 1$ or $\limsup \sqrt[n]{|a_n|} = \infty$, then $\sum_{n=1}^{\infty} a_n$ diverges.*

Note that the divergence criterion of the root test involves the limit superior, whereas the divergence criterion of the ratio test involves the limit inferior.

Example 23.16. Applying the root test to the series $\sum_{n=2}^{\infty} \frac{1}{(\ln(n))^n}$, we take $a_n = \frac{1}{(\ln(n))^n}$ and calculate

$$\lim_{n\to\infty} \sqrt[n]{a_n} = \lim_{n\to\infty} \frac{1}{\ln(n)} = 0.$$

As this limit is less than 1, we conclude that $\sum_{n=2}^{\infty} \frac{1}{(\ln(n))^n}$ converges.

Exercise 23.19. Prove the root test, Theorem 23.15.

Exercise 23.20. Use the root test to determine whether the given series converges absolutely or diverges.

(a) $\sum_{n=1}^{\infty} \frac{(-n)^3}{e^{2n}}$

(b) $\sum_{n=1}^{\infty} \frac{(-n)^n}{2^n}$

When $\limsup \sqrt[n]{|a_n|} = 1$, it is possible for the series $\sum_{n=1}^{\infty} a_n$ to either converge or diverge, so the root test is inconclusive.

Exercise 23.21. Let p be any positive real number and, for each natural number n, let $a_n = \frac{1}{n^p}$. Apply Exercise 10.19 to show that $\lim_{n\to\infty} \sqrt[n]{a_n} = 1$. Since the p-series $\sum_{n=1}^{\infty} a_n$ converges when $p > 1$ and diverges when $p \leq 1$, this demonstrates that when $\limsup \sqrt[n]{|a_n|} = 1$, both convergence and divergence of $\sum_{n=1}^{\infty} a_n$ are possible.

We use the symbols ∞ and $-\infty$ to describe specific ways in which the limit of a sequence, series, or function does not exist. When making statements about limiting behavior, it is sometimes convenient to think of ∞ and $-\infty$ as "values" a limit may possess, with the continued understanding that such limiting values are not real numbers. In such a context, we regard ∞ as being larger than any real number and $-\infty$ as being smaller than any real number. By doing so we can often be more succinct in our writing. For instance, if the inequality

$$\lim_{n\to\infty} \sqrt[n]{|a_n|} > 1$$

is interpreted to include the possibility that $\lim_{n\to\infty} \sqrt[n]{|a_n|} = \infty$, then the hypothesis for (2) of the root test, Theorem 23.15, could be abbreviated.

Another example is provided by the next theorem, where each limit appearing in the conclusion could be a nonnegative real number or ∞. This result indicates that when a conclusion regarding convergence or divergence of a series can be reached via the ratio test, the same conclusion would be reached via the root test.

Theorem 23.17. *For any sequence (a_n) of positive numbers,*

$$\liminf \frac{a_{n+1}}{a_n} \leq \liminf \sqrt[n]{a_n} \leq \limsup \sqrt[n]{a_n} \leq \limsup \frac{a_{n+1}}{a_n}.$$

Proof. Because all terms of the sequence (a_n) are positive, all of the limits in the inequality to be proved are nonnegative, though it is possible that one or more of them is ∞.

We first show that

$$\liminf \sqrt[n]{a_n} \leq \limsup \sqrt[n]{a_n}.$$

As the sequence $(\sqrt[n]{a_n})$ is bounded below by 0, it follows that $\inf\{\sqrt[k]{a_k} \mid k \geq n\}$ exists for every natural number n. If $(\sqrt[n]{a_n})$ is bounded above, then $\sup\{\sqrt[k]{a_k} \mid k \geq n\}$ also exists for every n, and as a set's infimum is always less than or equal to the set's supremum, we would have

$$\inf\{\sqrt[k]{a_k} \mid k \geq n\} \leq \sup\{\sqrt[k]{a_k} \mid k \geq n\}$$

for every n, from which we may then conclude that

$$\liminf \sqrt[n]{a_n} \leq \limsup \sqrt[n]{a_n}.$$

If $(\sqrt[n]{a_n})$ is not bounded above, then $\limsup \sqrt[n]{a_n} = \infty$, in which case it still follows that

$$\liminf \sqrt[n]{a_n} \leq \limsup \sqrt[n]{a_n}.$$

We now show that

$$\limsup \sqrt[n]{a_n} \leq \limsup \frac{a_{n+1}}{a_n}.$$

The desired inequality certainly holds if $\limsup \frac{a_{n+1}}{a_n} = \infty$. So suppose $\limsup \frac{a_{n+1}}{a_n} = L$ for some real number L that is necessarily nonnegative. Consider any real number r for which $r > L$. It suffices to show that

$$\limsup \sqrt[n]{a_n} \leq r.$$

As $\limsup \frac{a_{n+1}}{a_n} = L$, there exists a natural number N for which when $n \geq N$, it follows that

$$\frac{a_{n+1}}{a_n} < r.$$

An induction argument then shows that when $n > N$, we have

$$a_n < r^n \frac{a_N}{r^N},$$

so that

$$\sqrt[n]{a_n} < r\left(\sqrt[n]{\frac{a_N}{r^N}}\right).$$

Since $\frac{a_N}{r^N}$ is a positive constant, it follows by Exercise 10.20 that

$$\lim_{n\to\infty} \sqrt[n]{\frac{a_N}{r^N}} = 1.$$

Hence,

$$\limsup \sqrt[n]{a_n} \leq \limsup r\left(\sqrt[n]{\frac{a_N}{r^N}}\right) = \lim_{n\to\infty} r\left(\sqrt[n]{\frac{a_N}{r^N}}\right) = r \cdot 1 = r.$$

We leave the proof that $\liminf \frac{a_{n+1}}{a_n} \leq \liminf \sqrt[n]{a_n}$ as Exercise 23.22. □

Exercise 23.22. Complete the proof of Theorem 23.17 by demonstrating that $\liminf \frac{a_{n+1}}{a_n} \leq \liminf \sqrt[n]{a_n}$ for any sequence (a_n) of positive numbers.

Based on Theorem 23.17, it might seem that we should always try to apply the root test and never bother with the ratio test. However, algebraically simplifying roots is often much more challenging than simplifying ratios, so from a practical perspective the ratio test is easier to implement in some situations.

The root test is actually more powerful than the ratio test, as there are series whose convergence can be confirmed via the root test, but for which the ratio test is inconclusive.

Example 23.14 (Continued). Earlier, we determined that the ratio test would not tell us whether the series the series $\sum_{n=1}^{\infty} a_n$, where $a_n = \frac{1}{(5+(-1)^n)^n}$, converges or diverges. However, as

$$a_n = \begin{cases} (\frac{1}{4})^n, & \text{if } n \text{ is odd;} \\ (\frac{1}{6})^n, & \text{if } n \text{ is even;} \end{cases}$$

we see that

$$\sqrt[n]{a_n} = \begin{cases} \frac{1}{4}, & \text{if } n \text{ is odd;} \\ \frac{1}{6}, & \text{if } n \text{ is even;} \end{cases}$$

and it follows that

$$\limsup \sqrt[n]{a_n} = \frac{1}{4},$$

which is less than 1. Hence, the root test permits us to conclude that $\sum_{n=1}^{\infty} \frac{1}{(5+(-1)^n)^n}$ converges.

Exercise 23.23. For each natural number n, define a_n so that $a_n = 2^{-(n+1)/2}$ when n is odd and so that $a_n = 2^{-n/2}$ when n is even. Show that the ratio test does not help us determine whether the series $\sum_{n=1}^{\infty} a_n$ converges, but the root test does.

Raabe's test

If we find that both $\lim_{n\to\infty} \frac{a_{n+1}}{a_n} = 1$ and $\lim_{n\to\infty} \sqrt[n]{a_n} = 1$, so that neither the ratio test nor the root test is applicable, the following test, developed by the Swiss mathematician Joseph Ludwig Raabe (1801–1859), is sometimes helpful in resolving whether the series $\sum_{n=1}^{\infty} a_n$ converges.

Theorem 23.18 (Raabe's test). *Let (a_n) be a sequence of nonzero numbers and suppose $\lim_{n\to\infty} n(1 - |\frac{a_{n+1}}{a_n}|) = L$ for some real number L.*
(1) *If $L > 1$, then the series $\sum_{n=1}^{\infty} a_n$ converges absolutely.*
(2) *If $L < 1$, then the series $\sum_{n=1}^{\infty} a_n$ may converge conditionally, but does not converge absolutely.*

Proof. Suppose $L > 1$. Consider any number r for which $1 < r < L$. Then there is a natural number N such that when $n \geq N$, it follows that

$$n\left(1 - \left|\frac{a_{n+1}}{a_n}\right|\right) > r,$$

which via a bit of algebra can be rewritten as

$$(n-1)|a_n| - n|a_{n+1}| > (r-1)|a_n|.$$

Therefore, for any $n \geq N$, we have

$$\sum_{k=N}^{n} ((k-1)|a_k| - k|a_{k+1}|) > \sum_{k=N}^{n} (r-1)|a_k|,$$

which simplifies to

$$(N-1)|a_N| - n|a_{n+1}| > (r-1) \sum_{k=N}^{n} |a_k|.$$

But

$$(N-1)|a_N| > (N-1)|a_N| - n|a_{n+1}|$$

for any n, so it follows that

$$\sum_{k=N}^{n} |a_k| < \frac{(N-1)|a_N|}{r-1},$$

and, consequently, that

$$\sum_{k=1}^{n} |a_k| < \sum_{k=1}^{N-1} |a_k| + \frac{(N-1)|a_N|}{r-1},$$

when $n \geq N$, from which we may conclude that the partial sums of the series $\sum_{n=1}^{\infty} |a_n|$ are bounded. Thus, by Theorem 22.19, $\sum_{n=1}^{\infty} |a_n|$ converges, meaning $\sum_{n=1}^{\infty} a_n$ converges absolutely.

Now suppose $L < 1$ and consider any number r for which $\max\{L, 0\} < r < 1$. Then there is a natural number N such that when $n \geq N$, it follows that

$$n\left(1 - \left|\frac{a_{n+1}}{a_n}\right|\right) < r,$$

that is,

$$n|a_{n+1}| > (n-r)|a_n|.$$

But as $0 < r < 1$, it then follows that

$$(n-r)|a_n| > (n-1)|a_n|,$$

so we may conclude that

$$n|a_{n+1}| > (n-1)|a_n|$$

when $n \geq N$. Thus, the sequence $(n|a_{n+1}|)$ is increasing, at least from the index $N-1$ onward. Hence, for $n \geq N$, we have

$$n|a_{n+1}| > (N-1)|a_N|,$$

that is,

$$|a_{n+1}| > \frac{(N-1)|a_N|}{n},$$

and since the harmonic series $\sum_{n=1}^{\infty} \frac{1}{n}$ diverges, the comparison test permits us to conclude that $\sum_{n=1}^{\infty} |a_n|$ also diverges. Hence, $\sum_{n=1}^{\infty} a_n$ cannot converge absolutely. In Exercise 23.25, you are asked to find a series that converges conditionally and for which $L < 1$. □

In Exercises 23.17 and 23.21, you showed that the ratio and root tests are both inconclusive with respect to the convergence of p-series. Raabe's test, however, provides another means to verify that $\sum_{n=1}^{\infty} \frac{1}{n^p}$ converges absolutely when $p > 1$ and diverges when $0 < p < 1$, which you are asked to demonstrate in Exercise 23.24.

Exercise 23.24. Show that Raabe's test permits us to conclude that the p-series $\sum_{n=1}^{\infty} \frac{1}{n^p}$ converges absolutely when $p > 1$ and diverges when $0 < p < 1$. Show also that Raabe's test is inconclusive with respect to the harmonic series $\sum_{n=1}^{\infty} \frac{1}{n}$.

Exercise 23.25. Find a conditionally convergent series $\sum_{n=1}^{\infty} a_n$ with the property that $\lim_{n\to\infty} n(1 - |\frac{a_{n+1}}{a_n}|)$ is a real number less than 1.

Exercise 23.26. Let p and q be positive numbers, and for each natural number n, define

$$a_n = \frac{(p+1)(p+2)\cdots(p+n)}{(q+1)(q+2)\cdots(q+n)}.$$

(a) Show that the ratio test is inconclusive regarding the convergence of $\sum_{n=1}^{\infty} a_n$.
(b) Use Raabe's test to show that $\sum_{n=1}^{\infty} a_n$ converges when $q > p+1$ and diverges when $q < p+1$.
(c) Show that $\sum_{n=1}^{\infty} a_n$ diverges when $q = p+1$.

Exercise 23.27. Show that the ratio test does not help us to resolve the question of whether the series $\sum_{n=1}^{\infty} \frac{(2n)!}{4^n (n!)^2}$ converges, but Raabe's test does.

24 Regroupings and rearrangements of series

We consider what can happen with respect to the question of convergence if the summands of an infinite series are regrouped or rearranged. In the case of rearrangements of a conditionally convergent series, the possibilities are truly startling.

Regroupings of series

Recall that the associative property of addition tells us that for any real numbers a, b, and c, we have

$$(a + b) + c = a + (b + c).$$

This property implies that for any finite addition involving three or more summands, the calculated sum is the same, no matter how we use parentheses to group the summands. For example,

$$4 + (\pi + (e + 10)) + \left(\sqrt{5} + \frac{7}{8}\right) = ((4 + \pi) + e) + (10 + \sqrt{5}) + \frac{7}{8}.$$

Thus, as the grouping of summands is irrelevant, expressions such as

$$a + b + c$$

and

$$4 + \pi + e + 10 + \sqrt{5} + \frac{7}{8}$$

are unambiguous. It is reasonable to wonder whether a similar property holds for the "infinite additions" represented by infinite series. For instance, if we take the convergent geometric series

$$\sum_{n=1}^{\infty} \frac{1}{2^n} = \frac{1}{2} + \frac{1}{2^2} + \frac{1}{2^3} + \frac{1}{2^4} + \frac{1}{2^5} + \frac{1}{2^6} + \cdots$$

and regroup summands to create

$$\left(\frac{1}{2} + \frac{1}{2^2}\right) + \left(\frac{1}{2^3} + \frac{1}{2^4}\right) + \left(\frac{1}{2^5} + \frac{1}{2^6}\right) + \cdots = \sum_{n=1}^{\infty}\left(\frac{1}{2^{2n-1}} + \frac{1}{2^{2n}}\right)$$

or

$$\frac{1}{2} + \left(\frac{1}{2^2} + \frac{1}{2^3} + \frac{1}{2^4}\right) + \left(\frac{1}{2^5} + \frac{1}{2^6} + \frac{1}{2^7} + \frac{1}{2^8} + \frac{1}{2^9}\right) + \cdots = \sum_{n=1}^{\infty}\left(\frac{1}{2^{(n-1)^2+1}} + \cdots + \frac{1}{2^{n^2}}\right),$$

https://doi.org/10.1515/9783111429564-024

must these "new" series converge and, if so, must they converge to the same sum as the original series? The answer is affirmative, and is so in any case in which the series we begin with converges.

Theorem 24.1 (Regroupings of convergent series). *Suppose the series $\sum_{n=1}^{\infty} a_n$ converges with real number sum S. Let (k_n) be an increasing sequence of natural numbers and let $k_0 = 0$. Define the sequence (b_n) so that $b_n = \sum_{i=k_{n-1}+1}^{k_n} a_i$. Then the series $\sum_{n=1}^{\infty} b_n$ converges with sum S.*

In other words, any regrouping of the summands of a convergent series yields a series that converges to the same sum.

It is important to note that a regrouping of a series does not change the ordering of the summands. Regrouping simply inserts parentheses so that certain adjacent summands are added to form the summands of a new series, but the relative positioning of summands remains unchanged. For instance, using the notation given in the statement of Theorem 24.1, the series $\sum_{n=1}^{\infty} b_n$ is

$$(a_1 + a_2 + \cdots + a_{k_1}) + (a_{k_1+1} + a_{k_1+2} + \cdots + a_{k_2}) + (a_{k_2+1} + a_{k_2+2} + \cdots + a_{k_3}) + \cdots,$$

which is a regrouping of the original series $\sum_{n=1}^{\infty} a_n$, but preserves the relative ordering of the summands in the sense that if all the parentheses were removed, we would be back to the original series.

Exercise 24.1. Prove Theorem 24.1.

Exercise 24.2. Give an example of a regrouping of the summands of a divergent series that yields a convergent series.

Rearrangements of series

Having considered the effects of regrouping summands in an infinite series, we now turn to the question of rearranging the summands into a different order. Because the commutative property of addition tells us that

$$a + b = b + a$$

for all real numbers a and b, it may initially appear that it does not matter in what order the summands of a convergent series are added. But the story is not that simple.

First, though, let us make clear exactly what is meant by a rearrangement of a series. The series $\sum_{n=1}^{\infty} r_n$ is called a **rearrangement** of the series $\sum_{n=1}^{\infty} a_n$ if there is a bijection $k : \mathbb{N} \to \mathbb{N}$ such that for every natural number n we have $r_n = a_{k(n)}$. The bijection is actually a sequence that represents a specific reordering of the natural numbers, which then induces a specific reordering of the summands of the series.

Example 24.2. One example of a rearrangement of the series

$$\sum_{n=1}^{\infty}\left(-\frac{1}{2}\right)^{n-1} = 1 - \frac{1}{2} + \frac{1}{4} - \frac{1}{8} + \frac{1}{16} - \frac{1}{32} + \cdots$$

is the series

$$-\frac{1}{2} + 1 + \frac{1}{4} - \frac{1}{8} + \frac{1}{16} - \frac{1}{32} + \cdots,$$

obtained by switching the first two summands. In this case the bijection that reorders the indices is the function $k : \mathbb{N} \to \mathbb{N}$ where $k(1) = 2$, $k(2) = 1$, and $k(n) = n$ when $n \geq 3$; in other words, the new index sequence is $k = (2, 1, 3, 4, 5, 6, \ldots)$.

Another example of a rearrangement of $\sum_{n=1}^{\infty}(-\frac{1}{2})^{n-1}$ is the series

$$1 + \frac{1}{4} + \frac{1}{16} - \frac{1}{2} - \frac{1}{8} + \frac{1}{64} + \frac{1}{256} + \frac{1}{1024} - \frac{1}{32} - \frac{1}{128} + \cdots,$$

where we begin with the first three positive summands of the original series, followed by the first two negative summands, and continue to alternate in this way, the next three positive summands followed by the next two negative summands, forever. Here the reordering of the indices is $(1, 3, 5, 2, 4, 7, 9, 11, 6, 8, \ldots)$. Thus, if we let $a_n = (-\frac{1}{2})^{n-1}$, the original series is

$$a_1 + a_2 + a_3 + a_4 + a_5 + a_6 + a_7 + a_8 + a_9 + a_{10} + \cdots$$

and the rearrangement is

$$a_1 + a_3 + a_5 + a_2 + a_4 + a_7 + a_9 + a_{11} + a_6 + a_8 + \cdots.$$

Exercise 24.3. A rearrangement of a series $\sum_{n=1}^{\infty} a_n$ requires that each summand a_n eventually appears as a summand in the rearrangement. One of the summands in the series $\sum_{n=1}^{\infty}(-\frac{1}{2})^{n-1}$ from Example 24.2 is $-\frac{1}{2^{999}}$. What is the index of this summand in the second rearrangement described in this example?

There are certainly situations where a rearrangement of a convergent series also converges to the same sum. For instance, you establish in Exercise 24.4 that if only finitely many summands of a convergent series are displaced by a rearrangement of the series, the resulting series must converge to the same sum.

Exercise 24.4. Show that if a rearrangement of a convergent series moves only finitely many summands from their positions in the original series, the rearrangement converges and has the same sum.

We can also rearrange the summands of an absolutely convergent series in any way we would like and the result is still a series that converges to the same sum.

Theorem 24.3. *If an infinite series converges absolutely, then any rearrangement of the series also converges to the same sum.*

Proof. Suppose $\sum_{n=1}^{\infty} a_n$ converges absolutely with $\sum_{n=1}^{\infty} a_n = S$. Let $\sum_{n=1}^{\infty} r_n$ be a rearrangement of $\sum_{n=1}^{\infty} a_n$. We show $\sum_{n=1}^{\infty} r_n = S$.

Let (s_n) be the sequence of partial sums of $\sum_{n=1}^{\infty} a_n$ and let (t_n) be the sequence of partial sums of $\sum_{n=1}^{\infty} r_n$. Consider any positive number ε. Because $\sum_{n=1}^{\infty} a_n = S$ and $\sum_{n=1}^{\infty} |a_n|$, being convergent, is Cauchy, there exists a natural number N such that both

$$|s_n - S| < \frac{\varepsilon}{2}$$

and

$$\sum_{k=N+1}^{n} |a_k| < \frac{\varepsilon}{2},$$

whenever $n > N$.

Choose a natural number M large enough so that among the summands in $\sum_{k=1}^{M} r_k$ are all of $a_1, a_2, \ldots, a_{N+1}$. Then when $n \geq M$ it follows that

$$t_n - s_{N+1} = t_n - \sum_{k=1}^{N+1} a_k$$

is a sum of finitely many terms of the sequence (a_n) each having index larger than $N+1$. Hence,

$$|t_n - s_{N+1}| \leq \sum_{k=N+2}^{J} |a_k| < \frac{\varepsilon}{2},$$

for some natural number J for which $J > N+1$. Therefore, whenever $n > M$, which since $M \geq N+1$ means $n > N+1$, we may conclude that

$$|t_n - S| \leq |t_n - s_{N+1}| + |s_{N+1} - S| < \frac{\varepsilon}{2} + \frac{\varepsilon}{2} = \varepsilon.$$

Having shown the partial sums of $\sum_{n=1}^{\infty} r_n$ converge to S, it follows that $\sum_{n=1}^{\infty} r_n = S$. □

Exercise 24.5. Must every rearrangement of an absolutely convergent series also be absolutely convergent? Explain.

Rearrangements of conditionally convergent series are much less predictable relative to the question of convergence. The basis for the quite remarkable behavior we shall observe can be found in the following theorem. It states that a series that converges conditionally not only must have infinitely many positive summands and infinitely many negative summands, but the partial sums of just the positive summands must become unbounded, as must the partial sums of just the negative summands.

Theorem 24.4. *Suppose the series $\sum_{n=1}^{\infty} a_n$ converges conditionally.*

(1) *Infinitely many terms of the sequence (a_n) are positive and infinitely many terms are negative.*
(2) *If (p_n) is the subsequence of (a_n) consisting of all the positive terms of (a_n), then the series $\sum_{n=1}^{\infty} p_n$ diverges to ∞.*
(3) *If (q_n) is the subsequence of (a_n) consisting of all the negative terms of (a_n), then the series $\sum_{n=1}^{\infty} q_n$ diverges to $-\infty$.*

Proof. We leave the proof of (1) as Exercise 24.6.

To establish (2) and (3), obtain the sequence (b_n) by replacing all the negative terms of (a_n) with 0 and obtain the sequence (c_n) by replacing all the positive terms of (a_n) with 0. Observe that for every n it follows that

$$b_n + c_n = a_n$$

and

$$b_n - c_n = |a_n|.$$

Thus, as $\sum_{n=1}^{\infty} a_n$ converges, the convergence of either one of $\sum_{n=1}^{\infty} b_n$ or $\sum_{n=1}^{\infty} c_n$ would imply the convergence of the other, from which it would then follow that $\sum_{n=1}^{\infty} |a_n|$ converges, a contradiction as $\sum_{n=1}^{\infty} a_n$ is only conditionally convergent. Therefore, neither $\sum_{n=1}^{\infty} b_n$ or $\sum_{n=1}^{\infty} c_n$ can converge, and since (b_n) is nonnegative the partial sums of $\sum_{n=1}^{\infty} b_n$ are not bounded above, while since (c_n) is nonpositive the partial sums of $\sum_{n=1}^{\infty} c_n$ are not bounded below.

Now take (p_n) to be the subsequence of (b_n) consisting of the nonzero terms of (b_n), which also makes (p_n) the subsequence of (a_n) consisting of the positive terms of (a_n). Also take (q_n) to be the subsequence of (c_n) consisting of the nonzero terms of (c_n), which also makes (q_n) the subsequence of (a_n) consisting of the negative terms of (a_n). The only difference between the two series $\sum_{n=1}^{\infty} b_n$ and $\sum_{n=1}^{\infty} p_n$ is the presence of zero summands in $\sum_{n=1}^{\infty} b_n$ that are not included in $\sum_{n=1}^{\infty} p_n$. Hence, $\sum_{n=1}^{\infty} p_n = \sum_{n=1}^{\infty} b_n$ (see Exercise 24.7), so $\sum_{n=1}^{\infty} p_n$ diverges to ∞. In a similar fashion, $\sum_{n=1}^{\infty} q_n = \sum_{n=1}^{\infty} c_n$, so $\sum_{n=1}^{\infty} q_n$ diverges to $-\infty$. □

Exercise 24.6. Prove (1) of Theorem 24.4.

Exercise 24.7. Suppose the series $\sum_{n=1}^{\infty} a_n$ converges to the real number S and that infinitely many terms of the sequence (a_n) are nonzero. Let (a_{k_n}) be the subsequence of (a_n) consisting of the nonzero terms of (a_n). Show that $\sum_{n=1}^{\infty} a_{k_n} = S$.

It turns out that a conditionally convergent series can be rearranged to yield any desired sum. This result is stated formally in Theorem 24.6, but we first demonstrate the general process of the proof with a particular example.

Example 24.5. The series

$$\sum_{n=1}^{\infty} \frac{(-1)^{n+1}}{\sqrt{n}} = 1 - \frac{1}{\sqrt{2}} + \frac{1}{\sqrt{3}} - \frac{1}{\sqrt{4}} + \frac{1}{\sqrt{5}} - \frac{1}{\sqrt{6}} + \cdots$$

converges by the alternating series test, Theorem 23.10. Let S be the sum of this series and, for each natural number n, let $s_n = \sum_{k=1}^{n} \frac{(-1)^{k+1}}{\sqrt{k}}$ be the nth partial sum. Observe that $S < 1$ since (s_{2n-1}) is a decreasing sequence (you are asked to verify this in Exercise 24.8) and $s_1 = 1$.

Now suppose we want to try to rearrange the series in such a way that the rearrangement $\sum_{n=1}^{\infty} r_n$ has a different sum, say 3. This requires that the partial sums of the rearrangement converge to 3. We begin by adding up the minimum number of the initial positive summands of $\sum_{n=1}^{\infty} \frac{(-1)^{n+1}}{\sqrt{n}}$ to create a partial sum of the rearrangement $\sum_{n=1}^{\infty} r_n$ that is larger than 3. Since

$$1 + \frac{1}{\sqrt{3}} + \frac{1}{\sqrt{5}} + \frac{1}{\sqrt{7}} + \frac{1}{\sqrt{9}} \approx 2.736 < 3,$$

while

$$1 + \frac{1}{\sqrt{3}} + \frac{1}{\sqrt{5}} + \frac{1}{\sqrt{7}} + \frac{1}{\sqrt{9}} + \frac{1}{\sqrt{11}} \approx 3.037 > 3,$$

in our rearrangement $\sum_{n=1}^{\infty} r_n$, we take

$$r_1 = 1, \quad r_2 = \frac{1}{\sqrt{3}}, \quad r_3 = \frac{1}{\sqrt{5}}, \quad r_4 = \frac{1}{\sqrt{7}}, \quad r_5 = \frac{1}{\sqrt{9}}, \quad \text{and} \quad r_6 = \frac{1}{\sqrt{11}},$$

so that the partial sum $\sum_{k=1}^{6} r_k$ of the rearrangement is greater than 3.

Keeping in mind that we want the partial sums of $\sum_{n=1}^{\infty} r_n$ to converge to 3, we now add on the minimum number of the initial negative summands of the original series to achieve a partial sum that is less than or equal to 3. Since

$$1 + \frac{1}{\sqrt{3}} + \frac{1}{\sqrt{5}} + \frac{1}{\sqrt{7}} + \frac{1}{\sqrt{9}} + \frac{1}{\sqrt{11}} - \frac{1}{\sqrt{2}} \approx 2.330 < 3,$$

we take

$$r_7 = -\frac{1}{\sqrt{2}}$$

so that the partial sum $\sum_{k=1}^{7} r_k$ of the rearrangement is less than 3.

We continue in this way, alternately adding on just enough of the next positive summands to obtain a partial sum larger than 3, and then adding on just enough of the next negative summands to obtain a partial sum less than or equal to 3. Thus, as

$$1 + \frac{1}{\sqrt{3}} + \frac{1}{\sqrt{5}} + \frac{1}{\sqrt{7}} + \frac{1}{\sqrt{9}} + \frac{1}{\sqrt{11}} - \frac{1}{\sqrt{2}} + \frac{1}{\sqrt{13}} + \frac{1}{\sqrt{15}} \approx 2.866 < 3,$$

$$1 + \frac{1}{\sqrt{3}} + \frac{1}{\sqrt{5}} + \frac{1}{\sqrt{7}} + \frac{1}{\sqrt{9}} + \frac{1}{\sqrt{11}} - \frac{1}{\sqrt{2}} + \frac{1}{\sqrt{13}} + \frac{1}{\sqrt{15}} + \frac{1}{\sqrt{17}} \approx 3.108 > 3,$$

and

$$1 + \frac{1}{\sqrt{3}} + \frac{1}{\sqrt{5}} + \frac{1}{\sqrt{7}} + \frac{1}{\sqrt{9}} + \frac{1}{\sqrt{11}} - \frac{1}{\sqrt{2}} + \frac{1}{\sqrt{13}} + \frac{1}{\sqrt{15}} + \frac{1}{\sqrt{17}} - \frac{1}{\sqrt{4}} \approx 2.608 < 3,$$

we take

$$r_8 = \frac{1}{\sqrt{13}}, \quad r_9 = \frac{1}{\sqrt{15}}, \quad r_{10} = \frac{1}{\sqrt{17}}, \quad \text{and} \quad r_{11} = -\frac{1}{\sqrt{4}}$$

so that $\sum_{k=1}^{10} r_k > 3$ and $\sum_{k=1}^{11} r_k < 3$.

We are always able to carry out this process because Theorem 24.4 guarantees the partial sums determined by the positive summands are unbounded, as are the partial sums determined by the negative summands. We expect the partial sums being constructed for the rearrangement to converge to 3, because the summands $\frac{(-1)^{n+1}}{\sqrt{n}}$ from which they are fashioned converge to 0.

Exercise 24.8. Prove that the sequence (s_{2n-1}) from Example 24.5 is decreasing.

Exercise 24.9. Referring to Example 24.5, identify the values of r_{12} through r_{28}.

What makes it possible to rearrange a conditionally convergent series so that the rearrangement converges to any specified real number is the fact that the partial sums of both the sequence of positive summands and the sequence of negative summands are unbounded, as this is what permits us to move "back and forth" to either side of the proposed sum. The resulting rearranged partial sums eventually become arbitrarily close to the proposed sum, because both the sequence of positive summands and the sequence of negative summands of the original series must converge to 0. This process, illustrated in Example 24.5, is generalized in the proof of the following theorem.

Theorem 24.6. *If an infinite series converges conditionally and r is any real number, then there is a rearrangement of the series that converges r.*

Proof. Assume the series $\sum_{n=1}^{\infty} a_n$ converges conditionally and let r be a nonnegative real number; a straightforward modification of the following argument can be employed if r is negative. Without loss of generality, we may assume each summand a_n of the series is nonzero. Let (p_n) be the subsequence of (a_n) consisting of all the positive terms of (a_n) and let (q_n) be the subsequence of (a_n) consisting of all the negative terms of (a_n).

For convenience, we refer to any finite sum of one or more consecutive terms of either of the sequences (p_n) or (q_n) as a *block* from the sequence; for instance, a block from (p_n) is any sum of the form $\sum_{i=j}^{k} p_i$, where $j \le k$. We also say that a block B' *immediately follows* a block B if B' begins at the very next index after the last index of B; for instance, the block $\sum_{i=4}^{7} q_i$ from (q_n) immediately follows the block $\sum_{i=2}^{3} q_i$.

We now obtain a rearrangement of $\sum_{n=1}^{\infty} a_n$ that is a sum

$$P_1 + Q_1 + P_2 + Q_2 + P_3 + Q_3 + \cdots$$

of alternating blocks, each block P_n from (p_n) and each block Q_n from (q_n), where P_{n+1} always immediately follows P_n and Q_{n+1} always immediately follows Q_n, and with P_1 and Q_1 being blocks that begin with index 1. The blocks are chosen by appealing to Theorem 24.4, which says that the partial sums of $\sum_{n=1}^{\infty} p_n$ are not bounded above and the partial sums of $\sum_{n=1}^{\infty} q_n$ are not bounded below, meaning that whenever $x < y$, we are always able to find a block P from (p_n), with arbitrary starting index, for which $x + P > y$, as well as a block Q from (q_n), with arbitrary starting index, for which $y + Q < x$.

Specifically, we recursively construct the sequences (P_n) and (Q_n) of blocks as follows. First, let $P_1 = \sum_{i=1}^{j_1} p_i$, where j_1 is the unique natural number for which

$$P_1 - p_{j_1} \le r < P_1.$$

In other words, the block P_1 is formed by adding the minimum number of initial terms of (p_n) in order to achieve a sum greater than the number r. Now let $Q_1 = \sum_{i=1}^{k_1} q_i$, where k_1 is the unique natural number for which

$$P_1 + Q_1 < r \le P_1 + Q_1 - q_{k_1}.$$

Thus, the block Q_1 is formed by adding the minimum number of initial terms of (q_n) to P_1 in order to achieve a sum less than r. Continuing in this way, under the assumption that j_n, P_n, k_n, and Q_n have already been defined so that

$$\sum_{i=1}^{n-1}(P_i + Q_i) + P_n - p_{j_n} \le r < \sum_{i=1}^{n-1}(P_i + Q_i) + P_n, \tag{*}$$

and

$$\sum_{i=1}^{n}(P_i + Q_i) < r \le \sum_{i=1}^{n}(P_i + Q_i) - q_{k_n}, \tag{**}$$

we define j_{n+1} and P_{n+1} so that

$$\sum_{i=1}^{n}(P_i + Q_i) + P_{n+1} - p_{j_{n+1}} \leq r < \sum_{i=1}^{n}(P_i + Q_i) + P_{n+1}$$

and then define k_{n+1} and Q_{n+1} so that

$$\sum_{i=1}^{n+1}(P_i + Q_i) < r \leq \sum_{i=1}^{n+1}(P_i + Q_i) - q_{k_{n+1}}.$$

We now have a rearrangement

$$\begin{aligned} &P_1 + Q_1 + P_2 + Q_2 + P_3 + Q_3 + \cdots \\ &= \sum_{i=1}^{j_1} p_i + \sum_{i=1}^{k_1} q_i + \sum_{n=2}^{\infty}\left(\sum_{i=1+j_{n-1}}^{j_n} p_i + \sum_{i=1+k_{n-1}}^{k_n} q_i\right) \end{aligned}$$

of the original series $\sum_{n=1}^{\infty} a_n$ that we can show converges to r.

Define $U_1 = P_1$ and $V_1 = P_1 + Q_1$, and for each natural number n greater than 1, define

$$U_n = \sum_{i=1}^{n-1}(P_i + Q_i) + P_n$$

and

$$V_n = \sum_{i=1}^{n}(P_i + Q_i),$$

and note that the requirements (*) and (**) can then be expressed as

$$U_n - p_{j_n} \leq r < U_n$$

and

$$V_n < r \leq V_n - q_{k_n}.$$

So for every n we have

$$0 < U_n - r \leq p_{j_n}$$

and

$$0 < r - V_n \leq -q_{k_n}.$$

As (p_{j_n}) and (q_{k_n}) are both subsequences of (a_n), which itself converges to 0 as $\sum_{n=1}^{\infty} a_n$ converges, we conclude that $p_{j_n} \to 0$ and $q_{k_n} \to 0$. It then follows from the above inequalities and Theorem 10.15, the squeeze theorem, that $U_n \to r$ and $V_n \to r$.

Now consider any partial sum s_n of our rearranged series. If the last summand of this partial sum is positive, we have

$$V_m < s_n \leq U_{m+1}$$

for some m, whereas if the last summand is negative, we have

$$V_m \leq s_n < U_m$$

for some m. Because $U_n \to r$ and $V_n \to r$, it now follows that $s_n \to r$, meaning the rearranged series converges to r. □

Not only can a conditionally convergent series be rearranged so as to converge to any real number we would like, it can also be rearranged to diverge to either ∞ or $-\infty$.

Theorem 24.7. *If an infinite series converges conditionally, then there are rearrangements of the series that diverge to ∞ and to $-\infty$.*

Exercise 24.10. Prove that a conditionally convergent series can be rearranged so as to diverge to ∞.

---

25 Sequences and series of functions

We present a more extensive chapter opening than usual to provide motivation for taking up the study of sequences and series of functions.

The formula for the sum of a convergent geometric series tells us that

$$\sum_{n=0}^{\infty} x^n = \frac{1}{1-x}$$

when $-1 < x < 1$. Thus, if we define the function $h : (-1, 1) \to \mathbb{R}$ so that

$$h(x) = \frac{1}{1-x},$$

we have

$$h(x) = 1 + x + x^2 + x^3 + \cdots,$$

which is interesting for (at least) two reasons. First, the function h has been represented by an infinite series, which makes us wonder whether series representations might be achievable for other functions. Second, h is a non-polynomial function, yet the series representing it is a sort-of "infinite degree" polynomial, which gets us thinking about this specific type of series and its potential for representing a function.

The appearance of this polynomial-like series also hints at our work with Taylor polynomials in Chapter 17. There we learned that, given a function f, it is sometimes possible to use these *Taylor polynomials* P_n derived from f, where

$$P_n(x) = \sum_{k=0}^{n} \frac{f^{(k)}(x_0)}{k!}(x - x_0)^k,$$

to approximate values $f(x)$ of f, at least for x sufficiently close to the particular input x_0. We also found that in some cases Theorem 17.3, Taylor's theorem, would help us to measure the potential error involved in the resulting approximations. From the examples we considered, it appeared that the higher the degree of a Taylor polynomial, the better the approximations it produces.

Now a function f is said to be **infinitely differentiable** or to **have derivatives of all orders** at a point x_0 in its domain if $f^{(n)}(x_0)$ exists for every natural number n. Taken together, the utility of Taylor polynomials for obtaining approximations to function values and the recognition that there are non-polynomial functions that can be represented by infinite series that look like infinite degree polynomials, naturally raises the question of whether an infinitely differentiable function might be representable by an "infinite" version of the Taylor polynomial. That is, if the function f has derivatives of all orders, is it possible that

https://doi.org/10.1515/9783111429564-025

$$f(x) = \sum_{n=0}^{\infty} \frac{f^{(n)}(x_0)}{n!}(x - x_0)^n,$$

at least for values of x sufficiently close to x_0? (Note that the partial sums of the series on the right side of this equation are the Taylor polynomials for f at x_0.)

The use of polynomials and infinite series to approximate function values and represent functions goes all the way back to the pioneering work of Isaac Newton with the calculus in the 1660s. Some of Newton's results in this area were published in the 1690s, but in an incomplete form, and while other mathematicians such as James Gregory (1638–1675) also contributed to this work, it was Brook Taylor (1685–1731) who in 1715 published a general method for obtaining what are now called *Taylor series* representations of functions.

It is important to note that both series

$$1 + x + x^2 + x^3 + \cdots$$

and

$$\sum_{n=0}^{\infty} \frac{f^{(n)}(x_0)}{n!}(x - x_0)^n$$

discussed above are dependent on the value of x. In other words, each becomes a specific numerical series only when a particular number is assigned to x (and of course, in the latter case, we are assuming a function f and a number x_0 in the domain of f are already under consideration). Thus, in each case, before a value of x is specified, we may view the series as an infinite sum for which the summands are functions. In particular, if for each nonnegative integer n, we define the function $g_n : (-1, 1) \to \mathbb{R}$ so that

$$g_n(x) = x^n,$$

the infinite series

$$1 + x + x^2 + x^3 + \cdots = g_0(x) + g_1(x) + g_2(x) + g_3(x) + \cdots$$

is the value of the infinite sum of functions

$$g_0 + g_1 + g_2 + g_3 + \cdots = \sum_{n=0}^{\infty} g_n$$

at the input x.

We are therefore inevitably drawn toward a study of *infinite series of functions*, and since a series is really the sequence determined by its partial sums, a study of *sequences of functions*. By engaging in such a study, we are then able to investigate the question of representing functions by means of infinite series, and more specifically, by infinite series whose partial sums are Taylor polynomials.

Examples of sequences and series of functions

Recall that a **sequence** is any function whose domain is the set $\mathbb{N}$ of natural numbers. Most of the sequences we have encountered have been **sequences of real numbers**, functions with domain $\mathbb{N}$ and range a subset of the set $\mathbb{R}$ of real numbers. Thus, a sequence

$$a = (a_n) = (a_1, a_2, a_3, \dots)$$

of real numbers is actually the function $a : \mathbb{N} \to \mathbb{R}$, where for each natural number n we are writing a_n for $a(n)$.

A **sequence of functions** is a function with domain $\mathbb{N}$ whose range consists of functions. For instance,

$$f = (f_n) = (f_1, f_2, f_3, \dots)$$

is a sequence of functions if f_n is a function for each natural number n.

For our purposes, the functions in a sequence of functions are almost always real-valued and have a common domain that is some nonempty subset of $\mathbb{R}$. Thus, when we say that (f_n) **is a sequence of functions on** A (or **defined on** A), we intend that A is a nonempty subset of $\mathbb{R}$ and that for each natural number n, we have a function $f_n : A \to \mathbb{R}$ whose domain is A.

Example 25.1. Define the sequence (f_n) of functions on $\mathbb{R}$ so that

$$f_n(x) = \frac{x(-1)^{n+1}}{n}.$$

In other words, define the sequence of functions

$$f = (f_n) = (f_1, f_2, f_3, \dots)$$

so that, for each natural number n, the function $f_n : \mathbb{R} \to \mathbb{R}$ is defined by

$$f_n(x) = \frac{x(-1)^{n+1}}{n}.$$

Thus,

$$f_1(x) = x, \quad f_2(x) = -\frac{1}{2}x, \quad f_3(x) = \frac{1}{3}x, \quad f_4(x) = -\frac{1}{4}x, \quad \dots.$$

When it is convenient to do so, we allow the starting index for a sequence of functions to be an integer different from 1.

Example 25.2. Define the sequence of functions

$$g = (g_n) = (g_0, g_1, g_2, \ldots)$$

so that, for each nonnegative integer n, the function $g_n : (-1,1) \to \mathbb{R}$ is defined by

$$g_n(x) = x^n.$$

Thus, g is a sequence of functions on $(-1,1)$ and we have

$$g_0(x) = 1, \quad g_1(x) = x, \quad g_2(x) = x^2, \quad \ldots.$$

Here the indexing set is $\{0,1,2,\ldots\}$ rather than the set $\mathbb{N}$. Note that we are also permitting ourselves to abuse notation and regard $g_0(0) = 1$ even though the expression 0^0 is undefined.

Exercise 25.1. Referring to Example 25.1, find the numerical value of $f_6(12)$.

Exercise 25.2. Referring to Example 25.2, find the numerical value of $g_{10}(-0.5)$.

Exercise 25.3. Recursively define the sequence (f_n) of functions on $\mathbb{R}$ so that $f_1(x) = x^5$ and, for each natural number n, we have $f_{n+1} = f_n'$. Find the formulas for $f_5(x)$, for $f_6(x)$, and for $f_{100}(x)$.

Let A be a nonempty subset of $\mathbb{R}$ and suppose that for every natural number n, we have a function $f_n : A \to \mathbb{R}$. The **(infinite) series of functions**

$$f_1 + f_2 + f_3 + \cdots = \sum_{n=1}^{\infty} f_n$$

with summands $f_1, f_2, f_3, \ldots$ from the sequence (f_n) of functions is defined to be the sequence

$$(f_1, f_1 + f_2, f_1 + f_2 + f_3, \ldots)$$

of functions whose nth term is the function

$$f_1 + f_2 + \cdots + f_n = \sum_{k=1}^{n} f_k,$$

which is also referred to as the n**th partial sum** of the series.

Example 25.3. Take the sequence of functions (f_n) to be as in Example 25.1. Then the series $\sum_{n=1}^{\infty} f_n$ is the sequence whose nth term is the function $f_1 + f_2 + \cdots + f_n$ defined on $\mathbb{R}$ by

$$(f_1 + f_2 + \cdots + f_n)(x) = x - \frac{1}{2}x + \frac{1}{3}x - \cdots + \frac{x(-1)^{n+1}}{n} = \sum_{k=1}^{n} \frac{x(-1)^{k+1}}{k}.$$

For efficiency of communication, it is customary to refer to

$$\sum_{n=1}^{\infty} f_n(x)$$

as a series of functions, when in reality it is a series of real numbers obtained by evaluating each summand function f_n at the number x, with the actual series of functions being $\sum_{n=1}^{\infty} f_n$.

Example 25.3 (Continued). This convention allows us to speak about, for example, the series of functions

$$\sum_{n=1}^{\infty} \frac{x(-1)^{n+1}}{n},$$

rather than the more verbose, but formally correct, series of functions

$$\sum_{n=1}^{\infty} f_n,$$

where, for each natural number n, the function f_n is defined so that

$$f_n(x) = \frac{x(-1)^{n+1}}{n}.$$

If it is convenient, we allow the indexing of a series of functions to begin with an integer different from 1.

Example 25.4. Take the sequence of functions (g_n) to be as in Example 25.2. Then the series $\sum_{n=0}^{\infty} g_n$ is the sequence whose nth term is the function $g_0 + g_1 + \cdots + g_n$ defined on $(-1, 1)$ by

$$(g_0 + g_1 + \cdots + g_n)(x) = 1 + x + x^2 + \cdots + x^n = \sum_{k=0}^{n} x^k.$$

This series of functions would more commonly be expressed as simply

$$\sum_{n=0}^{\infty} x^n,$$

using the convention discussed above.

Exercise 25.4. Let s_n be the nth partial sum of the series $\sum_{n=1}^{\infty} f_n$ defined in Example 25.3. Evaluate $s_4'(12)$.

Pointwise convergence

Suppose that A is a nonempty subset of $\mathbb{R}$ and that for each natural number n, we have a function $f_n : A \to \mathbb{R}$. For each point x in A, the sequence (f_n) of functions gives rise to a sequence

$$(f_n(x))$$

of real numbers, which may or may not converge. If for each x in A the sequence $(f_n(x))$ converges, and we define $g(x)$ to be the limit of $(f_n(x))$, we have actually defined a function $g : A \to \mathbb{R}$. In this circumstance, we say that the sequence (f_n) of functions **converges pointwise** on A to the function g and we call g the **limit** of (f_n). Symbolically, we may indicate the pointwise convergence of (f_n) to the limit g by writing either

$$\lim f_n = g$$

or

$$f_n \to g.$$

We also sometimes write

$$\lim_{n\to\infty} f_n(x) = g(x),$$

especially if we want to make clear that the limiting operation is that pertaining to sequential convergence, not, for instance, a limit as the function inputs x tend toward some specified value.

In brief, then, a sequence (f_n) of functions converges pointwise to a function g if and only if

$$f_n(x) \to g(x)$$

for every x in the common domain of the functions f_n. In this way, we have used our prior understanding of what it means for a sequence of real numbers to converge to define a notion of convergence for sequences of real-valued functions.

Exercise 25.5. Explain why the limit function for a sequence of functions that converges pointwise must be unique.

Example 25.1 (Continued). For any real number x, we have

$$\lim_{n\to\infty} f_n(x) = \lim_{n\to\infty} \frac{x(-1)^{n+1}}{n} = x \lim_{n\to\infty} \frac{(-1)^{n+1}}{n} = x \cdot 0 = 0.$$

Therefore, if we define the function $g : \mathbb{R} \to \mathbb{R}$ by $g(x) = 0$, it follows that

$$\lim_{n\to\infty} f_n(x) = g(x)$$

for every x in $\mathbb{R}$. So, by definition, the sequence (f_n) of functions converges pointwise on $\mathbb{R}$ to the function g. As n increases the graphs of the functions in the sequence (f_n) become less steeply inclined and "flatten out" toward the x-axis, which is the graph of g (Figure 25.1).

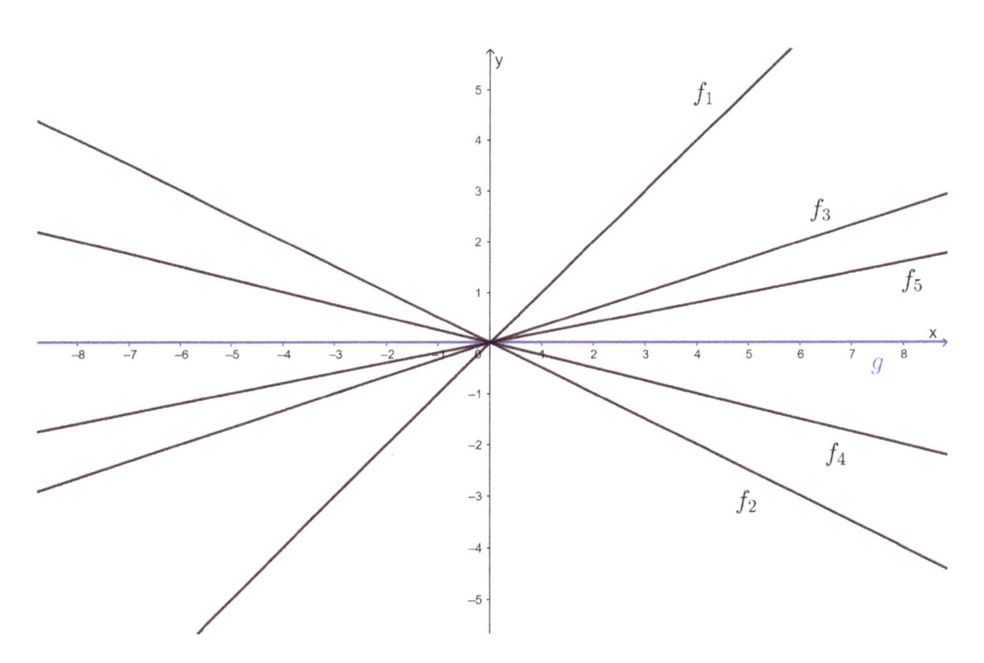

Figure 25.1: Graphs of the functions f_1, f_2, f_3, f_4, f_5, and g from Example 25.1.

Example 25.5. For each natural number n, define the function $f_n : [0,1] \to \mathbb{R}$ so that

$$f_n(x) = x^n.$$

Since

$$\lim_{n\to\infty} 1^n = 1$$

and, using the result from Example 10.20,

$$\lim_{n\to\infty} x^n = 0$$

when $0 \le x < 1$, it follows that

$$f_n \to g,$$

where the function $g : [0,1] \to \mathbb{R}$ is defined by

$$g(x) = \begin{cases} 0, & \text{if } 0 \le x < 1; \\ 1, & \text{if } x = 1. \end{cases}$$

Observe how the graphs of the functions f_n, with the exception of the point $(1,1)$ which remains fixed, collapse toward the x-axis as n increases, visually depicting the convergence of (f_n) to g (Figure 25.2).

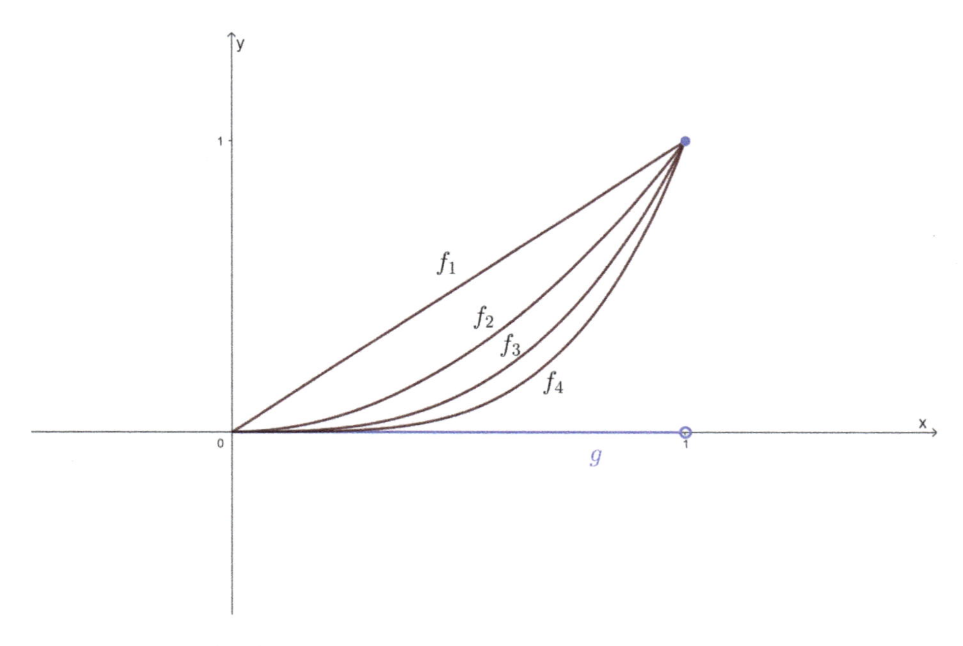

Figure 25.2: Graphs of the functions f_1, f_2, f_3, f_4, and g from Example 25.5.

Exercise 25.6. Find the limit function to which the given sequence of functions converges pointwise.

(a) The sequence (f_n) of functions defined on $\mathbb{R}$ so that $f_n(x) = \frac{x+nx^2}{n}$.

(b) The sequence (g_n) of functions defined on $\mathbb{R} - \{0\}$ so that $g_n(x) = ne^{-nx^2}$.

(c) The sequence (h_n) of functions defined on $\mathbb{R}$ so that $h_n(x) = \frac{x^{2n}}{1+x^{2n}}$.
(d) The sequence (i_n) of functions defined on $(0, \infty)$ so that $i_n(x) = \frac{1}{\sqrt{x}\ln(n)}$.
(e) The sequence (j_n) of functions defined on $[0, 1]$ so that $j_n(x) = \frac{1}{1+n^2 x}$.
(f) The sequence (k_n) of functions defined on $(-1, 1)$ so that $k_n(x) = \sqrt{x^2 + \frac{1}{n}}$.

Because an infinite series of functions is a special kind of sequence of functions, any definition or theorem concerning sequences of functions also applies to series of functions, and can be re-expressed in a fashion that is more directly applicable to series of functions.

For example, having defined the notion of pointwise convergence of a sequence of functions, it follows that a series $\sum_{n=1}^{\infty} f_n$ of functions **converges pointwise** to a function g if the sequence $(\sum_{k=1}^{n} f_k)$ of partial sums of $\sum_{n=1}^{\infty} f_n$ converges pointwise to g.

This definition is fine as it stands, but a more usable version of it can be achieved by directly incorporating within it the definitions of *pointwise convergence of a sequence of functions* and *convergence of a series of real numbers*. Doing so leads us to conclude that a series $\sum_{n=1}^{\infty} f_n$ of functions **converges pointwise** to a function g if, for each x in the common domain of the functions f_n, the real sequence $(\sum_{k=1}^{n} f_k(x))$ converges to $g(x)$, that is, if, for each x, the real series $\sum_{n=1}^{\infty} f_n(x)$ converges to (i. e., has sum) $g(x)$.

When the series $\sum_{n=1}^{\infty} f_n$ converges pointwise to g, we call g the **sum** of the series and we may write

$$\sum_{n=1}^{\infty} f_n = g.$$

Example 25.3 (Continued). Consider again the infinite series $\sum_{n=1}^{\infty} f_n$ of functions $f_n : \mathbb{R} \to \mathbb{R}$, where $f_n(x) = \frac{x(-1)^{n+1}}{n}$. In Exercise 23.9, you showed that

$$\sum_{n=1}^{\infty} \frac{(-1)^{n+1}}{n} = \ln(2).$$

Hence, for any real number x, we have

$$\sum_{n=1}^{\infty} f_n(x) = \sum_{n=1}^{\infty} \frac{x(-1)^{n+1}}{n} = x \sum_{n=1}^{\infty} \frac{(-1)^{n+1}}{n} = x\ln(2),$$

so we may conclude that the series $\sum_{n=1}^{\infty} f_n$ converges pointwise on $\mathbb{R}$ to the function $g : \mathbb{R} \to \mathbb{R}$, where $g(x) = x\ln(2)$, which is the sum of the series.

Example 25.4 (Continued). Recall that for each nonnegative integer n, we defined the function $g_n : (-1, 1) \to$ so that $g_n(x) = x^n$. Using the formula for a convergent geometric series, we conclude that

$$\sum_{n=0}^{\infty} g_n(x) = \sum_{n=0}^{\infty} x^n = \frac{1}{1-x}.$$

Thus, defining the function $h : (-1,1) \to \mathbb{R}$ so that $h(x) = \frac{1}{1-x}$, we may say that the series $\sum_{n=0}^{\infty} g_n$ converges pointwise on $(-1,1)$ to h and write $\sum_{n=0}^{\infty} g_n = h$.

Exercise 25.7. Find the sum function to which the given series of functions converges to pointwise, as well as the largest possible domain for the sum function.
(a) $\sum_{n=1}^{\infty} f_n$, where for each natural number n, the function f_n is defined so that $f_n(x) = \frac{(x-2)^n}{3^{n+1}}$;
(b) $\sum_{n=1}^{\infty} g_n$, where for each natural number n, the function g_n is defined so that $g_n(x) = \frac{x}{(1+x^2)^n}$.

Taylor series

Having introduced the notion of pointwise convergence of a sequence of functions, we return to the question raised at the beginning of the chapter, whether, given a function f having derivatives of all orders, the function's sequence

$$(P_0, P_1, P_2, P_3, \ldots)$$

of Taylor polynomials at x_0 converges pointwise to f on some subset of the domain of f that includes the point x_0.

Recall that for each nonnegative integer n, we have defined the **nth Taylor polynomial** P_n for a function f whose nth derivative is defined at x_0 (treating the zeroth derivative as f itself) so that

$$P_n(x) = \sum_{k=0}^{n} \frac{f^{(k)}(x_0)}{k!}(x - x_0)^k.$$

The nth Taylor polynomial P_n is a polynomial function for which

$$\begin{aligned}
P_n^{(0)}(x_0) &= P_n(x_0) = f(x_0) = f^{(0)}(x_0),\\
P_n^{(1)}(x_0) &= P_n'(x_0) = f'(x_0) = f^{(1)}(x_0),\\
P_n^{(2)}(x_0) &= P_n''(x_0) = f''(x_0) = f^{(2)}(x_0),\\
&\vdots\\
P_n^{(n)}(x_0) &= f^{(n)}(x_0),
\end{aligned}$$

meaning P_n and its first n derivatives, when evaluated at x_0, agree, respectively, with f and its first n derivatives, when evaluated at x_0.

When a function f is infinitely differentiable at x_0, we define the **Taylor series for (or of) f (centered) at x_0** to be the series of functions

$$\sum_{n=0}^{\infty} \frac{f^{(n)}(x_0)}{n!}(x - x_0)^n,$$

whose nth partial sum is the nth Taylor polynomial P_n for f at x_0. Note that there is the same abuse of notation here as we had with Taylor polynomials, in that 0^0 is undefined, but our intention is that $\frac{f^{(0)}(x_0)}{0!}(x_0 - x_0)^0 = f(x_0)$.

We can now reframe the question of convergence of the Taylor polynomials as one of trying to determine for which values of x the Taylor series for f converges to $f(x)$.

Example 25.6. Since the natural exponential function exp is defined on $\mathbb{R}$ so that

$$\exp(x) = e^x,$$

for every natural number n, we then have

$$\exp^{(n)}(x) = e^x,$$

from which it follows that

$$\exp^{(n)}(0) = e^0 = 1.$$

Hence, the Taylor series for exp at $x_0 = 0$ is

$$\sum_{n=0}^{\infty} \frac{\exp^{(n)}(0)}{n!}(x - 0)^n = \sum_{n=0}^{\infty} \frac{x^n}{n!}.$$

For which values of x the series $\sum_{n=0}^{\infty} \frac{x^n}{n!}$ converges to e^x is an unresolved question at this stage.

The Scottish mathematician Colin Maclaurin (1698–1746) explored in some depth the special case of a Taylor series with center 0, and so a Taylor series centered at $x_0 = 0$ is often referred to as a **Maclaurin series.**

Example 25.6 (Continued). The Taylor series $\sum_{n=0}^{\infty} \frac{x^n}{n!}$ for the natural exponential function exp at $x_0 = 0$ is the Maclaurin series for exp.

Example 25.7. The sine function sin is infinitely differentiable. Calculating its derivatives reveals the cyclical pattern where, for any nonnegative integer k, we have

$$\begin{aligned}
\sin^{(4k+1)}(x) &= \cos(x), \\
\sin^{(4k+2)}(x) &= -\sin(x), \\
\sin^{(4k+3)}(x) &= -\cos(x),
\end{aligned}$$

and

$$\sin^{(4k)}(x) = \sin(x).$$

So, as $\sin(0) = 0$ and $\cos(0) = 1$, the Maclaurin series for the sine function is

$$\begin{aligned}
\sum_{n=0}^{\infty} \frac{\sin^{(n)}(0)}{n!} x^n
&= \frac{\sin(0)}{0!} + \frac{\cos(0)}{1!} x + \frac{-\sin(0)}{2!} x^2 + \frac{-\cos(0)}{3!} x^3 + \frac{\sin(0)}{4!} x^4 + \frac{\cos(0)}{5!} x^5 \\
&\quad + \frac{-\sin(0)}{6!} x^6 + \frac{-\cos(0)}{7!} x^7 + \cdots \\
&= 0 + 1x + 0x^2 - \frac{1}{3!} x^3 + 0x^4 + \frac{1}{5!} x^5 + 0x^6 - \frac{1}{7!} x^7 + \cdots \\
&= x - \frac{x^3}{3!} + \frac{x^5}{5!} - \frac{x^7}{7!} + \cdots \\
&= \sum_{n=0}^{\infty} \frac{(-1)^n x^{2n+1}}{(2n+1)!}.
\end{aligned}$$

Again, we emphasize that we do not yet know for what values of x this series might converge to $\sin(x)$.

Example 25.8. The function $f : (0, \infty) \to \mathbb{R}$ defined by

$$f(x) = \sqrt{x}$$

is infinitely differentiable on its domain. Observe that $f'(x) = \frac{1}{2\sqrt{x}}$. In Exercise 25.8, you are asked to verify that

$$f^{(n)}(x) = \frac{(-1)^{n+1}(1 \cdot 3 \cdot 5 \cdot \ldots \cdot (2n-3))}{2^n} x^{0.5-n} = \frac{(-1)^{n+1} \prod_{i=1}^{n-1}(2i-1)}{2^n} x^{0.5-n}$$

when $n > 1$. Thus, the Taylor series for f centered at $x_0 = 4$ is

$$\begin{aligned}
\sum_{n=0}^{\infty} \frac{f^{(n)}(4)}{n!}(x-4)^n &= \frac{\sqrt{4}}{0!} + \frac{\frac{1}{2\sqrt{4}}}{1!}(x-4) + \sum_{n=2}^{\infty} \frac{\frac{(-1)^{n+1}\prod_{i=1}^{n-1}(2i-1)}{2^n} \cdot 4^{0.5-n}}{n!}(x-4)^n \\
&= 2 + \frac{1}{4}(x-4) + \sum_{n=2}^{\infty} \frac{(-1)^{n+1}\prod_{i=1}^{n-1}(2i-1)}{n! 2^{3n-1}}(x-4)^n.
\end{aligned}$$

As in Examples 25.6 and 25.7, no claim is yet being made about whether this Taylor series converges to the function f from which it has been generated.

Exercise 25.8. Let f be the function from Example 25.8. Use induction to prove that

$$f^{(n)}(x) = \frac{(-1)^{n+1}(1 \cdot 3 \cdot 5 \cdot \ldots \cdot (2n-3))}{2^n} x^{0.5-n} = \frac{(-1)^{n+1} \prod_{i=1}^{n-1}(2i-1)}{2^n} x^{0.5-n}$$

for each integer n larger than 1.

Exercise 25.9. Obtain the Taylor series for the given function f at the given point x_0.
(a) $f : \mathbb{R} \to \mathbb{R}$, where $f(x) = e^x$ at $x_0 = 1$
(b) $f : (-1, 1) \to \mathbb{R}$, where $f(x) = \sqrt[3]{1-x}$ at $x_0 = 0$
(c) $f : \mathbb{R} \to \mathbb{R}$, where $f(x) = \sin(x)$ at $x_0 = \frac{\pi}{2}$
(d) $f : (-1, 1) \to \mathbb{R}$, where $f(x) = \ln(x+1)$ at $x_0 = 0$

Exercise 25.10. Explain why a polynomial function of degree n is essentially its own Taylor series, regardless of the point x_0 chosen. Then, for the polynomial function f defined so that $f(x) = x^3$, write out the Taylor series for f centered at $x_0 = 5$ and show that it simplifies to x^3.

While we are always able to obtain the Taylor series for a function f that is infinitely differentiable at x_0, as suggested in our comments in Examples 25.6, 25.7, and 25.8, this Taylor series does not necessarily converge to $f(x)$ for any x other than x_0. The classic example of this phenomenon, given in the next example, is credited to the American mathematician Einar Hille (1894–1980) [4].

Example 25.9. Consider the function f defined on $\mathbb{R}$ so that

$$f(x) = \begin{cases} e^{-1/x^2}, & \text{if } x \neq 0; \\ 0, & \text{if } x = 0. \end{cases}$$

In Exercise 25.11, you show that f is infinitely differentiable on $\mathbb{R}$ and that $f^{(n)}(0) = 0$ for every natural number n. Hence, the Maclaurin series for f is

$$\sum_{n=0}^{\infty} \frac{f^{(n)}(0)}{n!}(x-0)^n = \sum_{n=0}^{\infty} \frac{0}{n!}(x-0)^n = 0,$$

the constant zero function on $\mathbb{R}$. Thus, as $f(x) = e^{-1/x^2} \neq 0$ for all nonzero real numbers x, the Maclaurin series converges to $f(x)$ for only $x = 0$, and does not represent f on any nontrivial interval about the center $x_0 = 0$. So even if a Taylor series converges, there is no guarantee it converges to the function from which it is generated.

Exercise 25.11. Here you are working with the function f defined in Example 25.9. The goal is to show that f is infinitely differentiable on $\mathbb{R}$ and that $f^{(n)}(0) = 0$ for every natural number n.

(a) Use the definition of the derivative and L'Hôpital's rule to show that $f'(0) = 0$.
First remark: Observe that, when $x \neq 0$, we have

$$f'(x) = \frac{2}{x^3} \cdot e^{-1/x^2} = p_1\left(\frac{1}{x}\right) \cdot e^{-1/x^2},$$

where p_1 is the polynomial function defined by $p_1(x) = 2x^3$.

(b) Show that the sum of any two polynomial functions p_1 and p_2 is also a polynomial function.

(c) Show that when c is a real number constant and p is a polynomial function, the constant multiple function cp is also a polynomial function.

(d) Show that the derivative p' of a polynomial function p is also a polynomial function.

(e) Show that if the function g is defined for nonzero x so that

$$g(x) = p_1\left(\frac{1}{x}\right) \cdot \frac{1}{x^m},$$

where p_1 is a polynomial function and m is a natural number, then

$$g(x) = p_2\left(\frac{1}{x}\right)$$

for some polynomial function p_2.

(f) Use the results from (b), (c), (d), and (e) to show that if the function g is defined for all nonzero x so that

$$g(x) = p_1\left(\frac{1}{x}\right) \cdot e^{-1/x^2},$$

where p_1 is a polynomial function, then for all nonzero x, we have

$$g'(x) = p_2\left(\frac{1}{x}\right) \cdot e^{-1/x^2}$$

for some polynomial function p_2.
Second remark: Since we noted above that for nonzero x, we have

$$f'(x) = p_1\left(\frac{1}{x}\right) \cdot e^{-1/x^2}$$

for a polynomial function p_1, the result from (f) immediately yields an induction argument that permits us to conclude that for nonzero x and any natural number n, we also have

$$f^{(n)}(x) = p_n\left(\frac{1}{x}\right) \cdot e^{-1/x^2}$$

for some polynomial function p_n. Thus, f is infinitely differentiable at all nonzero x, and these derivatives have a special form.

(g) Show that if g is a function defined on $\mathbb{R}$ so that

$$g(x) = \begin{cases} p(\frac{1}{x}) \cdot e^{-1/x^2}, & \text{if } x \neq 0; \\ 0, & \text{if } x = 0; \end{cases}$$

where p is a polynomial function, then $g'(0) = 0$.

Final remark: Our observations about the form $f^{(n)}(x)$ must take for nonzero x, together with the result from (g), immediately yield an induction argument that permits us to conclude that $f^{(n)}(0) = 0$ for every natural number n. Our goal has been achieved!

Power series

The Taylor series for a function that is infinitely differentiable at a particular point is defined as the limit of the sequence of Taylor polynomials constructed at that point. Taylor series are an example of a special type of series of functions known as *power series*. A **power series about** (or **centered at**) a real number x_0 is an infinite series of functions

$$\sum_{n=0}^{\infty} a_n(x - x_0)^n,$$

where

$$(a_n) = (a_0, a_1, a_2, \dots)$$

is a real sequence. The numbers $a_0, a_1, a_2, \dots$ are called the **coefficients** of the power series and the number x_0 is called the **center** of the power series. Power series are significant because not only do they include among their members the Taylor series, but, as we shall see, their behavior is very well regulated as compared to series of functions more generally. Once more, we are abusing notation as we shall interpret $a_0(x_0 - x_0)^0$ as representing a_0, even though the expression 0^0 is undefined.

Example 25.10. The series

$$\begin{aligned} \sum_{n=1}^{\infty} \frac{(-1)^{n+1}x^n}{n} &= 1x - \frac{1}{2}x^2 + \frac{1}{3}x^3 - \cdots \\ &= 0 + 1(x-0) + \left(-\frac{1}{2}\right)(x-0)^2 + \frac{1}{3}(x-0)^3 + \cdots \end{aligned}$$

is a power series centered at $x_0 = 0$. The coefficient of x^0 is $a_0 = 0$ and, for each natural number n, the coefficient of x^n is $a_n = \frac{(-1)^{n+1}}{n}$.

Example 25.11. The series

$$\sum_{n=0}^{\infty} 2^{-n}(x+1)^n = 1 + \frac{1}{2}(x-(-1)) + \frac{1}{4}(x-(-1))^2 + \frac{1}{8}(x-(-1))^3 + \cdots$$

is a power series with center $x_0 = -1$. For each nonnegative integer n, the coefficient of $(x+1)^n$ is $a_n = 2^{-n}$.

Example 25.4 (Continued). The series

$$\sum_{n=0}^{\infty} x^n$$

is a power series about $x_0 = 0$ for which all of the coefficients are 1.

Example 25.12. The series

$$\sum_{n=0}^{\infty} \sqrt[n+2]{x} = \sqrt{x} + \sqrt[3]{x} + \sqrt[4]{x} + \cdots$$

is not a power series as the summands involve roots of x rather than nonnegative integer powers of x.

Exercise 25.12. Determine which of the series from Exercise 25.7 are power series. For each that is, identify where it is centered and the coefficients.

A power series

$$\sum_{n=0}^{\infty} a_n(x-x_0)^n = a_0 + a_1(x-x_0) + a_2(x-x_0)^2 + a_3(x-x_0)^3 + \cdots$$

denotes what may be viewed informally as a "polynomial of infinite degree," which is especially obvious if we take $x_0 = 0$ to obtain

$$\sum_{n=0}^{\infty} a_n x^n = a_0 + a_1 x + a_2 x^2 + a_3 x^3 + \cdots.$$

Thus, results we prove about power series may be regarded, on an intuitive level, as results about infinite degree polynomials.

We have already mentioned that a Taylor series is a power series. Specifically, the Taylor series

$$\sum_{n=0}^{\infty} \frac{f^{(n)}(x_0)}{n!}(x-x_0)^n$$

is the power series

$$\sum_{n=0}^{\infty} a_n(x - x_0)^n$$

for which the coefficients are given by

$$a_n = \frac{f^{(n)}(x_0)}{n!}.$$

Hence, any results we prove for power series also apply to Taylor series.

Example 25.6 (Continued). We determined that the Maclaurin series for the natural exponential function exp is

$$\sum_{n=0}^{\infty} \frac{x^n}{n!}.$$

This series is the power series centered at $x_0 = 0$ for which the coefficient of x^n is $a_n = \frac{1}{n!}$.

Example 25.8 (Continued). The series

$$2 + \frac{1}{4}(x - 4) + \sum_{n=2}^{\infty} \frac{(-1)^{n+1} \prod_{i=1}^{n-1}(2i - 1)}{n!2^{3n-1}}(x - 4)^n$$

is the Taylor series for the principal square root function at $x_0 = 4$, but it is also a power series centered at 4. The coefficient of $(x - 4)^0$ is

$$a_0 = 2,$$

the coefficient of $(x - 4)^1$ is

$$a_1 = \frac{1}{4},$$

and, when $n \geq 2$, the coefficient of $(x - 4)^n$ is

$$a_n = \frac{(-1)^{n+1} \prod_{i=1}^{n-1}(2i - 1)}{n!2^{3n-1}}.$$

Exercise 25.13. Identify the coefficient of $(x - \frac{\pi}{2})^6$ in the Taylor series you obtained in Exercise 25.9(c).

If A is the set of values of x for which the power series $\sum_{n=0}^{\infty} a_n(x - x_0)^n$ converges, we may define a function $f : A \to \mathbb{R}$ so that

$$f(x) = \sum_{n=0}^{\infty} a_n (x - x_0)^n.$$

In this way, we may (and should) view a power series as representing a function. The set A here is known as the **domain of convergence** of the power series. The following theorem, whose statement uses the lowercase Greek letter λ (lambda), demonstrates that this set is always an interval, so for that reason it is usually referred to as the **interval of convergence** of the power series.

Theorem 25.13 (Convergence of a power series). *Given a power series* $\sum_{n=0}^{\infty} a_n (x - x_0)^n$, *let* $\lambda = \limsup \sqrt[n]{|a_n|}$.

(1) *If* $\lambda = 0$, *then* $\sum_{n=0}^{\infty} a_n (x - x_0)^n$ *converges absolutely for every real number* x.
(2) *If* $\lambda \in (0, \infty)$, *then* $\sum_{n=0}^{\infty} a_n (x - x_0)^n$ *converges absolutely when* $|x - x_0| < \frac{1}{\lambda}$ *and diverges when* $|x - x_0| > \frac{1}{\lambda}$.
(3) *If* $\lambda = \infty$, *then* $\sum_{n=0}^{\infty} a_n (x - x_0)^n$ *converges for only* $x = x_0$.

In establishing results about power series, it is usually sufficient to demonstrate the result holds for power series

$$\sum_{n=0}^{\infty} a_n x^n$$

about $x_0 = 0$, since a power series

$$\sum_{n=0}^{\infty} a_n (x - x_0)^n$$

centered at $x_0 \neq 0$ becomes a power series

$$\sum_{n=0}^{\infty} a_n y^n$$

centered at 0 via the substitution

$$y = x - x_0.$$

The explicit proof can then be applied to $\sum_{n=0}^{\infty} a_n y^n$ to obtain the result for $\sum_{n=0}^{\infty} a_n (x - x_0)^n$. We take this approach when it is feasible, including the following proof for Theorem 25.13.

Proof of Theorem 25.13. As discussed directly above, it suffices to explicitly present the argument for the case in which $x_0 = 0$.

Using the result of Exercise 4.18, we have

$$\limsup \sqrt[n]{|a_n x^n|} = (\limsup \sqrt[n]{|a_n|})|x| = \lambda |x|.$$

Thus, according to the root test, Theorem 23.15, the series $\sum_{n=0}^{\infty} a_n x^n$ converges absolutely when

$$\lambda|x| < 1$$

and diverges when

$$\lambda|x| > 1.$$

If $\lambda = 0$, the first of these two inequalities is true for all real x, thereby establishing (1).

If $\lambda \in (0, \infty)$, dividing the inequalities by λ yields (2).

If $\lambda = \limsup \sqrt[n]{|a_n|} = \infty$, then as $|x| > 0$ when $x \neq 0$, it follows that

$$\lambda|x| = (\limsup \sqrt[n]{|a_n|})|x| = \infty > 1,$$

and so the series diverges for $x \neq 0$. But when $x = 0$, the summands of the series are all 0, so the series converges to 0 for $x = 0$. Hence, (3) holds. □

This theorem tells us that the domain of convergence of a power series $\sum_{n=0}^{\infty} a_n(x - x_0)^n$ is always an interval centered at x_0. If (1) holds, this interval is

$$\mathbb{R} = (-\infty, \infty).$$

If (3) holds, it is

$$\{x_0\}.$$

If (2) holds, taking $\lambda = \limsup \sqrt[n]{|a_n|}$, it is one of the intervals

$$\left(x_0 - \frac{1}{\lambda}, x_0 + \frac{1}{\lambda}\right), \quad \left[x_0 - \frac{1}{\lambda}, x_0 + \frac{1}{\lambda}\right), \quad \left(x_0 - \frac{1}{\lambda}, x_0 + \frac{1}{\lambda}\right], \quad \text{or} \quad \left[x_0 - \frac{1}{\lambda}, x_0 + \frac{1}{\lambda}\right],$$

depending on whether the series converges at one, both, or neither of the endpoints, which cannot be determined from the theorem itself (and any of the possibilities listed here can occur, as demonstrated via examples and exercises to follow).

Based on this discussion, we define the **radius of convergence** of a power series to be

(1) ∞ if the interval of convergence is the entire set $\mathbb{R}$ of real numbers,
(2) 0 if the interval of convergence includes only the center x_0, and
(3) half the distance between the endpoints (i. e., half the length) of the interval of convergence, otherwise.

We denote the radius of convergence of a power series by R.

Note that when $\lambda = \limsup \sqrt[n]{|a_n|}$ is a positive real number, it follows that $R = \frac{1}{\lambda}$, and so the interval of convergence is one of the intervals

$$(x_0 - R, x_0 + R), \quad [x_0 - R, x_0 + R), \quad (x_0 - R, x_0 + R], \quad \text{or} \quad [x_0 - R, x_0 + R],$$

where, as mentioned above, an investigation of convergence at the endpoints must be separately conducted.

In the context of discussing the radius of convergence of a power series, it is customary to view 0 and ∞ as "reciprocals" of one another, as in doing so, it then follows that $R = \frac{1}{\lambda}$ for all possible values of R and λ, even 0 and ∞. In employing this convention, however, we are neither promoting the symbol ∞ to the status of being a real number, nor declaring that 0 has a multiplicative inverse.

The next theorem indicates that, provided $\lim_{n\to\infty} |\frac{a_n}{a_{n+1}}|$ exists or is infinite, its value must be the radius of convergence of the power series $\sum_{n=0}^{\infty} a_n(x - x_0)^n$. In practice, it is often easier to calculate the radius of convergence by means of this limit.

Theorem 25.14. *The radius of convergence of a power series $\sum_{n=0}^{\infty} a_n(x - x_0)^n$ is given by $\lim_{n\to\infty} |\frac{a_n}{a_{n+1}}|$ as long as this limit exists or is infinite.*

Proof. In this situation $\lim_{n\to\infty} |\frac{a_{n+1}}{a_n}|$ would also exist or be infinite, in which case

$$\lim_{n\to\infty}\left|\frac{a_{n+1}}{a_n}\right| = \liminf\left|\frac{a_{n+1}}{a_n}\right| = \limsup\left|\frac{a_{n+1}}{a_n}\right|.$$

Then, since we know from Theorem 23.17 that

$$\liminf\left|\frac{a_{n+1}}{a_n}\right| \le \limsup \sqrt[n]{|a_n|} \le \limsup\left|\frac{a_{n+1}}{a_n}\right|,$$

we would have

$$\lim_{n\to\infty}\left|\frac{a_{n+1}}{a_n}\right| = \limsup \sqrt[n]{|a_n|}.$$

Hence,

$$\lim_{n\to\infty}\left|\frac{a_n}{a_{n+1}}\right| = \frac{1}{\limsup \sqrt[n]{|a_n|}},$$

which is the radius of convergence of the power series. □

Example 25.4 (Continued). We have already used the formula for a convergent geometric series to determine that the power series

$$\sum_{n=0}^{\infty} x^n = 1 + x + x^2 + x^3 + \cdots$$

has interval of convergence $(-1, 1)$. Thus, the radius of convergence of this power series is $R = \frac{1-(-1)}{2} = 1$.

In most instances, though, we find the radius of convergence first and then use it to help us obtain the interval of convergence. To obtain the radius of convergence via Theorem 25.13, we observe that the coefficients of this series are $a_n = 1$ for all n, so that

$$\lambda = \limsup \sqrt[n]{|a_n|} = \limsup \sqrt[n]{|1|} = 1,$$

from which it follows that the radius of convergence is

$$R = \frac{1}{\lambda} = 1.$$

Taking this approach, since the series is centered at $x_0 = 0$, we would then conclude that the interval of convergence has left endpoint

$$x_0 - R = 0 - 1 = -1$$

and right endpoint

$$x_0 + R = 0 + 1 = 1.$$

When $x = -1$, the series $\sum_{n=0}^{\infty} x^n$ becomes

$$\sum_{n=0}^{\infty} (-1)^n,$$

which diverges as the partial sums alternate between 1 and 0, and when $x = 1$ the series becomes

$$\sum_{n=0}^{\infty} 1^n,$$

which diverges to ∞. Thus, neither endpoint is included in the interval of convergence, which is $(-1, 1)$.

A third approach to finding the radius of convergence is to apply the root test to the power series $\sum_{n=0}^{\infty} x^n$ itself, rather than to just the coefficients. According to the root test, the series $\sum_{n=0}^{\infty} x^n$ converges (absolutely) if

$$\limsup \sqrt[n]{|x^n|} = |x| < 1$$

and diverges if

$$\limsup \sqrt[n]{|x^n|} = |x| > 1,$$

from which we conclude that the radius of convergence is 1. From here, we would apply the same analysis as above to determine the interval of convergence as $(-1, 1)$, with neither endpoint included.

Example 25.10 (Continued). For the power series

$$\sum_{n=1}^{\infty} \frac{(-1)^{n+1} x^n}{n} = 1x - \frac{1}{2}x^2 + \frac{1}{3}x^3 - \cdots$$

we observe that the coefficients are

$$a_n = \frac{(-1)^{n+1}}{n},$$

so that, using the result of Exercise 10.19,

$$\lambda = \limsup \sqrt[n]{|a_n|} = \limsup \frac{1}{\sqrt[n]{n}} = 1.$$

Therefore, the radius of convergence is

$$R = \frac{1}{\lambda} = 1$$

and it follows that the endpoints of the interval of convergence are

$$x_0 - R = 0 - 1 = -1$$

and

$$x_0 + R = 0 + 1 = 1.$$

When $x = -1$, the series $\sum_{n=1}^{\infty} \frac{(-1)^{n+1} x^n}{n}$ becomes

$$\sum_{n=1}^{\infty} \frac{-1}{n},$$

the opposite of the harmonic series $\sum_{n=1}^{\infty} \frac{1}{n}$, and so diverges. When $x = 1$, the series $\sum_{n=1}^{\infty} \frac{(-1)^{n+1} x^n}{n}$ becomes the alternating harmonic series

$$\sum_{n=1}^{\infty} \frac{(-1)^{n+1}}{n},$$

and so converges. Thus, the interval of convergence for $\sum_{n=1}^{\infty} \frac{(-1)^{n+1} x^n}{n}$ is $(-1, 1]$, which includes the right endpoint but not the left endpoint.

Example 25.8 (Continued). To find the radius of convergence R of the power series

$$2 + \frac{1}{4}(x-4) + \sum_{n=2}^{\infty} \frac{(-1)^{n+1}\prod_{i=1}^{n-1}(2i-1)}{n!2^{3n-1}}(x-4)^n,$$

we note that when $n \geq 2$, the coefficients are $a_n = \frac{(-1)^{n+1}\prod_{i=1}^{n-1}(2i-1)}{n!2^{3n-1}}$, all of which are nonzero. Using Theorem 25.14, we calculate

$$R = \lim_{n\to\infty}\left|\frac{a_n}{a_{n+1}}\right| = \lim_{n\to\infty} \frac{\prod_{i=1}^{n-1}(2i-1)}{n!2^{3n-1}} \cdot \frac{(n+1)!2^{3(n+1)-1}}{\prod_{i=1}^{n}(2i-1)} = \lim_{n\to\infty} \frac{(n+1)\cdot 2^3}{2n-1} = 4.$$

As $x_0 = 4$, the endpoints of the interval of convergence are

$$x_0 - R = 4 - 4 = 0$$

and

$$x_0 + R = 4 + 4 = 8.$$

You are asked to investigate whether the series converges at either endpoint in Exercise 25.14.

Exercise 25.14. Determine the interval of convergence of the power series from Example 25.8.

Example 25.6 (Continued). For the power series

$$\sum_{n=0}^{\infty} \frac{x^n}{n!},$$

the coefficients are $a_n = \frac{1}{n!}$. Applying Theorem 25.14, the radius of convergence is

$$R = \lim_{n\to\infty}\left|\frac{a_n}{a_{n+1}}\right| = \lim_{n\to\infty}(n+1) = \infty,$$

and we conclude that the series converges for all real numbers x, and thus has interval of convergence $\mathbb{R}$.

Example 25.15. The coefficients of the power series

$$\sum_{n=0}^{\infty} n!x^n$$

are $a_n = n!$, so as

$$\lim_{n\to\infty}\left|\frac{a_n}{a_{n+1}}\right| = \lim_{n\to\infty}\frac{1}{n+1} = 0,$$

we conclude via Theorem 25.14 that the radius of convergence is $R = 0$ and the series converges only at its center $x_0 = 0$, thus making the interval of convergence $\{0\}$.

Exercise 25.15. Find the radius of convergence of the given power series.

(a) $\sum_{n=1}^{\infty} \frac{1}{n^n} x^n$

(b) $\sum_{n=1}^{\infty} (\sqrt{n})^n x^n$

(c) $\sum_{n=1}^{\infty} \frac{n^{10}}{n!} (x-1)^n$

(d) $\sum_{n=1}^{\infty} \frac{4^n}{n^4} (x+3)^n$

(e) $\sum_{n=2}^{\infty} \frac{(-1)^{n+1}}{\ln(n)} x^n$

(f) $\sum_{n=1}^{\infty} \frac{1}{(\sqrt{n})^{\sqrt{n}}} x^n$

(g) $\sum_{n=1}^{\infty} n^2 (\frac{x}{3})^{2n}$

Exercise 25.16. Find the interval of convergence for each power series in Exercise 25.15.

Exercise 25.17. Use induction to prove that for any natural number n, we have

$$\frac{(2n)!}{(n!)^2 \cdot 4^n} \le \frac{1}{\sqrt{3n+1}}.$$

Hint: In the inductive step, use the fact that $\frac{(2n)!}{(n!)^2 \cdot 4^n} = \frac{1\cdot 3\cdot 5\cdot \ldots \cdot (2n-1)}{2\cdot 4\cdot 6\cdot \ldots \cdot (2n)}$, which you should algebraically verify.

Exercise 25.18. Consider the power series $\sum_{n=0}^{\infty} \frac{(2n)!}{(n!)^2} x^n$.

(a) Verify that the radius of convergence is $\frac{1}{4}$.

(b) Use Raabe's test to show the series diverges when $x = \frac{1}{4}$.

(c) Use the alternating series test to show the series converges when $x = -\frac{1}{4}$. *Hint*: Make use of the result from Exercise 25.17.

Exercise 25.19. Prove the sequence $((1+\frac{1}{n})^{n+0.5})$ is decreasing and converges to e. *Hint*: It is straightforward to prove the sequence converges to e. To prove it is decreasing, first argue that it is sufficient to show $(1+\frac{1}{n(n+2)})^{2n+1} \cdot (1-\frac{1}{n+2})^2 > 1$ for every natural number n. Then apply Bernoulli's inequality (see Exercise 3.14) to $(1+\frac{1}{n(n+2)})^{2n+1}$.

Exercise 25.20. Consider the power series $\sum_{n=1}^{\infty} \frac{n^n}{n!} x^n$.

(a) Verify the radius of convergence is $\frac{1}{e}$.

(b) Show the series diverges when $x = \frac{1}{e}$. *Hint*: First verify that for any natural number greater than 1, we have

$$n = \prod_{k=1}^{n-1}\left(1 + \frac{1}{k}\right).$$

Use this fact to then verify that

$$n! = \frac{n^n}{\prod_{k=1}^{n-1}(1 + \frac{1}{k})^k}.$$

Then use this fact and the result of Exercise 25.19 to show that

$$n! \leq \frac{n^{n+0.5}}{e^{n-1}}.$$

Finally, apply the comparison test with the 0.5-series, which we know diverges. (In Exercise 27.14, you show that the series converges when $x = -\frac{1}{e}$.)

Exercise 25.21. Suppose the interval of convergence of a power series has two distinct endpoints. Explain why it is not possible for the power series to converge absolutely at one of the endpoints but not at the other.

Exercise 25.22. Find the radius of convergence of the power series $\sum_{n=0}^{\infty} a_n x^n$ if there exist positive numbers b and c for which $b < |a_n| < c$ for every nonnegative integer n.

26 Uniform convergence

Pointwise convergence of a sequence or series of functions is a natural extension of the notion of convergence of a sequence or series of real numbers. Unfortunately, though, the pointwise limit of a sequence of functions need not inherit any of the key properties of continuity, differentiability, or integrability even when shared by all of the functions in the sequence. Hence, we are drawn to introduce a stronger form of convergence, *uniform convergence*, that interacts more favorably with these properties. This new form of convergence also plays a significant role when we continue our exploration of power series in the next chapter.

Deficiencies of pointwise convergence

The sum rule for derivatives tells us that if f_1 and f_2 are differentiable functions, so is $f_1 + f_2$, with

$$(f_1 + f_2)' = f_1' + f_2'.$$

This property can be extended by induction to any finite sum of differentiable functions. Can it be extended to infinite sums of the type represented by an infinite series? That is, if (f_n) is a sequence of differentiable functions for which $\sum_{n=1}^{\infty} f_n = g$, does it necessarily follow that the function g is differentiable and that

$$g' = \sum_{n=1}^{\infty} f_n'?$$

For instance, as Example 25.4 demonstrated that

$$1 + x + x^2 + x^3 + \cdots = \frac{1}{1 - x}$$

when $-1 < x < 1$, can we differentiate summand-by-summand on the left side of this equation to obtain the derivative of the function on the right side? In other words, is it true that the series

$$(1)' + (x)' + (x^2)' + (x^3)' + \cdots = 1 + 2x + 3x^2 + \cdots$$

converges to

$$\left(\frac{1}{1 - x}\right)' = \frac{-1}{(1 - x)^2},$$

at least when $-1 < x < 1$?

https://doi.org/10.1515/9783111429564-026

The sum rule for integrals raises a similar question: If (f_n) is a sequence of functions each of which is integrable on the interval $[a,b]$ and we know that $\sum_{n=1}^{\infty} f_n = g$, does it necessarily follow that the function g is integrable on $[a,b]$ and that $\int_a^b g = \sum_{n=1}^{\infty} \int_a^b f_n$?

More broadly, since a series is a sequence of partial sums, it is natural to ask how the fundamental notions of differentiability, integrability, and continuity relate to a convergent *sequence* of functions. Specifically, given a sequence (f_n) of functions for which $f_n \to g$,

(1) if each function f_n is continuous, must g also be continuous?
(2) if each function f_n is differentiable, must g also be differentiable and must $f_n' \to g'$?
(3) if each function f_n is integrable on $[a,b]$, must g also be integrable on $[a,b]$ and must $\int_a^b f_n \to \int_a^b g$?

Regrettably, the answer to each of these questions is no, as the following examples illustrate.

Example 26.1. In Example 25.5, we showed that the sequence (f_n) of functions defined on $[0,1]$ so that

$$f_n(x) = x^n$$

converges to the function g defined on $[0,1]$ so that

$$g(x) = \begin{cases} 0, & \text{if } 0 \le x < 1; \\ 1, & \text{if } x = 1. \end{cases}$$

Each function f_n, being a polynomial function, is continuous, but the limit function g is not continuous at $x = 1$ since $g(1) = 1$ but $\lim_{x \to 1} g(x) = 0$. Thus, *the pointwise limit of continuous functions need not be continuous.*

Furthermore, each function f_n, being a polynomial function, is differentiable, but the fact that the limit function g is discontinuous at 1 also means g is not differentiable at 1. Thus, *the pointwise limit of differentiable functions need not be differentiable.*

Example 26.2. For each natural number n, define the function $f_n : [0,1] \to \mathbb{R}$ so that

$$f_n(x) = \frac{\sin(nx)}{n}.$$

As

$$-1 \le \sin(nx) \le 1$$

for all x, it follows that

$$-\frac{1}{n} \le \frac{\sin(nx)}{n} \le \frac{1}{n},$$

and so, by Theorem 10.15, the squeeze theorem, we have

$$\lim_{n\to\infty} \frac{\sin(nx)}{n} = 0$$

for all x. Thus, we may conclude that $f_n \to g$, where the function $g : [0,1] \to \mathbb{R}$ is the constant zero function defined by

$$g(x) = 0.$$

Each function f_n is differentiable, with

$$f_n'(x) = \cos(nx),$$

and g is differentiable, with

$$g'(x) = 0.$$

Thus,

$$g'(0) = 0$$

and, for every n, we have

$$f_n'(0) = \cos(n \cdot 0) = 1,$$

and it follows that

$$f_n'(0) \nrightarrow g'(0),$$

so that

$$f_n' \nrightarrow g'$$

on $[0,1]$. Thus, *even when the pointwise limit of differentiable functions is differentiable, the limit of the derivatives need not be the derivative of the limit.*

Example 26.3. The rational numbers in the interval $[0,1]$ form a countably infinite set, so there is a bijection $q : \mathbb{N} \to \mathbb{Q} \cap [0,1]$. As q is a sequence we shall write q_n for $q(n)$. For each natural number n, define the function $f_n : [0,1] \to \mathbb{R}$ so that

$$f_n(x) = \begin{cases} 1, & \text{if } x \in \{q_1, q_2, \dots, q_n\}; \\ 0, & \text{otherwise.} \end{cases}$$

The function f_n has only finitely many discontinuities, at each of $q_1, q_2, \dots, q_n$, and so is integrable. However, $f_n \to g$, where g is the Dirichlet function defined on $[0,1]$ so that

$$g(x) = \begin{cases} 1, & \text{if } x \text{ is rational;} \\ 0, & \text{if } x \text{ is irrational;} \end{cases}$$

and our work in Example 18.11 shows that g is not integrable. Thus, *the pointwise limit of integrable functions need not be integrable.*

Example 26.4. For each natural number n, define the function $f_n : [0,1] \to \mathbb{R}$ so that

$$f_n(x) = nx^3(1-x^4)^n.$$

You show in Exercise 26.1 that $f_n \to g$, where g is the constant zero function defined on $[0,1]$. Each f_n, being a polynomial function, is integrable on $[0,1]$, as is g. However,

$$\lim_{n\to\infty} \int_0^1 f_n = \lim_{n\to\infty} -\frac{n(1-x^4)^{n+1}}{4(n+1)}\bigg|_{x=0}^1 = \lim_{n\to\infty} \frac{n}{4(n+1)} = \frac{1}{4},$$

whereas

$$\int_0^1 g = 0.$$

Thus, *even when the pointwise limit of integrable functions is integrable, the limit of the integrals need not be the integral of the limit.*

Exercise 26.1. Referring to Example 26.4, show that $f_n \to g$.

Exercise 26.2. For each natural number n, define the function $f_n : [0,1] \to \mathbb{R}$ so that $f_n(x) = \frac{\sin(n^2 x)}{n}$. Show that the sequence $(f_n'(0))$ does not converge.

Exercise 26.3. For each natural number n, define the function $f_n : \mathbb{R} \to \mathbb{R}$ so that

$$f_n(x) = \begin{cases} |x|, & \text{if } |x| \ge \frac{1}{n}; \\ \frac{1}{4}(5nx^2 - n^5x^6), & \text{if } |x| < \frac{1}{n}. \end{cases}$$

(a) Prove that $f_n \to g$ where g is the absolute value function defined on $\mathbb{R}$.
(b) Prove that each f_n is differentiable on $\mathbb{R}$ and find a (piecewise) formula for $f_n'(x)$.
(c) Show that $\lim_{n\to\infty} f_n'(0)$ exists, even though $g'(0)$ is undefined.

Exercise 26.4. For each natural number n, define the function $f_n : [0,1] \to \mathbb{R}$ so that

$$f_n(x) = \begin{cases} 1, & \text{if } x \in \{\frac{1}{2^n}, \frac{2}{2^n}, \ldots, \frac{2^n}{2^n} = 1\}; \\ 0, & \text{otherwise.} \end{cases}$$

Also, define the function $g : [0,1] \to \mathbb{R}$ so that

$$g(x) = \begin{cases} 1, & \text{if } x = \frac{k}{2^n} \text{ for some natural numbers } k \text{ and } n; \\ 0, & \text{otherwise.} \end{cases}$$

(a) Show that (f_n) converges pointwise on $[0,1]$ to g.
(b) Indicate why each f_n is integrable on $[0,1]$ and evaluate $\int_0^1 f_n$. Then find the limit of the sequence $(\int_0^1 f_n)$.
(c) Show that the function g is not integrable on $[0,1]$.

Exercise 26.5. For each natural number n, define the function $f_n : [0,1] \to \mathbb{R}$ so that

$$f_n(x) = \begin{cases} n - n^2 x, & \text{if } 0 < x < \frac{1}{n}; \\ 0, & \text{if } x = 0 \text{ or } \frac{1}{n} \le x \le 1. \end{cases}$$

(a) Identify the limit function g to which the sequence (f_n) converges pointwise on $[0,1]$.
(b) Each of the functions f_n is integrable on $[0,1]$, as is the function g. Indicate why.
(c) Show that $\int_0^1 f_n \nrightarrow \int_0^1 g$.

The preceding examples and exercises indicate that the properties of continuity, differentiability, and integrability, even when shared by all the functions in a convergent sequence of functions, need not be inherited by the limit function. Ultimately, we shall look for conditions under which the inheritance of one of these properties by the limit function is guaranteed, but before doing so, we want to provide another perspective on the matter that can be instructive.

The "inheritance" or "preservation" question can be reframed as one concerning the interchanging of limiting processes. For example, when we ask whether the limit function g of a convergent sequence (f_n) of continuous functions is also continuous, we are asking whether

$$\lim_{n\to\infty}\left(\lim_{x\to p} f_n(x)\right)$$

is equal to

$$\lim_{x\to p}\left(\lim_{n\to\infty} f_n(x)\right).$$

To see why, note that the assumption of continuity for each f_n at p means that

$$\lim_{x\to p} f_n(x) = f_n(p)$$

for each n, while the pointwise convergence of (f_n) to g means that

$$\lim_{n\to\infty} f_n(x) = g(x)$$

for each x. Thus, the continuity of g at p, which requires that

$$\lim_{x\to p} g(x) = g(p),$$

may be translated as

$$\lim_{x\to p}\Big(\lim_{n\to\infty} f_n(x)\Big) = \lim_{n\to\infty}\Big(\lim_{x\to p} f_n(x)\Big),$$

where the order in which the two limiting processes are being applied has been interchanged.

Similarly, assuming differentiability of all the functions involved, the question of whether $\lim_{n\to\infty} f_n = g$ implies that $\lim_{n\to\infty} f_n' = g'$ is really that of asking whether the equality

$$\lim_{n\to\infty} f_n' = \Big(\lim_{n\to\infty} f_n\Big)'$$

holds, the equality representing the interchanging of the limit process represented by differentiation with the limit process represented by convergence of a sequence of functions.

Exercise 26.6. Suppose (f_n) is a sequence of functions each of which is integrable on $[a,b]$ and g is also a function that is integrable on $[a,b]$. Indicate how the question of whether $\lim_{n\to\infty} f_n = g$ implies that $\int_a^b f_n \to \int_a^b g$ is really a question about interchanging two limiting processes.

As we suggested earlier, in the context of a convergent infinite series of functions, the inheritance questions related to continuity, differentiability, and integrability are usually framed with respect to passing from the summands to the sum (limit) of the series. Specifically, assuming (f_n) is a sequence of functions for which $\sum_{n=1}^{\infty} f_n = g$,

(1) if each function f_n is continuous, must g also be continuous?
(2) if each function f_n is differentiable, must g also be differentiable and must $g' = \sum_{n=1}^{\infty} f_n'$?
(3) if each function f_n is integrable on $[a,b]$, must g also be integrable on $[a,b]$ and must $\int_a^b g = \sum_{n=1}^{\infty} \int_a^b f_n$?

Each of these questions receives a negative answer as we can build counterexamples based on the failure of limits of convergent sequences of functions to necessarily inherit the properties of continuity, differentiability, and integrability.

To illustrate, let (f_n) be a sequence of continuous functions that converges to a function g that is not continuous (take these functions as in Example 26.1 if you want to see a concrete illustration). Define a new sequence (h_n) of functions, where

$$h_1 = f_1$$

and

$$h_n = f_n - f_{n-1}$$

when $n > 1$. As the difference of continuous functions is continuous, each h_n is continuous. Note that for each n, we have

$$\sum_{k=1}^{n} h_k = f_1 + (f_2 - f_1) + (f_3 - f_2) + \cdots + (f_n - f_{n-1}) = f_n,$$

which tells us that the nth partial sum of the series $\sum_{n=1}^{\infty} h_n$ is the function f_n. Since a series is, by definition, its sequence of partial sums, we conclude that the sum of the series $\sum_{n=1}^{\infty} h_n$ is g, the limit of the sequence (f_n). Thus, $\sum_{n=1}^{\infty} h_n = g$, where each summand h_n is continuous, but the sum g of the series is not continuous. So *the infinite summation of continuous functions need not be continuous.*

Be sure to convince yourself that this line of reasoning can be applied to show that neither differentiability nor integrability is necessarily inherited by the sum function of a convergent series of functions, and that, even if the property is inherited, the sum function's derivative (respectively, integral) need not be the sum of the series of derivatives (respectively, integrals) of the summands of the original series.

Uniform convergence

Because the domain of convergence of a power series is always an interval, when a power series converges at a point p other than its center, it must converge for all points closer to the center than p is.

Theorem 26.5. *If a power series $\sum_{n=0}^{\infty} a_n(x - x_0)^n$ converges at $x = p$, where $p \neq x_0$, and $0 < r < |p - x_0|$, then $\sum_{n=0}^{\infty} a_n(x - x_0)^n$ converges absolutely for every x in the interval $[x_0 - r, x_0 + r]$.*

The mathematical content of this theorem is, of course, already known to us via Theorem 25.13. In fact, since the newly-stated theorem does not provide a mechanism for identifying the radius of convergence, it actually conveys less information than the earlier theorem. We are restating the result here, and in this form, so that we can give a new proof of it that provides motivation for a new type of function convergence, *uniform convergence*, that interacts much more favorably with continuity, differentiability, and integrability than does pointwise convergence.

Proof of Theorem 26.5. Without loss of generality, we take the center of the power series to be $x_0 = 0$. Assume $\sum_{n=0}^{\infty} a_n p^n$ converges for some number p, where $p \neq 0$. Consider any number r such that $0 < r < |p|$ and suppose $x \in [-r, r]$; we show $\sum_{n=0}^{\infty} a_n x^n$ converges absolutely. To do so, it suffices to show that $\sum_{n=0}^{\infty} |a_n x^n|$, viewed as a sequence of partial sums, is a Cauchy sequence.

Consider any positive number ε. Since $\sum_{n=0}^{\infty} a_n p^n$ converges, the sequence $(a_n p^n)$ is bounded, so there is a positive number B such that

$$|a_n| \leq \frac{B}{|p|^n}$$

for every natural number n. Let $\rho = \frac{r}{|p|}$ and observe that $0 < \rho < 1$ (here ρ is the lowercase Greek letter *rho*). Since $|x| \leq r$, it follows that

$$0 \leq \frac{|x|}{|p|} \leq \rho,$$

so that

$$|x|^n \leq |p|^n \rho^n$$

for every n. Now as $\lim_{n\to\infty} \rho^n = 0$ and $\frac{\varepsilon(1-\rho)}{B} > 0$, we may choose a natural number N so that

$$\rho^N < \frac{\varepsilon(1-\rho)}{B}.$$

As long as $n \geq m \geq N$, and using the formula for a convergent geometric series, we then have

$$\begin{aligned}
\sum_{k=m}^{n} |a_k x^k| &\leq \sum_{k=m}^{n} \frac{B}{|p|^k} |p|^k \rho^k \\
&= B\rho^m \sum_{k=0}^{n-m} \rho^k \\
&< B\rho^N \sum_{k=0}^{\infty} \rho^k \\
&< B \cdot \frac{\varepsilon(1-\rho)}{B} \cdot \frac{1}{1-\rho} \\
&< \varepsilon.
\end{aligned}$$

□

Before formally introducing *uniform convergence*, we re-express the definition of *pointwise convergence* of a sequence of functions to incorporate the formal meaning of convergence of a sequence of real numbers, Definition 9.9. Our goal in doing so is to make apparent the distinction between these two forms of function convergence.

Definition 26.6 (Definition of pointwise convergence). A sequence (f_n) of functions defined on A **converges pointwise** on A to the function g if for every positive number ε and for every number x in A, there is a natural number index $N = N_{\varepsilon,x}$ that may depend on both ε and x, such that when $n \geq N$, it follows that $|f_n(x) - g(x)| < \varepsilon$.

Note in particular that pointwise convergence of a sequence of functions permits the value of N to vary with not only the value of ε, but also the value of x. It is for this reason that we wrote $N = N_{\varepsilon,x}$ in the definition.

However, looking back at the proof of Theorem 26.5, observe that the selection of the natural number N in this argument *does not depend* on the number x taken from the interval $[-r, r]$. The *same* value of N works for all numbers x in $[-r, r]$, though N may vary with the value of ε. Because of our particular interest in power series, the observation that N does not depend on x in this proof prompts us to make the following definition.

Definition 26.7 (Definition of uniform convergence). A sequence (f_n) of functions defined on A **converges uniformly** on A to the function g if for every positive number ε, there is a natural number index $N = N_\varepsilon$ that may depend on ε, but not on x, such that whenever $n \geq N$, it follows that $|f_n(x) - g(x)| < \varepsilon$ for all x in A.

For future reference, observe that if a sequence of functions converges uniformly on a set, the sequence also converges uniformly on any subset of the set.

Based on the definition of uniform convergence and the remark made immediately prior to its statement, the proof of Theorem 26.5 immediately yields the following result concerning uniform convergence of a power series.

Theorem 26.8 (Uniform convergence of power series). *A power series* $\sum_{n=0}^{\infty} a_n(x - x_0)^n$ *having radius of convergence R converges uniformly on any closed bounded interval that is a subset of the interval* $(x_0 - R, x_0 + R)$.

The only difference between pointwise and uniform convergence, but a highly significant one, is that pointwise convergence allows for N to be a function of both ε and x, while uniform convergence requires N to be a function of ε only. Thus, uniform convergence is stronger than pointwise convergence in the sense that any sequence of functions that converges uniformly on a set automatically converges pointwise, to the same limiting function, on that set. Even though the logic behind this observation is trivial, the result itself is important enough to be stated as a theorem.

Theorem 26.9 (Uniform convergence implies pointwise convergence). *If a sequence* (f_n) *of functions converges uniformly on A to the function g, then* (f_n) *also converges pointwise on A to g.*

Graphically, in order for (f_n) to converge uniformly to g on A, the graphs of all of the functions in the sequence beyond a certain index must be able to be trapped within a band of arbitrarily small positive width about the graph of g (Figure 26.1).

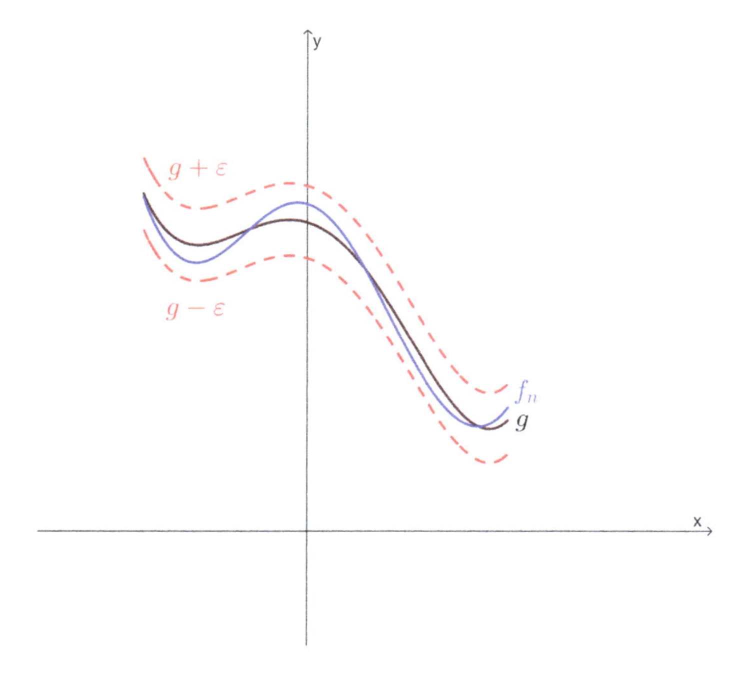

Figure 26.1: For sufficiently large n, uniform convergence requires that the entire graph of f_n lies between the dashed curves enclosing the graph of g.

Once more, we stress that when a sequence of functions converges uniformly to a function, this limit function is the same function to which the sequence converges pointwise. The additional knowledge that the convergence is uniform tells us something about *how* the convergence is happening, not what function is serving as the limit. Conceptually, when (f_n) converges uniformly to g, the sequence $f_n(x)$ converges to $g(x)$ at the same relative rate, no matter what the value of x, whereas when (f_n) converges pointwise to g, but not uniformly, the convergence of $f_n(x)$ to $g(x)$ is considerably more rapid for some points x than for others.

Example 26.1 (Continued). Consider once again the sequence (f_n) of functions $f_n : [0,1] \to \mathbb{R}$, where

$$f_n(x) = x^n.$$

We saw earlier that (f_n) converges pointwise on $[0,1]$ to the function $g : [0,1] \to \mathbb{R}$, where

$$g(x) = \begin{cases} 0, & \text{if } 0 \le x < 1; \\ 1, & \text{if } x = 1. \end{cases}$$

That the convergence is not uniform is suggested visually by examining the right side of Figure 25.2, near $x = 1$, and noting that the graphical criterion for uniform convergence described above and illustrated in Figure 26.1 is not achieved. Analytically, if we take $\varepsilon = 0.9$ and $x_n = \sqrt[n]{0.9}$ for each natural number n, then

$$|f_n(x_n) - g(x_n)| = |(\sqrt[n]{0.9})^n - 0| = 0.9 \not< 0.9 = \varepsilon,$$

so there is no natural number N for which the inequality $|f_n(x) - g(x)| < 0.9 = \varepsilon$ holds for all x in $[0,1]$ and all natural numbers n at least as large as N. Hence, (f_n) does not converge uniformly on $[0,1]$ to g.

Example 26.10. Choose any number b with $0 < b < 1$, and define the sequence (f_n) of functions $f_n : [0,b] \to \mathbb{R}$ so that

$$f_n(x) = x^n$$

and the function $g : [0,b] \to \mathbb{R}$ so that

$$g(x) = 0.$$

To show (f_n) converges uniformly on $[0,b]$ to g, begin by considering any positive number ε. Since $b^n \to 0$, there is a natural number N such that $b^N < \varepsilon$ (note that the selection of N here does not depend on x). Thus, for all x in $[0,b]$ and for all n for which $n \ge N$, we have

$$|f_n(x) - g(x)| = |x^n - 0| = x^n \le b^n < \varepsilon.$$

So (f_n) converges uniformly on $[0,b]$ to g.

The argument presented in Example 26.1, showing that (f_n) does not converge uniformly on $[0,1]$ to g, is based on the logical negation of the definition of uniform convergence. Formally negating this definition produces the following theorem, which is typically used to show that convergence of a sequence of functions is not uniform.

Theorem 26.11. *Given a sequence (f_n) of functions defined on A and a function g defined on A, the following are equivalent:*

(i) *the sequence (f_n) does not converge uniformly on A to g;*

(ii) *there exists a positive number ε, a subsequence (f_{k_n}) of (f_n), and a sequence (x_n) in A for which $|f_{k_n}(x_n) - g(x_n)| \ge \varepsilon$ for every natural number n.*

Example 26.1 (Continued). In showing (f_n) does not converge uniformly on $[0,1]$ to g, we essentially applied Theorem 26.11 by taking $\varepsilon = 0.9$, by taking the subsequence (f_{k_n}) of (f_n) to be (f_n) itself, and by taking the sequence (x_n) in $[0,1]$ to be the sequence $(\sqrt[n]{0.9})$.

Example 26.12. In Example 25.1, we showed that the sequence (f_n) of functions $f_n : \mathbb{R} \to \mathbb{R}$ defined by

$$f_n(x) = \frac{x(-1)^{n+1}}{n}$$

converges pointwise on $\mathbb{R}$ to the function $g : \mathbb{R} \to \mathbb{R}$, where

$$g(x) = 0.$$

To see that this convergence is not uniform on $\mathbb{R}$, note that for each natural number n, we have

$$|f_{2n-1}(2n-1) - g(2n-1)| = \left| \frac{(2n-1)(-1)^{2n}}{2n-1} - 0 \right| = 1,$$

so condition (ii) of Theorem 26.11 is satisfied with $\varepsilon = 1$.

Exercise 26.7. Show that the sequence (f_n) of functions defined on $[0,1]$ so that $f_n(x) = \frac{x+nx^2}{n}$ converges uniformly on $[0,1]$ to the function g, where $g(x) = x^2$.

Exercise 26.8. Prove the convergence in Exercise 25.6(f) is uniform. *Hint*: Make use of the algebraic identity $a^2 - b^2 = (a+b)(a-b)$.

Exercise 26.9. Show that the convergence in each of parts (a) through (e) of Exercise 25.6 is not uniform.

Exercise 26.10. Show that the convergence of (f_n) to g in Example 26.3 is not uniform.

Exercise 26.11. Suppose (f_n) and (g_n) are sequences of functions that converge uniformly on A to, respectively, the functions F and G. Prove that $(f_n + g_n)$ converges uniformly on A to $F + G$.

Exercise 26.12. Give an example of a sequence (f_n) of functions that converges uniformly on $\mathbb{R}$, but for which (f_n^2) does not converge uniformly on $\mathbb{R}$, thereby demonstrating that the product of uniformly convergent sequences of functions need not be uniformly convergent.

Exercise 26.13. Suppose that for each natural number n, we have a function f_n that is defined on the interval $[a,b]$. Show that if the sequence (f_n) converges uniformly on the interval (a,b) and the sequences $(f_n(a))$ and $(f_n(b))$ both converge, then (f_n) converges uniformly on $[a,b]$.

The uniform norm

Recall that a real-valued function is said to be **bounded** if its range is a bounded subset of $\mathbb{R}$.

Exercise 26.14. Let A be a nonempty subset of $\mathbb{R}$ and let (f_n) be a sequence of bounded real-valued functions defined on A.

(a) Prove that if (f_n) converges uniformly on A to a function g, then g is bounded.

(b) Give an example to show that if (f_n) converges pointwise on A to a function g, then g need not be bounded.

Given a nonempty subset A of $\mathbb{R}$ and a bounded real-valued function defined on A, the **uniform norm** of f, denoted $\|f\|$, is defined so that

$$\|f\| = \sup\{|f(x)| \mid x \in A\}.$$

Note that as the function f is assumed to be bounded, the completeness property of $\mathbb{R}$ guarantees the value of $\|f\|$ is a nonnegative real number.

Example 26.1 (Continued). The nth power function f_n is increasing, hence bounded, on $[0,1]$, with

$$\|f_n\| = \sup\{x^n \mid x \in [0,1]\} = 1^n = 1.$$

The function $g : [0,1] \to \mathbb{R}$ defined so that

$$g(x) = \begin{cases} 0, & \text{if } 0 \le x < 1; \\ 1, & \text{if } x = 1. \end{cases}$$

has range $\{0,1\}$, which is a bounded set, and we may then conclude that

$$\|g\| = \sup\{0,1\} = 1.$$

Example 26.10 (Continued). Since g is the constant zero function on $[0,b]$, we have

$$\|g\| = \sup\{0\} = 0.$$

Since the nth power function f_n is increasing on $[0,b]$, we have

$$\|f_n\| = \sup\{x^n \mid x \in [0,b]\} = b^n.$$

In mathematics generally, a *norm* may be viewed as a generalization of *absolute value* that is intended to measure the distance of an object of a certain type from a specified *zero object* of the same type. Given a nonempty subset A of $\mathbb{R}$, the objects to which the uniform norm can be applied are all of the functions $f : A \to \mathbb{R}$ that are bounded. These functions form the **function space of bounded real-valued functions defined on** A, which we shall denote by $B(A)$. One specific function in $B(A)$ is the constant zero function on A, which serves as the zero object in $B(A)$. For any function f in $B(A)$, the uniform norm $\|f\|$ of f represents one way of measuring the distance between f and the constant zero function in the function space $B(A)$.

Exercise 26.15. Show that the uniform norm $\|\|$ for the function space $B(A)$ of bounded real-valued functions having domain A possesses the following properties.
(a) For any function f in $B(A)$, we have $\|f\| \geq 0$, with equality holding only for the constant zero function z.
(b) For any real number r and any function f in $B(A)$, we have $\|rf\| = |r|\|f\|$.
(c) For any functions f and g in $B(A)$, we have $\|f + g\| \leq \|f\| + \|g\|$. (This property is known as the *triangle inequality*.)

When trying to determine whether a sequence of bounded functions converges uniformly, an approach based on the following theorem is often easier to apply than the definition of uniform convergence. The proof of the theorem is assigned as Exercise 26.19.

Theorem 26.13 (Uniform norm criterion for uniform convergence). *Given a sequence (f_n) of bounded functions defined on A and a function g defined on A, the following are equivalent:*
(i) *(f_n) converges uniformly on A to g;*
(ii) $\|f_n - g\| \to 0$.

Example 26.10 (Continued). Recalling that $f_n(x) = x^n$ and $g(x) = 0$ for all x in $[0, b]$, where $0 < b < 1$, we observe that

$$\|f_n - g\| = \sup\{|f_n(x) - g(x)| \mid 0 \leq x \leq b\} = \sup\{|x^n - 0| \mid 0 \leq x \leq b\} = b^n \to 0.$$

Therefore, according to the uniform norm criterion, the sequence (f_n) converges uniformly on $[0, b]$ to g. (We reached the same conclusion earlier by directly applying the definition of uniform convergence.)

Example 26.1 (Continued). Here $f_n(x) = x^n$ for all x in $[0, 1]$, while $g(x) = 0$ for all x in $[0, 1)$, but $g(1) = 1$. Thus, for each n we have

$$|f_n(x) - g(x)| = \begin{cases} x^n, & \text{if } 0 \leq x < 1; \\ 0, & \text{if } x = 1; \end{cases}$$

from which it follows that

$$\|f_n - g\| = \sup\{|f_n(x) - g(x)| \mid 0 \leq x \leq 1\} = 1.$$

Therefore, by the uniform norm criterion, the sequence (f_n) does not converge uniformly on $[0,1]$ to g. (We reached the same conclusion earlier by applying Theorem 26.11.)

Exercise 26.16. Use the uniform norm criterion to show that the sequence (f_n) of functions defined on $[1,4)$ so that $f_n(x) = \frac{x^2 - n^2x^3}{n^2}$ converges uniformly on $[1,4)$ to the function h where $h(x) = -x^3$.

Exercise 26.17. Use the uniform norm criterion to show that the sequence (f_n) of functions defined on $[0,1]$ so that $f_n(x) = x^n - x^{n+1}$ converges uniformly on $[0,1]$ to the constant zero function. *Hint*: Use the derivative to find the location of the maximum value of f_n on $[0,1]$.

Exercise 26.18. Use the uniform norm criterion to show that the convergence of (f_n) to g in Exercise 26.5 is not uniform.

Exercise 26.19. Prove Theorem 26.13.

Exercise 26.20. Suppose (f_n) and (g_n) are sequences of bounded functions that converge uniformly on A to, respectively, the functions F and G. Prove that (f_ng_n) converges uniformly on A to FG. (Compare with Exercise 26.12.)

Exercise 26.21. Suppose (f_n) is a sequence of functions that converges uniformly on A to the function F and that there is a positive number B for which $|f_n(x)| \leq B$ for all natural numbers n and all points x in A. Also suppose that the function g is continuous on $[-B, B]$. Prove that the sequence $(g \circ f_n)$ converges uniformly on A to $g \circ F$. *Hint*: Use the fact that a function that is continuous on a closed bounded interval is uniformly continuous on that interval (Theorem 14.20).

Just as

$$|a - b|$$

represents the distance between the real numbers a and b, the value of

$$\|f - g\|$$

can be taken as representing the distance between the functions f and g in the function space $B(A)$ of bounded real-valued functions defined on A. With this interpretation, the uniform norm criterion given in Theorem 26.13 is telling us that uniform convergence of a sequence (f_n) of functions in $B(A)$ to a function g is equivalent to the convergence of the sequence of distances between f_n and g to the real number 0.

Exercise 26.22. Show that the uniform norm $\|\|$ for the function space $B(A)$ of bounded real-valued functions having domain A possesses the following distance-like properties.

(a) For any functions f and g in $B(A)$, we have $\|f - g\| \geq 0$, with equality holding if and only if $f = g$. (Hence, the distance between functions in $B(A)$ is always nonnegative, and is zero precisely when the functions are identical.)

(b) For any functions f and g in $B(A)$, we have $\|f - g\| = \|g - f\|$. (Hence, the distance is the same, whether going from f to g or from g to f.)

(c) For any functions f, g, and h in $B(A)$, we have $\|f - g\| \leq \|f - h\| + \|h - g\|$. (Hence, there is no advantage, from the point of view of minimizing distance, to go out of our way to visit any particular function h if we are trying to get from f to g; this is another manifestation of the triangle inequality.)

The Cauchy criterion for uniform convergence and the Weierstrass M-test

Sometimes it is not clear what function might serve as the limit of a sequence of functions, but we would still like to know that the sequence converges uniformly to *some* function. The following theorem, obtained by modifying the Cauchy criterion for convergence of real sequences so that the index N not depend on the function input x, is often useful in such circumstances. A formal proof is assigned as an exercise.

Theorem 26.14 (The Cauchy criterion for uniform convergence). *Given a sequence (f_n) of functions defined on A, the following are equivalent:*

(i) *(f_n) converges uniformly on A;*

(ii) *for every positive number ε, there is an index N that may depend on ε, but not on x, such that whenever $m \geq N$ and $n \geq N$, it follows that $|f_m(x) - f_n(x)| < \varepsilon$ for all x in A.*

Example 26.15. Let a be a positive number and define the sequence (f_n) of functions on $[-a, a]$ so that

$$f_n(x) = \frac{x(-1)^{n+1}}{n}.$$

Given a positive number ε, we may use the Archimedean property to obtain a natural number N such that $\frac{1}{N} < \frac{\varepsilon}{2a}$. Assuming that $m \geq N$ and $n \geq N$ and $x \in [-a, a]$, we then have

$$|f_m(x) - f_n(x)| = \left|\frac{x(-1)^{m+1}}{m} - \frac{x(-1)^{n+1}}{n}\right| \leq |x|\left(\frac{1}{m} + \frac{1}{n}\right) \leq \frac{2a}{N} < \varepsilon.$$

Thus, as the sequence (f_n) satisfies the Cauchy criterion for uniform convergence (note that N does not depend on x) it converges uniformly on $[-a, a]$. In fact, it is straightforward to show that the limit is the constant zero function defined on $[-a, a]$.

Exercise 26.23. Prove Theorem 26.14.

Because an infinite series is the sequence whose terms are the partial sums of the series, an infinite series of functions **converges uniformly** provided that the sequence of partial sums of the series converges uniformly. This definition is usually difficult to apply, because it can be challenging to obtain a simple expression for the partial sums. Often, as was the case for infinite series of real numbers, this in turn leads to difficulty determining the limit, that is, the sum, of a convergent series of functions.

Thus, it is convenient to have uniform convergence tests devised specifically for series of functions. The first such test we put forth is just a translation of the Cauchy criterion for uniform convergence of a sequence of functions into the setting of an infinite series of functions.

Theorem 26.16 (The Cauchy criterion for uniform convergence of a series of functions). *Given a sequence (f_n) of functions defined on A, the following are equivalent:*
(i) *the infinite series $\sum_{n=1}^{\infty} f_n$ converges uniformly on A;*
(ii) *for every positive number ε, there is an index N that may depend on ε, but not on x, such that for any natural numbers m and n for which $n \geq m \geq N$, it follows that $|\sum_{k=m}^{n} f_k(x)| < \varepsilon$ for all x in A.*

Even this Cauchy criterion, though, can be difficult to apply because getting a handle on the expression $\sum_{k=m}^{n} f_k(x)$, which is similar to a partial sum, may not be feasible. Much more useful from a practical perspective is the following result whose proof relies on the Cauchy criterion.

Theorem 26.17 (The Weierstrass M-test). *Suppose (f_n) is a sequence of functions defined on A. If, for each n, there is a positive number M_n such that $|f_n(x)| \leq M_n$ for all x in A, and if the series $\sum_{n=1}^{\infty} M_n$ converges, then the series $\sum_{n=1}^{\infty} f_n$ converges uniformly on A, and for each x in A, the series $\sum_{n=1}^{\infty} f_n(x)$ converges absolutely.*

Proof. Consider any positive number ε. As $\sum_{n=1}^{\infty} M_n$ converges, the Cauchy criterion, Theorem 22.10, for convergence of a real series implies there is an index N such that whenever $n \geq m \geq N$, it follows that $\sum_{k=m}^{n} M_k < \varepsilon$. Therefore, assuming $n \geq m \geq N$, we have

$$\left| \sum_{k=m}^{n} f_k(x) \right| \leq \sum_{k=m}^{n} |f_k(x)| \leq \sum_{k=m}^{n} M_k < \varepsilon,$$

for all x in A. So, by the Cauchy criterion, Theorem 26.16, for uniform convergence of a series of functions, we may conclude that the series $\sum_{n=1}^{\infty} f_n$ converges uniformly on A. Furthermore, as we have determined that $\sum_{k=m}^{n} |f_k(x)| < \varepsilon$ when $n \geq m \geq N$, it follows using the Cauchy criterion that the series $\sum_{n=1}^{\infty} |f_n(x)|$ converges for each x in A, thereby implying that $\sum_{n=1}^{\infty} f_n(x)$ converges absolutely. □

Example 26.18. Suppose $f_n : [-1,1] \to \mathbb{R}$ is defined for each natural number n so that

$$f_n(x) = \frac{x(-1)^{n+1}}{n^2}.$$

Note that, as $-1 \leq x \leq 1$, we have

$$|f_n(x)| = \frac{|x|}{n^2} \leq \frac{1}{n^2}.$$

Thus, taking $M_n = \frac{1}{n^2}$, since

$$\sum_{n=1}^{\infty} M_n = \sum_{n=1}^{\infty} \frac{1}{n^2}$$

is the convergent 2-series, the Weierstrass M-test permits us to conclude that the series $\sum_{n=1}^{\infty} \frac{x(-1)^{n+1}}{n^2}$ converges uniformly on $[-1,1]$ and absolutely for each x in $[-1,1]$.

It is still possible for a series of functions to converge uniformly even in situations in which the Weierstrass M-test may not apply.

Example 26.19. If we define functions $f_n : [-1,1] \to \mathbb{R}$ so that

$$f_n(x) = \frac{x(-1)^{n+1}}{n},$$

the Weierstrass M-test is not helpful in the examination of $\sum_{n=1}^{\infty} f_n$ because here $|f_n(x)| = \frac{|x|}{n}$, which achieves a maximum value of $\frac{1}{n}$ at $x = \pm 1$, and the harmonic series $\sum_{n=1}^{\infty} \frac{1}{n}$ diverges.

However, because we know from Exercise 23.9 that the alternating harmonic series $\sum_{n=1}^{\infty} \frac{(-1)^n}{n}$ sums to $\ln(2)$, it follows that

$$\sum_{n=1}^{\infty} f_n(x) = \sum_{n=1}^{\infty} \frac{x(-1)^{n+1}}{n} = x\ln(2)$$

for each x in $[-1,1]$. Also, as $\sum_{n=1}^{\infty} \frac{1}{n}$ diverges, we may conclude that the convergence is only conditional, not absolute, when $x \neq 0$.

Now suppose $\varepsilon > 0$. Since $\sum_{n=1}^{\infty} \frac{(-1)^n}{n}$ converges, the Cauchy criterion from Theorem 22.10 tells us there is a natural number N such that when $n \geq m \geq N$, it follows that $|\sum_{k=m}^{n} \frac{(-1)^k}{k}| < \varepsilon$. Hence, assuming $n \geq m \geq N$, since $|x| \leq 1$, it follows that

$$\left|\sum_{k=m}^{n} \frac{x(-1)^k}{k}\right| = |x|\left|\sum_{k=m}^{n} \frac{(-1)^k}{k}\right| < |x|\varepsilon \leq \varepsilon,$$

and so the Cauchy criterion from Theorem 26.16 allows us to conclude that $\sum_{n=1}^{\infty} \frac{x(-1)^{n+1}}{n}$ converges uniformly on $[-1,1]$.

Example 26.19 demonstrates that uniform convergence does not guarantee absolute convergence. Neither does absolute convergence guarantee uniform convergence, as the next example illustrates.

Example 26.20. The Weierstrass M-test is not applicable to the series $\sum_{n=1}^{\infty} f_n$ of functions where f_n is defined on $\mathbb{R}$ so that

$$f_n(x) = \frac{x(-1)^{n+1}}{n^2}$$

because each function f_n is unbounded on $\mathbb{R}$. You are asked to show this series converges absolutely for each real number x, but not uniformly on $\mathbb{R}$, in Exercise 26.24.

Exercise 26.24. Consider the series $\sum_{n=1}^{\infty} f_n$ from Example 26.20.
(a) Show that this series converges absolutely for each real number x.
(b) Use the Cauchy criterion from Theorem 26.16 to show that this series does not converge uniformly on $\mathbb{R}$.

Exercise 26.25. Use the Weierstrass M-test to show the given series of functions converges uniformly on the specified domain.
(a) the series from Exercise 25.15(a), on $[-1, 1]$
(b) the series from Exercise 25.15(d), on $[-\frac{13}{4}, -\frac{11}{4}]$
(c) the series $\sum_{n=1}^{\infty} \frac{\sin(nx)}{n^2\sqrt{n}}$, on $\mathbb{R}$

Exercise 26.26. Show that if the real series $\sum_{n=0}^{\infty} a_n$ converges absolutely, then the power series $\sum_{n=0}^{\infty} a_n x^n$ converges uniformly on $[-1, 1]$.

Uniform convergence and continuity

One of the primary reasons for introducing uniform convergence is its relatively favorable relationship with the important properties of continuity, integrability, and differentiability. For instance, while pointwise convergence of a sequence of continuous functions does not guarantee continuity of the limit function, as demonstrated in Example 26.1, the uniform limit of continuous functions is always continuous.

Theorem 26.21 (Uniform convergence and continuity). *If a sequence (f_n) of continuous functions converges uniformly on A to the function g, then g is also continuous.*

Proof. Assume (x_n) is a sequence in A that converges to a point p in A. Let ε be an arbitrary positive number. As (f_n) converges uniformly on A to g, there is an index N such that when $n \geq N$, it follows that

$$|f_n(x) - g(x)| < \frac{\varepsilon}{3}$$

for all x in A. Then, as f_N is continuous on A and $x_n \to p$, we may conclude that $f_N(x_n) \to f_N(p)$, so there is an index M at least as large as N for which

$$|f_N(x_n) - f_N(p)| < \frac{\varepsilon}{3}$$

when $n \geq M$. Therefore, when $n \geq M$, it follows that

$$\begin{aligned} |g(x_n) - g(p)| &\leq |g(x_n) - f_N(x_n)| + |f_N(x_n) - f_N(p)| + |f_N(p) - g(p)| \\ &< \frac{\varepsilon}{3} + \frac{\varepsilon}{3} + \frac{\varepsilon}{3} \\ &= \varepsilon, \end{aligned}$$

so that $g(x_n) \to g(p)$. As g preserves convergent sequences, g is continuous. □

Example 26.1 (Continued). Because the functions in the sequence (f_n) defined on $[0,1]$ so that $f_n(x) = x^n$ are continuous, but the limit function $g : [0,1] \to \mathbb{R}$, where

$$g(x) = \begin{cases} 0, & \text{if } 0 \leq x < 1; \\ 1, & \text{if } x = 1; \end{cases}$$

is not continuous, it follows from Theorem 26.21 that the convergence of (f_n) to g is not uniform. (We established this conclusion earlier by other means.)

Exercise 26.27. Assume (f_n) is a sequence of continuous functions that converges uniformly on A to a function g. Prove that if (x_n) is a sequence in A that converges to a point p in A, then the sequence $(f_n(x_n))$ converges to $g(p)$.

The following corollary expresses Theorem 26.21 in the context of infinite series of functions.

Corollary 26.22. *If (f_n) is a sequence of continuous functions and the series $\sum_{n=1}^{\infty} f_n$ converges uniformly on A to a function g, then g is also continuous.*

Exercise 26.28. Prove Corollary 26.22.

Example 26.23. In Exercise 26.25(c), you showed that the series $\sum_{n=1}^{\infty} \frac{\sin(nx)}{n^2\sqrt{n}}$ converges uniformly on $\mathbb{R}$ to a (not explicitly identified) function g, which permits us to write

$$g(x) = \sum_{n=1}^{\infty} \frac{\sin(nx)}{n^2\sqrt{n}}$$

for every real number x. Because each summand function f_n, where

$$f_n(x) = \frac{\sin(nx)}{n^2\sqrt{n}},$$

is continuous, it follows via Corollary 26.22 that the sum function g is also continuous.

Exercise 26.29. Use Corollary 26.22 to show that the convergence of the series in Exercise 25.7(b) is not uniform.

Although we have seen that pointwise convergence of a sequence of continuous functions does not guarantee continuity of the limit function, it can happen that the limit of a sequence of continuous functions that is pointwise, but not uniformly, convergent is continuous.

Example 26.12 (Continued). We have seen that the sequence (f_n) of continuous functions defined on $\mathbb{R}$ so that $f_n(x) = \frac{x(-1)^{n+1}}{n}$ converges pointwise on $\mathbb{R}$ to the continuous function $g : \mathbb{R} \to \mathbb{R}$, where $g(x) = 0$, even though we have also shown that the convergence is not uniform.

Exercise 26.30. Give an example of a sequence (f_n) of functions on $[0, 1]$ for which each function f_n is discontinuous at every point of $[0, 1]$ but (f_n) converges uniformly on $[0, 1]$ to a continuous function g.

A sequence (f_n) of functions defined on a nonempty subset A of $\mathbb{R}$ is called **monotone** if, for every x in A, the sequence $(f_n(x))$ is nonincreasing or, for every x in A, the sequence $(f_n(x))$ is nondecreasing. Although continuity of the limit of a sequence of continuous functions does not guarantee the convergence is uniform, Ulisse Dini (1845–1918) established that the convergence must be uniform if the sequence is also monotone and the domain of the functions is a closed bounded interval.

Theorem 26.24 (Dini's theorem). *If a monotone sequence (f_n) of continuous functions converges pointwise on a closed bounded interval $[a, b]$ to a continuous function, then the convergence is actually uniform on $[a, b]$.*

Recall that any closed bounded interval is compact (Theorem 8.10). In Exercise 26.31 you will prove a more general version of Dini's theorem that replaces $[a, b]$ with an arbitrary compact subset of $\mathbb{R}$.

Exercise 26.31. Let A be a compact subset of $\mathbb{R}$ and assume (f_n) is a monotone sequence of continuous functions that converges pointwise on A to a continuous function g. More specifically, assume $(f_n(x))$ is nondecreasing for every x in A, meaning $f_n(x) \leq f_{n+1}(x)$ for all natural numbers n and all points x in A. (The case where $(f_n(x))$ is nonincreasing for every x is similar.) For each natural number n, define the function F_n on A so that

$$F_n(x) = g(x) - f_n(x).$$

(a) Explain why, for each x in A, the sequence $(F_n(x))$ is nonincreasing, nonnegative, and converges to 0.

(b) Consider any positive number ε and for each natural number n, let $U_n = F_n^{-1}[(-\infty, \varepsilon)]$. Show that $\mathcal{U} = \{U_n \mid n \in \mathbb{N}\}$ is an open cover of A. Then use the compactness of A to argue why there is a natural number N for which $F_n(x) < \varepsilon$ for all x in A whenever $n \geq N$.

Uniform convergence and integrability

In Examples 26.3 and 26.4, we demonstrated that the pointwise limit of integrable functions need not be integrable, and even when integrability is inherited by the limit function, the limit of the sequence of integrals of the functions need not be the integral of the limit function. However, the uniform limit of integrable functions is always integrable, and the integral of the limit function must be the limit of the sequence of integrals of the functions.

Theorem 26.25 (Uniform convergence and the integral). *If a sequence (f_n) of functions that are integrable on $[a, b]$ converges uniformly on $[a, b]$ to the function g, then g is also integrable on $[a, b]$ and $\int_a^b g = \lim_{n\to\infty} \int_a^b f_n$.*

In other words, under the stated hypotheses, $\int_a^b \lim_{n\to\infty} f_n$ exists and is equal to $\lim_{n\to\infty} \int_a^b f_n$.

Proof. First, we show that g is integrable on $[a, b]$. Consider any positive number ε. Since (f_n) converges uniformly on $[a, b]$ to g, there exists a natural number N such that

$$f_N(x) - \frac{\varepsilon}{6(b-a)} < g(x) < f_N(x) + \frac{\varepsilon}{6(b-a)}$$

for all x in $[a, b]$. As a consequence, for any tagged partition $\dot{P}$ of $[a, b]$, we have

$$S(f_N, \dot{P}) - \frac{\varepsilon}{6} < S(g, \dot{P}) < S(f_N, \dot{P}) - \frac{\varepsilon}{6}.$$

Then, since f_N is integrable on $[a, b]$, Theorem 19.9, the Cauchy criterion for integrability, tells us there is a positive number δ such that when $\dot{P}$ and $\dot{Q}$ are tagged partitions of $[a, b]$ whose norms are less than δ, it follows that

$$|S(f_N, \dot{P}) - S(f_N, \dot{Q})| < \frac{\varepsilon}{3}.$$

Thus, for any tagged partitions $\dot{P}$ and $\dot{Q}$ of $[a, b]$ whose norms are less than δ, we have

$$\begin{aligned}
&|S(g, \dot{P}) - S(g, \dot{Q})| \\
&\quad \leq |S(g, \dot{P}) - S(f_N, \dot{P})| + |S(f_N, \dot{P}) - S(f_N, \dot{Q})| + |S(f_N, \dot{Q}) - S(g, \dot{Q})| \\
&\quad < \frac{\varepsilon}{3} + \frac{\varepsilon}{3} + \frac{\varepsilon}{3} \\
&\quad = \varepsilon.
\end{aligned}$$

Hence, via the Cauchy criterion, g is integrable on $[a, b]$.

Now we show $\int_a^b g = \lim_{n\to\infty} \int_a^b f_n$. Again, consider any positive number ε. Since (f_n) converges uniformly on $[a,b]$ to g, there exists a natural number N such that, when $n \geq N$, it follows that

$$f_n(x) - \frac{\varepsilon}{4(b-a)} < g(x) < f_n(x) + \frac{\varepsilon}{4(b-a)},$$

for all x in $[a,b]$. Using the monotonicity of the integral, we may then conclude that when $n \geq N$, we have

$$\int_a^b f_n - \frac{\varepsilon}{4} \leq \int_a^b g \leq \int_a^b f_n + \frac{\varepsilon}{4}$$

so that

$$\left| \int_a^b f_n - \int_a^b g \right| \leq \frac{\varepsilon}{2} < \varepsilon.$$

Hence, $\int_a^b f_n \to \int_a^b g$. □

Example 26.26. You showed in Exercise 26.1 that the sequence (f_n) of functions defined on $[0,1]$ so that $f_n(x) = nx^3(1-x^4)^n$ converges pointwise to the constant zero function g. But as we also discovered in Example 26.4 that $\lim_{n\to\infty} \int_0^1 f_n = \frac{1}{4}$, while $\int_0^1 g = 0$, Theorem 26.25 now permits us to conclude that the convergence of (f_n) to g is not uniform.

Exercise 26.32. Verify the result of Theorem 26.25 using the sequence of functions given in Exercise 26.17.

Exercise 26.33. Suppose (f_n) is a sequence of functions that are integrable on $[a,b]$ and that (f_n) converges uniformly on $[a,b]$ to the function g. For each natural number n, define the function $F_n : [a,b] \to \mathbb{R}$ so that $F_n(x) = \int_a^x f_n$, and also define the function $G : [a,b] \to \mathbb{R}$ so that $G(x) = \int_a^x g$. Show that (F_n) converges uniformly on $[a,b]$ to G.

Exercise 26.34. Suppose (f_n) is a sequence of functions that are integrable on $[a,b]$ and that (f_n) converges uniformly on $[a,b]$ to the function g. Let h be a function that is integrable on $[a,b]$. Prove that $(\int_a^b f_n h)$ converges to $\int_a^b gh$. *Hint*: Make use of the result from Exercise 26.20.

The conclusion of Theorem 26.25 sometimes holds even when the convergence of the series of functions is not uniform. This scenario is illustrated in Exercise 26.35.

Exercise 26.35. Consider the sequence (f_n) of functions defined on $[0,1]$ so that $f_n(x) = \frac{nx}{1+nx}$. Show that (f_n) converges pointwise, but not uniformly, on $[0,1]$ to a function g that is integrable on $[0,1]$ and that $\int_0^1 g = \lim_{n\to\infty} \int_0^1 f_n$.

The version of Theorem 26.25 for series is stated in the following corollary. It follows immediately by applying the theorem to the partial sums of the series $\sum_{n=1}^{\infty} f_n$.

Corollary 26.27. *If (f_n) is a sequence of functions that are integrable on $[a,b]$ and the series $\sum_{n=1}^{\infty} f_n$ converges uniformly on $[a,b]$ to a function g, then g is also integrable on $[a,b]$ and $\int_a^b g = \sum_{n=1}^{\infty} \int_a^b f_n$.*

In other words, under the stated hypotheses, $\int_a^b \sum_{n=1}^{\infty} f_n$ exists and $\int_a^b \sum_{n=1}^{\infty} f_n = \sum_{n=1}^{\infty} \int_a^b f_n$.

When $\int_a^b \sum_{n=1}^{\infty} f_n = \sum_{n=1}^{\infty} \int_a^b f_n$, such as is the case when the hypotheses of Corollary 26.27 are satisfied, it is customary to say that **the integral of $\sum_{n=1}^{\infty} f_n$ on $[a,b]$ can be computed summand-by-summand.**

Example 26.23 (Continued). Consider again the function g determined by the series

$$g(x) = \sum_{n=1}^{\infty} \frac{\sin(nx)}{n^2\sqrt{n}},$$

which we already know converges uniformly on $\mathbb{R}$. Since each summand function is continuous on $[0,\pi]$, it is integrable on $[0,\pi]$. Thus, by Corollary 26.27, we are able to compute the integral of g on $[0,\pi]$ summand-by-summand, so that

$$\begin{aligned}
\int_0^{\pi} g &= \int_0^{\pi} \left(\sum_{n=1}^{\infty} \frac{\sin(nx)}{n^2\sqrt{n}} \right) dx \\
&= \sum_{n=1}^{\infty} \int_0^{\pi} \frac{\sin(nx)}{n^2\sqrt{n}}\, dx \\
&= \sum_{n=1}^{\infty} \frac{-\cos(nx)}{n^3\sqrt{n}} \Bigg|_{x=0}^{\pi} \\
&= \sum_{n=1}^{\infty} \frac{1-\cos(n\pi)}{n^3\sqrt{n}} \\
&= \sum_{n=1}^{\infty} \frac{2}{(2n-1)^3\sqrt{2n-1}}.
\end{aligned}$$

Exercise 26.36. Show that if the series $\sum_{n=0}^{\infty} a_n$ converges absolutely, then $\int_0^1 (\sum_{n=0}^{\infty} a_n x^n) dx = \sum_{n=0}^{\infty} \frac{a_n}{n+1}$.
Hint: Make use of the result from Exercise 26.26.

Uniform convergence and differentiability

Uniform convergence does not behave quite as nicely with respect to differentiability as it does with respect to continuity and integrability. For instance, while we might hope that the uniform limit of differentiable functions would have to be differentiable, this is not always the case.

Example 26.28. Consider again the sequence (k_n) of functions defined on $(-1, 1)$ so that

$$k_n(x) = \sqrt{x^2 + \frac{1}{n}}$$

from Exercise 25.6(f). Each of these functions is differentiable on $(-1, 1)$ with

$$k_n'(x) = \frac{x}{\sqrt{x^2 + \frac{1}{n}}}.$$

In Exercise 26.8, you showed that (k_n) converges uniformly on $(-1, 1)$ to the function K, where $K(x) = |x|$, a function which is not differentiable at $x = 0$ (see Example 15.7).

Actually, even when the uniform limit of a sequence of differentiable functions is itself differentiable, the derivative of the limit need not equal the limit of the derivatives.

Example 26.2 (Continued). The convergence of the sequence (f_n) of functions defined on $[0, 1]$ so that

$$f_n(x) = \frac{\sin(nx)}{n}$$

to the constant zero function g is uniform because, given any positive number ε, the Archimedean property can be used to obtain a natural number N such that

$$\frac{\sin(nx)}{n} \leq \frac{1}{n} \leq \frac{1}{N} < \varepsilon$$

for all $n \geq N$ and all x in $[0, 1]$. Since

$$f_n'(0) = \cos(n \cdot 0) = 1,$$

whereas

$$g'(0) = 0,$$

we see that $f_n'(0) \nrightarrow g'(0)$ even though (f_n) converges uniformly on $[0, 1]$ to g.

What we can prove is that the convergence of a sequence (f_n) of functions to a function h implies the convergence of the sequence (f_n') of derivatives to the function h' provided that we already know (f_n') converges uniformly to some function and $(f_n(x_0))$ converges for at least one input x_0.

Theorem 26.29 (Uniform convergence and the derivative). *Suppose (f_n) is a sequence of functions defined and differentiable on an interval I.*

If there is a point x_0 in I for which the sequence $(f_n(x_0))$ converges, and if the sequence (f_n') converges uniformly on I to a function g, then (f_n) converges uniformly on I to a function h that is differentiable on I and for which $h' = g$.

In other words, under the stated hypotheses, $\lim_{n\to\infty} f_n$ exists and is differentiable, with $(\lim_{n\to\infty} f_n)' = \lim_{n\to\infty} f_n'$.

Proof. We first show (f_n) converges uniformly on I. Consider any positive number ε. As $(f_n(x_0))$ converges, it is Cauchy (Theorem 11.16), so there is an index N_1 for which, when $m \geq N_1$ and $n \geq N_1$, it follows that

$$|f_m(x_0) - f_n(x_0)| < \frac{\varepsilon}{2}.$$

As (f_n') converges uniformly on I, the Cauchy criterion, Theorem 26.16, for uniform convergence tells us there is an index N_2 for which, when $m \geq N_2$ and $n \geq N_2$, it follows that

$$|f_m'(x) - f_n'(x)| < \frac{\varepsilon}{2(b-a)}$$

for all x in I. Let $N = \max\{N_1, N_2\}$, take $m \geq N$ and $n \geq N$, and consider any x in I for which $x \neq x_0$. Since the function $f_m - f_n$ is differentiable, hence continuous, on the closed bounded interval whose endpoints are x_0 and x, Theorem 16.7, the mean value theorem, can be applied to obtain a number c between x_0 and x such that

$$f_m'(c) - f_n'(c) = \frac{f_m(x) - f_n(x) - (f_m(x_0) - f_n(x_0))}{x - x_0}.$$

It now follows that

$$\begin{aligned}
|f_m(x) - f_n(x)| &= |f_m(x_0) - f_n(x_0) + (x - x_0)(f_m'(c) - f_n'(c))| \\
&\leq |f_m(x_0) - f_n(x_0)| + (b-a)|f_m'(c) - f_n'(c)| \\
&< \frac{\varepsilon}{2} + (b-a) \cdot \frac{\varepsilon}{2(b-a)} \\
&= \varepsilon.
\end{aligned}$$

Thus, according to the Cauchy criterion for uniform convergence, we may conclude that (f_n) converges uniformly on I to a function we shall call h.

To complete the proof, we must show h is differentiable on I and that $h'(x) = g(x)$ for all x in I. Consider any positive number ε. Using the hypothesis that (f_n') converges uniformly on I, we obtain, via the Cauchy criterion, an index N_1 for which, when $m \geq N_1$ and $n \geq N_1$, it follows that

$$|f_m'(x) - f_n'(x)| < \frac{\varepsilon}{3}$$

for all x in I. Take $m \geq N_1$ and $n \geq N_1$, consider any x and p in I for which $x \neq p$, and apply the mean value theorem to obtain a number d between x and p such that

$$f_m'(d) - f_n'(d) = \frac{f_m(x) - f_n(x) - (f_m(p) - f_n(p))}{x - p}.$$

It follows that

$$\left|\frac{f_m(x) - f_m(p)}{x - p} - \frac{f_n(x) - f_n(p)}{x - p}\right| = |f_m'(d) - f_n'(d)| < \frac{\varepsilon}{3}$$

so that

$$\frac{f_m(x) - f_m(p)}{x - p} - \frac{\varepsilon}{3} < \frac{f_n(x) - f_n(p)}{x - p} < \frac{f_m(x) - f_m(p)}{x - p} + \frac{\varepsilon}{3}.$$

Since the sequence whose nth term is $\frac{f_n(x)-f_n(p)}{x-p}$ converges to $\frac{h(x)-h(p)}{x-p}$, it follows that

$$\frac{f_m(x) - f_m(p)}{x - p} - \frac{\varepsilon}{3} < \frac{h(x) - h(p)}{x - p} < \frac{f_m(x) - f_m(p)}{x - p} + \frac{\varepsilon}{3},$$

that is,

$$\left|\frac{h(x) - h(p)}{x - p} - \frac{f_m(x) - f_m(p)}{x - p}\right| < \frac{\varepsilon}{3}.$$

Also, because $f_n'(p) \to g(p)$, there is an index N larger than N_1 for which

$$|f_N'(p) - g(p)| < \frac{\varepsilon}{3}.$$

Then, as f_N is differentiable at p with derivative $f_N'(p)$, there is a positive number δ for which

$$\left|\frac{f_N(x) - f_N(p)}{x - p} - f_N'(p)\right| < \frac{\varepsilon}{3}$$

when x is in I and $0 < |x - p| < \delta$. Putting all our conclusions here together, we find that when x is in I and $0 < |x - p| < \delta$, we have

$$\begin{aligned}\left|\frac{h(x) - h(p)}{x - p} - g(p)\right| &\leq \left|\frac{h(x) - h(p)}{x - p} - \frac{f_N(x) - f_N(p)}{x - p}\right| + \left|\frac{f_N(x) - f_N(p)}{x - p} - f_N'(p)\right| \\ &\quad + |f_N'(p) - g(p)| \\ &< \frac{\varepsilon}{3} + \frac{\varepsilon}{3} + \frac{\varepsilon}{3} \\ &= \varepsilon.\end{aligned}$$

Thus, by definition, h is differentiable at p and $h'(p) = g(p)$. As p was arbitrarily chosen from I, we may conclude that $h'(x) = g(x)$ for all x in I. □

Example 26.30. In Exercise 26.16, you showed the sequence (f_n) of functions defined on $[1, 4)$ so that

$$f_n(x) = \frac{x^2 - n^2x^3}{n^2}$$

converges uniformly on $[1, 4)$ to the function h where

$$h(x) = -x^3.$$

Each function f_n is differentiable on $[1, 4)$ with derivative

$$f_n'(x) = \frac{2x - 3n^2x^2}{n^2}.$$

Define the function g on $[1, 4)$ so that

$$g(x) = -3x^2.$$

Since

$$\|f_n' - g\| = \sup\{|f_n'(x) - g(x)| \mid 1 \leq x < 4\} = \sup\left\{\frac{2x}{n^2} \;\middle|\; 1 \leq x < 4\right\} = \frac{8}{n^2} \to 0,$$

we may conclude via the uniform norm criterion from Theorem 26.13 that (f_n') converges uniformly on $[1, 4)$ to g. Hence, by Theorem 26.29, it follows that $h' = g$, which we can readily see is the case.

Exercise 26.37. Give an example of a sequence (f_n) of functions defined on the interval $(0, 1)$ for which the sequence (f_n') converges uniformly on $(0, 1)$, but for which the sequence $(f_n(x))$ does not converge for any x in $(0, 1)$. (Thus, including convergence of $(f_n(x_0))$ for some point x_0 as part of the hypotheses of Theorem 26.29 is necessary.)

Exercise 26.38. This exercise illustrates that pointwise, rather than uniform, convergence of the sequence of derivatives does not guarantee the conclusion of Theorem 26.29 holds. Define the sequence (f_n) of functions on $[0, 1]$ so that $f_n(x) = \frac{x^n}{n}$.

(a) Show that (f_n) converges uniformly on $[0, 1]$ to the function g where $g(x) = 0$.

(b) Show that (f_n') converges pointwise on $[0, 1]$, but that $(f_n'(1))$ does not converge to $g'(1)$.

Given a series $\sum_{n=1}^{\infty} f_n$ of differentiable functions, the series $\sum_{n=1}^{\infty} f_n'$ whose summands are the derivatives of the summands of $\sum_{n=1}^{\infty} f_n$ is called the **derived series** determined by $\sum_{n=1}^{\infty} f_n$. Corollary 26.31, which converts Theorem 26.29 to the language of series, articulates conditions under which the derivative of a series of differentiable functions is the derived series.

Corollary 26.31. *Suppose (f_n) is a sequence of differentiable functions defined on an interval I.*

If there is a point x_0 in I for which the series $\sum_{n=1}^{\infty} f_n(x_0)$ converges, and if the derived series $\sum_{n=1}^{\infty} f_n'$ converges uniformly on I to a function g, then the series $\sum_{n=1}^{\infty} f_n$ converges uniformly on I to a function h that is differentiable on I and for which $h' = g$.

In other words, under the stated hypotheses, $\sum_{n=1}^{\infty} f_n$ is differentiable, with $(\sum_{n=1}^{\infty} f_n)' = \sum_{n=1}^{\infty} f_n'$.

When $(\sum_{n=1}^{\infty} f_n)' = \sum_{n=1}^{\infty} f_n'$, such as is the case when the hypotheses of Corollary 26.31 are satisfied, it is customary to say that **the derivative of $\sum_{n=1}^{\infty} f_n$ can be computed summand-by-summand.**

Example 26.23 (Continued). The uniform convergence on $\mathbb{R}$ of the series $\sum_{n=1}^{\infty} \frac{\sin(nx)}{n^2\sqrt{n}}$ to a function g means that

$$g(x) = \sum_{n=1}^{\infty} \frac{\sin(nx)}{n^2\sqrt{n}}$$

for all x in $\mathbb{R}$. Note that each summand function f_n, where

$$f_n(x) = \frac{\sin(nx)}{n^2\sqrt{n}},$$

is differentiable on $\mathbb{R}$, with

$$f_n'(x) = \frac{\cos(nx)}{n\sqrt{n}}.$$

As

$$\left|\frac{\cos(nx)}{n\sqrt{n}}\right| \leq \frac{1}{n\sqrt{n}}$$

for all n and all x, and as $\sum_{n=1}^{\infty} \frac{1}{n\sqrt{n}}$ is the convergent 1.5-series, the Weierstrass M-test tells us that the derived series

$$\sum_{n=1}^{\infty} f_n'(x) = \sum_{n=1}^{\infty} \frac{\cos(nx)}{n\sqrt{n}}$$

converges uniformly on $\mathbb{R}$. Therefore, Corollary 26.31 permits us to calculate the derivative of the function g summand-by-summand, so that

$$g'(x) = \left(\sum_{n=1}^{\infty} \frac{\sin(nx)}{n^2\sqrt{n}}\right)' = \sum_{n=1}^{\infty}\left(\frac{\sin(nx)}{n^2\sqrt{n}}\right)' = \sum_{n=1}^{\infty} \frac{\cos(nx)}{n\sqrt{n}}$$

for every real number x. In other words, the derivative of g is the derived series determined from g.

Exercise 26.39. Confirm that Corollary 26.31 can be applied to the specified series on the given interval and obtain the derived series representation of the derivative of the series.

(a) the series from Exercise 25.15(a), on $(-1, 1)$

(b) the series $\sum_{n=1}^{\infty} \frac{1}{1+n^2x}$, on $(1, \infty)$

A function that is differentiable on a nonempty subset A of $\mathbb{R}$ is said to be **continuously differentiable** on A if its derivative is continuous on A (see Example 16.10 for a situation in which this is not the case). It is much easier to obtain the conclusion of Theorem 26.29 if we assume that all of the functions in the sequence (f_n) are continuously differentiable, not just differentiable (see Exercise 26.40).

Exercise 26.40. Use both parts of the fundamental theorem of calculus to prove a version of Theorem 26.29 in which, in addition to the hypotheses stated there, the functions in the sequence (f_n) are continuously differentiable on I.

---

27 More about power series and Taylor series

In this chapter, we complete our study of power series, identifying the highly-regulated structure they must possess. We then return to the question of convergence of a Taylor series, describing several forms the Taylor remainder may take, and showing the natural exponential function is represented by its Maclaurin series. We also prove that e is irrational and extend the notion of binomial coefficient to obtain a generalization of the binomial theorem.

Power series revisited

Given a real number x_0, recall that a **power series with center** x_0 is an infinite series of functions

$$\sum_{n=0}^{\infty} a_n (x - x_0)^n,$$

where $(a_n) = (a_0, a_1, a_2, \dots)$ is a real sequence whose terms are referred to as the **coefficients** of the power series. We learned in Chapter 25 that the set of values of x for which a power series converges always forms an interval, called the **interval of convergence**, which can be

$$\{x_0\}$$

or

$$\mathbb{R} = (-\infty, \infty)$$

or one of

$$(x_0 - R, x_0 + R), \quad [x_0 - R, x_0 + R), \quad (x_0 - R, x_0 + R], \quad \text{or} \quad [x_0 - R, x_0 + R],$$

for some positive real number R, which is referred to as the **radius of convergence** of the power series. We also agreed to take $R = 0$ if the interval of convergence is $\{x_0\}$ and $R = \infty$ if the interval of convergence is $\mathbb{R}$. The radius of convergence R can always be computed as

$$R = \frac{1}{\limsup \sqrt[n]{|a_n|}},$$

and can also be computed as

$$R = \lim_{n \to \infty} \left| \frac{a_n}{a_{n+1}} \right|$$

https://doi.org/10.1515/9783111429564-027

provided this limit exists or is infinite. If I is the interval of convergence of the power series $\sum_{n=0}^{\infty} a_n(x-x_0)^n$, the power series is identified with the function $f : I \to \mathbb{R}$ defined by

$$f(x) = \sum_{n=0}^{\infty} a_n(x-x_0)^n.$$

We also indicated that because a general power series $\sum_{n=0}^{\infty} a_n(x-x_0)^n$ can be obtained from a power series $\sum_{n=0}^{\infty} a_n y^n$ centered at 0 via the substitution $y = x - x_0$, it is usually possible to give an explicit proof for a result that holds for power series in general for only the case where the power series has the form $\sum_{n=0}^{\infty} a_n x^n$, where $x_0 = 0$.

The **interior** of an interval I is the open interval that includes all points of I except any endpoints of I.

Example 27.1. The interior of each of $[0,1]$, $(0,1)$, $[0,1)$, and $(0,1]$ is $(0,1)$. The interior of each of $[0,\infty)$ and $(0,\infty)$ is $(0,\infty)$. The interior of $\mathbb{R}$ is $\mathbb{R}$.

In Chapter 26, we demonstrated that the convergence of a power series on any closed bounded interval contained within the interior of the interval of convergence is uniform. For convenience, we restate this theorem here.

Theorem 27.2 (Uniform convergence of a power series). *A power series*

$$\sum_{n=0}^{\infty} a_n(x-x_0)^n$$

with radius of convergence R converges uniformly on any closed bounded interval that is a subset of the interval $(x_0 - R, x_0 + R)$.

We leave the proof of the following corollary concerning continuity and integrability of a power series as Exercise 27.1.

Corollary 27.3 (Continuity and integrability of a power series). *Let the function f be the power series defined so that*

$$f(x) = \sum_{n=0}^{\infty} a_n(x-x_0)^n$$

and suppose the radius of convergence is R.

Then f is continuous on the interval $(x_0 - R, x_0 + R)$. Also, f is integrable on any closed bounded interval $[a,b]$ that is a subset of $(x_0 - R, x_0 + R)$, with

$$\int_a^b f = \sum_{n=0}^{\infty} \int_a^b a_n(x-x_0)^n \, dx.$$

Example 27.4. Let f be the power series defined so that

$$f(x) = \sum_{n=1}^{\infty} \frac{4^n}{n^4}(x+3)^n,$$

which you showed in Exercise 25.16(d) has interval of convergence $[-3.25, -2.75]$. Thus, f is continuous on $(-3.25, -2.75)$ and, for instance,

$$\int_{-3}^{-2.9} f = \sum_{n=1}^{\infty} \int_{-3}^{-2.9} \frac{4^n}{n^4}(x+3)^n dx = \sum_{n=1}^{\infty} \frac{4^n}{n^4} \cdot \frac{(x+3)^{n+1}}{n+1}\bigg|_{x=-3}^{-2.9} = \sum_{n=1}^{\infty} \frac{4^n}{n^4} \cdot \frac{(0.1)^{n+1}}{n+1}.$$

Exercise 27.1. Prove Corollary 27.3.

Does a power series have to be continuous at an endpoint of the interval of convergence if the power series converges at this endpoint? The next theorem provides an affirmative answer to this question.

Theorem 27.5 (Abel's theorem). *A power series is continuous on its entire interval of convergence.*

Proof. Continuity of a power series at all points in the interior of the interval of convergence is part of Corollary 27.3. So we need only show that if a power series $\sum_{n=0}^{\infty} a_n x^n$ has a positive number R for its radius of convergence, it is continuous at R if R is included within the interval of convergence, and it is continuous at $-R$ if $-R$ is included within the interval of convergence. The arguments are similar, so we demonstrate only the former.

Suppose $\sum_{n=0}^{\infty} a_n x^n$ has radius of convergence R, where R is a positive number, and assume $\sum_{n=0}^{\infty} a_n R^n$ converges with $\sum_{n=0}^{\infty} a_n R^n = S$. Define the function f on $(-R, R)$ so that

$$f(x) = \sum_{n=0}^{\infty} a_n x^n.$$

To show the power series $\sum_{n=0}^{\infty} a_n x^n$ is continuous at R, we show that $\lim_{x\to R^-} f(x) = S$. Thus, in what follows, we assume that $0 < x < R$.

For each natural number n, let

$$s_n = \sum_{k=0}^{n} a_k R^k,$$

and take

$$s_{-1} = 0.$$

According to Abel's partial summation formula, Lemma 23.7,

$$\begin{aligned}\sum_{k=0}^{n} a_k x^k &= \sum_{k=0}^{n} (a_k R^k)\left(\frac{x}{R}\right)^k \\ &= \sum_{k=0}^{n-1} s_k\left(\left(\frac{x}{R}\right)^k - \left(\frac{x}{R}\right)^{k+1}\right) + s_n\left(\frac{x}{R}\right)^n - s_{-1}\left(\frac{x}{R}\right)^0 \\ &= \left(1 - \frac{x}{R}\right)\sum_{k=0}^{n-1} s_k\left(\frac{x}{R}\right)^k + s_n\left(\frac{x}{R}\right)^n.\end{aligned}$$

We know $s_n \to S$. Also, because $0 < x < R$, it follows that $(\frac{x}{R})^n \to 0$. Hence, taking the limit as n approaches ∞ yields

$$\sum_{n=0}^{\infty} a_n x^n = \left(1 - \frac{x}{R}\right)\sum_{n=0}^{\infty} s_n\left(\frac{x}{R}\right)^n + S \cdot 0,$$

so that for $0 < x < R$, we have

$$f(x) = \left(1 - \frac{x}{R}\right)\sum_{n=0}^{\infty} s_n\left(\frac{x}{R}\right)^n.$$

Consider any positive number ε. As $s_n \to S$, there is a natural number N such that, when $n > N$, it follows that

$$|s_n - S| < \frac{\varepsilon}{2}.$$

Let

$$D = \sum_{n=1}^{N} |s_n - S|$$

and then let

$$\delta = \frac{R\varepsilon}{2D + 1}.$$

Note that as $D \geq 0$, it follows that $\delta > 0$. Also note that

$$D = \frac{R\varepsilon}{2\delta} - \frac{1}{2} < \frac{R\varepsilon}{2\delta}.$$

Assume now that $\max\{0, R - \delta\} < x < R$, so that

$$0 < 1 - \frac{x}{R} < \frac{\delta}{R}$$

and, using the formula for a convergent geometric series,

$$\sum_{n=0}^{\infty}\left(\frac{x}{R}\right)^n = \frac{1}{1-\frac{x}{R}}.$$

We may now conclude that

$$\begin{aligned}
|f(x) - S| &= \left|\left(1-\frac{x}{R}\right)\sum_{n=0}^{\infty} s_n\left(\frac{x}{R}\right)^n - S\left(1-\frac{x}{R}\right)\sum_{n=0}^{\infty}\left(\frac{x}{R}\right)^n\right| \\
&= \left|\left(1-\frac{x}{R}\right)\sum_{n=0}^{\infty}(s_n - S)\left(\frac{x}{R}\right)^n\right| \\
&\le \left(1-\frac{x}{R}\right)\sum_{n=1}^{N}|s_n - S|\left(\frac{x}{R}\right)^n + \left(1-\frac{x}{R}\right)\sum_{n=N+1}^{\infty}|s_n - S|\left(\frac{x}{R}\right)^n \\
&< \frac{\delta}{R}\sum_{n=1}^{N}|s_n - S|\cdot 1 + \frac{\varepsilon}{2}\left(1-\frac{x}{R}\right)\sum_{n=N+1}^{\infty}\left(\frac{x}{R}\right)^n \\
&< \frac{\delta}{R}D + \frac{\varepsilon}{2}\left(1-\frac{x}{R}\right)\sum_{n=0}^{\infty}\left(\frac{x}{R}\right)^n \\
&< \frac{\delta}{R}\cdot\frac{R\varepsilon}{2\delta} + \frac{\varepsilon}{2}\left(1-\frac{x}{R}\right)\cdot\frac{1}{1-\frac{x}{R}} \\
&= \varepsilon.
\end{aligned}$$

Hence, by definition, $\lim_{x\to R^-} f(x) = S$. □

Example 27.4 (Continued). Thus, according to Abel's lemma, the power series f defined so that

$$f(x) = \sum_{n=1}^{\infty}\frac{4^n}{n^4}(x+3)^n$$

is continuous on its interval of convergence $[-3.25, -2.75]$.

Exercise 27.2. In Exercise 23.9, you showed the sum of the alternating harmonic series $\sum_{n=0}^{\infty}\frac{(-1)^{n+1}}{n}$ is $\ln(2)$. Here is another way to reach this conclusion.

(a) Verify that the interval of convergence for the power series $\sum_{n=0}^{\infty}(-1)^n x^n$ is $(-1, 1)$ and demonstrate that $\sum_{n=0}^{\infty}(-1)^n x^n = \frac{1}{1+x}$ for all x in $(-1, 1)$.

(b) Use (a) to show that, for each x in $(-1, 1)$, we have $\ln(1+x) = \sum_{n=1}^{\infty}\frac{(-1)^{n+1}x^n}{n}$.

(c) Use Abel's theorem to explain how it follows from (b) that $\ln(2) = \sum_{n=1}^{\infty}\frac{(-1)^{n+1}}{n}$.

(d) Find the interval of convergence of the series $\sum_{n=1}^{\infty}\frac{(-1)^{n+1}x^n}{n}$.

A power series is always differentiable within the interior of its interval of convergence and may be differentiated summand-by-summand in order to obtain the derivative. In other words, the derived series of a power series is the derivative of the power series.

Theorem 27.6 (Differentiation theorem for power series). *Let the function f be the power series defined so that*

$$f(x) = \sum_{n=0}^{\infty} a_n (x - x_0)^n.$$

If R is the radius of convergence of f, then f is differentiable on the interval $(x_0 - R, x_0 + R)$ and

$$f'(x) = \sum_{n=1}^{\infty} n a_n (x - x_0)^{n-1},$$

so that the function f' is the derived series of f, and is also a power series with the same radius of convergence R as f.

Proof. Define the function g so that

$$g(x) = \sum_{n=1}^{\infty} n a_n (x - x_0)^{n-1}$$

for all x for which this power series converges. From Exercise 10.19, we recall that $\lim_{n\to\infty} \sqrt[n]{n} = 1$. Let $\lambda = \limsup \sqrt[n]{|a_n|}$. If λ is a real number, then

$$\limsup(\sqrt[n]{n}\sqrt[n]{|a_n|}) = \lim_{n\to\infty} \sqrt[n]{n} \cdot \limsup \sqrt[n]{|a_n|} = 1 \cdot \lambda = \lambda,$$

so that the radius of convergence of g is $\frac{1}{\lambda}$, identical to the radius of convergence of f. If $\lambda = \infty$, the quantity $\sqrt[n]{|a_n|}$ is unbounded, but since $\lim_{n\to\infty} \sqrt[n]{n} = 1$, the quantity $\sqrt[n]{n}\sqrt[n]{|a_n|}$ is also unbounded, meaning

$$\limsup(\sqrt[n]{n}\sqrt[n]{|a_n|}) = \infty = \lambda,$$

and the radius of convergence of g is once again $\frac{1}{\lambda}$, the same as the radius of convergence of f.

Consider any closed bounded interval I that is a subset of $(x_0 - R, x_0 + R)$. Now that we know the power series g has the same radius of convergence as the power series f, we can apply Theorem 27.2 to conclude that g converges uniformly on I. Since g is the series whose summands are the derivatives of the summands of f, we can then apply Corollary 26.31 to conclude that f is differentiable on I and that f' can be obtained through summand-by-summand differentiation, meaning

$$f'(x) = \sum_{n=1}^{\infty} n a_n (x - x_0)^{n-1}$$

for all x in I. Since every point of $(x_0 - R, x_0 + R)$ is included in some closed bounded interval that is a subset of $(x_0 - R, x_0 + R)$, we conclude that f is differentiable on $(x_0 - R, x_0 + R)$ and the equation

$$f'(x) = \sum_{n=1}^{\infty} na_n(x - x_0)^{n-1}$$

is valid for all x in $(x_0 - R, x_0 + R)$. □

Example 27.4 (Continued). Thus, the power series f defined by

$$f(x) = \sum_{n=1}^{\infty} \frac{4^n}{n^4}(x + 3)^n$$

is differentiable on $(-3.25, -2.75)$, with

$$f'(x) = \sum_{n=1}^{\infty} \frac{4^n}{n^4} \cdot n(x + 3)^{n-1} = \sum_{n=1}^{\infty} \frac{4^n}{n^3}(x + 3)^{n-1} = \sum_{n=0}^{\infty} \frac{4^{n+1}}{(n + 1)^3}(x + 3)^n.$$

By repeatedly applying Theorem 27.6, we obtain the following corollary telling us that a power series has derivatives of all orders (i. e., is infinitely differentiable) within the interior of its interval of convergence.

Corollary 27.7. *Let the function f be the power series defined so that*

$$f(x) = \sum_{n=0}^{\infty} a_n(x - x_0)^n.$$

If R is the radius of convergence of f, then for every natural number k, the derivative $f^{(k)}$ exists on the interval $(x_0 - R, x_0 + R)$ and is a power series with the same radius of convergence R as f. Moreover, for every natural number k, we have

$$f^{(k)}(x) = \sum_{n=k}^{\infty} \frac{n!}{(n - k)!} a_n(x - x_0)^{n-k},$$

a series that converges absolutely for every x in $(x_0 - R, x_0 + R)$ and uniformly on any closed bounded interval that is a subset of $(x_0 - R, x_0 + R)$.

Example 27.4 (Continued). It follows that the power series f defined by

$$f(x) = \sum_{n=1}^{\infty} \frac{4^n}{n^4}(x + 3)^n$$

is infinitely differentiable on $(-3.25, -2.75)$, with higher-order derivatives given by

$$f''(x) = \sum_{n=2}^{\infty} \frac{4^n}{n^3}(n - 1)(x + 3)^{n-2},$$

$$f'''(x) = \sum_{n=3}^{\infty} \frac{4^n}{n^3}(n-1)(n-2)(x+3)^{n-3},$$

and, more generally,

$$f^{(k)}(x) = \sum_{n=k}^{\infty} \frac{4^n (n-1)!}{n^3 (n-k)!}(x+3)^{n-k}.$$

Taking $x = -3$, we may then further conclude that

$$f^{(k)}(-3) = \frac{4^k (k-1)!}{k^3}.$$

Exercise 27.3. Use induction to prove Corollary 27.7. (Note that the base step is already established via Theorem 27.6.)

From Corollary 27.7, we obtain the following formula for the coefficients of a power series.

Theorem 27.8 (Formula for the coefficients of a power series). *Let the function f be the power series defined so that*

$$f(x) = \sum_{n=0}^{\infty} a_n (x-x_0)^n,$$

and suppose the radius of convergence R is nonzero. Then, for every natural number n, we have

$$a_n = \frac{f^{(n)}(x_0)}{n!}.$$

Proof. We are assuming the radius of convergence is nonzero so that there are points at which the power series f is differentiable. Setting $x = x_0$ in the formula

$$f^{(k)}(x) = \sum_{n=k}^{\infty} \frac{n!}{(n-k)!} a_n (x-x_0)^{n-k}$$

from Corollary 27.7 yields

$$f^{(k)}(x_0) = \sum_{n=k}^{\infty} \frac{n!}{(n-k)!} a_n (x_0-x_0)^{n-k} = k! a_k,$$

and consequently, $a_k = \frac{f^{(k)}(x_0)}{k!}$. □

Example 27.4 (Continued). Here f is defined by

$$f(x) = \sum_{n=1}^{\infty} \frac{4^n}{n^4}(x+3)^n$$

for x in $[-3.25, -2.75]$, so the coefficients are

$$a_n = \frac{4^n}{n^4}$$

for natural numbers n, with $a_0 = 0$. From the coefficient formula given in Theorem 27.8, we may conclude that, for each natural number n, we have

$$f^{(n)}(-3) = \frac{4^n \cdot n!}{n^4} = \frac{4^n(n-1)!}{n^3},$$

which agrees with the direct calculation we performed earlier.

Thus, when a function f is defined by a power series with center x_0, the coefficients of this power series are precisely the coefficients for the Taylor series of f with center x_0. This observation tells us that power series with the same center but different coefficients must converge to different functions.

Theorem 27.9 (Uniqueness theorem for power series). *If the power series $\sum_{n=0}^{\infty} a_n(x-x_0)^n$ and $\sum_{n=0}^{\infty} b_n(x-x_0)^n$ both converge to the same function on an interval $(x_0 - R, x_0 + R)$, where $R > 0$, then $a_n = b_n$ for every natural number n.*

Proof. Suppose $\sum_{n=0}^{\infty} a_n(x-x_0)^n$ and $\sum_{n=0}^{\infty} b_n(x-x_0)^n$ both converge to the function f on the interval $(x_0 - R, x_0 + R)$, where $R > 0$. Then, according to the coefficient formula from Theorem 27.8,

$$a_n = \frac{f^{(n)}(x_0)}{n!} = b_n,$$

for every natural number n. □

The uniqueness theorem can be applied to obtain the special forms taken by power series representations of *even* and *odd* functions. Recall that a function $f : I \to \mathbb{R}$, where I is either $\mathbb{R}$ or an interval having the form $(-a, a)$ or $[-a, a]$ for some positive number a, is **even** if $f(-x) = f(x)$ for all x in I and **odd** if $f(-x) = -f(x)$ for all x in I.

Theorem 27.10. *Let the function f be the power series defined so that*

$$f(x) = \sum_{n=0}^{\infty} a_n x^n,$$

and suppose the radius of convergence R is nonzero.

(1) *If f is an even function, then $a_n = 0$ for every odd natural number n.*
(2) *If f is an odd function, then $a_n = 0$ for every even natural number n.*

Exercise 27.4. Prove (1) of Theorem 27.10.

Convergence of Taylor series

When $x = x_0$, the Taylor series

$$\sum_{n=0}^{\infty} \frac{f^{(n)}(x_0)}{n!}(x - x_0)^n$$

for an infinitely differentiable function f is equal to $f(x_0)$, hence, converges to $f(x_0)$. However, as mentioned when we introduced Taylor series in Chapter 25, it is not guaranteed that the Taylor series converges to $f(x)$ for any other value of x, even when x lies in the interval of convergence of the series (we observed an extreme instance of this behavior in Example 25.9). Knowing p is in the interval of convergence of the Taylor series for f only tells us that when x is assigned the value p, the Taylor series converges to *some* number, not that this number must be $f(p)$.

Thus, we are interested in determining the values of x for which the Taylor series for f converges to $f(x)$ since, for such values of x, not only does the Taylor series provide us with an alternate representation of the function, but the partial sums of the Taylor series, that is, the Taylor polynomials, present us with a means to approximate $f(x)$.

Sometimes we can apply already established facts about convergent series to demonstrate convergence of a Taylor series to the function from which it was generated. In particular, we may draw upon our knowledge of convergent geometric series, as well as the ability to differentiate or integrate a given power series to obtain a new power series with the same radius of convergence.

Example 27.11. In Example 25.4, we used the formula for a convergent geometric series to determine that

$$\sum_{n=0}^{\infty} x^n = \frac{1}{1-x},$$

when x is in $(-1, 1)$. Thus, we obtained a power series representation $\sum_{n=0}^{\infty} x^n$ for the function $h : (-1, 1) \to \mathbb{R}$ defined so that

$$h(x) = \frac{1}{1-x}.$$

The uniqueness theorem for power series guarantees this power series is the Maclaurin series (i. e., Taylor series centered at 0) for h. What we have learned about the convergence of geometric series guarantees the convergence of this Taylor series to the function h on the interval $(-1, 1)$.

Then, because a power series is differentiable within the interior of the interval of convergence and the derivative of a power series is the derived series of the power series, we may conclude that

$$h'(x) = \frac{1}{(1-x)^2} = \sum_{n=0}^{\infty}(x^n)' = \sum_{n=1}^{\infty} nx^{n-1} = \sum_{n=0}^{\infty}(n+1)x^n,$$

again when x is in $(-1, 1)$. The power series $\sum_{n=0}^{\infty}(n+1)x^n$ is the Maclaurin series for h' and it converges to h' on $(-1, 1)$ because of what we know about the derivatives of power series via the differentiation theorem for power series.

The parallel result for integrals tells us that for any x in $(-1, 1)$, we must have

$$\int_0^x h = \sum_{n=0}^{\infty} \int_0^x t^n dt.$$

Since

$$\int_0^x h = \int_0^x \frac{1}{1-t} dt = -\ln(1-x) = \ln((1-x)^{-1}) = \ln\left(\frac{1}{1-x}\right),$$

while

$$\sum_{n=0}^{\infty} \int_0^x t^n dt = \sum_{n=0}^{\infty} \frac{x^{n+1}}{n+1} = \sum_{n=1}^{\infty} \frac{x^n}{n},$$

we may then conclude that

$$\ln\left(\frac{1}{1-x}\right) = \sum_{n=1}^{\infty} \frac{x^n}{n},$$

for all x in $(-1, 1)$. Thus, $\sum_{n=1}^{\infty} \frac{x^n}{n}$ is the Maclaurin series for the function $g : (-1, 1) \to \mathbb{R}$, where $g(x) = \ln(\frac{1}{1-x})$, and this series converges to g on $(-1, 1)$.

Example 27.12. Your work in Exercise 27.2 shows that

$$\ln(1+x) = \sum_{n=1}^{\infty} \frac{(-1)^{n+1}x^n}{n}$$

for x in $(-1, 1]$. Thus, the series here is the Maclaurin series for the function $y = \ln(1+x)$ and this series converges on $(-1, 1]$ to $\ln(1+x)$.

More generally, to determine whether the Taylor series

$$\sum_{n=0}^{\infty} \frac{f^{(n)}(x_0)}{n!}(x - x_0)^n$$

at x_0 for an infinitely differentiable function f converges to f itself, we make use of the **Taylor remainder** R_n, defined for each nonnegative integer n and for each x in the interval of convergence I of the Taylor series, so that

$$R_n(x) = f(x) - P_n(x),$$

where $P_n(x)$ is the nth partial sum $\sum_{k=0}^{n} \frac{f^{(k)}(x_0)}{k!}(x - x_0)^k$ of the Taylor series, that is, the nth Taylor polynomial. The next theorem is simply the observation that the Taylor series converges to f if and only if the Taylor remainder converges to zero on the interval of convergence.

Theorem 27.13 (Convergence of a Taylor series). *Let f be a function that is infinitely differentiable on an open interval I that includes x_0. For each x in I, the Taylor series $\sum_{n=0}^{\infty} \frac{f^{(n)}(x_0)}{n!}(x - x_0)^n$ converges to $f(x)$ if and only if the sequence $(R_n(x))$ of Taylor remainders converges to 0.*

We first encountered the Taylor remainder R_n in our study of Taylor polynomial approximations to a function in Chapter 17. There we also proved a result, known as *Taylor's theorem*, which provides a specific form the remainder must take under certain hypotheses. Here we present an extended version of Taylor's theorem that also includes additional information about the form of the remainder in the presence of somewhat stronger hypotheses.

Theorem 27.14 (Taylor's theorem). *Let n be a nonnegative integer, let I be an open interval that includes x_0, and let f be a function for which $f^{(n+1)}$ exists on I.*

(1) *For each x in $I - \{x_0\}$, there exists a number c between x and x_0 such that the Taylor remainder $R_n(x)$ is given by*

$$R_n(x) = \frac{f^{(n+1)}(c)}{(n+1)!}(x - x_0)^{n+1}.$$

(2) *If $f^{(n+1)}$ is integrable on every closed bounded interval that is a subset of I, then for each x in I, the Taylor remainder $R_n(x)$ is given by*

$$R_n(x) = \frac{1}{n!}\int_{x_0}^{x} f^{(n+1)}(t)(x - t)^n\,dt.$$

(3) *If $f^{(n+1)}$ is continuous on I, there exists a number c between x and x_0 such that the Taylor remainder $R_n(x)$ is given by*

$$R_n(x) = \frac{f^{(n+1)}(c)}{n!}(x - x_0)(x - c)^n.$$

Note that (1) of Theorem 27.14 is the original formulation of Taylor's theorem stated and proved as Theorem 17.3. You proved (2) in Exercise 20.7. The proof of (3) is assigned as Exercise 27.5.

The form of the remainder given in (1) is sometimes called the **Lagrange form of the remainder**, while the form in (2) is the **integral form of the remainder**, and the form in (3) is the **Cauchy form of the remainder**. The remainder formulas assist us in determining whether the remainder converges to zero and in obtaining error bounds when using a Taylor polynomial to approximate a function value (see Examples 17.1 and 20.25 for illustrations of the latter). The particular form of the remainder employed varies based on the information at hand.

Exercise 27.5. Prove (3) of Theorem 27.14. *Hint*: Make use of the integral form (2) of the Taylor remainder and the mean value theorem for Integrals from Exercise 19.3.

Exercise 27.6. Explain why, for $n = 0$, the result (1) from Theorem 27.14 is a conclusion we could reach via Theorem 16.7, the mean value theorem.

Exercise 27.7. Use the series representation of $\ln(1 + x)$ given in Example 27.12, along with the Lagrange form (1) of the remainder given in Taylor's theorem, to show that if $0 < x < 1$, we have

$$\sum_{k=1}^{2n} \frac{(-1)^{k+1}x^k}{k} < \ln(1 + x) < \sum_{k=1}^{2n+1} \frac{(-1)^{k+1}x^k}{k}$$

for each natural number n. Then use this inequality to accurately approximate $\ln(1.2)$ to within 0.01 of its true value.

Example 27.15. Consider the infinitely differentiable function $f : (0, \infty) \to \mathbb{R}$ defined by

$$f(x) = \sqrt{x}.$$

In Exercise 25.8, you verified that

$$f^{(n)}(x) = \frac{(-1)^{n+1}\prod_{i=1}^{n-1}(2i - 1)}{2^n}x^{0.5-n},$$

which we used in Example 25.8 to derive the Taylor series

$$2 + \frac{1}{4}(x - 4) + \sum_{n=2}^{\infty} \frac{(-1)^{n+1}\prod_{i=1}^{n-1}(2i - 1)}{n!2^{3n-1}}(x - 4)^n$$

for f about $x_0 = 4$. We also found the interval of convergence of this series to be $(0, 8]$. We now show that the Taylor series converges on $(0, 8]$ to f. The convergence to $f(4) = \sqrt{4}$ at the center $x_0 = 4$ is, of course, automatic. The convergence to $f(8) = \sqrt{8}$ at the endpoint 8 follows via Abel's theorem once we establish convergence to $f(x)$ at all other points x of the interval of convergence.

In the case where $4 < x < 8$, fix n and use the Lagrange form of the remainder given in (1) of Taylor's theorem to obtain a number c for which $4 < c < x$ and

$$|R_n(x)| = \left|\frac{f^{(n+1)}(c)}{(n+1)!}(x-4)^{n+1}\right| = \frac{\prod_{i=1}^{n}(2i-1)}{(n+1)! \cdot 2^{n+1}} c^{-0.5-n}(x-4)^{n+1}.$$

Define the function g on $(4, 8)$ by

$$g(t) = t^{-0.5-n}$$

and note that

$$g'(t) = \frac{-0.5-n}{t^{1.5+n}} < 0$$

for all t, so g is decreasing on $(4, 8)$. Thus, as $4 < c$, it follows that

$$c^{-0.5-n} < 4^{-0.5-n}.$$

Hence,

$$\begin{aligned} 0 &\le |R_n(x)| \\ &\le \frac{\prod_{i=1}^{n}(2i-1)}{(n+1)! \cdot 2^{n+1}} \cdot 4^{-0.5-n} \cdot 4^{n+1} \\ &= \frac{1 \cdot 3 \cdot 5 \cdot \ldots \cdot (2n-1)}{2 \cdot 4 \cdot 6 \cdot \ldots \cdot (2n+2)} \cdot 2 \\ &= \frac{1}{2} \cdot \frac{3}{4} \cdot \frac{5}{6} \cdot \ldots \cdot \frac{2n-1}{2n} \cdot \frac{1}{2n+2} \cdot 2 \\ &< \frac{2}{2n+2}. \end{aligned}$$

Thus, since $\frac{2}{2n+2} \to 0$, it follows that $R_n(x) \to 0$, and we may conclude, via Theorem 27.13, that the Taylor series for f converges to f when $4 < x < 8$.

For the case where $0 < x < 4$, fix n and use the Cauchy form of the remainder given in (3) of Taylor's theorem to obtain a number c for which $x < c < 4$ and

$$|R_n(x)| = \left|\frac{f^{(n+1)}(c)}{n!}(x-4)(x-c)^n\right| = \frac{\prod_{i=1}^{n}(2i-1)}{n! \cdot 2^{n+1}} c^{-0.5-n}(4-x)(c-x)^n.$$

Hence,

$$|R_n(x)| = \frac{1}{2} \cdot \frac{3}{4} \cdot \frac{5}{6} \cdot \ldots \cdot \frac{2n-1}{2n} \cdot \frac{1}{2} \cdot \left(\frac{c-x}{c}\right)^n \cdot \frac{4-x}{\sqrt{c}} < \left(1 - \frac{x}{c}\right)^n \cdot \frac{4-x}{\sqrt{c}}.$$

Define the function h on $(0,4)$ by

$$h(t) = 1 - \frac{x}{t}$$

and note that

$$h'(t) = \frac{x}{t^2} > 0$$

for all t, so h is increasing on $(0,4)$. Thus, as $c < 4$, it follows that

$$1 - \frac{x}{c} < 1 - \frac{x}{4}.$$

Also, as $0 < x < c$, we have

$$\frac{4-x}{\sqrt{c}} < \frac{4-x}{\sqrt{x}}.$$

Hence,

$$0 \leq |R_n(x)| < \left(1 - \frac{x}{4}\right)^n \cdot \frac{4-x}{\sqrt{x}}.$$

But since $0 < x < 4$, it follows that

$$0 < 1 - \frac{x}{4} < 1,$$

so $(1 - \frac{x}{4})^n \to 0$ and, as a consequence, $R_n(x) \to 0$. Therefore, the Taylor series for f also converges to f when $0 < x < 4$.

The following result can be useful in analyzing the Taylor remainder. It says that even when $|a|$ is large, the quantity $n!$ grows much faster than the exponential expression a^n.

Theorem 27.16. *For any real number a, the sequence $(\frac{a^n}{n!})$ converges to 0.*

Exercise 27.8. Prove Theorem 27.16.

If all the derivatives of an infinitely differentiable function are bounded on an open interval by the same positive number, then a Taylor series for the function converges to the function on that interval.

Theorem 27.17. *Let f be a function that is infinitely differentiable on an open interval I that includes x_0. If there exists a positive number B such that*

$$|f^{(n)}(x)| \le B$$

for all x in I and all natural numbers n, then

$$f(x) = \sum_{n=0}^{\infty} \frac{f^{(n)}(x_0)}{n!}(x - x_0)^n$$

for all x in I.

Exercise 27.9. Prove Theorem 27.17. *Hint*: Make use of Theorem 27.16.

Exercise 27.10. Let f be a function that is infinitely differentiable on an open interval I that includes x_0. Show that if there exists a positive number M such that $|f^{(n)}(x)| \le M^n$ for all x in I and all natural numbers n, then $f(x) = \sum_{n=0}^{\infty} \frac{f^{(n)}(x_0)}{n!}(x - x_0)^n$ for all x in I. *Hint*: Make use of Theorem 27.16.

The natural exponential function revisited

As another application of Taylor's theorem, we prove the Maclaurin series for the natural exponential function converges to e^x for every real number x.

Theorem 27.18 (Series representation of the natural exponential function and Euler's number). *For every real number x, we have*

$$\exp(x) = e^x = 1 + x + \frac{x^2}{2} + \frac{x^3}{3!} + \frac{x^4}{4!} + \cdots = \sum_{n=0}^{\infty} \frac{x^n}{n!}.$$

In particular, then

$$e = 1 + 1 + \frac{1}{2} + \frac{1}{3!} + \frac{1}{4!} + \cdots = \sum_{n=0}^{\infty} \frac{1}{n!}.$$

Proof. For the natural exponential function exp, we have $\exp^{(n)}(x) = e^x$ for every natural number n. We obtained the Maclaurin series

$$\sum_{n=0}^{\infty} \frac{x^n}{n!}$$

for exp in Example 25.6. According to (1) of Taylor's theorem, given a nonzero number x and a natural number n, there is a number c between x and $x_0 = 0$ such that

$$R_n(x) = \frac{\exp^{(n+1)}(c)}{(n+1)!}x^{n+1} = \frac{e^c}{(n+1)!}x^{n+1}.$$

If $x < 0$, then $x < c < 0$, and it follows that $e^c < e^0 = 1$, and thus

$$0 < |R_n(x)| = e^c \frac{(-x)^{n+1}}{(n+1)!} < \frac{(-x)^{n+1}}{(n+1)!}.$$

As $\frac{(-x)^{n+1}}{(n+1)!} \to 0$ via Theorem 27.16, we may conclude that $R_n(x) \to 0$ when $x < 0$.
If $x > 0$, then $0 < c < x$, which then implies $e^c < e^x$, and consequently,

$$0 < |R_n(x)| = e^c \frac{x^{n+1}}{(n+1)!} < e^x \frac{x^{n+1}}{(n+1)!}.$$

Once again applying Theorem 27.16, we are able to conclude that $R_n(x) \to 0$ when $x > 0$.

As $R_n(x) \to 0$ for every real number x, it follows using Theorem 27.13 that the Maclaurin series for exp converges to e^x for every real number x. Hence,

$$e^x = \sum_{n=0}^{\infty} \frac{x^n}{n!}$$

for all x in $\mathbb{R}$. □

Exercise 27.11. It is possible to *define* the natural exponential function exp by the equation

$$\exp(x) = \sum_{n=0}^{\infty} \frac{x^n}{n!},$$

rather than as the inverse of the natural logarithmic function, the approach we took in Chapter 20. Using the series definition for $\exp(x)$ given here, verify each of the following.

(a) $\exp(0) = 1$.
(b) $\exp'(x) = \exp(x)$.
(c) $\lim_{x\to\infty} \exp(x) = \infty$.

Exercise 27.12. Use Theorem 27.17 to provide an alternate proof of Theorem 27.18.

We can use the series representation for Euler's number e to show e is irrational. The proof originated with Joseph Fourier (1768–1830).

Theorem 27.19 (Irrationality of e). *Euler's number e is irrational.*

Proof. Suppose to the contrary that $e = \frac{m}{n}$ for some natural numbers m and n. From Theorem 27.18, we know $e = \sum_{k=0}^{\infty} \frac{1}{k!}$, so that

$$e - \sum_{k=0}^{n} \frac{1}{k!} = \sum_{k=n+1}^{\infty} \frac{1}{k!},$$

which implies

$$n!\left(e - \sum_{k=0}^{n} \frac{1}{k!}\right) = n! \sum_{k=n+1}^{\infty} \frac{1}{k!}. \tag{*}$$

Hence,

$$n!\left(e - \sum_{k=0}^{n} \frac{1}{k!}\right) = n!\left(\frac{m}{n} - \sum_{k=0}^{n} \frac{1}{k!}\right) = m(n-1)! - \sum_{k=0}^{n} \frac{n!}{k!},$$

which is a positive integer, since $n!(e - \sum_{k=0}^{n} \frac{1}{k!}) > 0$ and $m(n-1)! - \sum_{k=0}^{n} \frac{n!}{k!}$ is an integer. However, using comparison of series and the formula for the sum of a convergent geometric series, we now have

$$\begin{aligned}
n!\left(e - \sum_{k=0}^{n} \frac{1}{k!}\right) &= n! \sum_{k=n+1}^{\infty} \frac{1}{k!} \\
&= \frac{1}{(n+1)} + \frac{1}{(n+1)(n+2)} + \frac{1}{(n+1)(n+2)(n+3)} + \cdots \\
&< \frac{1}{(n+1)} + \frac{1}{(n+1)^2} + \frac{1}{(n+1)^3} + \cdots \\
&= \frac{\frac{1}{n+1}}{1 - \frac{1}{n+1}} \\
&= \frac{1}{n} \\
&\leq 1,
\end{aligned}$$

a contradiction, as there is no positive integer less than 1. Therefore, e is irrational. □

Exercise 27.13. In this exercise, you eventually show that $\lim_{n\to\infty} \frac{n^n}{n! e^n} = 0$.

(a) Use the fact that the natural logarithmic function is increasing on its domain to show that

$$\int_{k-1}^{k} \ln(x)dx < \ln(k) < \int_{k}^{k+1} \ln(x)dx$$

for any natural number k.

(b) Indicate how it follows from (a) that

$$\int_{0}^{n} \ln(x)dx < \ln(n!) < \int_{1}^{n+1} \ln(x)dx$$

for any natural number n.

(c) Verify that

$$\int_{0}^{n} \ln(x)dx = n\ln(n) - n$$

and

$$\int_1^{n+1} \ln(x)dx = (n+1)\ln(n+1) - n.$$

(d) Define the sequence a so that $a_n = \ln(n!) - (n+0.5)\ln(n) + n$. Show that

$$a_n - a_{n+1} = (n+0.5)\ln\left(\frac{n+1}{n}\right) - 1$$

for any natural number n.

(e) Verify that $\frac{n+1}{n} = \frac{1+\frac{1}{2n+1}}{1-\frac{1}{2n+1}}$.

(f) Use the Maclaurin series representation for $\ln(1+x)$ obtained in Example 27.12 to obtain the Maclaurin series representation for $\ln(\frac{1+x}{1-x})$ on $(-1,1)$.

(g) Use (d), (e), and (f), along with the formula for a convergent geometric series, to verify that

$$0 < a_n - a_{n+1} = \sum_{k=1}^{\infty} \frac{1}{2k+1} \cdot \left(\frac{1}{2n+1}\right)^{2k} < \frac{1}{3}\sum_{k=1}^{\infty}\left(\frac{1}{2n+1}\right)^{2k} = \frac{1}{12}\left(\frac{1}{n} - \frac{1}{n+1}\right).$$

(h) Define the sequence b so that $b_n = a_n - \frac{1}{12n}$. Use the result from (g) to explain why the sequence a is decreasing and the sequence b is increasing.

(i) Explain how it follows from (h) that the sequence a must converge to a positive number p. Show that as a consequence we must have $e^p = \lim_{n\to\infty} \frac{n!e^n}{n^n\sqrt{n}}$.

(j) Indicate how we may now conclude that $\lim_{n\to\infty} \frac{n^n}{n!e^n} = 0$.

Exercise 27.14. Use the alternating series test to show that the series $\sum_{n=1}^{\infty} \frac{(-1)^n n^n}{n!e^n}$ converges. Doing so, together with the results from Exercise 25.20, shows that the interval of convergence of the power series $\sum_{n=1}^{\infty} \frac{n^n}{n!}x^n$ is $[-\frac{1}{e}, \frac{1}{e})$. *Hint*: Make use of the limit established in Exercise 27.13.

Binomial coefficients and the binomial series

In our statement of Theorem 3.23, the binomial theorem, we defined the *binomial coefficient* $\binom{n}{k}$ for natural numbers n and nonnegative integers k no larger than n. More generally, for any real number a, we define the **binomial coefficient** $\binom{a}{k}$ so that when k is a natural number, we have

$$\binom{a}{k} = \frac{a(a-1)(a-2)\cdots(a-k+1)}{k!},$$

and when $k=0$, we have

$$\binom{a}{0} = 1.$$

For example,

$$\binom{-0.5}{4} = \frac{(-0.5)(-0.5-1)(-0.5-2)(-.5-3)}{4!}.$$

Exercise 27.15. Find the numerical value of each of $\binom{0.1}{2}$ and $\binom{2}{4}$.

Exercise 27.16. Show that $\binom{a}{k} = \frac{a!}{k!(a-k)!}$ when a and k are nonnegative integers for which $k \leq a$.

For any real number a, the series $\sum_{n=0}^{\infty} \binom{a}{n} x^n$, which we show converges when $-1 < x < 1$, is called a **binomial series**. The following theorem, due to Newton, is a kind of extension of the binomial theorem.

Theorem 27.20 (Convergence of the binomial series). *Let a be any real number and let x be any real number for which $-1 < x < 1$. Then*

$$(1+x)^a = \sum_{n=0}^{\infty} \binom{a}{n} x^n.$$

Proof. Define the function $f : (-1,1) \to \mathbb{R}$ so that

$$f(x) = (1+x)^a.$$

Observe that, for each natural number n, we have

$$f^{(n)}(x) = a(a-1)(a-2)\cdots(a-n+1)(1+x)^{a-n} = \binom{a}{n} n!(1+x)^{a-n}$$

and, therefore,

$$f^{(n)}(0) = \binom{a}{n} n!.$$

Hence, the Maclaurin series for f is

$$\sum_{n=0}^{\infty} \frac{f^{(n)}(0)}{n!} x^n = \sum_{n=0}^{\infty} \binom{a}{n} x^n.$$

The rest of the proof is devoted to showing this series converges on $(-1,1)$ to f.

First, note that if $a = 0$, we have

$$f(x) = (1+x)^0 = 1,$$

and if $a = 1$, we have

$$f(x) = (1+x)^1 = 1 + x,$$

both of which are power series, hence, by the uniqueness theorem, must be the respective Maclaurin series for f. So the convergence to f holds when $a = 0$ and when $a = 1$. We now present the proof when $a < 1$ and $a \neq 0$, leaving the proof when $a > 1$ as Exercise 27.18.

As all the derivatives of f are continuous on $(-1, 1)$, they are integrable on any closed bounded subinterval of $(-1, 1)$, so applying (2) of Taylor's theorem, we know that for each x in $(-1, 1)$, the Taylor remainder is given by

$$R_n(x) = \frac{1}{n!} \int_0^x f^{(n+1)}(t)(x-t)^n \, dt.$$

Using the formula we derived for $f^{(n)}$, we may then conclude that

$$\begin{aligned} R_n(x) &= \frac{1}{n!} \int_0^x \binom{a}{n+1}(n+1)!(1+t)^{a-n-1}(x-t)^n \, dt \\ &= (n+1)\binom{a}{n+1} \int_0^x \left(\frac{x-t}{1+t}\right)^n (1+t)^{a-1} \, dt. \end{aligned}$$

Suppose x is in $(-1, 1)$. The convergence of the Maclaurin series to f is necessary at the center $x = 0$, so we need only consider the possibilities $-1 < x < 0$ and $0 < x < 1$.

First, assume $0 < x < 1$. Then, when $0 < t < x$, we have

$$0 < \frac{x-t}{1+t} < x.$$

Also, because $1 + t > 1$ and $a - 1 < 0$, we have

$$0 < (1+t)^{a-1} < 1.$$

Thus,

$$\begin{aligned} |R_n(x)| &= (n+1)\left|\binom{a}{n+1}\right| \left| \int_0^x \left(\frac{x-t}{1+t}\right)^n (1+t)^{a-1} \, dt \right| \\ &\leq (n+1)\left|\binom{a}{n+1}\right| \left| \int_0^x x^n \, dt \right| \\ &= (n+1)\left|\binom{a}{n+1}\right| x^{n+1}. \end{aligned}$$

Applying the result from Exercise 27.17 below, we are able to conclude that

$$\lim_{n\to\infty}(n+1)\binom{a}{n+1}x^{n+1} = 0,$$

and it follows that $R_n(x) \to 0$ when $0 < x < 1$.

Proceeding to the case in which $-1 < x < 0$, note that when $x < t < 0$, we have

$$0 < \frac{t-x}{1+t} < -x = |x|.$$

Thus, keeping in mind that $a \neq 0$, it follows that

$$\begin{aligned}
|R_n(x)| &= (n+1)\left|\binom{a}{n+1}\right|\left|\int_x^0 \left(\frac{t-x}{1+t}\right)^n (1+t)^{a-1}dt\right| \\
&\leq (n+1)\left|\binom{a}{n+1}\right||x|^n \int_x^0 (1+t)^{a-1}dt \\
&= (n+1)\left|\binom{a}{n+1}\right||x|^n \frac{1-(1+x)^a}{a} \\
&= (n+1)\left|\binom{a}{n+1}\right||x|^{n+1} \frac{1-(1+x)^a}{a|x|}.
\end{aligned}$$

Since $\frac{1-(1+x)^a}{a|x|}$ is a constant, we may again apply the result from Exercise 27.17 to conclude that $R_n(x) \to 0$ when $-1 < x < 0$.

Since the Taylor remainder converges to zero for all x in $(-1,1)$ when $a < 1$, the binomial series $\sum_{n=0}^{\infty}\binom{a}{n}x^n$ converges on $(-1,1)$ to $(1+x)^a$ when $a < 1$. □

Exercise 27.17. Use the result from Exercise 10.27 to show that $\lim_{n\to\infty} n\binom{a}{n}x^n = 0$ when $|x| < 1$.

Exercise 27.18. Complete the proof of Theorem 27.20 by establishing the desired convergence holds for the case in which $a > 1$.

---

28 The trigonometric functions

We show that the trigonometric functions can be analytically defined via power series, provide a rigorous definition for the number π, and show both the cosine and sine functions are periodic of period 2π. We also prove π is irrational. Thus, in this chapter we complete an analytic study of the trigonometric functions that is consistent with the geometric approach familiar from earlier mathematical experiences.

The cosine and sine functions

Consider the power series

$$1 - \frac{x^2}{2} + \frac{x^4}{4!} - \frac{x^6}{6!} + \cdots = \sum_{n=0}^{\infty} a_n x^n,$$

where

$$a_n = \begin{cases} \frac{(-1)^{n/2}}{n!}, & \text{if } n \text{ is even;} \\ 0, & \text{if } n \text{ is odd.} \end{cases}$$

For each natural number n and each real number x, we may observe that

$$|a_n x^n| \leq \left|\frac{x^n}{n!}\right|,$$

so as the series $\sum_{n=0}^{\infty} \frac{x^n}{n!}$ converges (Theorem 27.18), it follows by comparison that the series $\sum_{n=0}^{\infty} a_n x^n$ does as well. For convenience, we re-index the series and define the function $f : \mathbb{R} \to \mathbb{R}$ so that

$$f(x) = 1 - \frac{x^2}{2} + \frac{x^4}{4!} - \frac{x^6}{6!} + \cdots = \sum_{n=0}^{\infty} \frac{(-1)^n x^{2n}}{(2n)!},$$

which is differentiable, being a power series. Differentiating yields

$$f'(x) = -x + \frac{x^3}{3!} - \frac{x^5}{5!} + \frac{x^7}{7!} + \cdots = \sum_{n=0}^{\infty} \frac{(-1)^{n+1} x^{2n+1}}{(2n+1)!}$$

and

$$f''(x) = -1 + \frac{x^2}{2!} - \frac{x^4}{4!} + \frac{x^6}{6!} + \cdots = \sum_{n=0}^{\infty} \frac{(-1)^{n+1} x^{2n}}{(2n)!},$$

so that

https://doi.org/10.1515/9783111429564-028

$$f''(x) + f(x) = 0$$

for every real number x, which means f is a solution to the differential equation

$$y'' + y = 0.$$

Actually, since we also have

$$f(0) = 1$$

and

$$f'(0) = 0,$$

the function f is a solution to the initial value problem

$$\begin{cases} y'' + y = 0, \\ y(0) = 1, \\ y'(0) = 0. \end{cases}$$

Theorem 16.18 states that this initial value problem has a unique solution, but at that time, we presented only a partial proof of the theorem, with a commitment to later demonstrate the genuine existence of a solution. With what has been shown here about the function f, that commitment is now fulfilled. Using the terminology also introduced in the statement of Theorem 16.18, the function f is revealed to be the *cosine* function. The theorem also defines the *sine* function by

$$\sin(x) = -\cos'(x),$$

and so it follows that

$$\sin(x) = -f'(x) = x - \frac{x^3}{3!} + \frac{x^5}{5!} - \frac{x^7}{7!} + \cdots = \sum_{n=0}^{\infty} \frac{(-1)^n x^{2n+1}}{(2n+1)!}.$$

Definition 28.1 (Definitions of cosine and sine). The **cosine** and **sine** functions, abbreviated cos and sin, respectively, are defined for every real number x so that

$$\cos(x) = 1 - \frac{x^2}{2} + \frac{x^4}{4!} - \frac{x^6}{6!} + \cdots = \sum_{n=0}^{\infty} \frac{(-1)^n x^{2n}}{(2n)!}$$

and

$$\sin(x) = x - \frac{x^3}{3!} + \frac{x^5}{5!} - \frac{x^7}{7!} + \cdots = \sum_{n=0}^{\infty} \frac{(-1)^n x^{2n+1}}{(2n+1)!}.$$

Exercise 28.1. Use the series definition of the sine function to verify directly that the sine function is a solution to the initial value problem

$$\begin{cases} y'' + y = 0, \\ y(0) = 0, \\ y'(0) = 1. \end{cases}$$

For convenience, we consolidate into a single theorem several results concerning the cosine and sine functions already established in Theorems 16.18 through 16.21. We can draw upon these facts to deduce further properties of the cosine and sine.

Theorem 28.2 (Elementary properties of the cosine and sine functions).

(1) $\cos(0) = 1$.

(2) $\sin(0) = 0$.

(3) *For every real number x, we have*

$$\cos'(x) = -\sin(x)$$

and

$$\sin'(x) = \cos(x).$$

(4) *For every real number x, we have*

$$\cos(-x) = \cos(x)$$

and

$$\sin(-x) = \sin(x).$$

(5) (Pythagorean identity) *For every real number x, we have*

$$\cos^2(x) + \sin^2(x) = 1.$$

(6) *For all real numbers x and y, we have*

$$\cos(x + y) = \cos(x)\cos(y) - \sin(x)\sin(y)$$

and

$$\sin(x + y) = \sin(x)\cos(y) + \cos(x)\sin(y).$$

(7) *Each of the cosine and sine functions has range $[-1, 1]$. Hence,*

$$-1 \le \cos(x) \le 1$$

and

$$-1 \leq \sin(x) \leq 1.$$

Exercise 28.2. Use term-by-term differentiation of the series defining the cosine and sine functions to verify the differentiation formulas $\cos'(x) = -\sin(x)$ and $\sin'(x) = \cos(x)$.

Periodic functions

Probably the most characteristic feature of the cosine and sine functions is their periodicity, manifested in the repeating images their graphs produce. Our next goal is to establish that these functions are periodic, each having period 2π. To obtain the numerical value of the period requires a definition for π, and we intend to offer an analytic one that is easily accepted based on geometric considerations.

Assuming A is a nonempty subset $\mathbb{R}$, a function $f : A \to \mathbb{R}$ is **periodic with period** p if there is a nonzero number p for which $f(x + p) = f(x)$ for all x in A. Note that if f is periodic with period p, it is also periodic with period kp for any nonzero integer k (Exercise 28.3). When p is positive and f is periodic with period p, but is not periodic with period q for any positive number q that is less than p, we call p the **fundamental period** of f. A function is **periodic** if it is periodic with some nonzero period.

Graphically, a periodic function repeats the same image over and over again as we look from left-to-right across the xy-plane (Figure 28.1) and the fundamental period, if it

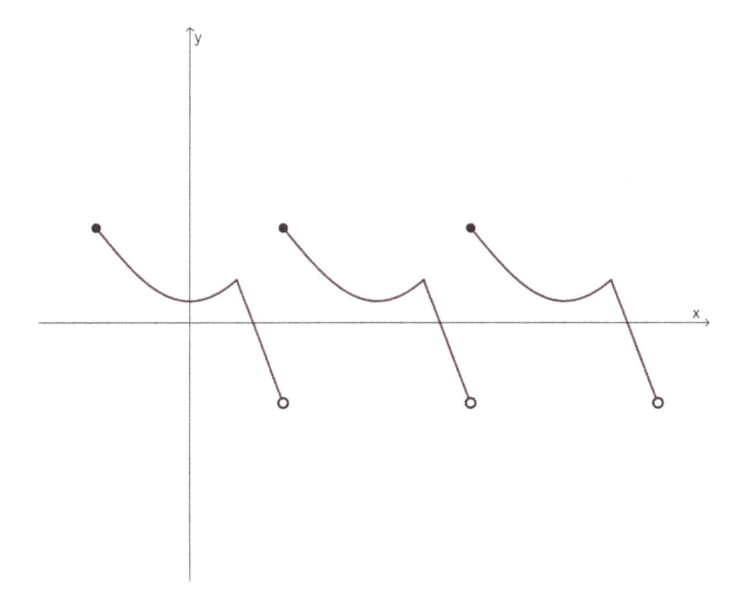

Figure 28.1: Graph of a periodic function.

exists, can be interpreted as the smallest length of an input interval over which we can see one complete image of what is being repeatedly traced out.

Exercise 28.3. Show that if the function f is periodic with period p, then f is also periodic with period kp for any nonzero integer k.

Exercise 28.4. Give an example of a periodic function that does not have a fundamental period.

A **zero** of a real-valued function f is an element a of the domain of f for which $f(a) = 0$.

Theorem 28.3. *The cosine function has a least positive zero. If this least positive zero of the cosine function is denoted by z, then* $\sin(z) = 1$*, and both the cosine and sine functions are periodic with fundamental period* $4z$*.*

Proof. As the sine function is differentiable, we may apply Theorem 16.7, the mean value theorem, to obtain a number c for which $0 < c < 2$ and

$$\sin'(c) = \cos(c) = \frac{\sin(2) - \sin(0)}{2}.$$

Consequently, as $\sin(0) = 0$, we have

$$2|\cos(c)| = |\sin(2)| \leq 1,$$

from which it follows that

$$\cos^2(c) \leq \frac{1}{4}.$$

Using the sum formula for the cosine and the Pythagorean identity, both given in Theorem 28.2, we may then conclude that

$$\cos(2c) = \cos(c + c) = \cos^2(c) - \sin^2(c) = 2\cos^2(c) - 1 \leq -\frac{1}{2} < 0.$$

As $\cos(0) = 1$ and $\cos(2c) < 0$, by Theorem 14.1, the intermediate value theorem, there exists a zero of the cosine function between 0 and $2c$. Having determined that the cosine has a positive zero, we can use Exercise 28.5 below to conclude that the cosine has a least positive zero z.

According to Exercise 14.15, the continuity of the cosine and the sine means that when we are able to conclude either of these functions is strictly monotone on an open interval (a, b), the function is actually strictly monotone on the closed interval $[a, b]$. We apply this fact several times in what follows.

As $\cos(z) = 0$ and $\cos^2(z) + \sin^2(z) = 1$, it follows that $\sin(z) = \pm 1$. But because $\sin'(x) = \cos(x) > 0$ for all x in $(0, z)$, it follows that the sine function is increasing on $[0, z]$, so as $\sin(0) = 0$ as well, we may conclude that $\sin(z) = 1$.

That both the cosine and sine functions must be periodic with period $4z$ follows from Exercise 28.6 below. Note that as a consequence, we must have

$$\cos(4z) = \cos(0) = 1.$$

To complete the proof, we must show neither the cosine nor the sine has period t for some positive number t that is less than $4z$. We do so for the cosine, leaving the argument for the sine as Exercise 28.8.

To show the cosine has fundamental period $4z$, it suffices to show that $\cos(x) \neq 1$ for all x in $(0, 4z)$. We show the cosine is decreasing on $[0, 2z]$, which, as $\cos(0) = 1$, implies that $\cos(x) \neq 1$ for all x in $(0, 2z]$. We also show the cosine is increasing on $[2z, 4z]$, which, as $\cos(4z) = 1$, implies that $\cos(x) \neq 1$ for all x in $(2z, 4z]$. By Theorem 16.9, the monotonicity theorem, to reach these conclusions, it is enough to show that $\cos'$ is negative on $(0, 2z)$ and positive on $(2z, 4z)$.

In demonstrating that $\sin(z) = 1$, we determined that the sine function is increasing on $[0, z]$, from the value 0 at $x = 0$ to the value 1 at $x = z$. Thus, $\sin(x) > 0$ for all x in $(0, z)$, which implies

$$\cos'(x) = -\sin(x) < 0$$

for all x in $(0, z)$. Also note that

$$\cos'(z) = -\sin(z) = -1 < 0.$$

Furthermore, if x is in $(z, 2z)$, we can use the result from Exercise 28.6(b) to conclude that

$$\sin(x) = \sin\big((x - z) + z\big) = \cos(x - z) > 0,$$

since $x - z$ is in $(0, z)$, and as a consequence, we have

$$\cos'(x) = -\sin(x) < 0.$$

We have now established that $\cos'(x) < 0$ and $\sin(x) > 0$ for all x in $(0, 2z)$.

Now if x is in $(2z, 4z)$, we can use the results from Exercise 28.6(b) and 28.6(a) to conclude that

$$\begin{aligned}\cos'(x) &= -\sin(x) \\ &= -\sin\big((x - 2z) + z + z\big) \\ &= -\cos\big((x - 2z) + z\big)\end{aligned}$$

$$= \sin(x - 2z)$$
$$> 0,$$

since $x - 2z$ is in $(0, 2z)$.

Having demonstrated that $\cos'$ is negative on $(0, 2z)$ and positive on $(2z, 4z)$, the plan described earlier permits us to conclude that the cosine function does not take the value 1 at any point in $(0, 4z)$. So, as the cosine is periodic with period $4z$, it follows that $4z$ is the fundamental period of the cosine. □

Exercise 28.5. Suppose the function f is continuous on the interval $[0, b]$, with $f(0) \neq 0$ and $f(c) = 0$ for some c in $(0, b]$. Prove that there is a least positive number z for which $f(z) = 0$.

Exercise 28.6. Take z to be the least positive zero of the cosine function. Show that the following equations are true for every real number x. *Hint*: Note that $x + 2z = (x + z) + z$ and $x + 4z = (x + 2z) + 2z$.

(a) $\cos(x + z) = -\sin(x)$.
(b) $\sin(x + z) = \cos(x)$.
(c) $\cos(x + 2z) = -\cos(x)$.
(d) $\cos(x + 4z) = \cos(x)$.

Exercise 28.7. Again, let z be the least positive zero of the cosine function. Verify each of the following.

(a) $\cos(2z) = -1$.
(b) $\sin(2z) = 0$.
(c) $\cos(3z) = 0$.
(d) $\sin(3z) = -1$.

Exercise 28.8. Among other things, your work in Exercises 28.6 and 28.7 permits us to deduce that the sine function is periodic with period $4z$, where z is the least positive zero of the cosine function. Show precisely how and then show that $4z$ is the fundamental period of the sine function.

The number π

From our geometrically-oriented study of the trigonometric functions in precalculus and calculus courses, we recognize the numerical value of the period of the cosine and sine functions as being 2π (see Appendix D). Thus, taking z to be the least positive zero of the cosine, we want to prove that $4z = 2\pi$ or, equivalently, $z = \frac{\pi}{2}$. Doing so requires a definition for π.

The usual way of defining π, as the ratio of a circle's circumference to its diameter, is unsuitable for our purposes, being expressed in terms of the geometric notions of *circle*, *diameter of a circle*, and *circumference of a circle*, which are outside the scope of mathematical analysis. We want an analytic definition, one that is constructed using only results we have established based on the axioms we formulated for the real number system all the way back in Chapters 1 and 4.

At the same time, we would like our definition to be consistent with geometric principles with which we are familiar. For instance, we learn in geometry that the area A of a circle is given by the formula

$$A = \pi r^2,$$

where r is the radius of the circle, and so the area of a circle having radius 1 is π. Because the integral can be interpreted as representing area, we shall define π by means of an expression that represents the area of a circle of radius 1 by means of an integral. The unit circle, the circle of radius 1 centered at the origin, does nicely in this regard. Its equation is

$$x^2 + y^2 = 1,$$

and so

$$y = \sqrt{1 - x^2},$$

describes the points lying on the upper half of the unit circle (Figure 28.2).

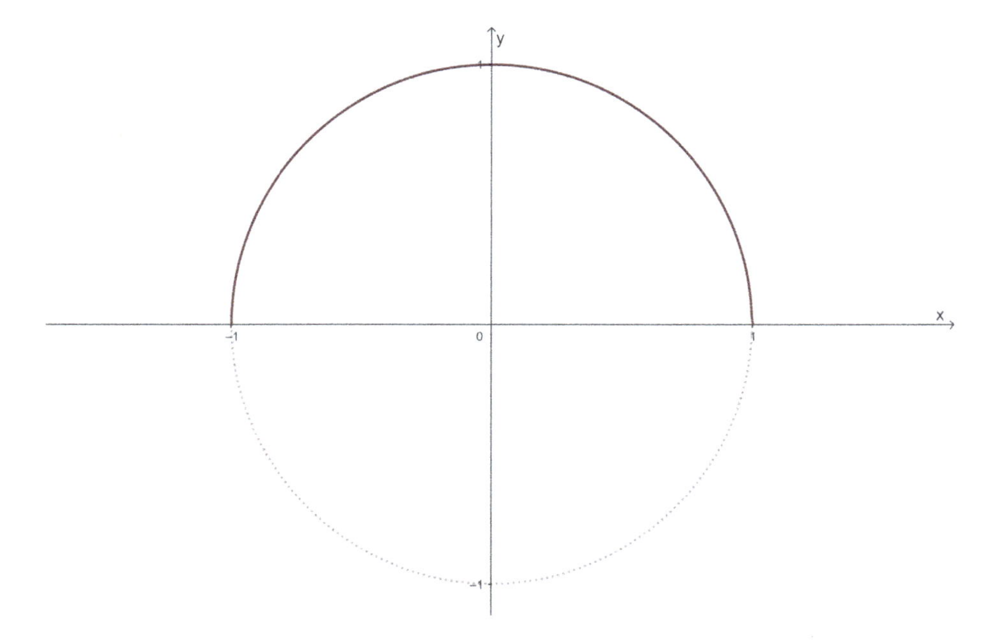

Figure 28.2: Graph of the unit circle, with the upper half highlighted.

The following definition takes into account that to obtain the entire area of the unit circle, we may calculate twice the area between the upper half and the x-axis.

Definition 28.4 (Definition of π). We define the real number π so that

$$\pi = 2\int_{-1}^{1} \sqrt{1-x^2}dx.$$

This definition, while informed by geometry, is purely analytic because it is expressed using only the analytic concept of the Riemann integral. The continuity of the integrand guarantees the integral in the definition exists.

The proof of the following theorem is assigned as Exercise 28.9.

Theorem 28.5. *The fundamental period of both the cosine and sine functions is* 2π.

Exercise 28.9. To prove Theorem 28.5, it suffices to establish that

$$\int_{-1}^{1} \sqrt{1-x^2}dx = z,$$

where z is the least positive zero of the cosine function.

(a) Taking $x = \cos(t)$, verify that all of the hypotheses of Theorem 20.28, the change of variables theorem, are satisfied, resulting in

$$\int_{-1}^{1} \sqrt{1-x^2}dx = \int_{2z}^{0} \sqrt{1-\cos^2(t)}\big(-\sin(t)\big)dt.$$

(b) Show that $\int_{2z}^{0} \sqrt{1-\cos^2(t)}(-\sin(t))dt = z$.

Exercise 28.10. The graph of the cosine function along the interval $[0, 2\pi]$ is displayed in Figure 28.3. Discuss how the graph is assembled based on conclusions we reached across Theorem 28.2, the proof of Theorem 28.3, and Exercise 28.7, along with an investigation of the subintervals of $[0, 2\pi]$ on which the graph is concave up and on which it is concave down (see Exercise 16.20 for the definitions of *concave up* and *concave down*).

Exercise 28.11. The graph of the sine function along the interval $[0, 2\pi]$ is displayed in Figure 28.4. Discuss how the graph is assembled, including an investigation of those subintervals of $[0, 2\pi]$ on which the function is increasing and those on which it is decreasing, as well as those subintervals on which the graph is concave up and those on which it is concave down.

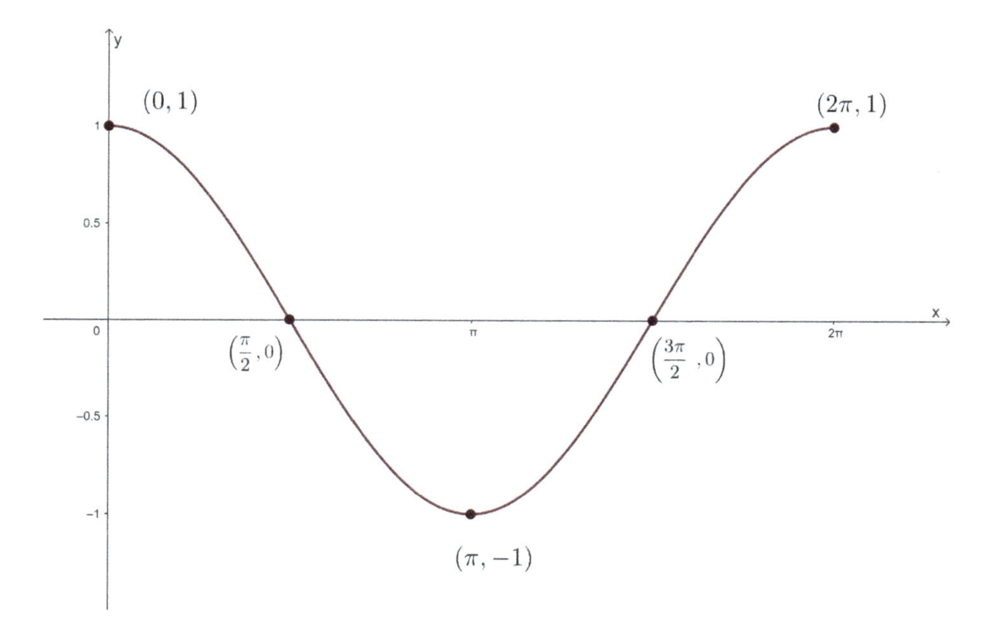

Figure 28.3: Graph of one period of the cosine function.

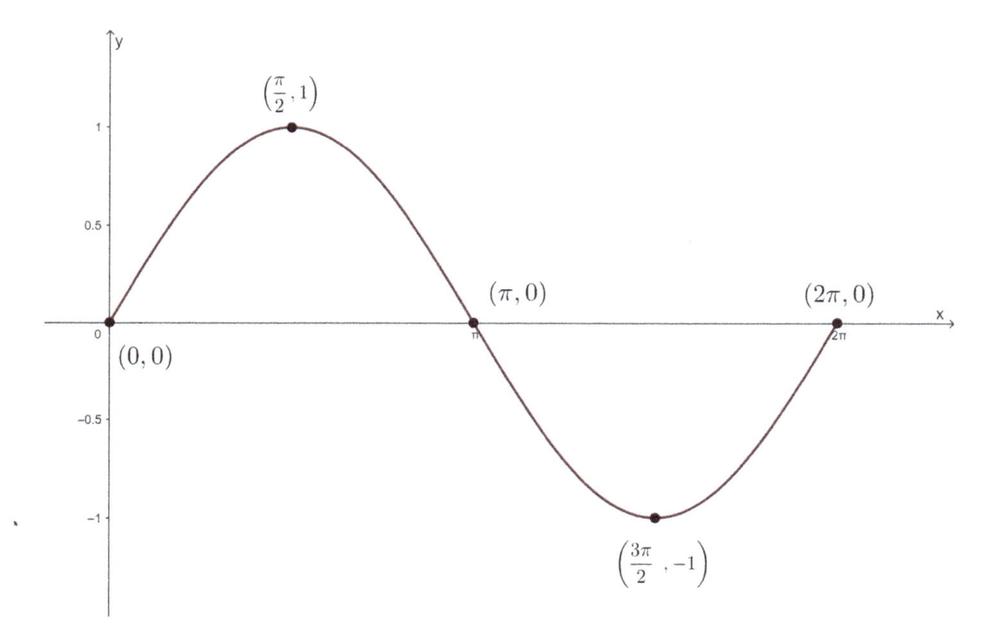

Figure 28.4: Graph of one period of the sine function.

Exercise 28.12. Show that the number a is a zero of the cosine function if and only if $a = \frac{\pi}{2} + n\pi$ for some integer n. Then show that the number a is a zero of the sine function if and only if $a = n\pi$ for some integer n.

Exercise 28.13. Prove that $\sin(\frac{\pi}{4}) = \cos(\frac{\pi}{4}) = \frac{1}{\sqrt{2}}$.

Exercise 28.14. Verify that $\int_0^\pi \sin = 2$.

The number π is often approximated as 3.14. We want to show this approximation is consistent with our definition for π. One approach is to apply one or another of the numerical integration techniques discussed in Chapter 21, for example, a trapezoidal approximating sum or Simpson's rule, to approximate the value of the integral

$$\int_{-1}^{1} \sqrt{1-x^2}dx$$

employed in Definition 28.4 to define π. While this is feasible, it is complicated by the fact that the integrand is not differentiable at ± 1, so we cannot easily obtain estimates for the error introduced by such approximations, because the error estimate formulas (see Theorems 21.5, 21.7, and 21.10) we obtained all require differentiability of the integrand over the entire interval of integration.

Instead, we use an infinite series to verify the approximate value 3.14 for π, but not until after introducing the inverse trigonometric functions. Here is the specific theorem we shall soon prove.

Theorem 28.6. *The real number π satisfies the inequality*

$$3.141 < \pi < 3.142$$

Thus, to the nearest hundredth, $\pi \approx 3.14$.

The tangent function

You are familiar with other trigonometric functions beyond the cosine and sine. In particular, the **tangent** function, abbreviated tan, is defined so that

$$\tan(x) = \frac{\sin(x)}{\cos(x)},$$

for every real number x such that $x \neq \frac{\pi}{2} + n\pi$ for any integer n (to avoid division by zero).

Theorem 28.7 (Elementary properties of the tangent function).
(1) $\tan(0) = 0$.
(2) *For every x in the domain of the tangent,* $\tan(-x) = -\tan(x)$.
(3) *The tangent function is differentiable on its domain, with* $\tan'(x) = \frac{1}{\cos^2(x)}$.
(4) *The range of the tangent function is* $\mathbb{R}$.
(5) *The tangent function is periodic with fundamental period* π.

Proof. The proofs of (1), (2), (3), and (5) are left as exercises. To demonstrate (4), first note that from (1) it is apparent that 0 is in the range of the tangent function. It also follows from (2) that whenever a positive number is in the range of the tangent, so is its opposite. So we need only show that every positive number is in the range.

From (3) we see that the derivative of the tangent is positive on $(0, \frac{\pi}{2})$, so the tangent is increasing on this interval. Moreover, as the tangent is differentiable on this interval, it is also continuous. Thus, because of Theorem 14.1, the intermediate value theorem, and the fact that $\tan(0) = 0$, it suffices to show that

$$\lim_{x \to \frac{\pi}{2}^-} \tan(x) = \infty.$$

Consider any positive number B. As the sine function is increasing on the interval $[1, \frac{\pi}{2})$ and the cosine function takes only positive values on this interval, using the fact that $1 < \frac{\pi}{2}$ (can you provide an analytic argument justifying this inequality?), it follows that, for all x in $[1, \frac{\pi}{2})$, we have

$$\tan(x) = \frac{\sin(x)}{\cos(x)} \geq \frac{\sin(1)}{\cos(x)}.$$

Furthermore, as $\lim_{x \to \frac{\pi}{2}^-} \cos(x) = 0$ and $\sin(1) > 0$, we can choose t in $[1, \frac{\pi}{2})$ so that $\cos(t) < \frac{\sin(1)}{B}$. It then follows that

$$\tan(t) = \frac{\sin(t)}{\cos(t)} \geq \frac{\sin(1)}{\cos(t)} > B.$$

Thus, $\lim_{x \to \frac{\pi}{2}^-} \tan(x) = \infty$. □

Exercise 28.15. Complete the proof of Theorem 28.7.

Exercise 28.16. Find a formula for $f'(x)$ if the function f is defined by $f(x) = \sqrt[3]{x \tan(\pi x)}$.

Exercise 28.17. Show that the tangent function is increasing on the interval $(-\frac{\pi}{2}, \frac{\pi}{2})$. Then, as a consequence of (4) and (5) of Theorem 28.7, we may conclude that the tangent function restricted to $(-\frac{\pi}{2}, \frac{\pi}{2})$ is a bijection onto $\mathbb{R}$.

The inverse trigonometric functions

Because the trigonometric functions sine, cosine, and tangent are periodic, none of them is one-to-one. However, suitable restrictions of the domains of these functions produce one-to-one functions which do have inverses that are functions. Specifically, we define

(1) the **arcsine function**, denoted arcsin, to be the inverse of the restriction of the sine function to the interval $[-\frac{\pi}{2}, \frac{\pi}{2}]$;
(2) the **arccosine function**, denoted arccos, to be the inverse of the restriction of the cosine function to the interval $[0, \pi]$;
(3) and the **arctangent function**, denoted arctan, to be the inverse of the restriction of the tangent function to the interval $(-\frac{\pi}{2}, \frac{\pi}{2})$.

As an informal exercise, you should look back to see exactly how our earlier work with the cosine, sine, and tangent functions shows that they are one-to-one on the intervals specified here.

Theorem 28.8 (Derivatives of the inverse trigonometric functions).

(1) *The arcsine function is differentiable on the interval* $(-1, 1)$ *and*

$$\arcsin'(x) = \frac{1}{\sqrt{1-x^2}}.$$

(2) *The arccosine function is differentiable on the interval* $(-1, 1)$ *and*

$$\arccos'(x) = \frac{-1}{\sqrt{1-x^2}}.$$

(3) *The arctangent function is differentiable on* $\mathbb{R}$ *and*

$$\arctan'(x) = \frac{1}{1+x^2}.$$

Proof. We prove (1), leaving the proofs of (2) and (3) as exercises. Define $f : [-\frac{\pi}{2}, \frac{\pi}{2}] \to \mathbb{R}$ so that $f(x) = \sin(x)$. Then $f^{-1}(x) = \arcsin(x)$ and $f'(x) = \cos(x)$. Thus, if we let $t = \arcsin(x)$, it follows that $\sin(t) = x$. Applying Theorem 15.15 concerning the derivatives of inverse functions, we then have

$$\arcsin'(x) = (f^{-1})'(x) = \frac{1}{f'(f^{-1}(x))} = \frac{1}{\cos(\arcsin(x))} = \frac{1}{\cos(t)} = \frac{1}{\sqrt{1-\sin^2(t)}} = \frac{1}{\sqrt{1-x^2}},$$

when $-1 < x < 1$. □

Exercise 28.18. Complete the proof of Theorem 28.8.

We now obtain the Maclaurin series for the arctangent function. It is this series that we can use to verify the accuracy of the approximation 3.14 for the value of π.

Theorem 28.9 (Series representation of the arctangent). *For any x in the interval $[-1,1]$, we have*

$$\arctan(x) = x - \frac{x^3}{3} + \frac{x^5}{5} - \frac{x^7}{7} + \cdots = \sum_{n=0}^{\infty} \frac{(-1)^n x^{2n+1}}{2n+1}.$$

Proof. Using the formula for the sum of a convergent geometric series, we have

$$\frac{1}{1+t^2} = \frac{1}{1-(-t^2)} = 1 - t^2 + t^4 - t^6 + \cdots = \sum_{n=0}^{\infty} (-1)^n t^{2n}$$

when $-1 < -t^2 < 1$, that is, when $-1 < t < 1$. The series here is a power series, hence, by Corollary 27.3, for any x in $(-1,1)$, it follows that

$$\int_0^x \frac{1}{1+t^2}\,dt = \sum_{n=1}^{\infty} \int_0^x (-1)^n t^{2n}\,dt.$$

Upon evaluating the integrals, we conclude that

$$\arctan(x) = x - \frac{x^3}{3} + \frac{x^5}{5} - \frac{x^7}{7} + \cdots = \sum_{n=0}^{\infty} \frac{(-1)^n x^{2n+1}}{2n+1}$$

when x is in $(-1,1)$. Also, if $x = \pm 1$, the series $\sum_{n=0}^{\infty} \frac{(-1)^n x^{2n+1}}{2n+1}$ converges by Theorem 23.10, the alternating series test, and so since Theorem 27.5, Abel's theorem, tells us that a power series is continuous on its entire interval of convergence, the equality

$$\arctan(x) = x - \frac{x^3}{3} + \frac{x^5}{5} - \frac{x^7}{7} + \cdots = \sum_{n=0}^{\infty} \frac{(-1)^n x^{2n+1}}{2n+1}$$

may be extended to all x in $[-1,1]$. The uniqueness of power series representations tells us that this series is the Maclaurin series for the arctangent function. □

In Exercise 28.13, you showed that $\sin(\frac{\pi}{4}) = \cos(\frac{\pi}{4})$, and so it follows that

$$\tan\left(\frac{\pi}{4}\right) = \frac{\sin(\frac{\pi}{4})}{\cos(\frac{\pi}{4})} = 1,$$

that is,

$$\arctan(1) = \frac{\pi}{4}.$$

Using the series representation of the arctangent function obtained in Theorem 28.9, we see that

$$\arctan(1) = 1 - \frac{1}{3} + \frac{1}{5} - \frac{1}{7} + \cdots = \sum_{n=0}^{\infty} \frac{(-1)^n}{2n+1}$$

and we may conclude that

$$\pi = 4 \sum_{n=0}^{\infty} \frac{(-1)^n}{2n+1}.$$

Thus, the partial sums of the series $\sum_{n=0}^{\infty} \frac{(-1)^n}{2n+1}$ can be used to help us approximate π. For each nonnegative integer n, let $s_n = \sum_{k=0}^{n} \frac{(-1)^k}{2k+1}$, the nth partial sum of this series. Observe that

$$s_{2(n+1)+1} = s_{2n+1} + \frac{1}{4n+5} - \frac{1}{4n+7} > s_{2n+1},$$

so the subsequence (s_{2n+1}) of (s_n) is increasing. Also observe that

$$s_{2(n+1)} = s_{2n} - \frac{1}{4n+3} + \frac{1}{4n+5} < s_{2n},$$

so the subsequence (s_{2n}) of (s_n) is decreasing. The monotone convergence of both sequences (s_{2n+1}) and (s_{2n}) to $\frac{\pi}{4}$ has therefore revealed that

$$s_1 < s_3 < s_5 < \cdots < \frac{\pi}{4} < \cdots < s_4 < s_2 < s_0.$$

Computer calculations reveal that

$$s_{2454} \approx 0.785296288911$$

and

$$s_{2455} \approx 0.785499996387,$$

so that

$$\frac{3.141}{4} = 0.78525 < s_{2455} < \frac{\pi}{4} < s_{2454} < 0.78550 = \frac{3.142}{4},$$

from which it follows that

$$3.141 < \pi < 3.142.$$

We have thus proved Theorem 28.6 and established the legitimacy of the approximation 3.14 for π.

The irrationality of π

The first proof that the number π is irrational was offered by Johann Lambert in 1761. We present Ivan Niven's proof, published in the Bulletin of the American Mathematical Society in 1947 [5], asking you to fill in some of the details.

Theorem 28.10 (Irrationality of π). *The number π is irrational.*

Proof. Suppose to the contrary that $\pi = \frac{a}{b}$ for some integers a and b, which we may assume are positive. For each natural number n, define the function $f_n : \mathbb{R} \to \mathbb{R}$ so that

$$f_n(x) = \frac{1}{n!}x^n(a - bx)^n.$$

First, we show that for every natural number m, the number $f_n^{(m)}(0)$ is an integer. Keep in mind that as f_n is a polynomial function, so are all of its derivatives, and the value of a polynomial function at the input 0 is just the constant term of the polynomial. Applying Theorem 3.23, the binomial theorem, we note that

$$x^n(a - bx)^n = x^n \sum_{k=0}^{n} \binom{n}{k} a^{n-k}(-bx)^k = \sum_{k=0}^{n} c_k x^{n+k},$$

where $c_k = \binom{n}{k} a^{n-k}(-b)^k$ is an integer. Therefore,

$$f_n(x) = \frac{1}{n!} \sum_{k=0}^{n} c_k x^{n+k} = \frac{1}{n!}(c_0 x^n + c_1 x^{n+1} + c_2 x^{n+2} + \cdots + c_n x^{2n}).$$

Because the degree of f_n is $2n$, it follows that $f_n^{(m)}$ is the zero polynomial when $m > 2n$, in which case $f_n^{(m)}(0) = 0$. Also, as f_n has no nonzero term for which x occurs to a power less than n, when $m < n$, it follows that the constant term of $f_n^{(m)}$ is zero, in which case $f_n^{(m)}(0) = 0$. All that remains is the situation where $n \leq m \leq 2n$, which we leave as Exercise 28.19(a). After completing that exercise, we will have demonstrated that $f_n^{(m)}(0)$ is an integer for every natural number m.

Next, for each natural number n, define the function $F_n : \mathbb{R} \to \mathbb{R}$ so that

$$F_n(x) = f_n(x) - f_n^{(2)}(x) + f_n^{(4)}(x) - \cdots + (-1)^n f_n^{(2n)}(x) = \sum_{k=0}^{n} (-1)^k f_n^{(2k)}(x).$$

We want to show that $F_n(\pi) + F_n(0)$ is an integer. Since $f_n(0) = 0$ and $f_n^{(m)}(0)$ is an integer for every natural number m, it is certainly true that $F_n(0)$ is an integer. In Exercise 28.19(b), you are asked to show that for every real number x, we have

$$f_n(\pi - x) = f_n(x).$$

Hence, for every natural number m, it follows via the chain rule for calculating derivatives that

$$f_n^{(m)}(\pi - x) = (-1)^m f_n^{(m)}(x).$$

In particular, then, we have

$$f_n(\pi) = f_n(0)$$

and

$$f_n^{(2k)}(\pi) = f_n^{(2k)}(0)$$

for every natural number k. Therefore,

$$F_n(\pi) = \sum_{k=0}^{n} (-1)^k f_n^{(2k)}(\pi) = \sum_{k=0}^{n} (-1)^k f_n^{(2k)}(0) = F_n(0),$$

and as we already showed that $F_n(0)$ is an integer, so is $F_n(\pi)$. Thus, we deduce that $F_n(\pi) + F_n(0)$ is an integer.

In Exercise 28.19(c), we ask you to verify that

$$F_n''(x) + F_n(x) = f_n(x)$$

for every real number x. As a result, it follows that

$$\frac{d}{dx}[F_n'(x)\sin(x) - F_n(x)\cos(x)] = F_n''(x)\sin(x) + F_n(x)\sin(x) = f_n(x)\sin(x).$$

So, using the fundamental theorem of calculus (Theorem 20.3),

$$\int_0^{\pi} f_n(x)\sin(x)dx = [F_n'(x)\sin(x) - F_n(x)\cos(x)]_{x=0}^{\pi} = F_n(\pi) + F_n(0),$$

and as we determined earlier that $F_n(\pi) + F_n(0)$ is an integer, we now know that $\int_0^{\pi} f_n(x)\sin(x)dx$ is an integer.

Because π is the least positive zero of the sine function, it follows that

$$0 < \sin(x) \leq 1$$

for all x in the interval $(0, \pi)$. Also, for each such x, as $\pi = \frac{a}{b}$, we may conclude that

$$0 < a - bx < a,$$

from which it follows that

$$0 < \frac{1}{n!}x^n(a-bx)^n = f_n(x) < \frac{(\pi a)^n}{n!}.$$

Thus, when x is in $(0,\pi)$, we have

$$0 < f_n(x)\sin(x) \le \frac{(\pi a)^n}{n!}$$

and, hence,

$$0 < \int_0^\pi f_n(x)\sin(x)dx \le \int_0^\pi \frac{(\pi a)^n}{n!}dx = \frac{(\pi a)^n}{n!}\cdot\pi,$$

where we have applied the monotonicity of the integral and the result from Exercise 19.1(a).

We now know that $\int_0^\pi f_n(x)\sin(x)dx$ must be a *positive* integer. However, noting that

$$\sum_{n=0}^{\infty}\frac{(\pi a)^n}{n!} = e^{\pi a},$$

the convergence of this series then implies, via the divergence test, Theorem 22.12, that $\frac{(\pi a)^n}{n!} \to 0$, so that, for a large enough value of n, the integral

$$\int_0^\pi f_n(x)\sin(x)dx$$

takes a value less than 1, hence, cannot be a positive integer, thereby producing a contradiction. This contradiction emerged from the assumption that π is rational, so it must be that π is irrational. □

Exercise 28.19. This exercise pertains to the proof of Theorem 28.10, the irrationality of π. The functions f_n and F_n are defined within the proof.

(a) Show that $f_n^{(m)}(0)$ is an integer in the case where $n \le m \le 2n$.

(b) Show that $f_n(\pi - x) = f_n(x)$ for every real number x.

(c) Verify that $F_n''(x) + F_n(x) = f_n(x)$ for every real number x.

---

29 Two remarkable results of Weierstrass

In this closing chapter, we explore two fairly astonishing mathematical facts, both originally proved by Karl Weierstrass. The first is the existence of functions that are continuous at all real numbers but not differentiable at any real number. The second is the ability to approximate arbitrarily closely any function continuous on a closed bounded interval by means of polynomial functions.

A nowhere differentiable continuous function

It seemed inconceivable to the mathematics community of the mid-19th century that a function could be everywhere continuous yet nowhere differentiable. But in 1872, Karl Weierstrass constructed the first example of such a function. Here we present a different example, one that is easier to prove is nowhere differentiable, developed by John McCarthy in 1953 [6].

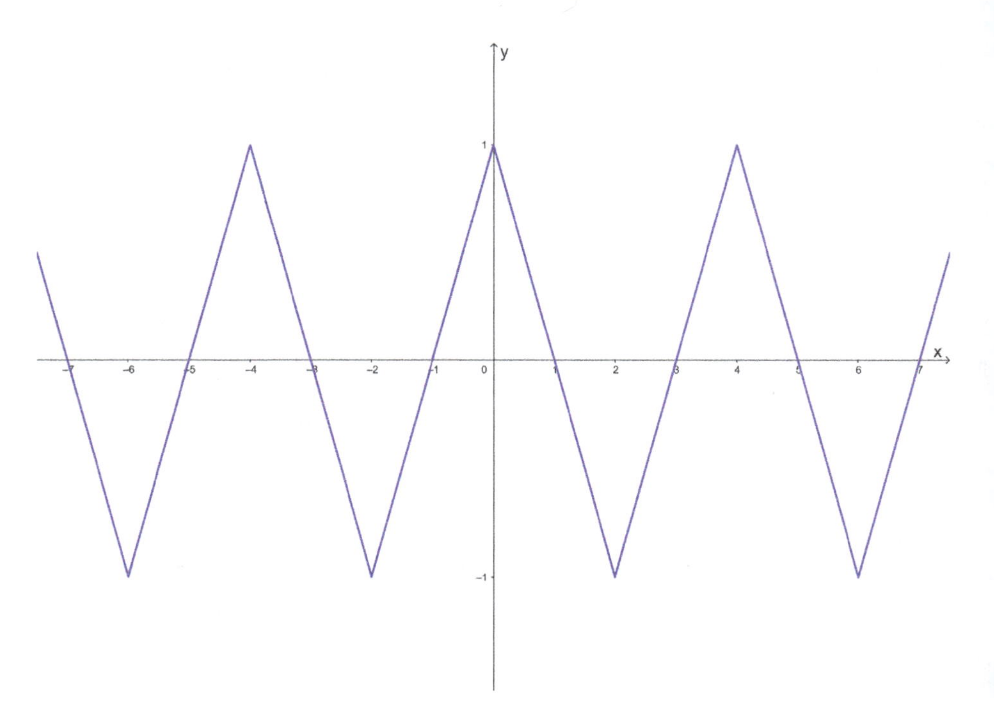

Figure 29.1: Graph of the saw-tooth function h defined in the proof of Theorem 29.1.

Recalling that the absolute value function is not differentiable at 0 because of the sharp bend in its graph, the idea behind McCarthy's function is to modify the "saw-

https://doi.org/10.1515/9783111429564-029

tooth" function, whose graph is depicted in Figure 29.1, by adding to it ever more saw-tooth functions, whose amplitudes and periods shrink to produce microscopically more jagged sharp bends. Figure 29.2 displays part of the graph of the second partial sum of McCarthy's function.

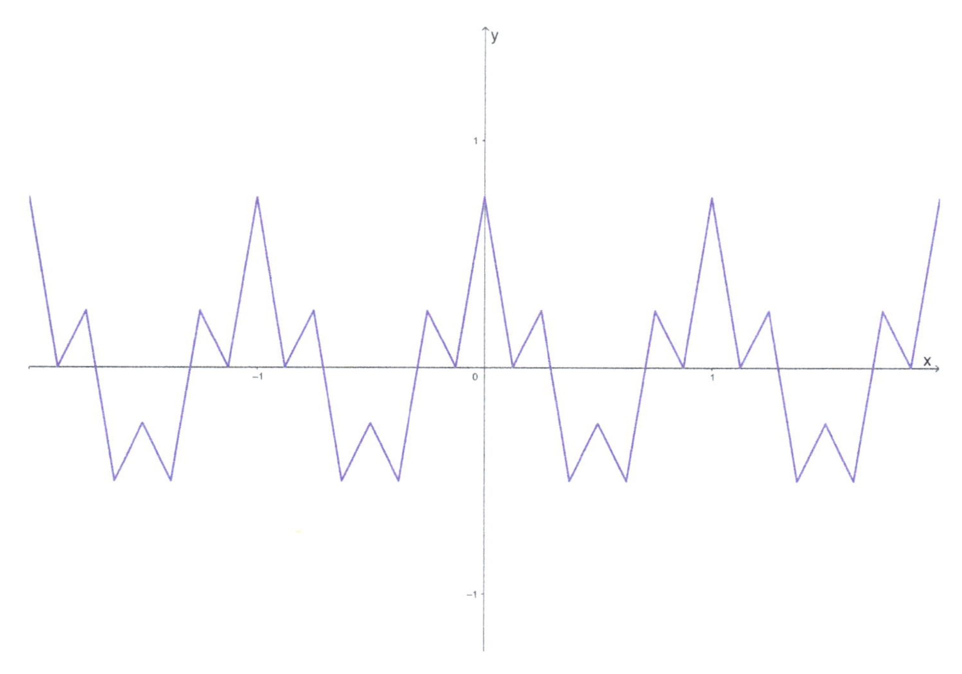

Figure 29.2: Graph of the partial sum $\sum_{k=1}^{2} \frac{1}{2^k} g_k$ of McCarthy's function.

Theorem 29.1. *There exists a function that is continuous at every point of* $\mathbb{R}$*, but which is not differentiable at any point of* $\mathbb{R}$*.*

Proof. First, define the function $h : \mathbb{R} \to [-1, 1]$ so that

$$h(x) = \begin{cases} 1 + x, & \text{if } -2 \le x \le 0; \\ 1 - x, & \text{if } 0 < x \le 2; \end{cases}$$

and, more generally, so that

$$h(x + 4) = h(x).$$

The function h is piecewise linear, continuous, and periodic with fundamental period 4. Its graph (Figure 29.1) consists of linear segments extending along input intervals $[2m, 2(m + 1)]$, where m is an integer. Because the slopes of these segments are either

1 or -1, it follows that if $(p, h(p))$ and $(q, h(q))$ lie on the same segment, then

$$|h(p) - h(q)| = |p - q|,$$

and more generally, for any real numbers p and q, that

$$|h(p) - h(q)| \leq |p - q|.$$

Next, for each natural number n, define the function $g_n : \mathbb{R} \to [-1, 1]$ so that

$$g_n(x) = h(2^{2^n} x).$$

Note that g_n is like the function h except that the fundamental period has been adjusted to $\frac{4}{2^{2^n}} = 2^{2-2^n}$. As a consequence, g_n is a piecewise linear continuous function whose graph consists of segments extending along input intervals $[\frac{2m}{2^{2^n}}, \frac{2(m+1)}{2^{2^n}}]$ having length $\frac{2}{2^{2^n}} = 2^{1-2^n}$, where m is an integer.

McCarthy's function f is then defined on $\mathbb{R}$ so that

$$f(x) = \sum_{n=1}^{\infty} \frac{1}{2^n} g_n(x).$$

Because

$$\left| \frac{1}{2^n} g_n(x) \right| \leq \frac{1}{2^n}$$

for every natural number n, and $\sum_{n=1}^{\infty} \frac{1}{2^n}$ is a convergent geometric series, the Weierstrass M-test (Theorem 26.17) tells us that $\sum_{n=1}^{\infty} \frac{1}{2^n} g_n(x)$ converges uniformly on $\mathbb{R}$. Then as $\frac{1}{2^n} g_n$ is continuous for each n and the uniform limit of continuous functions is also continuous (Theorem 26.21 and Corollary 26.22), we may conclude that f is continuous on $\mathbb{R}$.

Now consider any real number x. To show that f is not differentiable at x, we must show $\lim_{t \to x} \frac{f(t)-f(x)}{t-x}$ does not exist. It suffices, via the sequential characterization of function limits, to find a sequence (a_n) that converges to 0 and for which

$$\left| \frac{f(x + a_n) - f(x)}{a_n} \right| \to \infty.$$

Given any natural number n, the point $(x, g_n(x))$ lies on one of the segments S_n of the graph of g_n (it is possible that this point is the intersection of two adjacent segments, in which case, we can arbitrarily select one of the segments to serve as S_n). Since

$$|(x - 2^{-2^n}) - (x + 2^{-2^n})| = 2 \cdot 2^{-2^n} = 2^{1-2^n},$$

which is precisely the horizontal distance covered by S_n, at least one of the points $(x + 2^{-2^n}, g_n(x + 2^{-2^n}))$ or $(x - 2^{-2^n}, g_n(x - 2^{-2^n}))$ must also lie on the segment S_n. For each n, choose $a_n = \pm 2^{-2^n}$ so that $(x + a_n, g_n(x + a_n))$ lies on S_n.

Consider any natural numbers n and k. In Exercise 29.1, we ask you to show that if $k > n$, then

$$g_k(x + a_n) - g_k(x) = 0,$$

and that if $k = n$, then

$$|g_n(x + a_n) - g_n(x)| = 1.$$

For the case in which $k < n$, note that

$$\begin{aligned}
|g_k(x + a_n) - g_k(x)| &= |h(2^{2^k}(x + a_n)) - h(2^{2^k}x)| \\
&\le |(2^{2^k}x + 2^{2^k}a_n) - 2^{2^k}x| \\
&= 2^{2^k} \cdot 2^{-2^n} \\
&\le 2^{2^{n-1}} \cdot 2^{-2^n} \\
&= 2^{-2^{n-1}}.
\end{aligned}$$

For future reference, then, we see that

$$\begin{aligned}
\left|\sum_{k=1}^{n-1} \frac{1}{2^k}(g_k(x + a_n) - g_k(x))\right| &\le \sum_{k=1}^{n-1} \frac{1}{2^k}|g_k(x + a_n) - g_k(x)| \\
&\le \sum_{k=1}^{n-1} |g_k(x + a_n) - g_k(x)| \\
&\le (n - 1) \cdot 2^{-2^{n-1}} \\
&< 2^{n-1} \cdot 2^{-2^{n-1}} \\
&\le 1.
\end{aligned}$$

Thus, for fixed n, applying Theorem 2.3, the reverse triangle inequality, and the fact that the series defining f is uniformly convergent, we may conclude that

$$\begin{aligned}
\left|\frac{f(x + a_n) - f(x)}{a_n}\right| &= 2^{2^n}\left|\sum_{k=1}^{\infty} \frac{1}{2^k} g_k(x + a_n) - \sum_{k=1}^{\infty} \frac{1}{2^k} g_k(x)\right| \\
&= 2^{2^n}\left|\sum_{k=1}^{\infty} \frac{1}{2^k}(g_k(x + a_n) - g_k(x))\right| \\
&= 2^{2^n}\left|\frac{1}{2^n} \cdot 1 + \sum_{k=1}^{n-1} \frac{1}{2^k}(g_k(x + a_n) - g_k(x))\right|
\end{aligned}$$

$$\geq 2^{2^n}\left(\frac{1}{2^n} - \left|\sum_{k=1}^{n-1} \frac{1}{2^k}(g_k(x+a_n) - g_k(x))\right|\right)$$
$$> 2^{2^n}\left(\frac{1}{2^n} - 2^{n-1}\cdot 2^{-2^{n-1}}\right).$$

In Exercise 29.1(c), you show

$$2^{2^n}\left(\frac{1}{2^n} - 2^{n-1}\cdot 2^{-2^{n-1}}\right) \geq 2^{2^{n-1}}$$

when n is a natural number for which $n \geq 4$. Hence, for such n, we may conclude that

$$\left|\frac{f(x+a_n) - f(x)}{a_n}\right| > 2^{2^{n-1}},$$

which implies

$$\left|\frac{f(x+a_n) - f(x)}{a_n}\right| \to \infty.$$

Therefore, we have shown that f is not differentiable at x, and as x was arbitrarily chosen among the real numbers, f is not differentiable at any real number. ☐

Exercise 29.1. Here we ask you to fill in some details of the proof of Theorem 29.1 that were left to the reader.
(a) Show $g_k(x + a_n) - g_k(x) = 0$ when $k > n$. *Hint*: Keep in mind that h has period 4.
(b) Show $|g_n(x + a_n) - g_n(x)| = 1$. *Hint*: Use the fact that both $(x, g_n(x))$ and $(x + a_n, g_n(x + a_n))$ lie on the same segment of the graph of g_n.
(c) Show $2^{2^n}(\frac{1}{2^n} - 2^{n-1}\cdot 2^{-2^{n-1}}) \geq 2^{2^{n-1}}$ when $n \geq 4$.

The Weierstrass approximation theorem

Theorems 27.13 and 27.14 delineate criteria for an infinitely differentiable function to be represented by its Taylor series within some nontrivial open interval of the function's domain. Because the partial sums of a Taylor series are polynomials, the Taylor polynomials introduced in Chapter 17, we have therefore demonstrated that under appropriate circumstances an infinitely differentiable function may be locally approximated to any desired degree of accuracy by a polynomial function of sufficiently large degree.

What is truly remarkable, however, is that *any function that is continuous on a closed bounded interval* can be approximated to any desired degree of accuracy throughout the interval by a polynomial function of sufficiently large degree. This fact was first established by Weierstrass in 1872 and is called the Weierstrass approximation theorem. We present a version of the proof of this result by Sergei Bernstein published in 1912 [7].

Theorem 29.2 (The Weierstrass approximation theorem). *A function that is continuous on a closed bounded interval can be approximated arbitrarily closely on the interval by a polynomial function.*

That is, given a continuous function $f : [a,b] \to \mathbb{R}$ and a positive number ε, there is a polynomial function $p : [a,b] \to \mathbb{R}$ such that

$$|f(x) - p(x)| < \varepsilon$$

for all x in $[a,b]$.

Unlike Taylor's theorem, the hypothesis of the Weierstrass approximation theorem does not assume any differentiability criteria for the function to which it is applied, only that the function be continuous on a closed bounded interval. So, for instance, even a function such as McCarthy's function from the proof of Theorem 29.1, which is continuous but nowhere differentiable on every closed bounded interval $[a,b]$, can be approximated on $[a,b]$ to any prescribed tolerance by some polynomial function.

Bernstein's approach to the proof introduced the now-famous polynomials that bear his name. Specifically, a **Bernstein polynomial** for a continuous function f defined on $[a,b]$ is a function $B : [a,b] \to \mathbb{R}$ where

$$B(x) = \sum_{k=0}^{n} f\left(\frac{k}{n}\right)\binom{n}{k}x^k(1-x)^{n-k}$$

for some natural number n. It is this type of polynomial function that Bernstein used to approximate the continuous function f in his proof of the Weierstrass approximation theorem. In Exercise 29.2, we ask you to establish several preliminary results needed for the proof.

Exercise 29.2. Let n be a natural number and let x be a real number.

(a) Use Theorem 3.23, the binomial theorem, to show $\sum_{k=0}^{n}\binom{n}{k}x^k(1-x)^{n-k} = 1$.

(b) Show $\frac{1}{n}\sum_{k=0}^{n} k\binom{n}{k}x^k(1-x)^{n-k} = x$.

(c) Show $\frac{1}{n(n-1)}\sum_{k=0}^{n} k(k-1)\binom{n}{k}x^k(1-x)^{n-k} = x^2$ when $n \geq 2$.

(d) Show $n\sum_{k=0}^{n}(x - \frac{k}{n})^2\binom{n}{k}x^k(1-x)^{n-k} = x(1-x)$ when $n \geq 2$.

Proof of Theorem 29.2. We demonstrate the result for functions continuous on the interval $[0,1]$ and you show in Exercise 29.3 how it then follows that the conclusion of the theorem holds for functions continuous on an arbitrary closed bounded interval $[a,b]$.

Suppose the function f is continuous on $[0,1]$ and let ε be a positive number. By Theorem 14.14, the extreme value theorem, there is a positive number B for which

$$|f(x)| \leq B$$

for all x in $[0,1]$. Because a function continuous on a closed bounded interval is uniformly continuous on that interval (Theorem 14.20), f is uniformly continuous on $[0,1]$, so there is a positive number δ such that whenever x and y are in $[0,1]$ with $|x-y|<\delta$, it follows that

$$|f(x)-f(y)|<\frac{\varepsilon}{2}.$$

We may then use the Archimedean property to obtain a natural number n greater than 1 for which

$$\frac{1}{n}<\frac{\delta^2\varepsilon}{B}.$$

Take the function $p:[0,1]\to\mathbb{R}$ to be the Bernstein polynomial defined by

$$p(x)=\sum_{k=0}^{n}f\left(\frac{k}{n}\right)\binom{n}{k}x^k(1-x)^{n-k}.$$

Consider any x in $[0,1]$. Our goal is to show

$$|f(x)-p(x)|<\varepsilon,$$

but in order to do this we separately consider those integers k in $K=\{0,1,2,\dots,n\}$ for which $\frac{k}{n}$ is relatively close to x and those for which that is not the case. By "relatively close" we mean $|x-\frac{k}{n}|<\delta$. Let $C=\{k\in K\mid |x-\frac{k}{n}|<\delta\}$ and let $D=K-C$.

First, suppose $k\in C$. Then the uniform continuity of f on $[0,1]$ permits us to conclude that $|f(x)-f(\frac{k}{n})|<\frac{\varepsilon}{2}$.

Next, suppose $k\in D$. As B is an upper bound for the values f achieves on $[0,1]$, it follows that $|f(x)-f(\frac{k}{n})|\le 2B$. Because k is in D, we have $|x-\frac{k}{n}|\ge\delta$, and since $\frac{1}{n}<\frac{\delta^2\varepsilon}{B}$, it follows that $2B<2n\delta^2\varepsilon$. Hence,

$$\left|f(x)-f\left(\frac{k}{n}\right)\right|\le 2B<2n\delta^2\varepsilon\le 2n\left(x-\frac{k}{n}\right)^2\varepsilon.$$

In Exercise 29.2(a), you showed that

$$\sum_{k=0}^{n}\binom{n}{k}x^k(1-x)^{n-k}=1,$$

where we may observe that all the terms in the summation are nonnegative, as here x is in $[0,1]$, and in Exercise 29.2(d) you showed that

$$n\sum_{k=0}^{n}\left(x-\frac{k}{n}\right)^2\binom{n}{k}x^k(1-x)^{n-k}=x(1-x)$$

when $n \geq 2$. Therefore,

$$
\begin{aligned}
&|f(x) - p(x)| \\
&= \left| f(x) \sum_{k=0}^{n} \binom{n}{k} x^k (1-x)^{n-k} - \sum_{k=0}^{n} f\left(\frac{k}{n}\right) \binom{n}{k} x^k (1-x)^{n-k} \right| \\
&\leq \sum_{k=0}^{n} \left| f(x) - f\left(\frac{k}{n}\right) \right| \binom{n}{k} x^k (1-x)^{n-k} \\
&= \sum_{k \in C} \left| f(x) - f\left(\frac{k}{n}\right) \right| \binom{n}{k} x^k (1-x)^{n-k} + \sum_{k \in D} \left| f(x) - f\left(\frac{k}{n}\right) \right| \binom{n}{k} x^k (1-x)^{n-k} \\
&< \sum_{k \in C} \frac{\varepsilon}{2} \binom{n}{k} x^k (1-x)^{n-k} + \sum_{k \in D} 2n \left(x - \frac{k}{n} \right)^2 \varepsilon \binom{n}{k} x^k (1-x)^{n-k} \\
&\leq \frac{\varepsilon}{2} \sum_{k=0}^{n} \binom{n}{k} x^k (1-x)^{n-k} + 2n\varepsilon \sum_{k=0}^{n} \left(x - \frac{k}{n} \right)^2 \binom{n}{k} x^k (1-x)^{n-k} \\
&= \frac{\varepsilon}{2} \cdot 1 + 2n\varepsilon \cdot \frac{1}{n} x(1-x) \\
&= \frac{\varepsilon}{2} + 2\varepsilon \cdot \frac{1}{4} \\
&= \varepsilon,
\end{aligned}
$$

where we have observed that as x is in $[0, 1]$, we have $x(1-x) \leq \frac{1}{4}$. □

Exercise 29.3. Complete the proof of Theorem 29.2, the Weierstrass approximation theorem, by extending the conclusion to functions continuous on an arbitrary closed bounded interval $[a, b]$. *Hint*: Make use of the function $L : [0, 1] \to [a, b]$ defined so that $L(y) = a + y(b - a)$.

A Sets, relations, and functions

It is customary to employ the language of set theory as a means to facilitate mathematical communication. Here we summarize some fundamental terminology and notation from set theory that we shall regularly draw upon, including the important notions of *relation* and *function*.

Sets and set membership

The notion of *set* is undefined. We think of sets as being made up of "elements" or "members." Thus, we would think of Barack Obama as an element or member of the set of all former presidents of the United States.

A set is sometimes specified by listing its members within "curly brackets." For example, the set whose members are precisely the numbers 6, 8, and 10 is $\{6, 8, 10\}$.

It is often inconvenient to list all the members of a set, and this becomes impossible if the set has an infinite number of members. If there is a straightforward pattern by which members of a set are related to one another, we may be able to describe the set by initiating the pattern and then using the ellipsis notation $\ldots$ to indicate a gap in the list. For instance, $\{-1, -2, -3, -4, \ldots, -50\}$ is the set whose members are the fifty largest negative integers, while $\{-1, -2, -3, -4, \ldots\}$ is the set of all negative integers.

A set having exactly one member is called a **singleton** (in this book, boldface type usually indicates a term is being defined) and a set $\{a, b\}$ having a and b as its only members is called a **doubleton**. Thus, the set whose only member is the number 5 is the singleton $\{5\}$ and the set whose only members are the numbers -1 and 1 is the doubleton $\{-1, 1\}$.

We write $x \in A$ to indicate that x is a member of a set A and $x \notin A$ to indicate that x is *not* a member of a set A. For instance, $6 \in \{6, 8, 10\}$ and $7 \notin \{6, 8, 10\}$. We agree that, given any set A and any object x, either $x \in A$ or $x \notin A$, but not both. When $x \in A$, we may say that x **is in** A and when $x \notin A$ that x **is not in** A.

Set equality

Sets are **equal** when they have exactly the same members. For example, $\{6, 8, 10\} = \{8, 6, 10\}$, but $\{6, 8, 10\} \neq \{6, -8, 10\}$.

Set-builder notation

The set of all members of a specified set A possessing a certain property P is often expressed using **set-builder notation** as

https://doi.org/10.1515/9783111429564-030

$$\{x \in A \mid x \text{ has property } P\},$$

which can be read as *the set of all members x of the set A such that x has property P* or simply as *the set of all members of A with property P*. For example, if D is the set of all odd counting numbers, then $D = \{1, 3, 5, 7, 9, \ldots\}$ and

$$\{1, 3, 5\} = \{x \in D \mid x < 7\}.$$

Subsets

A set A is a **subset** of a set B, denoted $A \subseteq B$, provided that whenever x is in A, it follows that x is in B. For example, $\{6, 8, 10\} \subseteq \{6, 8, -8, 10\}$, but $\{6, 8, -8, 10\} \nsubseteq \{6, 8, 10\}$. When $A \subseteq B$, we may also write $B \supseteq A$.

When A is a subset of B, we can also say any of the following: the set A is **included in** the set B, the set A is **contained in** the set B, the set B **includes** the set A, the set B **contains** the set A, and the set B is a **superset** of the set A. So, as $\{6, 8, 10\}$ is a subset of $\{6, 8, -8, 10\}$, it follows that $\{6, 8, 10\}$ is included in $\{6, 8, -8, 10\}$, that $\{6, 8, 10\}$ is contained in $\{6, 8, -8, 10\}$, that $\{6, 8, -8, 10\}$ includes $\{6, 8, 10\}$, that $\{6, 8, -8, 10\}$ contains $\{6, 8, 10\}$, and that $\{6, 8, -8, 10\}$ is a superset of $\{6, 8, 10\}$. The subset relation $\subseteq$ can be referred to as either **set inclusion** or **set-theoretic containment**.

If $A \subseteq B$ and $A \neq B$ we sometimes write $A \subset B$ and we call A a **proper** subset of B. The **improper subset** of a set is the set itself. For example, because $\{6, 8, 10\}$ is a subset of $\{6, 8, -8, 10\}$, but $\{6, 8, -8, 10\}$ includes an element, namely -8, which is not among the elements of $\{6, 8, 10\}$, it follows that $\{6, 8, 10\}$ is a proper subset of $\{6, 8, -8, 10\}$, which can be expressed by writing $\{6, 8, 10\} \subset \{6, 8, -8, 10\}$. Note, however, that $\{6, 8, 10\} \not\subset \{6, 10, 8\}$.

Set operations

Given sets A and B, we define the **union** $A \cup B$, the **intersection** $A \cap B$, and the **difference** $A - B$ so that

$$A \cup B = \{x \mid x \in A \text{ or } x \in B\},$$
$$A \cap B = \{x \mid x \in A \text{ and } x \in B\},$$

and

$$A - B = \{x \mid x \in A \text{ and } x \notin B\}.$$

For example,

$$\{6, 8, 10\} \cup \{5, 6, 7, 8\} = \{5, 6, 7, 8, 10\},$$

$$\{6, 8, 10\} \cap \{5, 6, 7, 8\} = \{6, 8\},$$

and

$$\{6, 8, 10\} - \{5, 6, 7, 8\} = \{10\}.$$

The set-theoretic difference $A - B$ is often read as "A without B" and is sometimes called the **relative complement** of B in A or the **complement of B relative to** A.

The empty set

There is exactly one set $\emptyset$ having no members; it is called the **empty set** or **null set**. For instance, $\{6, 8, 10\} - \{5, 6, 7, 8, 9, 10\} = \emptyset$.

Disjoint sets

When the intersection of two sets is empty, the sets are said to be **disjoint**. For example, the sets $\{6, 8, 10\}$ and $\{5, 9\}$ are disjoint because $\{6, 8, 10\} \cap \{5, 9\} = \emptyset$.

Ordered pairs

The **first coordinate** of the ordered pair (a, b) is a and the **second coordinate** is b. We consider ordered pairs to be **equal** precisely when their corresponding coordinates are equal; that is, $(a, b) = (c, d)$ if and only if both $a = c$ and $b = d$. For example,

$$(1, -1) = (2 - 1, 1 - 2),$$

but

$$(1, -1) \neq (-1, 1).$$

Cartesian products

Given sets A and B, the **Cartesian product** $A \times B$ is the set $\{(a, b) \mid a \in A \text{ and } b \in B\}$ of all ordered pairs with first coordinates in A and second coordinates in B. Thus, for instance,

$$\{1, 2\} \times \left\{\pi, 33\frac{1}{3}, 1\right\} = \left\{(1, \pi), \left(1, 33\frac{1}{3}\right), (1, 1), (2, \pi), \left(2, 33\frac{1}{3}\right), (2, 1)\right\}.$$

Relations

By a **relation** we mean a set of ordered pairs. More specifically any subset of $A \times B$ is referred to as a **relation on** $A \times B$, and any subset of $A \times A$ as a **relation on** A. For example, $\{(1,2),(2,3),(1,3)\}$ can be viewed as a relation on $\{1,2,4\} \times \{0,1,2,3\}$, as a relation on $\{1,2\} \times \{2,3\}$, or as a relation on $\{1,2,3\}$.

The **domain** of a relation is the set of all first coordinates of the ordered pairs in the relation, and the **range** is the set of all second coordinates. Thus, the domain of the relation $\{(1,2),(2,3),(1,3)\}$ is $\{1,2\}$ and the range is $\{2,3\}$.

Functions

A relation is a **function** if each first coordinate is paired with exactly one second coordinate. Thus, for instance, the relation $\{(1,2),(2,3),(3,2)\}$ is a function, while the relation $\{(1,2),(2,3),(1,4)\}$ is not. Note that a function is permitted to pair the same second coordinate with different first coordinates, but is not allowed to pair different second coordinates with the same first coordinate.

Each member of the domain of a function is called an **input** to the function and each member of the range is called an **output**. For a given function f, the unique output corresponding to a particular input x is denoted by $f(x)$. The output $f(x)$ of the function f at the input x is also called the **value** of f at x. For each x in the domain of f, we say the input x is **mapped to** (**taken to**, **sent to**) the output (value) $f(x)$, and can indicate this by writing $x \mapsto f(x)$.

To illustrate, consider the function $f = \{(1,2),(2,3),(3,2)\}$. As $(1,2) \in f$, the function f assigns the output 2 to the input 1; that is, $f(1) = 2$. Thus, we can say that f maps (takes, sends) 1 to 2, which can be expressed symbolically by writing

$$1 \overset{f}{\mapsto} 2^{\cdot}$$

When a function f has the set X as its domain and some subset of the set Y as its range, we say that f is a **function from** X **into** Y, denoted

$$f : X \to Y.$$

In this context, the set Y is referred to as the **codomain** of f.

Note that the codomain must contain the range, but the codomain may have other members that are not in the range. One reason for not forcing the set Y in the notation $f : X \to Y$ to be the range of f, but rather just a set containing the range, is that in some situations it is only the type of output, not the precise set of outputs, that must be made clear.

For example, in this book we are often interested in functions $f : A \to \mathbb{R}$, where $\mathbb{R}$ is the set of real numbers and the domain A is some subset of $\mathbb{R}$. We are precise about the domain because we do not want to input a number for which the function is undefined (for example, we would not want to divide by zero). We may not, however, need to be as precise about the range since simply knowing the outputs are real numbers may be sufficient for our purposes (i. e., we may not always need to know precisely which real numbers the function takes as values). The range of a function is sometimes referred to as the function's **image**.

B Fundamentals of logic

Logic is the study of truth and reasoning. The undefined notions in this study are *statement* and *truth*. *True* (T) and *false* (F) are the possible **truth values** a statement may possess.

Axioms are statements that are assumed to be true. The following axiom expresses the assumption that a statement is a sentence that is unambiguously true or unambiguously false.

Axiom B.1. *A statement is either true or false, but cannot be both true and false.*

The sentence

Paris is the capital of France.

is a true statement, whereas

The integer 5 *is even.*

is a false statement.

New statements can be created from existing statements through the use of the **logical connectives** *not*, *and*, *or*, *If ... then...*, and *if and only if*. In the next axiom, we give the formal names and notations for these sorts of statements and also summarize the circumstances under which they are true and under which they are false.

Axiom B.2. *For any statements P and Q,*

(1) *the **negation** of P, denoted $\sim P$ and read "Not P," is true when P is false and is false when P is true;*
(2) *the **conjunction** of P and Q, denoted $P \wedge Q$ and read "P and Q," is true precisely when both P and Q are true;*
(3) *the **disjunction** of P and Q, denoted $P \vee Q$ and read "P or Q," is false precisely when both P and Q are false;*
(4) *the **conditional** or **implication** with **hypothesis** P and **conclusion** Q, denoted $P \Rightarrow Q$ and read "If P, then Q," is false precisely when P is true and Q is false;*
(5) *the **biconditional** of P and Q, denoted $P \Leftrightarrow Q$ and read "P if and only if Q," is true precisely when P and Q have the same truth value, both true or both false.*

We give a few examples to suggest the "correctness" of the truth assignments described here.

Example B.3. The negation of a given statement has the "opposite" truth value from the statement itself. For instance, as the statement

The integer 11 *is prime*

https://doi.org/10.1515/9783111429564-031

is true, its negation

The integer 11 *is not prime*

is false. However, as the statement

The integer 5 *is an even number*

is false, its negation

The integer 5 *is not an even number*

is true.

Example B.4. Based on our everyday interpretation of the English word *and* (and assuming we know to whom the pronoun *I* refers), the conjunction

Today is Friday and this weekend I will be going to a baseball game

would be considered true precisely when both of its components

Today is Friday

and

This weekend I will be going to a baseball game

are true. If either component, or both, is false the conjunction would be considered false.

In logic generally, and in mathematics more specifically, the connective *or* is interpreted *inclusively*. This means that a disjunction is considered true as long as at least one of the component statements is true (so knowing $P \vee Q$ is true *includes* the possibility that both P and Q are true).

Example B.5. A graduation requirement such as

Students must take a math course or a physics course

would be fulfilled if a student takes a math course but not a physics course, takes a physics course but not a math course, or takes both a math course and a physics course. The only way the requirement is not satisfied is if a student does not take a math course and also does not take a physics course.

Thus, if our intention is to create a statement in which *or* is to be interpreted *exclusively*, we must write out something like *either… or… but not both* to make this clear. Note that the *exclusive or* can be represented symbolically as $(P \vee Q) \wedge \sim(P \wedge Q)$.

A conditional statement $P \Rightarrow Q$ is considered true unless the hypothesis P is true and the conclusion Q is false. Not everyone immediately agrees to this, so you may find the next example helpful.

Example B.6. Consider the conditional statement

If you are a math major, then you must take a programming course.

Note that finding a non-math major who does or does not have to take a programming course would not provide evidence against the truth of this statement; it is for this reason that an *If ..., then...* statement with a false hypothesis is automatically taken to be true. In order to be convinced that the statement is false, we would have to find an instance of a math major who does not have to take a programming course, that is, an instance where the hypothesis is true, but the conclusion is false.

A biconditional is really the conjunction of two conditional statements.

Example B.7. The biconditional

I will go to the baseball game if and only if I get my homework done

is the conjunction of the statements

I will go to the baseball game if I get my homework done

and

I will go to the baseball game only if I get my homework done,

where the latter statement is just another way of saying

If I go to the baseball game, then I did get my homework done.

This example helps us to see that

$$P \Leftrightarrow Q$$

may be regarded as a condensed way of expressing

$$(P \Rightarrow Q) \wedge (Q \Rightarrow P).$$

Note that in either form, it is precisely when the statements P and Q have the same truth value, both true or both false, that the biconditional is true. This way of assigning truth values matches our intuitive sense that the statement

I will go to the baseball game if and only if I get my homework done

is believable in the event that I go to the game having finished my homework or in the event that I do not go to the game having not finished my homework, but is a lie in the event that I go to the game having not finished my homework, or the event that I do not go the game having finished my homework.

Statement forms and truth tables

A **statement variable** is a letter, such as P, that can be replaced by any statement. A **statement form** is an expression created out of statement variables and logical connectives that becomes a statement once actual statements are substituted for the variables; for example, each of the expressions $\sim P$, $P \wedge Q$, $P \vee Q$, $P \Rightarrow Q$, and $P \Leftrightarrow Q$ is a statement form. A **truth table** for a statement form exhibits the truth value assigned to the form by each possible assignment of truth values to the form's variables. The truth tables displayed in Tables B.1 and B.2 provide an efficient means of representing the assumptions stated in Axiom B.2.

Table B.1: Truth assignments for negation.

P	$\sim P$
T	F
F	T

Table B.2: Truth assignments for conjunction, disjunction, conditional, and biconditional.

P	Q	$P \wedge Q$	$P \vee Q$	$P \Rightarrow Q$	$P \Leftrightarrow Q$
T	T	T	T	T	T
T	F	F	T	F	F
F	T	F	T	T	F
F	F	F	F	T	T

The first column in the first of these truth tables gives the possible truth values for the statement assigned to the variable P, using T for *true* and F for *false*, while the second column shows the resulting truth values for $\sim P$. In the second truth table, the first two columns give the four different ways truth values can be assigned to the variables P and Q, while the remaining columns show the resulting truth values for the forms heading the respective columns.

A truth table can be constructed for any statement form.

Example B.8. Table B.3 is a truth table for the statement form $P \Rightarrow \sim(Q \Leftrightarrow R)$.

Table B.3: A truth table for a statement form involving three variables.

P	Q	R	$Q \Leftrightarrow R$	$\sim(Q \Leftrightarrow R)$	$P \Rightarrow \sim(Q \Leftrightarrow R)$
T	T	T	T	F	F
T	T	F	F	T	T
T	F	T	F	T	T
T	F	F	T	F	F
F	T	T	T	F	T
F	T	F	F	T	T
F	F	T	F	T	T
F	F	F	T	F	T

Besides including a column at the very right for the form of interest, we have included columns for the forms $Q \Leftrightarrow R$ and $\sim(Q \Leftrightarrow R)$ to make it easier to obtain the truth values for $P \Rightarrow \sim(Q \Leftrightarrow R)$. Each row gives a different assignment of truth values for the variables P, Q, and R, along with the resulting truth assignments to each of $Q \Leftrightarrow R$, $\sim(Q \Leftrightarrow R)$, and $P \Rightarrow \sim(Q \Leftrightarrow R)$. For instance, the first row ultimately tells us that when all of P, Q, and R are replaced by true statements, the statement $P \Rightarrow \sim(Q \Leftrightarrow R)$ is false.

When more than one logical connective appears in a statement form, we usually must employ grouping symbols to make clear the order in which the connectives are to be applied. We did this in the previous example with $P \Rightarrow \sim(Q \Leftrightarrow R)$.

It is generally agreed, though, that the negation operator $\sim$ is given priority so as to reduce the quantity of grouping symbols that would otherwise be needed. So $\sim P \Rightarrow Q$ should be interpreted as $(\sim P) \Rightarrow Q$, and if we want to form the negation of $P \Rightarrow Q$, we would have to write $\sim(P \Rightarrow Q)$.

Tautologies and contradictions

A statement form is a
(1) **tautology** if every possible assignment of truth values to its statement variables yields a truth value of T to the form;
(2) **contradiction** if every possible assignment of truth values to its statement variables yields a truth value of F to the form.

Note that since it is always the case that either a statement or its negation must be true, the form $P \vee \sim P$ is a tautology, but since it is never the case that both a statement and its negation can be true, the form $P \wedge \sim P$ is a contradiction.

Whenever a statement has the form of a tautology, that statement can be used in a proof. Thus, as $P \vee \sim P$ is a tautology, we could include the statement

The positive integer n is either prime or not prime

in a proof if we believe it would help us build our argument.

Logical equivalence

Two statement forms are **logically equivalent** if any given assignment of truth values to the forms' variables yields the same truth value for both forms, meaning that when a truth table includes columns for each of the two forms, the truth values in these two columns match each other in every row. Two statements are **logically equivalent** provided they possess forms that are logically equivalent.

Example B.9. Note that in Table B.4

Table B.4: A truth table demonstrating a logical equivalence.

P	Q	$P \Rightarrow Q$	$Q \Rightarrow P$	$\sim P$	$\sim Q$	$\sim Q \Rightarrow \sim P$
T	T	T	T	F	F	T
T	F	F	T	F	T	F
F	T	T	F	T	F	T
F	F	T	T	T	T	T

the columns under $P \Rightarrow Q$ and $\sim Q \Rightarrow \sim P$ match row by row. The forms $P \Rightarrow Q$ and $\sim Q \Rightarrow \sim P$ are true at the same time (in all but the second row) and they are false at the same time (in the second row only). Hence, we may conclude that $\sim Q \Rightarrow \sim P$ is logically equivalent to $P \Rightarrow Q$.

Having verified the logical equivalence of the forms $P \Rightarrow Q$ and $\sim Q \Rightarrow \sim P$, it follows that any pair of statements created from these forms would be logically equivalent. For instance, the statements

If you are a math major, then you must take a programming course

and

If you do not have to take a programming course, then you are not a math major

are logically equivalent.

However, also observe that the columns in the truth table under $P \Rightarrow Q$ and $Q \Rightarrow P$ do not match in every row; for instance, they have different truth values in the second row, F for $P \Rightarrow Q$, but T for $Q \Rightarrow P$. Thus, $Q \Rightarrow P$ is not logically equivalent to $P \Rightarrow Q$, and neither are the statements

If you are a math major, then you must take a programming course

and

If you must take a programming course, then you are a math major.

For a conditional statement $P \Rightarrow Q$, the statement $Q \Rightarrow P$ is called the **converse** of $P \Rightarrow Q$, and the statement $\sim Q \Rightarrow \sim P$ is called the **contrapositive** of $P \Rightarrow Q$. Our work in Example B.9 demonstrates that a conditional statement is logically equivalent to its contrapositive but not to its converse. The following theorem lists the logical equivalences that most often arise in mathematical settings.

Theorem B.10 (Some commonly employed logical equivalences). *The following pairs of statement forms are logically equivalent to each other:*

(1) $P \wedge Q$ *and* $Q \wedge P$*;*
(2) $P \vee Q$ *and* $Q \vee P$*;*
(3) $P \Rightarrow Q$ *and* $\sim Q \Rightarrow \sim P$*;*
(4) $P \vee Q$ *and* $\sim P \Rightarrow Q$*;*
(5) $P \Leftrightarrow Q$ *and* $(P \Rightarrow Q) \wedge (Q \Rightarrow P)$*;*
(6) $\sim(P \wedge Q)$ *and* $\sim P \vee \sim Q$ *(this is one of DeMorgan's laws);*
(7) $\sim(P \vee Q)$ *and* $\sim P \wedge \sim Q$ *(this is the other of DeMorgan's laws);*
(8) $\sim(P \Rightarrow Q)$ *and* $P \wedge \sim Q$*;*
(9) $P \wedge (Q \vee R)$ *and* $(P \wedge Q) \vee (P \wedge R)$ *(this is a distributive law);*
(10) $P \vee (Q \wedge R)$ *and* $(P \vee Q) \wedge (P \vee R)$ *(this is another distributive law).*

Valid arguments and proofs

Generally speaking, an **argument** consists of one or more statements, called **premises**, that are assumed to be true, along with a statement, called the **consequence**, whose truth allegedly follows from the premises. The only arguments we are willing to accept in mathematics are those that are **valid**, meaning they exhibit an underlying logical form that forces the consequence to be true when all the premises are true. A **proof** may then be regarded as a sequence of valid arguments that demonstrates how a new result follows logically from given assumptions.

So, what exactly does it mean for an argument to exhibit an underlying logical form that forces the consequence to be true when all the premises are true? Using the symbol $\therefore$ as logical shorthand for *therefore*, and given statement forms $\mathbb{P}_1, \mathbb{P}_2, \ldots, \mathbb{P}_n$, and $\mathbb{C}$, we refer to

$$\mathbb{P}_1, \mathbb{P}_2, \ldots, \mathbb{P}_n \therefore \mathbb{C}$$

as an **argument form** having **premises** $\mathbb{P}_1, \mathbb{P}_2, \ldots, \mathbb{P}_n$ and **consequence** $\mathbb{C}$. An argument form is **valid** if every truth assignment that makes all the premises true also results in the consequence being true. An argument is then taken to be **valid** if it exhibits a valid argument form. Each deduction in a mathematical proof must be the consequence of a valid argument, whose premises have appeared earlier in the proof (either explicitly or implicitly).

Example B.11. Table B.5

Table B.5: Truth assignments to the conditional.

P	Q	$P \Rightarrow Q$
T	T	T
T	F	F
F	T	T
F	F	T

reveals that the argument form

$$P \Rightarrow Q, P \therefore Q$$

is valid because there is only one row in which the premises $P \Rightarrow Q$ and P are both true, the first row, and in the first row the consequence Q is also true. Thus, if we are writing a proof and have already determined that

If the function f is differentiable, then the function f is continuous

and that

The function f is differentiable,

we would be able to deduce that

The function f is continuous,

as this conclusion would result from applying the valid argument form

$$P \Rightarrow Q, P \therefore Q.$$

On the other hand, the same truth table reveals that the argument form

$$P \Rightarrow Q, Q \therefore P$$

is not valid because in the third row both of the premises $P \Rightarrow Q$ and Q are true, but the consequence P is false. This means that if we are writing a proof and we already know that

If the function f is differentiable, then the function f is continuous

and that

The function f is continuous

we would *not* be able to deduce that

The function f is differentiable.

Perhaps, depending on what other information might be available to us, there might be another way to reach this conclusion, but it does not follow as a logical consequence of the other two statements.

The next theorem lists the argument forms that are most often applied in mathematical proofs.

Theorem B.12 (Some commonly employed argument forms). *Each of the following is a valid argument form:*

(1) $P \Rightarrow Q, P \therefore Q;$
(2) $P \Rightarrow Q, \sim Q \therefore \sim P;$
(3) $P \vee Q, \sim P \therefore Q;$
(4) $P \wedge Q \therefore Q;$
(5) $P \Rightarrow Q, Q \Rightarrow P \therefore P \Leftrightarrow Q;$
(6) $P \Leftrightarrow Q, P \therefore Q;$
(7) $P \Leftrightarrow Q \therefore \sim P \Leftrightarrow \sim Q;$
(8) $P \Rightarrow Q, Q \Rightarrow R \therefore P \Rightarrow R;$
(9) $P \Leftrightarrow Q, Q \Leftrightarrow R \therefore P \Leftrightarrow R.$

A proof does not usually include explicit mention of the valid argument forms it employs, mainly because the same few argument forms are used over and over again so they become easily recognizable.

Open statements and quantifiers

A sentence that has at least one variable in it and that becomes a statement once each variable is assigned a particular value is called an **open statement**.

Example B.13. The equation

$$x + y = x$$

is an open statement. It becomes the false statement $4 + 2 = 4$ if $x = 4$ and $y = 2$, but it becomes the true statement $4 + 0 = 4$ if $x = 4$ and $y = 0$.

The logical phrase *for all* is referred to as the **universal quantifier**. It may be rendered symbolically as $\forall$ and also translated verbally by other phrases such as *for every* and *for each*.

The logical phrase *there exists* is referred to as the **existential quantifier**. It may be rendered symbolically as $\exists$ and also translated verbally by other phrases such as *for some* and *there is at least one*.

An open statement becomes a statement when quantifiers are applied to the variables appearing in it.

Example B.13 (Continued). Applying universal quantification to both variables x and y in the open statement

$$x + y = x,$$

and assuming these variables represent real numbers, yields the statement

For all real numbers x and y, we have $x + y = x$.

This statement is false, since taking $x = 4$ and $y = 2$, for instance, produces the false equation $4 + 2 = 4$.

If we instead apply existential quantification to y and *afterward* apply universal quantification to x, we obtain the statement

There exists a real number y such that for every real number x, we have $x + y = x$.

This statement is true because taking $y = 0$ the equation $x + y = x$ becomes

$$x + 0 = x,$$

which is true no matter what real number is assigned as the value of x.

It is customary to avoid using logical symbols such as $\forall$, $\exists$, and $\Rightarrow$ in most formal mathematical writing, so we typically use verbal phrases for quantifiers and logical operators rather than their symbolic representations. In less formal writing, for instance, when outlining ideas for a proof on “scratch” paper, mathematicians often choose to use such symbols, so we point out that

$$\forall x \in A,\ P(x),$$

is an abbreviation for

For every x, *if* $x \in A$, *then* $P(x)$,

and

$$\exists x \in A,\ P(x),$$

is an abbreviation for

There exists x *such that* $x \in A$ *and* $P(x)$.

Example B.13 (Continued). Thus, in our "private" notes, where we are thinking through some ideas, we might write the statement

There exists a real number y *such that for every real number* x, *we have* $x + y = x$

as

$$\exists y \in \mathbb{R}, \forall x \in \mathbb{R}, x + y = x.$$

The one exception to the suppression of logical symbols in formal mathematical writing is when the topic is logic itself. That is why we are using such symbols in the statements of results in this appendix. Since mathematical proof also falls under the heading of logic, we also sometimes use logical symbols when stating proof strategies.

When something is not always true, that means there exists an instance for which it is false. Also, when there does not exist an instance for which something is true, that means it is always false. These principles are formalized in the following axiom.

Axiom B.14 (Negating statements containing quantifiers). *Let* $P(x)$ *be an open statement that depends on a variable* x.
(1) *The statement* $\sim(\forall x, P(x))$ *is logically equivalent to the statement* $\exists x, \sim P(x)$.
(2) *The statement* $\sim(\exists x, P(x))$ *is logically equivalent to the statement* $\forall x, \sim P(x)$.

This axiom essentially tells us that when logical negation moves past one of the quantifiers *for all* or *there exists*, that quantifier must be changed to the other one.

Example B.15. The negation of

Every student in a certain class is a math major

is the statement

There exists a student in the class who is not a math major.

The negation of

There exists a real number y *such that for every real number* x*, we have* $x + y = x$

is the statement

For every real number y *there exists a real number* x *such that* $x + y \neq x$.

A great many of the definitions in the study of real analysis involve one or both of the universal and existential quantifiers.

C Essential proof strategies

The proof strategies recorded here are the ones that are used most frequently throughout mathematics. There is usually some discussion of a strategy at the point in the text where it is first introduced, so we have included each strategy's chapter and item number in that chapter so that you are able to easily locate it. Also included is a reference to at least one proof in which the strategy is employed.

Unless the subject is logic itself, which is not the case in this book, good mathematical writing style dictates that we should not use logical symbols such as $\forall$, $\exists$, $\Rightarrow$, $\Leftrightarrow$, $\sim$, $\wedge$, and $\vee$ in our proofs, instead writing out verbal equivalents to these symbols. The reason these symbols appear in the statements below is that they are proof strategies, and the study of proof is part of logic.

Also, keep in mind that

$$\forall x \in A, P(x),$$

is an abbreviated form of

$$\forall x, x \in A \Rightarrow P(x),$$

while

$$\exists x \in A, P(x),$$

is an abbreviated form of

$$\exists x, x \in A \wedge P(x).$$

Proving an *if... then...* statement. To prove a statement of the form

$$P \Rightarrow Q,$$

assume P and then show how to deduce Q from this assumption.

This is Proof Strategy 1.4. See the proof of Theorem 1.3.

Proving an *if and only if* statement. To prove a statement of the form

$$P \Leftrightarrow Q,$$

prove each of $P \Rightarrow Q$ and $Q \Rightarrow P$.

This is Proof Strategy 1.23. See the proof of Theorem 1.24.

Proving a *for all* statement. To prove a statement of the form

$$\forall x \in A, P(x),$$

https://doi.org/10.1515/9783111429564-032

consider an anonymous element x of the set A and then show the statement $P(x)$ is true for this anonymous x.

This is Proof Strategy 1.6. See the proof of Theorem 1.5. Note also that to disprove a *for all* statement, we need only exhibit a single *counterexample*; see Example 1.14.

Proving existence. To prove a statement of the form

$$\exists x \in A, P(x),$$

identify, define, or construct an element t of the set A and show that the statement $P(t)$ is true for this particular object t.

This is Proof Strategy 1.19. See the proofs of Theorems 1.18 and 1.20.

Proving uniqueness. To prove there is only one object having a certain property, assume there are two objects possessing the property and show they really are the same.

This is Proof Strategy 1.7. See the proof of Theorem 1.8.

Proving an *or* statement. To prove a statement of the form

$$P \vee Q,$$

assume P is not true and then deduce Q.

This is Proof Strategy 1.26. See the proof of Theorem 1.25.

Proof by contradiction. To prove a statement P using the method of proof by contradiction, assume its negation $\sim P$ is true and deduce a contradiction.

This is Proof Strategy 1.16. See the proof of Theorem 1.17.

Proof by contraposition. To prove the *if... then...* statement

$$P \Rightarrow Q,$$

by contraposition, assume $\sim Q$ and then deduce $\sim P$.

This is Proof Strategy 3.10. See the proof of Theorem 3.11.

Proof by induction. Suppose that for each natural number n, we have a statement $S(n)$. To prove that all of the statements

$$S(1), \quad S(2), \quad S(3), \quad S(4), \quad \ldots,$$

are true, it is enough to do both of the following:

Base step: Prove that $S(1)$ is true.

Inductive step: Prove that whenever $S(k)$ is true, it follows that $S(k+1)$ is true.

This is Proof Strategy 3.2. See the proofs of Theorems 3.3, 3.4, and 3.5.

Proving set inclusion. To prove the subset relationship

$$A \subseteq B,$$

consider an anonymous member x of the set A and show x is a member of the set B.

This is Proof Strategy 3.29. See the proofs of Theorems 3.30 and 3.31.

Proving sets are equal. To prove the set equality

$$A = B,$$

prove both $A \subseteq B$ and $B \subseteq A$.

This is Proof Strategy 3.32. See the proof given in Example 3.33.

Proving functions are equal. To prove that functions f and g are equal, show f and g have the same domain and also show that $f(x) = g(x)$ for each x in this shared domain.

This is Proof Strategy 5.2. See Example 5.3.

Proving a function is one-to-one, onto, or a bijection. To prove a function $f : X \to Y$ is

(1) *one-to-one*, assume x_1 and x_2 are in X with $f(x_1) = f(x_2)$ and show $x_1 = x_2$;
(2) *onto*, assume y is in Y and show there exists x in X with $f(x) = y$;
(3) a *bijection*, prove f is one-to-one and prove f is onto.

This is Proof Strategy 5.6. See Examples 5.7 and 5.8.

D A geometric approach to the cosine and sine functions

In earlier courses, the trigonometric functions *cosine* and *sine* are defined geometrically by means of right triangles and the unit circle. We lay an analytic foundation for these functions in Chapters 16 and 28, where we also rigorously prove some of the fundamental properties they possess without recourse to geometric ideas such as length, angle, and so on. However, it is convenient to have the cosine and sine functions available for use in examples prior to the analytic study we eventually undertake. Moreover, it is both satisfying and instructive for readers of this book to see how their prior experience with these functions is ultimately consistent with a more formal analytic treatment of them.

For these reasons, we use this appendix to remind you of the geometric construction of the cosine and sine functions, and to geometrically derive some of their elementary properties. The material presented here is no doubt familiar to you, but we emphasize that the analytic approach available to us once we have developed some of the theory of the *derivative* and *convergence of series of functions* ultimately supersedes the informal geometric approach taken here, and will thereby put the trigonometric functions on a firmly rigorous foundation.

Recall that the **unit circle** is the circle of radius 1 centered at the origin of an xy-coordinate system (Figure D.1). Any real number t can be identified with a **standard position arc** on the unit circle as follows: the arc always has $(1, 0)$ as its initial point,

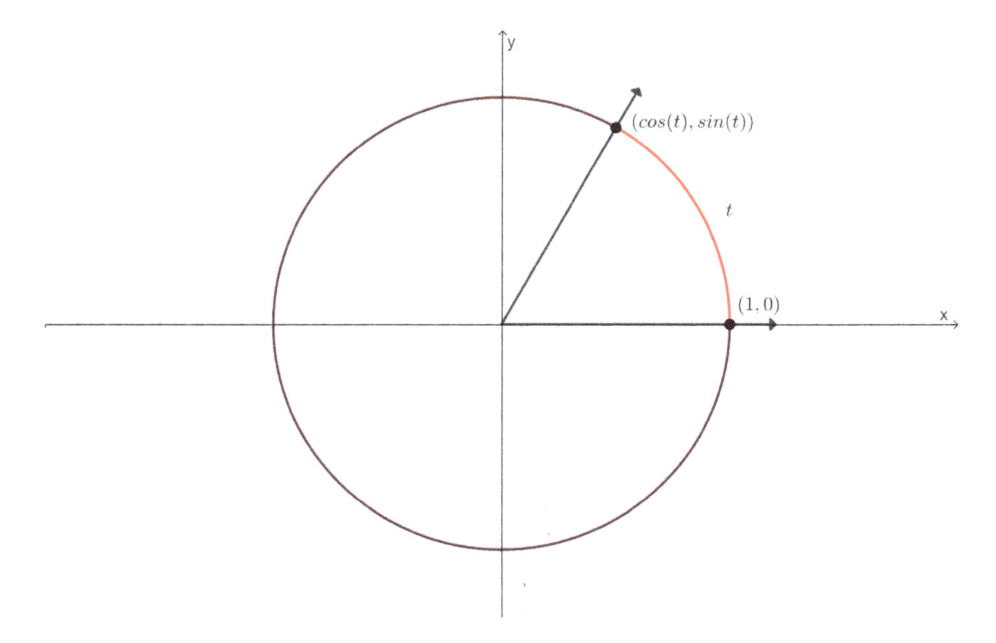

Figure D.1: Unit circle definitions of cosine and sine.

https://doi.org/10.1515/9783111429564-033

its length is equal to $|t|$, and it is oriented counterclockwise if $t > 0$, clockwise if $t < 0$, and consists of the single point $(1, 0)$ if $t = 0$. We can then define functions $\cos : \mathbb{R} \to \mathbb{R}$ and $\sin : \mathbb{R} \to \mathbb{R}$, so that for each real number t, the numbers $\cos(t)$ and $\sin(t)$ are, respectively, the x-coordinate and the y-coordinate of the point on the unit circle where the arc identified with t terminates. The function cos is called the **cosine function** and the function sin is called the **sine function**.

Since the coordinates of the points on the unit circle consist of precisely the numbers in the interval $[-1, 1]$, it follows that the range of both the cosine and sine functions is $[-1, 1]$.

Now, as the trivial arc $t = 0$ terminates where it begins, at $(1, 0)$, we have

$$\cos(0) = 1$$

and

$$\sin(0) = 0.$$

Then, because the circumference of the unit circle is 2π, Figure D.2 indicates that the arc consisting of the quarter of the unit circle lying in the first (upper right) quadrant, which is identified with the real number

$$t = \frac{1}{4} \cdot 2\pi = \frac{\pi}{2},$$

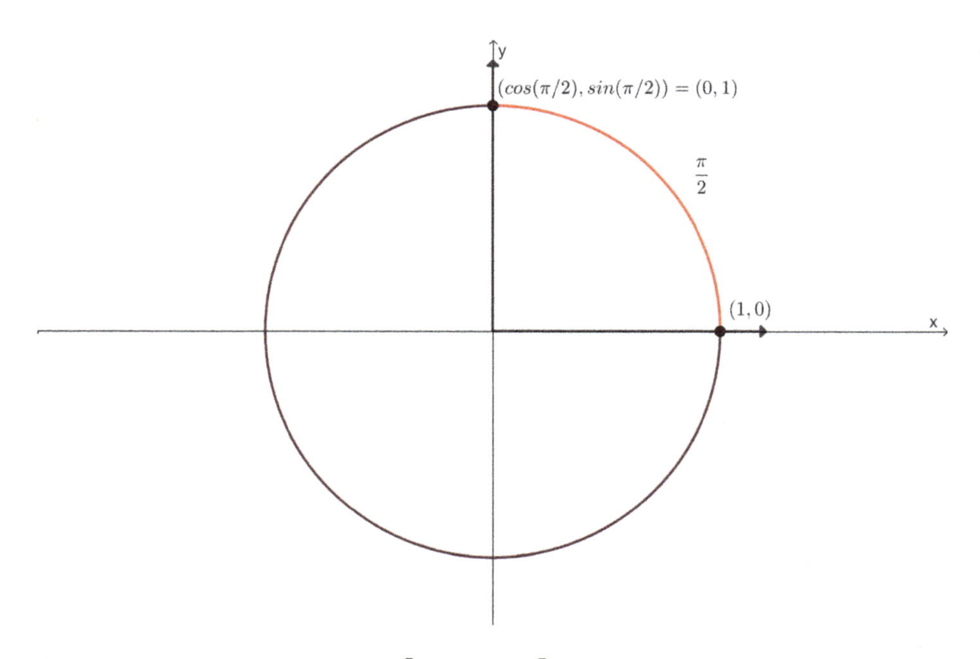

Figure D.2: Geometric explanation for $\cos(\frac{\pi}{2}) = 0$ and $\sin(\frac{\pi}{2}) = 1$.

terminates at $(0,1)$, from which it follows that

$$\cos\left(\frac{\pi}{2}\right) = 0$$

and

$$\sin\left(\frac{\pi}{2}\right) = 1.$$

In a similar manner, we may conclude that

$$\begin{aligned} \cos(\pi) &= -1, \\ \sin(\pi) &= 0, \\ \cos\left(\frac{3\pi}{2}\right) &= 0, \end{aligned}$$

and

$$\sin\left(\frac{3\pi}{2}\right) = -1.$$

If we imagine a satellite that is traveling at a constant speed of 1 unit per second counterclockwise around the unit circle and that is located at the point $(1,0)$ at time $t = 0$ seconds, then the horizontal position of the satellite at an arbitrary time t is given by $\cos(t)$ and the vertical position is given by $\sin(t)$. Once again using the fact that the circumference of the unit circle is 2π, we determine that the satellite returns to the same position on the unit circle every 2π seconds, meaning

$$\cos(t + 2\pi n) = \cos(t)$$

and

$$\sin(t + 2\pi n) = \sin(t)$$

for every real number t and every integer n.

The satellite scenario can also be used to provide a geometric/physical explanation for the derivative formulas

$$\cos'(t) = -\sin(t)$$

and

$$\sin'(t) = \cos(t)$$

for the cosine and since functions. In Figure D.3, the satellite's position at time t seconds, indicated by the blue position vector, is given by

$$\langle \cos(t), \sin(t) \rangle.$$

Also, at any given moment the satellite's direction of motion, indicated by the red velocity vector, is tangent to the circle.

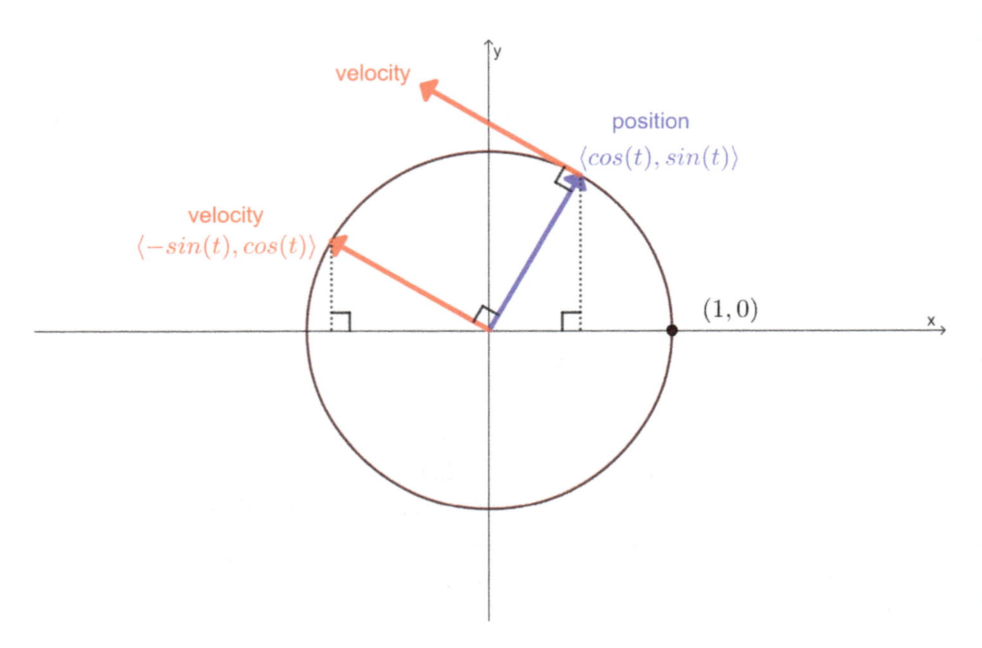

Figure D.3: Position and velocity vectors for a satellite traversing the unit circle.

As the only relevant attributes of a vector are its *length* and *direction*, we can slide the red velocity vector, maintaining its length and the direction in which it points, so that it begins at the origin. Because the velocity vector is tangent to the circle, the angle between the position and velocity vectors is a right angle. We also know the position and velocity vectors both have length 1. Hence, the two right triangles shown in the diagram with these vectors lying along the triangles' respective hypotenuses are congruent, so their corresponding sides have the same length.

Now as the horizontal leg of the "position triangle" has length $\cos(t)$ and corresponds to the vertical leg of the "velocity triangle," the vertical leg of the velocity triangle must have length $\cos(t)$. And as the vertical leg of the position triangle has length $\sin(t)$ and corresponds to the horizontal leg of the velocity triangle, the horizontal leg of the velocity triangle must have length $\sin(t)$. Since the velocity vector is pointing to the left and up, its horizontal component is negative, while its vertical component is positive, which is why the negative sign is introduced in the horizontal component of the velocity.

By definition velocity is the derivative of position, so it naturally follows that the derivative of an object's horizontal position is its horizontal velocity and the derivative of its vertical position is its vertical velocity. So, as the horizontal position of the satellite is $\cos(t)$ and its horizontal velocity is $-\sin(t)$, we may conclude that the derivative of $\cos(t)$ is $-\sin(t)$. And as the vertical position of the satellite is $\sin(t)$ and its vertical velocity is $\cos(t)$, it follows that the derivative of $\sin(t)$ is $\cos(t)$.

Bibliography

[1] Guyer P, Wood A (editors). The Cambridge edition of the works of Immanuel Kant. Cambridge, England, Cambridge University Press, 1998.
[2] Cauchy A-L. Cours d'analyse: an annotated translation. New York, NY, USA, Springer, 2009.
[3] Lebesgue H. Leçons sur l'integration et la recherche des fonctions primitive. Paris, France, Gauthier-Villars, 1904.
[4] Hille E. Analytic function theory. Waltham, MA, USA, Ginn and Company, 1959.
[5] Niven I. A simple proof that π is irrational. Bulletin of the American Mathematical Society USA 1947, 53:6, 509.
[6] McCarthy J. An everywhere continuous nowhere differentiable function. The American Mathematical Monthly USA, 1953, 60, 709.
[7] Bernstein S. Demonstration of a theorem of Weierstrass based on the calculus of probabilities. Communications of the Kharkov Mathematical Society Ukraine 1912, 8, 1–2.

https://doi.org/10.1515/9783111429564-034

Index

https://doi.org/10.1515/9783111429564-035

www.ingramcontent.com/pod-product-compliance
Lightning Source LLC
Jackson TN
JSHW060133050825
88763JS00010BA/35
* 9 7 8 3 1 1 1 4 2 9 2 8 1 *